The Diatoms
Applications for the Environmental and Earth Sciences

EDITED BY E. F. STOERMER AND JOHN P. SMOL

CAMBRIDGE
UNIVERSITY PRESS

PUBLISHED BY THE PRESS SYNDICATE OF THE UNIVERSITY OF CAMBRIDGE
The Pitt Building, Trumpington Street, Cambridge, United Kingdom

CAMBRIDGE UNIVERSITY PRESS
The Edinburgh Building, Cambridge CB2 2RU, UK
40 West 20th Street, New York, NY 10011-4211, USA
10 Stamford Road, Oakleigh, VIC 3166, Australia
Ruiz de Alarcón 13, 28014 Madrid, Spain
Dock House, The Waterfront, Cape Town 8001, South Africa

http://www.cambridge.org

First published 1999
Reprinted 2000
First paperback edition 2001

Printed in the United Kingdom at the University Press, Cambridge

Typeface Lexicon No. 2 (*The Enschedé Font Foundry*) 9.25/12.75   *System* QuarkXPress™   [SE]

A catalogue record for this book is available from the British Library

Library of Congress Cataloguing in Publication data

The diatoms: applications for the environmental and earth sciences /
    [edited by] Eugene F. Stoermer and John P. Smol
        p.   cm.
    Includes index.
    ISBN 0 521 58281 4 (hardbound)
    1. Diatoms.   2. Diatoms – Ecology.   3. Plant indicators.
    I. Stoermer, Eugene F., 1934–   .   II. Smol, J. P. (John P.)
    QK569.D54D536   1999
    579.8′5–dc21   98-29506 CIP

ISBN 0 521 58281 4 hardback
ISBN 0 521 00412 8 paperback

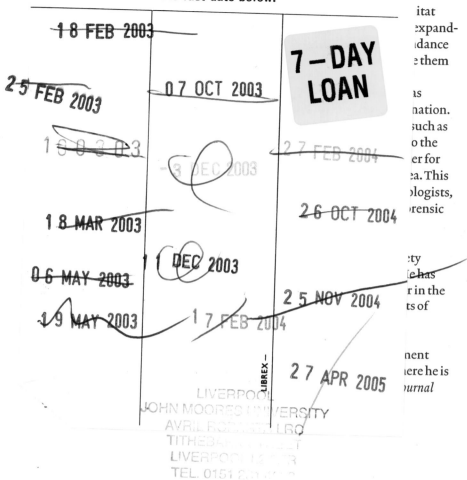
...itat
...expand-
...dance
...e them

...as
...nation.
...such as
...o the
...er for
...a. This
...ologists,
...rensic

...ty
...e has
...r in the
...ts of

...nent
...ere he is
...ournal

'We stand on the shoulders of giants.'

This book is dedicated to Dr John D. Dodd, Dr Ruth Patrick, and Dr Charles W. Reimer, whose varied contributions so greatly expanded and improved diatom studies in North America.

Contents

Contributors

JOHN P. SMOL
Paleoecological Environmental Assessment and Research Laboratory (PEARL), Department of Biology, Queen's University, Kingston, Ontario K7L 3N6, Canada
smolj@biology.queensu.ca

EUGENE F. STOERMER
Center for Great Lakes and Aquatic Sciences, University of Michigan, Ann Arbor, MI 48109-2099, USA
stoermer@umich.edu

RICHARD W. BATTARBEE
Environmental Change Research Centre, University College London, 26 Bedford Way, London WC1H 0AP, UK
rbattargb@geography.ucl.ac.uk

NIGEL CAMERON
Environmental Change Research Centre, University College London, 26 Bedford Way, London WC1H 0AP, UK

J. PLATT BRADBURY
US Geological Survey, MS 919 Box 25046, Federal Center, Denver, CO 80225, USA
jbradbur@usgs.gov

DONALD F. CHARLES
Patrick Center for Environmental Research, The Academy of Natural Sciences, 1900 Benjamin Franklin Parkway, Philadelphia, PA 19103-1195, USA

SHERRI R. COOPER
Duke Wetlands Center, SOE, Box 90333, Duke University, Durham, NC 27708-0333, USA
slcooper@acpub.duke.edu

BRIAN F. CUMMING
Paleoecological Environmental Assessment and Research Lab (PEARL), Department of Biology, Queen's University, Kingston, Ontario K7L 3N6, Canada

LUC DENYS
De Lescluzestraat 68, B-2600 Berchem (Antwerpen), Belgium
LUDE@ruca.ua.ac.be

HEIN DE WOLF
Netherlands Institute of Applied Geoscience TNO-National Geological Survey, Geo-mapping Support and Development, Postbox 157, NL-2000 AD Haarlem, The Netherlands

SUSHIL S. DIXIT
Paleoecological Environmental and Assessment Laboratory, Department of Biology, Kingston, Ontario K7L 3N6, Canada

MARIANNE S. V. DOUGLAS
Department of Geology, University of Toronto, 22 Russell Street, Toronto, Ontario M5S 3B1, Canada
msvd@opal.geology.utoronto.ca

HAMISH C. DUTHIE
Department of Biology, University of Waterloo, Ontario N2L 3Q1, Canada

SHERILYN C. FRITZ
Department of Geosciences, University of Nebraska, Lincoln, NE 68588, USA

GRETA A. FRYXELL
Department of Botany, University of Texas, Austin, TX 78712-7640, USA
gfryxell@bongo.cc.utexas.edu

FRANÇOISE GASSE
Cerege, Université Aix-Marseille III – CNRSFU 017, Pôle d'activité commerciale de l'Arbois, BP 80, 13545 Aix-En-Provence Cedex 4, France

ROLAND I. HALL
Climate Impacts Research Centre & Department of Environmental Health, Umeå University, Abisko Scientific Research Station, Box 62, s-981 07 Abisko, Sweden
Roland.Hall@ans.kiruna.se

MARGARET A. HARPER
Department of Geology, Research School of Earth Sciences, Victoria University of Wellington, PO Box 600, Wellington, New Zealand
Margaret.Harper@vuw.ac.nz

DAVID M. HARWOOD
Department of Geology, University of Nebraska, Lincoln, NE 68588-0340, USA
dharwood@unlinfo.unl.edu

JEFFREY R. JOHANSEN
Department of Biology, John Carroll University, University Heights, OH 44118, USA
johansen@jcvaxa.jcu.edu

STEVEN JUGGINS
Department of Geography, University of Newcastle, Newcastle upon Tyne NE1 7RU, UK
stephen.juggins@newcastle.ac.uk

WILLIAM N. KREBS
AMOCO Production Co., 501 Westlake Park Blvd., Houston, TX 77079, USA
bill.n.krebs@amoco.com

KATHLEEN R. LAIRD
Paleoecological Environmental Assessment and Research Lab (PEARL), Department of Biology, Queen's University, Kingston, Ontario K7L3N6, Canada

ANDRÉ F. LOTTER
EAWAG, CH-8600 Duebendorf, Switzerland
lotter@eawag.ch

DIANE M. MCKNIGHT
Institute of Arctic and Alpine Research, University of Colorado, Boulder, CO 80303, USA

YANGDON PAN
Environmental Sciences and Resources, Portland State University, Portland, OR 97207, USA

ANTHONY J. PEABODY
Metropolitan Forensic Science Laboratory, 109 Lambeth Road, London SE1 7LP, UK

REINHARD PIENITZ
Centre d'Études Nordiques & Département de Géographie, Université Laval, Québec G1K 7P4, Canada

INGEMAR RENBERG
Department of Environmental Health, Umeå University, s-90187 Umeå, Sweden

CONSTANCE A. SANCETTA
National Science Foundation, 4201 Wilson Blvd., Arlington, VA 22230, USA
csancett@nsf.gov

CLAIRE L. SCHELSKE
Department of Fisheries and Aquatic Sciences, University of Florida, 7922 NW 71st Street, Gainesville, FL 32653, USA
schelsk@nervm.nerdc.ufl.edu

ROLAND SCHMIDT
Institut für Limnologie, Österreichische Akademie der Wissenschaften, Gaisberg 116, A-5310 Mondsee, Austria

PAULI SNOEIJS
Institute of Ecological Botany, Uppsala University, Box 559, S-751 22 Uppsala, Sweden
pauli.Snoeijs@vaxtbio.uu.se

SARAH A. SPAULDING
Diatom Section, Invertebrate Zoology and Geology, California Academy of Sciences, Golden Gate Park, San Francisco, CA 94118-4599, USA
spauldin@CAS.calacademy.org

R. JAN STEVENSON
Department of Biology, University of Louisville, Louisville, KY 40292, USA
rjstevo1@ulkyvm.louisville.edu

MICHAEL J. SULLIVAN
Department of Biological Sciences, Mississippi State University, PO Box GY, Mississippi State, MS 39762, USA
mjs2@ra.msstate.edu

MARIA C. VILLAC
Universidade Federal do Rio de Janeiro, Instituto de Biologia, Departamento de Biologia Marinha, Cidade Universitaria CCS, Blocco A, Rio de Janeiro, Brazil

JULIE A. WOLIN
Department of Biology, Indiana University of Pennsylvannia, Indiana, PA 15705, USA
jwolin@grove.iup.edu

Preface

Diatoms are being used increasingly in a wide range of applications, and the number of diatomists and their publications continues to increase rapidly. Although a number of books have dealt with various aspects of diatom biology, ecology, and taxonomy, to our knowledge, no volume exists that summarizes the many applications and uses of diatoms.

Our overall goal was to collate a series of review chapters, which would cover most of the key applications and uses of diatoms to the environmental and earth sciences. Due to space limitations, we could not include all types of applications, but we hope to have covered the main ones. Moreover, many of the chapters could easily have been double in size, and in fact several chapters could have been expanded to the size of books. Nonetheless, we hope the material has been reviewed in sufficient breadth and detail to make this a valuable reference book for a wide spectrum of scientists, managers, and other users. In addition, we hope that researchers who occasionally use diatoms in their work, or at least read about how diatoms are being used by their colleagues (e.g., archeologists, forensic scientists, climatologists, etc.), will also find the book useful.

The volume is broadly divided into six parts. Following our brief Prologue, Part II contains seven chapters that review how diatoms can be used as indicators of environmental change in flowing waters and lakes. Part III summarizes work completed on diatoms from extreme environments, such as the High Arctic, the Antarctic, and aerial habitats. These ecosystems are often considered to be especially sensitive bellwethers of environmental change. Part IV contains five chapters dealing with diatoms in marine and estuarine environments, whilst Part V summarizes some of the other applications and uses of diatoms (e.g., archeology, oil exploration and correlation, forensic studies, toxic effects, atmospheric transport, diatomites). We conclude with a short epilogue (Part VI), followed by a glossary (including acronyms) and an index.

Many individuals have helped with the preparation of this volume. We are especially grateful to Barnaby Willits (Cambridge University Press), who shepherded this project from its inception. As always, the reviewers provided excellent suggestions for improving the chapters. We are also grateful to our colleagues at the University of Michigan and at Queen's University, and elsewhere, who helped in many ways to bring this volume to its completion. And, of course, we thank the authors.

These are exciting times for diatom-based research. We hope that the following chapters effectively summarize how these powerful approaches can be used by a diverse group of users.

EUGENE F. STOERMER
Ann Arbor, Michigan, USA

JOHN P. SMOL
Kingston, Ontario, Canada

Part I
Introduction

1 Applications and uses of diatoms: prologue

EUGENE F. STOERMER AND JOHN P. SMOL

This book is about the uses of diatoms (Class Bacillariophyceae), a group of microscopic algae abundant in almost all aquatic habitats. There is no accurate estimate of the number of diatom species. Estimates on the order of 10^4 are often given (Guillard & Kilham, 1977), and Mann and Droop (1996) point out that this estimate would be raised to at least 10^5 by application of modern species concepts. Diatoms are characterized by a number of features, but are most easily recognized by their siliceous (opaline) cell walls, composed of two valves, that together form a frustule (Fig. 1.1). The size, shape, and sculpturing of diatom cell walls are taxonomically diagnostic. Moreover, because of their siliceous composition, they are often very well preserved in fossil deposits, and have a number of other industrial uses.

This book is not about the biology and taxonomy of diatoms. Other volumes, for example, Round *et al.* (1990) and the review articles and books cited in the following chapters, provide introductions to the biology, ecology, and taxonomy of diatoms. Instead, we focus on the applications and uses of diatoms, with a further focus on environmental and earth sciences. Although this book contains chapters on direct applications, such as uses of fossilized diatom remains in industry, oil exploration, and forensic applications, most of the book deals with using these indicators to decipher the effects of long-term ecological perturbations, such as climatic change, lake acidification, and eutrophication. As many others have pointed out (e.g., Dixit *et al.*, 1992), diatoms are almost ideal biological monitors. There are a very large number of ecologically sensitive species, which are abundant in nearly all habitats where water is at least occasionally present, and leave their remains preserved in the sediments of most lakes and many areas of the oceans, as well as in other environments.

Precisely when and how people first began to use the occurrence and abundance of diatom populations directly, and to sense environmental conditions and trends, is probably lost in the mists of antiquity. It is known that diatomites were used as a palliative food substitute during times of starvation (Taylor, 1929), and Bailey's notes attached to the type collection of *Gomphoneis herculeana* (Ehrenb.) Cleve (Stoermer & Ladewski, 1982) indicate that masses of this species were used by native Americans for some medicinal purpose, especially by women (J. W. Bailey, unpublished notes associated with the type

Fig 1.1. Frustules of the diatom *Stephanodiscus niagarae* Ehrenb., viewed in a scanning electron microscope. Larger frustule is complete, and is near maximum size range for this taxon. In the smaller frustule, which is near the minimum size range for the species, the valves are separated, and the interior of one valve is visible. Micrograph courtesy of M. B. Edlund.

gathering of *G. herculeana*, housed in the Humboldt Museum, Berlin). It is interesting to speculate how early peoples may have used the gross appearance of certain algal masses as indications of suitable water quality (or contra-indications of water suitability!), or the presence of desirable and harvestable fish or invertebrate communities.

However, two great differences separate human understanding of higher plants and their parallel understanding of algae, particularly diatoms. The first is direct utility. Anyone can quite quickly grasp the difference between having potatoes and not having potatoes. It is somewhat more difficult to establish the consequences of, for example, *Cyclotella americana* Fricke being extirpated from Lake Erie.

The second is perception. At this point in history, nearly any person living in temperate latitudes can correctly identify a potato. Some people whose existence has long been associated with potato culture can provide a wealth of information, even if they lack extended formal education. Almost any university will have individuals who have knowledge of aspects of potato biology or, at a minimum, know where this rich store of information may be obtained. Of course, knowledge is never perfect, and much research remains to be done before our understanding of potatoes approaches completeness.

Diatoms occupy a place near the opposite end of the spectrum of under-standing. Early peoples could not sense individual diatoms, and their only

E. F. STOERMER AND J. P. SMOL

knowledge of this fraction of the world's biota came from mass occurrences of either living (e.g., biofilms) or fossil (diatomites) diatoms. Even in today's world, it is difficult to clearly and directly associate diatoms with the perceived values of the majority of the world's population. The consequences of this history are that the impetus to study diatoms was not great. Hence, many questions concerning basic diatom biology remain to be addressed. Indeed, it is still rather rare to encounter individuals deeply knowledgeable about diatoms even amongst university faculties. This, however, is changing rapidly.

What is quite clear is that people began to compile and speculate upon the relationships between the occurrence of certain diatoms and other things which were useful to know, almost as soon as optical microscopes were developed. In retrospect, some of the theories developed from these early observations and studies may appear rather quaint in the light of current knowledge. For instance, Ehrenberg (see Jahn, 1995) thought that diatoms were animal-like organisms, and his interpretation of their cytology and internal structure was quite different from our modern understanding. Further, his interpretations of the origins of airborne diatoms (he thought they were directly associated with volcanoes) seem rather outlandish today. On the other hand, Ehrenberg did make phytogeographic inferences which are only now being rediscovered.

As will be pointed out in chapters following, knowledge about diatoms can help us know about the presence of petroleum, if and where a deceased person drowned, when storms over the Sahara and Sub-Saharan Africa were of sufficient strength to transport freshwater diatoms to the mid-Atlantic, and indeed to the most remote areas of Greenland, as well as many other applications and uses. As will be reflected in the depth of presentation in these chapters, diatoms provide perhaps the best biological index of annual to millennial changes in Earth's biogeochemistry. As it becomes increasingly evident that human actions are exercising ever greater control over the conditions and processes which allow our existence, fully exercising all the tools which may serve to infer the direction and magnitude of change, and indeed the limits of change, becomes increasingly imperative. This need has fueled a considerable increase in the number of studies that deal with diatoms, particularly as applied to the problems alluded to above.

The primary motivation for this book therefore, is to compile this rapidly accumulating and scattered information into a form readily accessible to interested readers. A perusal of the literature will show that the authors of the different chapters are amongst the world leaders in research on the topics addressed.

The perceptive reader will also note that, despite their great utility, the store of fundamental information concerning diatoms is not as great as might logically be expected for a group of organisms which constitute a significant fraction, Werner (1977) estimates 25%, of Earth's biomass. For example, readers will find few references to direct experimental physiological studies of the species discussed. Sadly, there are few studies to cite, and practically none of those available was conducted on the most ecologically sensitive freshwater species. Readers may also note that there are some differences of opinion concerning taxonomic limits, even of common taxa, and that naming conventions are presently in a state of flux. These uncertainties are real, and devolve from the history of diatom studies.

As already mentioned, study of diatoms started relatively late, compared with most groups of macroorganisms. Diatoms have only been studied in any organized fashion for about 200 years, and the period of effective study has only been about 150 years. It is also true that the history of study has been quite uneven. After the development of fully corrected optical microscopes, the study of microorganisms in general, and diatoms in particular, attracted immense interest and the attention of a number of prodigiously energetic and productive workers. This grand period of exploration and description produced a very substantial, but poorly assimilated, literature. Diatomists who worked toward the end of this grand period of growth produced remarkably advanced insights into cytology and similarly advanced theories of biological evolution (Mereschkowsky, 1903). This, and the fact that sophisticated and expensive optical equipment is required for their study, gained diatoms the reputation of a difficult group of organisms to study effectively. Partially for this reason, basic diatom studies entered a period of relative decline beginning *ca*. 1900, although a rich, if somewhat eclectic, amateur tradition flourished in England and North America. The area that remained most active was ecology. As Pickett-Heaps *et al.* (1984) have pointed out, Robert Lauterborn, an exceptionally talented biologist, well known for his studies of diatom cytology, could also be appropriately cited as one of the founders of aquatic ecology.

The people who followed generally did not command the degree of broad recognition enjoyed by their predecessors, and many of them operated at the margin of the academic world. As examples, Friedrich Hustedt, perhaps the best known diatomist in the period from 1900 to 1960 (Behre, 1970), supported himself and his family as a high school teacher for much of his career. B. J. Cholnoky (Archibald, 1973) was caught up in the vicissitudes of the Second World War, and produced his greatest works on diatom autecology, including his large summary work (Cholnoky, 1968), after he became an employee of the South African Water Resources Institute. Although many workers of this era produced notable contributions, they were peripheral to the main thrusts of academic ecological thought and theory, particularly in North America. Although this continent had numerous individuals who were interested in diatoms, and published on the group, most of them were either interested amateurs or isolated specialists working in museums or other non-university institutions. For example, when the senior author decided to undertake advanced degree work on diatoms in the late 1950s, there was no university in the United States with a faculty member specializing in the study of freshwater diatoms.

One of the most unfortunate aspects of separation of diatom studies from the general course of botanical research was substantial separation from the blossoming of new ideas. The few published general works on diatoms had a curious 'dated' quality, and relatively little new understanding, except for descriptions of new species. The main thing that kept this small branch of botanical science alive was applied ecology, and this was the area that, in our opinion, produced the most interesting new contributions.

This situation began to change in the late 1950s, partially as a result of the general expansion of scientific research in the post-Sputnik era, and partially as the result of technological advances, particularly in the area of electronics. The general availability of electron microscopes opened new orders of magnitude in resolution of diatom structure, which made it obvious that many of the older, radically condensed, classification schemes were untenable. This released a virtual flood of new, rediscovered, and reinterpreted entities (Round et al., 1990) which continues to grow today (e.g., Lange-Bertalot & Metzeltin, 1996; Krammer, 1997). At the same time, the general availability of high-speed digital computers made it possible to employ multivariate statistical techniques ideally suited to objective analysis of modern diatom communities and those contained in sediments.

The history of ecological studies centered on diatoms can be roughly categorized as consisting of three eras. The first is what we might term the 'era of exploration'. During this period (ca. 1830–1900), most research focused on diatoms as objects of study. Work during this period was largely descriptive, be the topic the discovery of new taxa, discovery of their life cycles and basic physiology, or observations of their geographic and temporal distributions. One of the hallmarks of this tradition was the 'indicator species concept'. Of course, the age of exploration is not over for diatoms. Taxa have been described at a rate of about 400 per year over the past three decades, and this rate appears to be accelerating in recent years (C. W. Reimer, personal communication). Basic information concerning cytology and physiology of some taxa continues to accumulate, although at a lesser rate than we might desire.

The second era of ecological studies can be termed the 'era of systematization' (ca. 1900–1970). During this period, many researchers attempted to reduce the rich mosaic of information and inference concerning diatoms to more manageable dimensions. The outgrowths of these efforts were the so-called systems and spectra (e.g., halobion, saprobion, pH, temperature, etc.). Such devices are still employed, and sometimes modified and improved. Indeed, there are occasional calls for simple indices as a means of conveying information to managers and the public.

We would categorize the current era of ecological studies focused on diatoms as the 'age of objectification'. Given the computational tools now generally available, it is possible to more accurately determine which variables affect diatom occurrence and growth and, more importantly, do so quantitatively, reproducibly, and with measurable precision. Thus, applied studies based on diatoms have been raised from a little understood art practised by a few extreme specialists, to a tool that more closely meets the general expectations of science and the users of this work, such as environmental managers.

The result is that we now live in interesting times. Diatoms have proven to be extremely powerful tools with which to explore and interpret many ecological and practical problems. The continuing flood of new information will, without doubt, make the available tools of applied ecology even sharper. It is also apparent that the maturation of this area of science will provide additional challenges. Gone are the comfortable days when it was possible to learn the characteristics of most freshwater genera in a few days and become familiar with the available literature in a few months. Although we might sometimes wish for the return of simpler days, it is clear that this field of study is rapidly expanding, and it is our conjecture that we are on the threshold of even larger changes. The motivation for producing this volume is to summarize accomplishments of the recent past and, thus, perhaps make the next step easier.

References

Archibald, R. E. M. (1973). Obituary: Dr. B. J. Cholnoky (1899–1972). *Revue Algologique*, N. S., **11**, 1–2.

Behre, K. (1970). Friedrich Hustedt's Leben und Werke. *Nova Hedwigia, Beiheft*, **31**, 11–22.

Cholnoky, B. J. (1968). *Die Ökologie der Diatomeen in Binnengewässern*. Verlag von J. Cramer: Lehre. 699 pp.

Dixit, S. S.; Smol, J. P., Kingston, J. C. & Charles, D. F. (1992). Diatoms: Powerful indicators of environmental change. *Environmental Science and Technology*, **26**, 22–33.

Guillard, R. R. L. & Kilham, P. (1977). The ecology of marine planktonic diatoms. In *The Biology of Diatoms. Botanical Monographs* vol. 13, ed. D. Werner, pp. 372–469. Oxford: Blackwell Scientific Publications.

Jahn, R. (1995). C. G. Ehrenberg's concept of the diatoms. *Archiv für Protistenkunde*, **146**, 109–16.

Krammer, K. (1997). Die cymbelloiden Diatomeen, Eine Monographie der weltweit bekannneten Taxa. Teil 21. *Encyonema* part., *Encyonopsis* and *Cymbellopsis*. *Bibliotheca Diatomologica* vol. 37. J. Cramer in der Gebrüder Borntraeger Verlagsbuchhandlung. 469 pp.

Lange-Bertalot, H. & Metzeltin, D. (1996). Indicators of oligotrophy – 800 taxa representative of three ecologically distinct lake types – carbonate buffered – oligodystrophic – weakly buffered soft water. *Iconographia Diatomologia – Annotated Diatom Micrographs*, ed. H. Lange Bertalot), Vol. 2, Ecology – diversity – taxonomy. Koeltz Scientific Books. 390 pp.

Mann, D. G. & Droop, J. M. (1996). Biodiversity, biogeography and conservation of diatoms. *Hydrobiologia*, **336**, 19–32.

Mereschkowsky, C. (1903). Nouvelles recherchés sur la structure et la division des Diatomées. *Bulletin Société Impériale des Naturalistes de Moscou*, **17**, 149–72.

Pickett-Heaps, J. D., Schmid, A-M. & Tippett, D. H. (1984). Cell division in diatoms: A translation of part of Robert Lauterborn's treatise of 1896 with some modern confirmatory observations. *Protoplasma*, **120**, 132–54.

Round, F. E., Crawford, R. M. & Mann, D. G. (1990). *The Diatoms: Biology and Morphology of the Genera*. Cambridge: Cambridge University Press. 747 pp.

Stoermer, E. F. & Ladewski, T. B. (1982). Quantitative analysis of shape variation in type and modern populations of *Gomphoneis herculeana*. *Nova Hedwigia, Beiheft*, **73**, 347–86.

Taylor, F. B. (1929). *Notes on Diatoms*. Bournemouth: Guardian Press. 269 pp.

Werner, D. (ed.). (1977). *The Biology of Diatoms*. Oxford: Blackwell Scientific Publications. 498 pp.

Part II
Diatoms as indicators of environmental change in flowing waters and lakes

2 Assessing environmental conditions in rivers and streams with diatoms

R. JAN STEVENSON AND YANGDONG PAN

Introduction

Assessments of environmental conditions in rivers and streams with diatoms have a long history, which has resulted in the development of the two basic conceptual and analytical approaches used today. First, based on the work of Kolkwitz and Marsson (1908), autecological indices were developed to infer levels of pollution based on the species composition of assemblages and the ecological preferences and tolerances of taxa (e.g., Butcher, 1947; Fjerdingstad, 1950; Zelinka & Marvan, 1961; Lowe, 1974; Lange-Bertalot, 1979). Second, Patrick's early monitoring studies (Patrick, 1949; Patrick et al., 1954; Patrick & Strawbridge, 1963) relied primarily on diatom diversity as a general indicator of river health (i.e., ecological integrity), because species composition of assemblages varied seasonally and species richness varied less. Thus, the concepts and tools for assessing ecosystem health and diagnosing causes of impairment in aquatic habitats, particularly rivers and streams, were established and developed between 50 and 100 years ago.

The many advances in the use of diatoms and other algae for monitoring stream and river quality have been reviewed by Patrick (1973) and, more recently, by Stevenson and Lowe (1986), Round (1991), Whitton et al. (1991), Coste et al. (1991), Whitton and Kelly (1995), Rosen (1995), and Lowe and Pan (1996). There are three major objectives for this chapter. First, we emphasize the importance of designing environmental assessments so that many approaches are used and results are based on rigorous statistical testing of hypotheses. Second, we review the many characteristics of diatom assemblages that could be used in assessments and the methods and indices of assessment. Finally, we develop the concept that assessments are composed of two processes, determining the ecological integrity of the habitat and inferring the causes of impairment of river ecosystems. We also describe approaches for developing a diatom index of biotic integrity (IBI).

Rationale for using diatoms

Rivers and streams are complex ecosystems in which many environmental factors vary on different spatial and temporal scales. These variables can range from climate, land use, and geomorphology in the watershed (e.g., Richards *et al.*, 1996) to the physical, chemical, and biological characteristics of rivers and streams. In most environmental studies, as many variables as possible should be measured to infer environmental conditions in a habitat (Barbour *et al.*, 1995; Norris & Norris, 1995). Measurement of all physical and chemical factors that could be important determinants of ecosystem integrity is impractical. Biological indicators respond to altered physical and chemical conditions that may not have been measured. Biological indicators, based on organisms living from 1 day to several years, provide an integrated assessment of environmental conditions in streams and rivers that are spatially and temporally highly variable. Biological indicators are important parts of environmental assessments because protection and management of these organisms are the objectives of most programs.

Using diatom indicators of environmental conditions in rivers and streams is important for three basic reasons: their importance in ecosystems, their utility as indicators of environmental conditions, and their ease of use. Diatom importance in river and stream ecosystems is based on their fundamental role in food webs (e.g., Mayer & Likens, 1987; for review see Lamberti, 1996), oxygenation of surface waters (J. P. Descy, personal communication), and linkage in biogeochemical cycles (e.g., Newbold *et al.*, 1981; Kim *et al.*, 1990; Mulholland, 1996). As one of the most species-rich components of river and stream communities, diatoms are important elements of biodiversity and genetic resources in rivers and streams (Patrick, 1961). In addition, diatoms are the source of many nuisance algal problems, such as taste and odor impairment of drinking water, reducing water clarity, clogging water filters, and toxic blooms (e.g., Palmer, 1962).

Diatoms are valuable indicators of environmental conditions in rivers and streams, because they respond directly and sensitively to many physical, chemical, and biological changes in river and stream ecosystems, such as temperature (Squires *et al.*, 1979; Descy & Mouvet, 1984), nutrient concentrations (Pringle & Bowers, 1984; Pan *et al.*, 1996), and herbivory (Steinman *et al.*, 1987a; b; McCormick & Stevenson, 1989). The species-specific sensitivity of diatom physiology to many habitat conditions is manifested in the great variability in biomass and species composition of diatom assemblages in rivers and streams (e.g., Patrick, 1961). This great variability is the result of complex interactive effects among a variety of habitat conditions that differentially affect physiological performance of diatom species, and thereby, diatom assemblage composition (Stevenson, 1997). Stevenson (1997) organizes these factors into a hierarchical framework in which higher level factors (e.g., climate and geology) can restrict effects of low-level factors. Low-level, proximate factors, such as resources (e.g., light, N, P) and stressors (e.g., pH, temperature, toxic substances), directly affect diatoms. At higher spatial and temporal levels, effects of resources and stressors on diatom assemblages can be constrained by climate, geology, and land use (Biggs, 1995; Stevenson, 1997). The sensitivity of diatoms to so many habitat conditions can make them highly valuable indicators, particularly if effects of specific factors can be distinguished. Knowing the hierarchical relations among factor effects will help to make diatom indicators more precise.

Diatoms occur in relatively diverse assemblages, and species are relatively easily distinguished when compared to other algae and invertebrates that also have diverse assemblages. Diatoms are readily distinguished to species and subspecies levels based on unique morphological features, whereas many other algal classes have more than one stage in a life cycle, and some of those stages are either highly variable ontogenically (e.g., blue-green algae), cannot be distinguished without special reproductive structures (e.g., Zygnematales), or cannot be distinguished without culturing (many unicellular green algae). Diverse assemblages provide more statistical power in inference models. Identification to species level improves precision and accuracy of indicators that could arise from autecological variability within genera. Diatoms are relatively similar in size (although varying many orders of magnitude in size) compared to variability among all groups of algae, so assemblage characterizations accounting for cell size (biovolume and relative biovolume) are not as necessary as when using all groups of algae together.

Diatoms have one of the shortest generation times of all biological indicators (Rott, 1991). They reproduce and respond rapidly to environmental change and provide early warning indicators of both pollution increases and habitat restoration success. Diatoms can be found in almost all aquatic habitats, so that the same group of organisms can be used for comparison of streams, lakes, wetlands, oceans, estuaries, and even some ephemeral aquatic habitats. Diatoms can be found on substrata in streams, even when the stream is dry; so they can be sampled at most times of the year. Diatom frustules are preserved in sediments and record habitat history, if undisturbed sediments can be found in lotic ecosystems, such as in reservoirs or deltas where rivers and streams drain (Amoros & van Urk, 1989). Combined cost of sampling and sample assay are relatively low when compared to other organisms. Samples can be archived easily for long periods of time for future analysis and long-term records.

Methods

RESEARCH DESIGN

The design of projects should be based on hypothesis testing, characterization of possible error variation in assessments, and a sound scientific approach, so that results provide reliable information for protection of natural resources. The choice of methods is highly dependent upon the habitat being assessed, the objectives of assessment, and frequently, the budget of the project. Therefore, the costs and benefits of different approaches should be considered with any project.

Approaches can be generally classified as observational and experimental.

Observational studies typically generate hypotheses for changes in ecosystems based on correlations between spatial or temporal changes in environmental conditions and in diatom communities. Observational approaches are particularly valuable in surveys of river and stream quality over broader regions or where experiments are not practical. In observation studies, assessment of variance associated with correlation of diatom assemblage attributes and environmental conditions or by duplicate samples in selected numbers of sites (e.g., quality assurance approach recommended in rapid bioassessment protocols, Plafkin *et al.*, 1989) is important to determine the precision of environmental inferences based on diatom assemblages. Experiments involve manipulation of specific environmental conditions to test the causes of ecological change. Because experiments that manipulate the presence and absence of power plants, sewage treatments plants, or land use in a watershed are not practical, experiments are usually conducted by manipulating specific physical and chemical conditions in microcosms or mesocosms.

A number of experimental systems have been developed to work with benthic or planktonic algae. These systems range from in-stream, stream-side, to laboratory systems. Within streams, benthic algal responses to anthropogenic impact can be tested with relatively great ecological realism in tubes oriented parallel to stream flow (Petersen *et al.*, 1983; Pringle, 1990) or with ion-diffusing substrata (e.g., Pringle & Bowers, 1984; Lowe *et al.*, 1986; McCormick & Stevenson, 1989; Chessman *et al.*, 1992; Pan & Lowe, 1994). However, in-stream systems are not practical in streams with frequent floods (personal experience). Stream-side facilities offer an alternative to in-stream channels, and they enable greater control of current, nutrient enrichment, light, and grazing pressure (Bothwell & Jasper, 1983; Bothwell, 1989; Peterson & Stevenson, 1992; Rosemond, 1994). Stream-side facilities can retain the natural environmental variability of many stream characteristics, but some large-scale features of natural environments, such as fish effects and habitat heterogeneity, may be lost. Laboratory streams provide even greater control (McIntire & Phinney, 1965; Lamberti & Steinman, 1993; Hoagland *et al.*, 1993), but environmental realism is the trade-off for the convenience, control, and ability to use toxic chemicals freely in laboratory streams. Sealed containers suspended in the water column, riverside pools, and laboratory microcosms have been used to experimentally manipulate the conditions in which plankton grow (Côté, 1983; Ghosh & Gaur, 1990; Thorp *et al.*, 1996).

A combination of survey (correlational) and experimental research provides the foundation for strong inference of the causes of ecological changes over broader regions and times. Surveys can produce ecologically realistic hypotheses based on quantitative analyses of complex and natural systems. These hypotheses can then be tested by manipulating specific environmental conditions in replicated experiments. Results of experiments in which specific physical, chemical, or biological conditions are manipulated, typically in only one ecosystem or laboratory, can then be compared to the correlations observed in large-scale surveys of many ecosystems. When several lines of evidence (e.g., diatom assemblage attributes) in experiments and surveys indicate the same ecological relationships, then results provide a reliable source of information for relating anthropogenic causes to environmental impacts and for managing our natural resources.

SAMPLING PLANKTON AND PERIPHYTON

The advantages of sampling plankton and periphyton vary with size of the river and objective of the research. Plankton should usually be sampled in large rivers and periphyton should be sampled in shallow streams, where each, respectively, are the most important sources of primary production (Vannote *et al.*, 1980). However, periphyton sampling could be more appropriate than phytoplankton in large rivers if assessing point sources of pollution and if high spatial resolution in water quality assessment are objectives. If greater spatial integration is desired, then plankton sampling even in small streams may be an appropriate approach. For example, fewer sampling stations should be necessary to assess pollution throughout a watershed if plankton were sampled instead of periphyton. Suspended algae originate from benthic algae in small rivers and streams and are transported downstream (Swanson & Bachman, 1976; Müller-Haeckel & Håkansson, 1978; Stevenson & Peterson, 1991). Therefore, plankton may provide a good spatially integrated sample of benthic algae in a stream. Further studies of the value of sampling plankton in small rivers and streams are warranted for assessment of watershed conditions.

Benthic algae on natural substrata and plankton should be sampled in stream assessments whenever objectives call for accurate assessment of ecosystem components (Aloi, 1990; Cattaneo & Amireault, 1992) or when travel costs to sites are high. Artificial substratum sampling is expensive because it requires two separate trips to the field, and artificial substrata are highly susceptible to vandalism and damage from floods. One problem with periphyton samples on natural substrata is that they can be highly variable. Composite sampling approaches have been used to reduce within habitat variability when sampling stream and river periphyton. Composite samples are collected by sampling periphyton on rocks at random locations along three or more random transects in a habitat and combining the samples into one composite sample (Porter *et al.*, 1993; Pan *et al.*, 1996).

Using artificial substrata is a valuable approach when objectives call for precise assessments in streams with highly variable habitat conditions, or when natural substrata are unsuitable for sampling. The latter may be the case in deep, channelized, or silty habitats. Benthic algal communities on artificial substrata are commonly different than those on natural substrata (e.g., Tuchman & Stevenson, 1980). However, when the ecology of the natural habitat is simulated, benthic diatom assemblages developing on artificial substrata can be similar to assemblages on natural substrata (for review see Cattaneo & Amireault, 1992). Cattaneo and Amireault recommend cautious use of artificial substrata, because algal quantity often differs and non-diatom algae are underrepresented on artificial substrata. When detecting changes in water quality is a higher priority than assessing effects of water quality on the specific natural assemblage of periphyton in that habitat, then the advantages of the high precision and sensitivity of diatoms on artificial substrata for assessing the physical and chemical conditions in the water may outweigh the disadvantages of questionable simulation of natural communities.

Assemblage characteristics used in assessment

Diatom community characteristics have been used to assess the ecological integrity of rivers and streams and to diagnose causes of degradation. Many diatom indices assess the ecological integrity of an ecosystem, i.e., the similarity of an assessed ecosystem to a reference ecosystem. Ecological integrity is legally established in the US in the Clean Water Act as 'chemical, physical, and biological integrity' (see Adler, 1995). We distinguish conceptually between ecological integrity and biotic integrity. Karr and Dudley (1981) define biotic integrity as 'the capability of supporting and maintaining a balanced, integrated, adaptive community of organisms having a species composition, diversity, and functional organization comparable to that of natural habitat of the region'. Ecological integrity, however, includes more than just biotic integrity; it includes the physical and chemical integrity of the habitat as well. In practice, biotic integrity has been defined as differences between community characteristics in assessed habitats and reference habitats. Reference habitats are typically upstream from the assessed habitat or are similarly sized streams and rivers with highly regarded attributes (little alteration of the watershed and stream channel by humans) within a region (see Hughes, 1995 for a discussion). In addition to knowing the degree of degradation, development of targeted remedial management strategies requires knowing the causes of degradation. Thus, diatom indices that infer specific causes of habitat degradation will be valuable for better diagnosis of the causes of ecosystem degradation.

Diatom assemblage characteristics are typically used in conjunction with the characteristics of entire periphyton or plankton assemblages in assessments, thus accounting for changes in other algae and microorganisms that occur in benthic and planktonic samples. These characteristics occur in two categories, structural and functional (Table 2.1). Species composition and biomass (measured as cell density or biovolume) are the only diatom assemblage characteristics that can be distinguished from other algae and microbes in periphyton and plankton samples. It is worthwhile to note that chlorophyll (chl) a, ash-free dry mass (AFDM), chemical composition, and functional characteristics of diatom assemblages cannot be distinguished from other algae, bacteria, and fungi in a periphyton or plankton sample. Little is known about the accuracy of diatom biomass assessments with chl c.

Table 2.1. *Characteristics of algal assemblages that could be used to assess the ecological integrity of a habitat*

Parameter	Assays	Citations
Biomass – structural	AFDM, chl *a* and other pigments, cell densities, cell biovolumes, other elements that are most common in microbial biomass (N or P)	APHA, 1992
Species composition – structural characteristic	relative abundances of species, composite diversity, species richness, species evenness, pigment ratios	common throughout the literature, e.g., Schoemann, 1976; Lange-Bertalot, 1979
Autecological indices – structural characteristic	Pollution Tolerance Index, SPI, GDI, DAIpo, DAI-pH, DAI-TP	Lange-Bertalot, 1979; Rumeau & Coste, 1988; Watanabe *et al.*, 1988; Prygiel & Coste, 1993; Kelly *et al.* 1995; Pan *et al.*, 1996
Morphology – structural characteristic	larger cells with UV effects, deformed frustules with metals	Bothwell *et al.*, 1993; McFarland *et al.*, 1997
Chemical ratios – structural characteristic	chl *a*:AFDM, chl *a*: phaeophytin, N:P, N:C, P:C, heavy metals: AFDM	Weber, 1973; Peterson & Stevenson, 1992; Humphrey & Stevenson, 1992; Biggs, 1995
Growth and dispersal rates – functional characteristic	reproduction rate, growth rate, accrual rate, immigration rate, emigration rate	Müller-Haeckel & Hakansson, 1978; Stevenson, 1983; 1984, 1986b, 1990, 1995, 1996; Biggs, 1990; McCormick & Stevenson, 1991; Stevenson & Pan, 1995
Metabolic rates – functional characteristic	photosynthetic rate, respiration rate, phosphatase activity	Blanck, 1985; Mullholland & Rosemond, 1992; Hill, 1997

Note:
SPI-Species Pollution Index; GDI, Indice Générique Diatomique; DAI, Diatom Autecological Index.

Periphyton and phytoplankton biomass can be estimated with assays of dry mass (DM), AFDM, chl a, cell densities, cell biovolumes, and cell surface area (Aloi, 1990; APHA, 1992). All these estimates have some bias in their measurement of algal biomass. Dry mass varies with the amount of inorganic as well as organic matter in samples. AFDM varies with the amount of detritus as well as the amount of bacteria, fungi, microinvertebrates, and algae in the sample. Chl a: algal C ratios can vary with light and N availability (Rosen & Lowe, 1984). Chl c density in habitats could be a good indicator of diatom biomass in a habitat (APHA, 1992). Cell density: algal C ratios vary with cell size and shape. Even cell volume: algal C ratios vary among species, particularly among some classes of algae, because vacuole size in algae varies (Sicko-Goad et al., 1977). Cell surface area may be a valuable estimate of algal biomass because most cytoplasm is adjacent to the cell wall. Elemental and chemical mass (other than chl a, such as μg P/cm^2, μg N/cm^2) per unit area could also be used to assess algal biomass, but many of these methods of assessing biomass have not been studied extensively.

We recommend using as many indicators of algal biomass as possible. We typically do not restrict our assays of algal biomass to diatom density and biovolume. We usually assess chl a and AFDM of samples, and count and identify all algae to the lowest possible taxonomic level in Palmer cells or wet mounts to determine algal cell density and biovolume. In ecological studies where distinguishing live and dead cells is important, diatoms are counted in syrup (Stevenson, 1984b) or high refractive-index media using vapor substitution (Stevenson & Stoermer, 1981; Crumpton, 1987). In assessment studies, where distinguishing live and dead diatoms has not been shown to be important, we count acid-cleaned diatoms in Naphrax® to ensure the best taxonomic assessments. Conceptually, counting dead diatoms that may have drifted into an area or persisted from the past should only increase the spatial and temporal scale of an ecological assessment.

Periphyton and phytoplankton biomass is highly variable in streams and rivers and, periphyton biomass in particular, has not been regarded as a reliable indicator of water quality (Whitton & Kelly, 1995; Leland, 1995). According to theories of community adaptation to stress (Stevenson, 1997), biomass should be less sensitive than species composition to environmental stress, because communities can adapt to environmental stress by changing species composition (see discussion on p. 20). Another problem with using algal biomass as an indicator of nutrient enrichment and toxicity is that low biomass may be the result of a recent natural physical and biotic disturbance (e.g., Tett et al., 1978; Steinman et al., 1987a) or toxicity (e.g., Gale et al., 1979). A more reliable indicator of environmental impacts on algal and diatom biomass in a habitat may be the peak biomass that can accumulate in a river or stream after a disturbance (Biggs, 1996; Stevenson, 1996). Peak biomass is the maximum biomass in the phytoplankton or the periphyton that accumulates after a disturbance. These maxima develop during low discharge periods, usually seasonally, for both phytoplankton and periphyton and, theoretically, should be highly correlated to nutrient and light availability in a system. Bothwell (1989) showed the clear relationship between phosphate concentration and peak biomass of periphytic diatoms. Peak biomass is also a valuable parameter because it indicates the potential for nuisance-levels of algal biomass accumulation.

DIVERSITY

Enumeration and identification of periphyton and plankton samples, and particularly diatom assemblages within them, provide the basis for many indices of ecological integrity and causes of degradation. Environmental inferences can be based on single indicator species and genera, such as *Gomphonema parvulum* (Kütz.) Kütz. and *Nitzschia amphibia* Grun. (e.g., Raschke, 1993), groups of indicator species (e.g., Schoeman, 1976), or the whole assemblage. Whole assemblage indicators include various diversity indices, indices of community similarity, and indices that infer environmental conditions in habitats that are based on the autecologies of diatoms in the habitat, i.e., autecological indices.

Many indices have been developed to characterize the number of species in a sample (species richness), the evenness of species abundances, and composite diversity. Composite diversity is represented in indices that respond to changes in both richness and evenness (e.g., Shannon, 1948; Simpson, 1949). High correlation between all of these indices has been observed (Archibald, 1972), probably because composite diversity and species richness measurements are highly dependent on evenness of species abundances in short counts (e.g., 500 valves).

Assessment of species richness in a habitat is particularly problematic because species numbers are highly correlated to species evenness in counts when a predetermined number of diatoms is counted (e.g., 500 valves). Better assessments of species richness can be determined by developing the relationship between species numbers observed and the number of cells that have been counted. Species richness can be defined as the number of species observed when no new species are observed with a specified additional counting effort. Alternatively, non-linear regression can be used to determine the asymptote of the relationship between number of species observed and number of cells counted (the asymptote is an estimate of the number of species in the sample). The precision of the asymptote, and thus the estimate of species richness, will be reported by most statistical programs. Stratified counting efforts can be employed to assess different community parameters, such as relative abundance of dominant taxa and species richness, by identifying and counting all cells until a prespecified number of cells is counted to determine relative abundance of the dominants; then only count cells to keep track of the number of cells observed and list species that have not been observed to estimate species richness. When the budget permits, species richness could be determined with long diatom counts (3000–8000 cells) and estimation using the assumption that the number of species in different density categories fit a log-normal curve (Patrick *et al.*, 1954). Assessments of species evenness do vary with the evenness parameter used (e.g., Hurlbert's vs. Alatalo's evenness, Hurlbert, 1971; Alatalo, 1981), but the utility of differential sensitivity of these characteristics has not been extensively investigated.

The best use of diversity-related indices in river and stream assessments is probably as an indicator of changes in species composition when comparing impacted and reference assemblages (Stevenson, 1984c; Jüttner *et al.*, 1996). Some investigators have found that diversity decreases with pollution (e.g., Rott & Pfister, 1988), that diversity can increase with pollution (e.g., Archibald, 1972; van Dam, 1982), and that diversity changes differently depending upon the type of pollution (Jüttner *et al.*, 1996). Patrick (1973) hypothesized ambiguity in diversity assessments of pollution when using composite diversity indices because of differing effects of pollutants on species richness and evenness. Patrick (1973) predicted that some pollutants (e.g., organic pollution) would differentially stimulate growth of some species and thereby decrease evenness of species abundances. She also predicted that toxic pollution could increase evenness and that severe pollution could decrease species numbers (Patrick, 1973). Therefore, depending upon the kind and severity of pollution, human alteration of river and stream conditions could decrease or increase the diversity that was characterized with composite indices that incorporate both the richness and evenness elements of diversity. Future research should test Patrick's theories more thoroughly.

SPECIES COMPOSITION

Changes in species composition tend to be the most sensitive responses of diatoms and other microbes to environmental change (van Dam, 1982; Niederlehner & Cairns, 1994). However, the temporal scale of the observation is important. In the very short term of a bioassay, algal metabolism responds sensitively to environmental stress (Blanck, 1985). Quickly, however, communities can adapt to many environmental stresses by changing species composition and, thereby, may achieve biomass and metabolic rates like those in unimpacted areas (Stevenson, 1997). Diatom assemblages in most field situations have had this time to adapt to environmental moderate stresses by changing species composition. Therefore, in most field sampling situations, when stresses have existed long enough for immigration of new species and accrual of rare taxa that are stress-tolerant, species composition should be more sensitive to changes in environmental conditions than changes in biomass or metabolic rates (e.g., Schindler, 1990).

Ordination, clustering, and community similarity indices are three approaches to assess variation in species composition among communities. Ordination (correspondence analysis, detrended correspondence analysis) is typically used to assess the multidimensional pattern in the relationships between assemblages based on species composition. Species and sample scores are related to ordination axes and can be used to determine which species were most important in groups of samples. Environmental conditions can also be related to the ordination axes by using canonical correspondence analysis and detrended canonical correspondence analysis (ter Braak, 1987–1992). Ordination and clustering can be used to show which assemblages are the most different from other assemblages, which may be caused by anthropogenic impacts (e.g., Chessman, 1986; Stevenson & White, 1995).

Community dissimilarity indices (for review see Pielou, 1984) can be used to test specific hypotheses about correlations between changes in species composition and the environment (Cairns & Kaesler, 1969; Moore & McIntire, 1977; Peterson & Stevenson, 1989, 1992). For example, Stevenson *et al.* (1991) used the relationship between periphytic diatom assemblage dissimilarity and difference in assemblage ages as an indicator of the rate of change in community composition, in this case an index of succession rate. Community dissimilarity indices can also be used to distinguish among groups of assemblages by testing the hypothesis that dissimilarity among assemblages within a group is significantly less than between groups of assemblages. Cluster analysis (e.g., TWINSPAN, Hill, 1979) groups assemblages based on the similarity in species composition between assemblages (Leland, 1995). Discriminant analysis can also be used to determine whether species composition of groups of assemblages differ significantly between clusters (ter Braak, 1986; Peterson & Stevenson, 1989).

AUTECOLOGICAL INDICES

Autecological indices use the relative abundance of species in assemblages and their ecological preferences, sensitivities, or tolerances to infer environmental conditions in an ecosystem. The sensitivity and tolerance of diatoms to a number of environmental characteristics, such as eutrophication, organic pollution, heavy metals, salinity, pH, and pesticides, are known to differ among species (Stevenson, 1996). These species-specific sensitivities and tolerances can be used to infer environmental conditions in a habitat (e.g., Lange-Bertalot, 1978).

Many diatom autecological indices of water pollution in rivers have been developed and are in widespread use. Diatom autecological indices can infer specific or general environmental conditions. Most are indicators of organic pollution of water (Palmer, 1969; Slàdecek, 1973; Descy, 1979; Lange-Bertalot, 1979; Watanabe *et al.*, 1986) and are reviewed by Coste *et al.* (1991) and in Whitton and Kelly (1995). The indices can be based on the detailed characterization of assemblages with many species (Prygiel, 1991 used 1550 species), or they can be simplified to only identify genera or a few species for use by non-specialists (Rumeau & Coste, 1988; Round, 1993 as cited in Whitton & Kelly, 1995).

Highly specific assessments of the environmental stressors affecting periphyton, and perhaps other organisms, can be made using relatively simple inference models. Clearly, identifying the stressors that affect aquatic ecosystems is important for watershed management so that effective strategies can be developed for solving environmental problems. For example, a diatom index for siltation is the percent of motile diatoms (e.g., *Cylindrotheca*, *Gyrosigma*, *Navicula*, *Nitzschia*, and *Surirella*) in a community (modified from Bahls, 1993). Based on *Epithemia* and *Rhopalodia* having blue-green endosymbionts (DeYoe et al., 1992) and responding when N is low in streams (Peterson & Grimm, 1992), the percentage of these two genera in assemblages could be used as an indicator of low N. The percentage of *Eunotia* in a habitat can be used to infer the pH of habitats, particularly the relative pH of two habitats when diatom assemblages are compared. The development of eutrophication indices has been advanced by Steinberg and Schiefele (1988). Similar efforts of development and testing of indices for pH and salinity have also been conducted with mixed results (for review see Schiefele and Schreiner, 1991). In most cases, these indices were based on autecological classification scales that were rather coarse ($<$ six levels) and on autecological characterizations for populations in lakes.

Remarkably precise autecological indices of environmental conditions that work across broad geographic scales can be developed based on the great wealth of autecological information in the literature and a simple formula (Zelinka & Marvan, 1961):

$$\Sigma p_i \Theta_i / \Sigma p_i$$

in which p_i is the proportional abundance of the ith taxon (p_i, for $i = 1, 2, \ldots, S$ species) and Θ_i is the autecological rank of a species for a specific stressor. Much autecological information in the literature (e.g., Hustedt, 1957; Cholnoky, 1968; Slàdecek, 1973; Lowe, 1974; Descy, 1979; Lange-Bertalot, 1979; Beaver, 1981; Fabri & Leclercq, 1984; van Dam et al., 1994) uses a ranking system (for reviews see Lowe, 1974; van Dam et al., 1994) with six or less ranks for a specific environmental stressor (pH, temperature, salinity, organic pollution, etc.). See van Dam et al. (1994) for additional references. Use of this autecological information in simple metrics of environmental stressors can be useful if more accurate information is not known about diatom autecologies.

The weighted-average inference model approach (ter Braak & van Dam, 1989) offers an opportunity for development of more accurate indicators of stream and river conditions if accurate characterizations of species autecologies along specific environmental gradients are known. Weighted average indices are based on the relative abundances of taxa in an assemblage, the optimum environmental condition for a taxon (v_i), and the tolerance of species to variation in environmental conditions (t_i; ter Braak & van Dam, 1989).

$$\Sigma p_i v_i t_i / \Sigma p_i t_i$$

The lack of information about the autecology of some species in samples may reduce the precision of these indices, but can be corrected by the denominator

R. J. STEVENSON AND Y. PAN

in the equation, which accounts for exclusion of some species in calculations of indices (perhaps because autecological information for those species is not known. This approach has been used widely in the development of diatom indices of lake pH, nutrient concentrations, and salinity (Kingston & Birks, 1990; Fritz, 1990; Fritz *et al.*, 1991; Sweets, 1992; Kingston *et al.*, 1992; Hall & Smol, 1992; Cumming *et al.*, 1992; Cumming & Smol, 1993; Reavie *et al.*, 1995; chapters in this book). Autecological indices have been developed to infer conductivity and TP in wetlands (Pan and Stevenson, 1996; Stevenson *et al.*, in press). Pan *et al.* (1996) have used this approach to develop and test diatom indices of pH and TP in streams.

The weighted-average approach requires assessing diatom assemblages and water chemistry characteristics in a large number of streams (Pan *et al.*, 1996). That information is then used to determine the optima and tolerances of diatoms to specific environmental gradients using a software program, WACALIB (Line *et al.*, 1994). Diatom indices can then be tested by a statistical technique called jackknifing with the software program, CALIBRATE (Juggins & ter Braak, 1992). By jackknifing, all sites are, in turn, left out of the development of the index when the environmental condition at that site is predicted. Even though the precision of weighted average diatom indices of stream conditions, such as pH and TP, can be relatively high for the region and for the time studied (Pan *et al.*, 1996), the transferability of indices to other regions and times needs to be tested.

Diatom-based autecological indices could be particularly valuable in stream and river assessments because one-time assay of species composition of diatom assemblages in streams could provide better characterizations of physical and chemical conditions than one-time measurement of those conditions. Precise characterization of environmental conditions in rivers and streams is difficult because of the high variability in discharge, water chemistry, temperature, and light availability associated with weather-related events. Charles (1985) showed that diatom inferred pH was a better characterization of mean annual pH than one-time sampling of pH for lakes. The RMSE for a weighted-average index of total phosphorus (TP) based on diatom species composition in streams of the Mid-Atlantic Highlands (USA) was 0.32 log(TP μg/l) (Pan *et al.*, 1996) and was substantially less than the range in TP concentration that is commonly observed in streams (1.0–4.0 log (TP μg/l) was observed over an 8-week period in Kentucky streams, R. J. Stevenson, personal observation). Future research should rigorously test the hypothesis that diatom-inferred environmental conditions are more accurate indicators of river and stream conditions than one-time sampling of water chemistry.

Considerable debate exists over how many diatoms to count and what level of taxonomic resolution is necessary to provide valuable autecological assessments. The answer to these questions again depends on the objectives and budget of the project. If one objective of a study is to determine the number and identity of most of the species in an assemblage, then 3000–8000 cells may need to be counted (Patrick *et al.*, 1954). Alternatively, if a precise characterization of just the dominant taxa is necessary, then between 500 and 1000 diatoms should be counted, depending upon the number and evenness of species in the community (Pappas & Stoermer, 1996). Some counting protocols require counting until at least ten cells of the ten dominant taxa have been observed, which should precisely characterize the relative abundance of the ten dominant taxa. Stratified counting approaches should be used when communities are dominated by one or a few species. For example, stop counting dominants after hundreds of them have been observed; record the initial proportion of sample observed; continue counting other taxa until precision for subdominants is established and record the final proportion of sample observed; then adjust counts of dominants by multiplying their counts by the final proportion of sample observed and dividing by the initial proportion of sample observed. Some program objectives and budgets may call for shorter counts ($\approx$300 cells) that only characterize the dominant taxa, with which coarse ecological inferences could be made.

Longer counts and identification to species level will surely provide more precise and accurate information than short counts and identification to the generic level. However, recent comparative studies show that generic level indices (GPI, Rumeau & Coste, 1988) perform nearly as well as species level indices (Coste *et al.*, 1991; Prygiel & Coste, 1993; Kelly *et al.*, 1995) with the autecological information that was used. More accurate autecological information is now being developed (e.g., Pan *et al.*, 1996) and shows important within-genus differences between species that provide valuable additional information for assessing specific environmental conditions.

The distinction between counting valves and cells should also be made. The objective of counts is to make enough observations of the dominant organisms to be able to precisely quantify their abundance in the habitat. In this case, an observation may be a frustule with protoplasm in live diatom counts or it could be one or two valves in a cleaned diatom sample, depending upon whether a whole frustule was observed or not. If counting only live diatoms in wet or syrup mounts (Stevenson, 1984*b*), then whole frustules are counted as a unit. If counting cleaned diatom samples, then valves must be counted because the valves of some frustules separate and the tendency to separate may be species-specific. When valves are broken, count pieces of the valve if it includes some distinctive part of the valve, like the central area. If we assume that half of the valves separate, then it would be necessary to count 750 valves (a complete frustule being counted as two valves) in a cleaned diatom sample to make the same number of observations as in a 500-cell count of whole frustules.

MORPHOLOGICAL CHARACTERISTICS

Little research has been conducted to evaluate the effects of stressors on diatom size, striae density, shape, and other morphological characteristics. Sexual reproduction and auxospore formation in high density periphyton assem-

　　　R. J. STEVENSON AND Y. PAN

blages after substantial colonization was hypothesized to be related to lower nutrient availability than immediately after rains and high discharge (Stevenson, 1990). UV radiation may cause an increase in cell size and abundance of stalked diatoms (Bothwell *et al.*, 1993). Aberrant diatom shape, such as indentations and unusually bending in frustules, has been shown to be related to heavy metal stress in streams (McFarland *et al.*, 1997). Bahls (personal communication) observed that 19% of diatoms in a metal contaminated stream had unusual shapes. More research is justified to pursue this potentially sensitive metric to important stream stressors. Cell size, striae density, and shape may also respond to environmental conditions.

CHEMICAL CHARACTERISTICS

Sediments and periphyton are important sinks for nutrients as well as many toxic inorganic and organic chemicals (Kelly & Whitton, 1989; Genter, 1996; Hoagland *et al.*, 1996). In addition to their potential as indicators of biomass, chemical characteristics of periphyton may provide valuable indications of the environmental conditions that affect periphytic diatoms. Total nitrogen (TN) and TP of periphyton communities have been used by Biggs (1995) to infer nutrient limitation and eutrophication in habitats (see also Humphrey & Stevenson, 1992). Harvey and Patrick (1968) used assays of radionuclides in periphyton to assess pollution. Kelly and Whitton (1989) demonstrate the accumulation of heavy metals in periphyton. Similarly, assays of particulate chemistry of the water column may provide insight into the chemical environment of phytoplankton that is not evident from assay of dissolved chemistry alone. For example, TP is used routinely to assess trophic status of lakes (Vollenweider & Kerekes, 1981).

FUNCTIONAL CHARACTERISTICS

Functional characteristics of diatom-dominated assemblages, such as photosynthesis and respiration rates, phosphatase activity, and growth rate have been used as indicators of environmental conditions in streams and rivers. Phosphatase activity is a valuable indicator of P limitation (Healey & Henzel, 1979; Mulholland & Rosemond, 1992). Photosynthesis and respiration can be used as measures of community productivity and health, but these assays are not commonly used in field surveys. Hill *et al.* (1997) uses the response of periphyton respiration rate to experimentally manipulated stressors as an indicator of those stressors in the habitat. Since assemblages may adapt to environmental stressors by changing species composition and maintaining functional ecological integrity (Stevenson, 1997), Hill *et al.* predict that respiration rates of assemblages will not be inhibited by exposure to that stressor.

Growth rate (Stevenson, 1996) has been used recently as an indicator of algal biomass production and can be assessed at population as well as

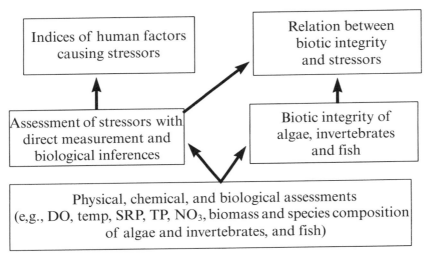

Indices of human factors causing stressors		Relation between biotic integrity and stressors
Assessment of stressors with direct measurement and biological inferences		Biotic integrity of algae, invertebrates and fish

Physical, chemical, and biological assessments
(e,g., DO, temp, SRP, TP, NO₃, biomass and species composition
of algae and invertebrates, and fish)

Fig 2.1. Schematic diagram of an approach that could be used to develop a diatom index of biotic integrity. Independent assessments of biotic integrity and environmental stressors are produced from determinations of species composition and other attributes of assemblages.

assemblage levels. Schoeman (1976) and Biggs (1990) used growth rates as an indicator of nutrient limitation in habitats by resampling habitats after a short time (3–7 d). Assessment of differing responses of species growth rates to environmental conditions may enhance the simple characterization of the autecology of species based on changes in their relative abundance.

Caution should be exercised when using metabolic rates as indicators of population performance or community health. Metabolic rates are affected greatly by areal algal biomass (e.g., cells/cm^2), whether metabolic rates are normalized based on area or biomass. Areal metabolic rates increase with algal biomass, but biomass-specific metabolic rates decrease with algal biomass (Stevenson, 1990; Hill & Boston, 1991). Future research should quantify the biomass effect on metabolism so that this sensitive measure of algal assemblages can be used in assessments.

Assessment of biotic integrity and diagnosis of environmental stressors

Managing stream and river ecosystems calls for an assessment of the ecological integrity of the ecosystem and a diagnosis of causes of degradation (Fig. 2.1). Indices of biotic integrity (IBI) of aquatic invertebrates, fish, and morphology are being widely used to characterize streams (Karr, 1981; Hilsenhoff, 1988; Plafkin *et al.*, 1989; Lenat, 1993). Similar indices of biotic integrity have been developed for diatoms, but have not been widely used or tested (Kentucky Division of Water, 1994; Hill, 1997). These indices are referred to as multimetric

Table 2.2. *Metrics and scoring ranges for Metzmeier's diatom bioassessment index (DBI)*

Score	Taxa richness	Diversity	DTI	RA(s)	PSc
1	<20	<1.5	1.0–1.5	<0.1	<10
2	20–30	1.5–2.5	1.5–2.0	0.1–1	10–30
3	30–50	2.5–3.5	2.0–2.5	1.0–5.0	30–50
4	50–70	3.5–4.5	2.5–3.0	5.0–20	50–75
5	>70	>4.5	>3.0	20–100	75–100

Notes:

(Kentucky Division of Water, 1994; Metzmeier, personal communication). Taxa richness is the number of taxa observed in a 500–1000 valve count. Diversity is Shannon (1948) diversity. DTI is the diatom tolerance index which is calculated with the equation $\Sigma p_i v_i / \Sigma p_i$ in which proportional abundances (p_i) of all taxa (for $i = 1,2,\ldots,S$, where S is the Taxa Richness) are known and all taxa are assigned a pollution tolerance rank (v_i) from lowest tolerance (4) to highest (1). RA(s) is the sum of the % relative abundance of all sensitive species (for which $v_i = 4$). PSc is the % similarity of the assessed and reference site assemblages ($100 - 0.5 \Sigma |r_{ia} - r_{ir}|$ where r_{ia} and r_{ir} are the % relative abundances (0–100%) for $i = 1,2,\ldots,S$ species in the assessed and reference sites, respectively). Scores are averaged for all metrics to calculate the DBI. Scores from 1–2 indicate severe impairment, from 2–3 indicate moderate impairment, from 3–4 indicate good ecological integrity, and from 4–5 indicate excellent ecological integrity.

indices because they are composed (an average or sum) of more than one metric (i.e, an attribute of an assemblage that responds to human alterations of the watershed, *sensu* Karr & Chu, 1997).

Metzmeier's diatom index of biotic integrity (Kentucky Division of Water, 1994) includes taxa richness, Shannon diversity (Shannon, 1948), the pollution tolerance index (Lange-Bertalot, 1979), and proportion of species sensitive to pollution. These metrics are then translated to a score ranging from 1 to 5, i.e., poor to excellent. The average score for the metrics is Metzmeier's Diatom Bioassessment Index (Table 2.2). Hill (1997) recently developed an index of biotic integrity with ten metrics that successfully distinguished reference and impacted stream sites. Hill's index of periphyton integrity used metrics such as relative genera richness, ratio of diatom cells/all algal cells, live diatoms/dead diatoms, and metrics that decreased as biomass approached nuisance levels.

Even though multimetric indices of biotic integrity only summarize the information that diatoms infer about environmental conditions in habitats, they are valuable management tools. Comparison of indices for different trophic levels and geomorphology provides a spatially, temporally, and ecologically integrated assessment of ecosystem integrity in streams (Plafkin *et al.*, 1989; Hannaford & Resh, 1995). The integrated assessment of composite indices facilitates communication of environmental information to the public and facilitates comparative risk assessment.

A diatom IBI can be constructed in many ways. We should expect that it will be necessary to develop different IBI for different stream types and ecoregions, or to have different expectations in different regions, as is necessary for invertebrate and fish IBI (e.g., Ohio Environmental Protection Agency, 1987). The basic steps for developing IBI have been outlined by Plafkin *et al.*(1989). A list of diatom assemblage attributes (characteristics) should be established, such as taxa richness, Shannon diversity (Shannon 1948), % pollution tolerant species, % pollution sensitive species, and % species or generic similarity to reference conditions, such as used in Metzmeier's Diatom Bioassessment Index (Kentucky Division of Water, 1994). Attributes as simple as % *Eunotia* or % *Nitzschia* and other generic characteristics of assemblages could correlate with human activity in the region. With the possibility that diatoms and other microorganisms will be included on endangered species list (Lange-Bertalot Rot List, distributed at International Diatom Meeting, Tokyo 1996), then rare and endangered species could be included as an IBI metric. Metrics (i.e., attributes that correlate with human activities in watersheds; *sensu* Karr & Chu, 1997) are then selected from the list of possible attributes, based on correlation to human impairment of ecosystems and on lack of correlation with other attributes, converted to a specified scale, and then summed. For example, an IBI developed with ten metrics with scores adjusted to a scale ranging from 0 to 10 would have a minimum score of 0 and a maximum score of 100.

Other monitoring frameworks are conceptually similar to the multimetric approach, but differ in the actual statistic approaches used. Ordination techniques can be used to identify patterns in species composition of diatom assemblages among sites; then sites with unusual species assemblages can be identified, as Stevenson and White (1995) did with river phytoplankton. Autecological indices of environmental conditions can be used to infer why unusual species assemblages occur. Alternatively, correspondence analysis can be used to relate patterns in species assemblages to environmental factors to assess human impacts on river ecosystems, as O'Connell *et al.* (1997) assessed water quality with epiphytic diatoms on *Cladophora*. The RIVPACs approach also uses ordination and classification techniques to characterize reference conditions and deviation of test assemblages from predicted reference assemblages as an assessment of biotic integrity (Wright *et al.*, 1993; Norris, 1995).

Indices of biotic integrity have been criticized because they do not accurately diagnose the causes of ecosystem impairment (Suter, 1984). Usually this problem is remedied by correlating IBI and measured environmental conditions, which results in hypotheses for causes of ecosystem impairment. Experiments could be run to test these hypotheses. Certainty for diagnoses of impaired conditions can be increased with results of past research showing response of ecosystems to specific stressors. Alternatively, additional autecological indices of environmental conditions and biological response to stressors could be produced during assessment and integrated into the assessment of ecological conditions. One approach is to develop an IBI that includes metrics that diagnose ecological conditions (Fig. 2.1). Another approach is to have different indices for biotic integrity and the stressors affecting biotic integrity.

A diatom IBI that diagnoses the causes of environmental degradation could be developed with autecological indices that inferred, for example, organic pollution, nitrogen enrichment, phosphorus enrichment, heavy metal contamination, sedimentation, and pH. Causes of environmental degradation could be determined by breaking down the IBI and assessing which metric or metrics, thus environmental factor or factors, caused a low IBI. This approach is convenient, because one summary metric is produced that can be decomposed to diagnose causes of impairment. However, developing the IBI based on autecological indices of stressors introduces the argument of circularity (dependence) between IBI and metric, i.e., the assessment of biotic integrity is low because the assessment of stressors is high.

The second approach for developing diagnostic relationships between the IBI and metrics that infer causes of ecological impairment emphasizes the independence of the calculation of the metrics. In this second approach, a suite of metrics is developed in which an IBI is related to measured environmental conditions and to autecological indices of environmental stressors that confirm impact of specific stressors of the assessed biological assemblages (Fig. 2.1; Stevenson, 1998). Thus the IBI would be based on such metrics as nuisance algal accumulation, % sensitive species, number of rare and endangered species, and deviation of species composition and diversity from reference conditions, which do specifically infer levels of environmental stressors. Then measured physical and chemical conditions, as well as conditions inferred with autecological indices of environmental conditions, can be used to independently diagnose conditions causing impaired biotic conditions. This later approach calls for a better understanding of the effect of environmental stressors on biotic integrity of streams and rivers.

Another important step in managing ecosystem health is determining the human activities that cause stressors that impair ecological integrity (Fig. 2.1), because these activities can be managed directly. These human activities are referred to as contributing factors (Stevenson, 1998). Multimetric diatom indices for contributing factors, such as industry, sewage contamination, agriculture, logging, and mining, could be constructed with suites of metrics that indicate stressors associated with specific contributing factors (Table 2.3). A similar approach has been developed for macroinvertebrate response signatures (Yoder & Rankin, 1995). Industrial pollution, for example, could be indicated by toxic effects causing low biomass, low richness, high evenness and resulting in low values for Lange-Bertalot's Pollution Tolerance Index and high values of diatom autecological indices for toxic organic and inorganic compounds. In some cases, contributing factors could be uniquely identified by using just the diatom IBI framework. For example, industry, sewage discharge, and agriculture have very different effects on diatom assemblages because they have different stressors (Table 2.3). In other cases, other information may be necessary to uniquely identify contributing factors, such as using macroinvertebrates or fish assemblages, to distinguish between effects of agriculture and forestry that have similar stressors. Other, more

Table 2.3. *Organization of indicators that could be used in a hierarchical index of diatom biotic integrity*

Indicator of contributing factor	Indicators of stressors
Industry	Low peak biomass, low richness, high evenness, low pollution tolerance index, high DAI for heavy metals, high DAI for toxic organic compounds
Sewage	High peak biomass, low evenness, low Chl a : AFDM ratio, high DAI for saprobic conditions, high DAI for organic N, high DAI for TP
Agriculture	High peak biomass, high richness, low evenness, high Chl a : AFDM ratio, high DAI for DIN, high DAI for organic N, high DAI for TP, high DAI for sediment
Forestry	High peak biomass, high richness, low evenness, high Chl a : AFDM ratio, high DAI for DIN, high DAI for TP, high DAI for sediment
Mining	Low peak biomass, low richness, low evenness, high DAI for heavy metals, low DAI for pH, high DAI for SO_4

specific contributing factors, such as feed lots for livestock, strip mining, or impermeable surfaces (paved roads and parking lots), could be inferred from suites of diatom indicators of stressors. Note that some diatom indicators of stressors, such as the diatom autecological index for TP, could be used in more than one diatom index for contributing factors (Table 2.3).

Concluding remarks

Developing approaches and indices for environmental assessment is an interactive process between scientists and policy makers. Interactions should focus on furthering the scientists' understanding of policy issues and environmental problems and on helping policy makers translate their goals into testable hypotheses and practical approaches to solve environmental problems.

Priorities of some policy makers have been directed toward understanding the relationship between land use, physical and chemical changes in streams and rivers, and ecological responses. Land use and zoning are important strategies for slowing environmental degradation in areas under development pressure from urban and suburban sprawl. Assessing watershed level changes in stream and river conditions could be valuable for inferring the land-use effects and the geological and climatic factors that make watersheds sensitive and tolerant to land-use changes (Robinson *et al.*, 1995; Richards *et al.*, 1996; Kutka & Richards, 1996). Plankton assemblages may reflect large-scale changes in watersheds better than periphyton. In addition, paleoecological approaches in basins where sediments accrue may provide valuable insight into the correlations between historic changes in water quality and land use.

In addition to greater use of weighted average indices of environmental conditions, other opportunities for developing diatom indicators of environmental conditions can be found in species and characteristics of species that we have not yet explored. For example, rare species may hold very valuable information about habitat characteristics because they are sensitive to environmental conditions. The tolerance of species for environmental variation may provide an indicator that improves characterizations of environmental variability as well as providing mean or median environmental conditions.

Ecological theory should become a more important foundation for the environmental assessments and the indices used in environmental assessments. For example, Patrick's work on species diversity as an indicator of water quality was well founded in the ecological theory that was being explored at the time (Patrick *et al.*, 1954; Patrick & Strawbridge, 1963). Lange-Bertalot's (1978) emphasis on the sensitivity and tolerance of species to pollution provided a better predictive concept of species responses to pollution than species preferences for polluted conditions. By placing research in an ecological context and testing a broader ecological theory, such as Odum's predictions for stressed ecosystems (Odum *et al.*, 1979; Odum, 1985) as well as specific diatom-based hypotheses, results of our research become more transferable to assessments with other organisms and to assessments of other habitats with diatoms.

Diatoms are valuable indicators of ecological integrity and the environmental factors that impair rivers and streams. Public concern often focuses on the biodiversity other organisms, but partly because they do not appreciate the diversity, beauty, and ecology of algae, particularly diatoms. Greater efforts should be made to inform the public and develop their appreciation for algae in aquatic ecosystems.

Summary

Diatoms have a long history of assessing the ecological integrity of streams. Diatom assemblages respond rapidly and sensitively to environmental change and provide highly informative assessments of the biotic integrity of streams and causes of ecosystem impairment. Periphytic diatoms from natural and artificial substrata are usually sampled from streams and small rivers, but plankton provide valuable assessments of conditions in large rivers. Structural and functional characteristics of diatom communities can be used in bioassessments, but relative abundances of diatom genera and species are usually used as the most valuable characteristics of diatom assemblages for bioassessment. Using these characteristics, multimetric indices of biotic integrity and environmental stressors have been developed that enable use of diatom assemblages in the risk assessment and management of stream and river ecosystems.

Acknowledgments

We thank E. F. Stoermer and J. Smol for providing the opportunity to write this paper and present our concepts of using diatoms to assess rivers and streams. We thank R. Lowe and the editors for thoughtful review. In particular, we want to thank the graduate students of Stevenson's lab and the University of Louisville, and R. Lowe, R. Sweets, D. Charles, L. Bahls, L. Metzmeier, and B. Hill for the many discussions that helped to develop these concepts of diatom bioassessments. Research enabling this synthesis was supported by grants from the United States Environmental Protection Agency (R 82167 and R 824783).

References

Adler, R. W. (1995). Filling the gaps in water quality standards: Legal perspectives on bio-criteria. In *Biological Assessment and Criteria: Tools for Water Resource Planning and Decision Making*, ed. W. S. Davis & T. P. Simon, pp. 31–47. Boca Raton, FL: Lewis Publishers.

Aloi, J. E. (1990). A critical review of recent freshwater periphyton methods. *Canadian Journal of Fisheries and Aquatic Sciences* 47, 656–70.

APHA. (1992). *Standard Methods for the Evaluation of Water and Wastewater*. 18th edn. Washington DC: American Public Health Association.

Alatalo, R. V. (1981). Problems in the measurement of evenness in ecology. *Oikos*, 37; 199–204.

Amoros, C. & Van Urk, G. (1989). Palaeoecological analyses of large rivers: Some principles and methods. In *Historical Change of Large Alluvial Rivers: Western Europe*, ed. G. E. Petts, H. Möller & A. L Roux pp. 143–65. Chichester: Wiley.

Archibald, R. E. M. (1972). Diversity in some South African diatom assemblages and its relation to water quality. *Water Research*, 6, 1229–38.

Bahls, L. L. (1993). *Periphyton Bioassessment Methods for Montana Streams*. Department of Health and Environmental Sciences, Helena, MT: Water Quality Bereau.

Barbour, M. T., Gerritsen, J., Snyder, B. D. & Stribling, J. B. revised draft. *Revision to Rapid Bioassessment Protocols for Use in Streams and Rivers: Periphyton, Benthic Macroinvertebrates, and Fish*. United States Environmental Protection Agency EPA 841-D-97-002, Washington, DC.

Barbour, M. T., Stribling, J. B & Karr, J. R. (1995). Multimetric approach for establishing bio-criteria and measuring biological condition. In *Biological Assessment and Criteria: Tools for Water Resource Planning and Decision Making*, ed. W. S. Davis & T. P. Simon, pp. 63–77. Boca Raton, FL: Lewis Publishers.

Beaver, J. (1981). *Apparent Ecological Characteristics of Some Common Freshwater Diatoms*. Ontario Ministry of the Environment, Rexdale, Ontario, Canada.

Biggs, B. J. F. (1990). Use of relative specific growth rates of periphytic diatoms to assess enrichment of a stream. *New Zealand Journal of Marine and Freshwater Research*, 24, 9–18.

(1995). The contribution of flood disturbance, catchment geology and land use to the habitat template of periphyton in stream ecosystems. *Freshwater Biology*, 33, 419–38.

(1996). Patterns of benthic algae in streams. In *Algal Ecology: Freshwater Benthic Ecosystems*, ed. R. J. Stevenson, M. Bothwell & R. L. Lowe, pp. 31–55. San Diego, CA: Academic Press.

Blanck, H. (1985). A simple, community level, ecotoxicological test system using samples of periphyton. *Hydrobiologia*, **124**, 251–61.

Bothwell, M. L. (1989). Phosphorus-limited growth dynamics of lotic periphytic diatom communities: Areal biomass and cellular growth rate responses. *Canadian Journal of Fisheries and Aquatic Sciences*, **46**, 1293–301.

Bothwell, M. L. & Jasper, S. (1983). A light and dark trough methodology for measuring rates of lotic periphyton settlement and net growth. In *Periphyton of Freshwater Ecosystems*, ed. R. G. Wetzel, pp. 253–65. The Hague: Dr. W. Junk Publishers.

Bothwell, M. L., Sherbot, D., Roberge, A. C. & Daley, R. J. (1993). Influence of natural ultraviolet radiation on lotic periphytic diatom community growth, biomass accrual, and species composition: Short-term versus long-term effects. *Journal of Phycology*, **29**, 24–35.

Butcher, R. W. (1947). Studies in the ecology of rivers. IV. The algae of organically enriched water. *Journal of Ecology*, **35**, 186–91.

Cairns, J., Jr. & Kaesler, R. L. (1969). Cluster analysis of Potomac River survey stations based on protozoan presence–absence data. *Hydrobiologia*, **34**, 414–32.

Cattaneo, A. & Amireault, M. C. (1992). How artificial are artificial substrata for periphyton? *Journal of the North American Benthological Society*, **11**, 244–56.

Charles, D. F. (1985). Relationships between surface sediment diatom assemblages and lake-waste characteristics in Adirondack lakes. *Ecology*, **66**, 994–1011.

Chessman, B. C. (1986). Diatom flora of an Australian river system: Spatial patterns and environmental relationships. *Freshwater Biology*, **16**, 805–19.

Chessman, B. C., Primrose, E. H. & Burch, J. M. 1992. Limiting nutrients for periphyton growth in sub-alpine, forest, agricultural, and urban streams. *Freshwater Biology*, **28**, 349–61.

Cholnoky, B. J. (1968). *Ökologie der Diatomeen in Binnegewässern*. Lehre: Cramer.

Coste, M., Bosca, C. & Dauta, A. (1991). Use of algae for monitoring rivers in France. In *Use of Algae for Monitoring Rivers*, ed. B. A. Whitton, E. Rott & G. Friedrich, pp. 75–88. Universität Innsbruck, Innsbruck, Austria: E. Rott, Publisher, Institut für Botanik.

Côté, R. (1983). Aspects toxiques du cuivre sur la biomasse et la productivité du phytoplankton de la rivière du Saguenay, Québec. *Hydrobiologia*, **98**, 85–95.

Crumpton, W. G. (1987). A simple and reliable method for making permanent mounts of phytoplankton for light and fluorescence microscopy. *Limnology and Oceanography*, **32**, 1154–9.

Cumming, B. F. & Smol, J. P. (1993). Development of diatom-based salinity models for paleoclimatic research from lakes in British Columbia (Canada). *Hydrobiologia*, **269/270**, 575–86.

Cumming, B. F., Smol, J. P., Kingston, J. C., Charles, D. F., Birks, H. J. B., Camburn, K. E., Dixit, S. S. Uutala, A. J. & Selle, A. R. (1992). How much acidification has occurred in Adirondack region lakes (New York, USA) since pre-industrial times? *Canadian Journal of Fisheries and Aquatic Sciences*, **49**, 128–41.

Descy, J. P. (1979). A new approach to water quality estimation using diatoms. *Nova Hedwigia*, **64**, 305–23.

Descy, J. P. & Mouvet, C. (1984). Impact of the Tihange nuclear power plant on the periphyton and the phytoplankton of the Meuse River (Belgium). *Hydrobiologia*, **119**, 119–28.

DeYoe, H. R., Lowe, R. L. & Marks, J. C. (1992). The effect of nitrogen and phosphorus on the endosymbiont load of *Rhopalodia gibba* and *Epithemia turgida* (Bacillariophyceae). *Journal of Phycology*, **23**, 773–7.

Fabri, R. & Leclercq, L. (1984). *Etude écologique des rivières du nord du massif Ardennais (Belgique): flore et végétation de diatomées et physico-chimie des eaux*. 1. Robertville: Station Scientifique des Hautes Fagnes. 379 pp.

Fjerdingstad, E. (1950). The microflora of the River Molleaa with special reference to the relation of benthic algae to pollution. *Folia Limnologica Scandanavica*, **5**, 1–123.

Fritz, S. C. (1990). Twentieth-century salinity and water-level fluctuations in Devil's Lake, North Dakota, test of a diatom-based transfer function. *Limnology and Oceanography*, **35**, 1771–81.

Fritz, S. C., Juggins, S., Battarbee, R. W. & Engstrum, D. R. (1991). Reconstruction of past changes in salinity and climate using a diatom-based transfer function. *Nature*, **352**; 706–8.

Gale, W. F., Gurzynski, A. J. & Lowe, R. L. (1979). Colonization and standing crops of epilithic algae in the Susquehanna River, Pennsylvania. *Journal of Phycology*, **15**, 117–23.

Genter, R. B. (1996). Ecotoxicology of inorganic chemical stress to algae. In *Algal Ecology: Freshwater Benthic Ecosystems*, ed. R. J. Stevenson, M. Bothwell & R. L. Lowe, pp. 403–68. San Diego, CA: Academic Press.

Ghosh, M. & Gaur, J. P. (1990). Application of algal assay for defining nutrient limitation in two streams at Shillong. *Proceedings of the Indian Academy of Sciences*, **100**, 361–8.

Hall, R. I. & Smol, J. P. (1992). A weighted-averaging regression and calibration model for inferring total phosphorus from diatoms from British Columbia (Canada) lakes. *Freshwater Biology*, **27**, 417–37.

Hannaford, M. J. & Resh, V. H. (1995). Variability in macroinvertebrate rapid bioassessment surveys and habitat assessments in a northern California stream. *Journal of the North American Benthological Society*, **14**, 430–9.

Harvey, R. S. & Patrick, R. (1968). Concentration of ^{137}Cs, ^{65}Zn, and ^{85}Sr by fresh-water algae. *Biotechnology and Bioengineering*, **9**, 449–56.

Healey, F. P., and Henzel, L. L. (1979). Fluorometric measurement of alkaline phosphatase activity in algae. *Freshwater Biology*, **9**, 429–39.

Hill, B. H. (1997). The use of periphyton assemblage data in an index of biotic integrity. *Bulletin of the North American Benthological Society*, **14**, 158.

Hill, B. H., Lazorchak, J. M., McCormick, F. H. & Willingham, W. T. (1997). The effects of elevated metals on benthic community metabolism in a Rocky Mountain stream. *Environmental Pollution*, **96**, 183–90

Hill, M. O. (1979). *TWINSPAN-A FORTRAN Program for Detrended Correspondence Analysis and Reciprocal Averaging*. Cornell University, Ithaca, New York, USA.

Hill, W. R. & Boston, H. L. (1991). Community development alters photosynthesis–irradiance relations in stream periphyton. *Limnology and Oceanography*, **36**, 1375–89.

Hilsenhoff, W. L. (1988). Rapid field assessment of organic pollution with a family level biotic index. *Journal of the North American Benthological Society*, **7**, 65–8.

Hoagland, K. D., Carder, J. P. & R. L. Spawn. (1996). Effects of organic toxic substances. In *Algal Ecology: Freshwater Benthic Ecosystems*. ed. R. J. Stevenson, M. Bothwell & R. L. Lowe, pp. 469–97. San Diego, CA: Academic Press.

Hoagland, K. D., Drenner, R. W., Smith, J. D. & Cross, D. R. (1993). Freshwater community responses to mixtures of agricultural pesticides: Effects of atrizine and bifenthrin. *Enviromental Toxicology and Chemistry*, **12**, 627–37.

Hughes, R. M. (1995). Defining acceptable biological status by comparing with reference conditions. In *Biological Assessment and Criteria: Tools for Water Resource Planning and Decision Making*, ed. W. S. Davis and T. P. Simon, pp. 31–47. Boca Raton, FL: Lewis Publishers.

Humphrey, K. P. & Stevenson, R. J. (1992). Responses of benthic algae to pulses in current and nutrients during simulations of subscouring spates. *Journal of the North American Benthological Society*, **11**, 37–48.

Hurlbert, S. H. (1971). The nonconcept of species diversity: A critique and alternative parameters. *Ecology*, **52**, 577–86.

Hustedt, F. (1957). Die Diatomeenflora des Flusssystems der Weser im Gebiet der Hansestadt Bremen. *Abhandlungen naturwissenschaftlichen Verein zu Bremen*, Bd. 34, Heft 3, S. 181–440, 1 Taf.

Juggins, S. & ter Braak, C. J. F. 1992. *CALIBRATE – a program for species–environment calibration by [weighted averaging] partial least squares regression*. Environmental Change Research Center, University College, London.

Jüttner, I., Rothfritz, H. & Omerod, S. J. (1996). Diatoms as indicators of river water quality in the Nepalese Middle Hills with consideration of the effects of habitat-specific sampling. *Freshwater Biology*, **36**, 475–86.

Karr, J. R. (1981). Assessment of biotic integrity using fish communities. *Fisheries*, **6**, 21–7.

Karr, J. R. & Chu, E. W. (1997). *Biological Assessment: Using Multimetric Indexes Effectively*. Washington, DC: United States Environmental Protection Agency.

Karr, J. R. & Dudley, D. R. (1981). Ecological perspective on water quality goals. *Environmental Management*, **5**, 55–68.

Kelly, M. G. & Whitton, B. A. (1989). Interspecific differences in Zn, Cd and Pb accumulation by freshwater algae and bryophytes. *Hydrobiologia*, **175**, 1–11.

Kelly, M. G., Penny, C. J. & Whitton, B. A. (1995). Comparative performance of benthic diatom indices used to assess river water quality. *Hydrobiologia*, **302**, 179–88.

Kentucky Division of Water. (1994). *Pond Creek Drainage (Ohio River – Oldham County) Biological and Water Quality Investigation*. Technical Report No. 51, Frankfort, Kentucky.

Kentucky Natural Resources and Environmental Protection Cabinet. (1997). *Kentucky Outlook 2000: A Strategy for Kentucky's Third Century. Executive Summary and Guide to the Technical Committee Reports*. Frankfort, Kentucky.

Kim, B. K., Jackman, A. P. & Triska, R. J. (1990). Modeling transient storage and nitrate uptake kinetics in a flume containing a natural periphyton community. *Water Resources Research*, **26**, 505–15.

Kingston, J. C. & Birks, H. J. B. (1990). Dissolved organic carbon reconstructions from diatom assemblages in PIRLA project lakes, North America. *Philosophical Transactions of the Royal Society of London*, **B 327**, 279–88.

Kingston, J. C., Birks, H. J. B. & Uutala, A. J., Cumming, B. F. & Smol, J. P. (1992). Assessing trends in fishery resources and lake water aluminum from paleolimnological analyses of siliceous algae. *Canadian Journal of Fisheries and Aquatic Sciences*, **49**, 127–38.

Kolkwitz, R. & Marsson, M. (1908). Ökologie der pflanzliche Saprobien. *Berichte der Deutsche Botanische Gesellschaften*, **26**, 505–19.

Kutka, F. J. & Richards, C. (1996). Relating diatom assemblage structure to stream habitat. *Journal of the North American Benthological Society*, **15**, 469–80.

Lamberti, G. A. (1996). The role of periphyton in benthic food webs. In *Algal Ecology: Freshwater Benthic Ecosystems*, ed. R. J. Stevenson, M. Bothwell, & R. L. Lowe, pp. 533–72. San Diego, CA: Academic Press.

Lamberti, G. A. & Steinman, A. D. (1993). Research in artificial streams: Applications, uses, and abuses. *Journal of the North American Benthological Society*, **12**, 313–84.

Lange-Bertalot, H. (1978). Diatomeen-Differentialarten anstelle von Leitformen: ein geeigneteres Kriterium der Gewässerbelastung. *Archiv für Hydrobiologie Supplement*, **51**, 393–427.

(1979). Pollution tolerance of diatoms as a criterion for water quality estimation. *Nova Hedwigia*, **64**, 285–304.

Leland, H. V. (1995). Distribution of phytobenthos in the Yakima River basin, Washington, in relation to geology, land use and other environmental factors. *Canadian Journal of Fisheries and Aquatic Sciences*, **52**, 1108–29.

Lenat, D. R. (1993). A biotic index for the southeastern United States: Derivation and list of tolerance values, with criteria for assigning water quality ratings. *Journal of the North American Benthological Society*, **12**, 279–90.

Line, J. M., ter Braak, C. J. F. & Birks, H. J. B. (1994). WACALIB version 3.3: A computer program to reconstruct environmental variables from fossil assemblages by weighted averaging and to derive sample-specific errors of prediction. *Journal of Paleolimnology*, **10**, 147–52.

Lowe, R. L. (1974). *Environmental Requirements and Pollution Tolerance of Freshwater Diatoms*. US Environmental Protection Agency, EPA-670/4–74–005. Cincinnati, Ohio, USA.

Lowe, R. L. & Pan, Y. (1996). Benthic algal communities and biological monitors. In *Algal Ecology: Freshwater Benthic Ecosystems*, ed. R. J. Stevenson, M. Bothwell & R. L. Lowe, pp. 705–39. San Diego, CA: Academic Press.

Lowe, R. L., Golladay, S. W. & Webster, J. R. (1986). Periphyton response to nutrient manipulation in streams draining clearcut and forested watersheds. *Journal of the North American Benthological Society*, **5**, 221–9.

Mayer, M. S. & Likens, G. E. (1987). The importance of algae in a shaded headwater stream as food for an abundant caddisfly (Trichoptera). *Journal of the North American Benthological Society*, **6**, 262–9.

McCormick, P. V. & Stevenson, R. J. (1989). Effects of snail grazing on benthic algal community structure in different nutrient environments. *Journal of the North American Benthological Society*, **82**, 162–72.

(1991). Mechanisms of benthic algal succession in different flow environments. *Ecology*, **72**, 1835–48.

McFarland, B H., Hill, B. H. & Willingham, W. T. (1997). Abnormal *Fragilaria* spp. (Bacillariophyceae) in streams impacted by mine drainage. *Journal of Freshwater Ecology*, **12**, 141–9.

McIntire, C. D. & Phinney, H. K. (1965). Laboratory studies of periphyton production and community metabolism in lotic environments. *Ecological Monographs*, **35**, 237–58.

Moore, W. W. & McIntire, C. D. (1977). Spatial and seasonal distribution of littoral diatoms in Yaquina Estuary, Oregon (USA). *Botanica Marina*, **20**, 99–109.

Mulholland, P. J. (1996). Role of nutrient cycling in streams. In *Algal Ecology: Freshwater Benthic Ecosystems*, ed. R. J. Stevenson, M. Bothwell & R. L. Lowe, pp. 609–39. San Diego, CA: Academic Press.

Mulholland, P. J & Rosemond, A. D. (1992). Periphyton response to longitudinal nutrient depletion in a woodland stream: Evidence of upstream–downstream linkage. *Journal of the North American Benthological Society*, **11**, 405–19.

Müller-Haeckel, A. & Håkansson, H. (1978). The diatom-flora of a small stream near Abisko (Swedish Lapland) and its annual periodicity, judged by drift and colonization. *Archiv für Hydrobiologie*, **84**, 199–217.

Newbold, J. D., Elwood, J. W., O'Neill, R. V. & Van Winkle, W. (1981). Measuring nutrient spiralling in streams. *Canadian Journal of Fisheries and Aquatic Sciences*, **38**, 680–3.

Niederlehner, B. R. & Cairns, J. C., Jr. (1994). Consistency and sensitivity of community level endpoints in microcosm tests. *Journal of Aquatic Ecosystem Health*, **3**, 93–99.

Noris, R. H. (1995). Biological monitoring: The dilemma of data analysis. *Journal of North American Benthological Society*, **14**, 440–50.

Norris, R. H. & Norris, K. R. (1995). The need for biological assessment of water quality: Australian perspective. *Australian Journal of Ecology*, **20**, 1–6.

O'Connell, J. M., Reavie, E. D. & Smol, J. P. (1997). Assessment of water quality using epiphytic diatom assemblages on *Cladophora* from the St. Lawrence River (Canada). *Diatom Research*, **12**, 55–70.

Odum, E. P. (1985). Trends expected in stressed ecosystems. *BioScience*, **35**, 412–22.

Odum, E. P., Finn, J. T. & Franz, E. H. 1979. Perturbation theory and the subsidy–stress gradient. *BioScience*, **29**, 349–52.

Ohio Environmental Protection Agency. (1987). Biological criteria for the protection of aquatic life: volumes I–III. Ohio Environmental Protection Agency, Columbus, Ohio, USA.

Palmer, C. M. (1962). *Algae in Water Supplies*. Washington, DC, USA: US Department of Health, Education and Welfare.

(1969). A composite rating of algae tolerating organic pollution. *Journal of Phycology*, **5**, 78–82.

Pan, Y. & Lowe, R. L. (1994). Independent and interactive effects of nutrients and grazers on benthic algal community structure. *Hydrobiologia*, **291**, 201–9.

Pan, Y. & Stevenson, R. J. (1996). Gradient analysis of diatom assemblages in western Kentucky wetlands. *Journal of Phycology*, **32**, 222–32.

Pan, Y., Stevenson, R. J., Hill, B. H., Herlihy, A. T. & Collins, G. B. (1996). Using diatoms as indicators of ecological conditions in lotic systems: A regional assessment. *Journal of the North American Benthological Society*, **15**, 481–95.

Pappas, J. L. & Stoermer, E. F. (1996). Quantitative methods for determining a representative algal count. *Journal of Phycology*, **32**, 693–6.

Patrick, R. (1949). A proposed biological measure of stream conditions based on a survey of the Conestoga Basin, Lancaster County, Pennsylvania. *Proceedings of the Academy of Natural Sciences of Philadelphia*, **101**, 277–341.

(1961). A study of the numbers and kinds of species found in rivers of the Eastern United States. *Proceedings of the Academy of Natural Sciences of Philadelphia*, **113**, 215–58.

(1973). Use of algae, especially diatoms, in the assessment of water quality. In *Biological Methods for the Assessment of Water Quality*, ASTM STP 528, pp. 76–95. Philadelphia, PA: American Society for Testing and Materials.

Patrick, R. & Strawbridge, D. (1963). Variation in the structure of natural diatom communities. *The American Naturalist*, **97**; 51–7.

Patrick, R., Hohn, M. H. & Wallace, J. H. 1954. A new method for determining the patter of the diatom flora. *Notulae Naturae*, No. 259. 12 pp.

Peterson, B. J., Hobbie, J. E. & Corliss, T. L. 1983. A continuous-flow periphyton bioassay: Tests of nutrient limitation in a tundra stream. *Limnology and Oceanography*, **28**, 583–91.

Peterson, C. G. & Grimm, N. B. (1992). Temporal variation in enrichment effects during periphyton succession in a nitrogen-limited desert stream ecosystem. *Journal of the North American Benthological Society*, **11**, 20–36.

Peterson, C. G. & Stevenson, R. J. (1989). Seasonality in river phytoplankton: multivariate analyses of data from the Ohio River and six Kentucky tributaries. *Hydrobiologia*, **182**, 99–114.

(1992). Resistance and recovery of lotic algal communities: Importance of disturbance timing, disturbance history, and current. *Ecology*, **73**, 1445–61.

Pielou, E. C. 1984. *The Interpretation of Ecological Data, A Primer on Classification and Ordination*. New York. John Wiley.

Plafkin, J. L., Barbour, M. T., Porter, K. D., Gross, S. K. & Hughes, R. M. (1989). *Rapid Bioassessment Protocols for Use in Streams and Rivers: Benthic Macroinvertebrates and Fish*. EPA/444/4–89–001. Washington, DC, US EPA Office of Water.

Porter, S. D., Cuffney, T. F., Gurtz, M. E. & Meador, M. R. (1993). *Methods for Collecting Algal Samples as Part of the National Water-Quality Assessment Program*. Report 93–409. Raleigh, NC: US Geological Survey.

Pringle, C. M. (1990). Nutrient spatial heterogeneity: Effects on community structure, physiognomy, and diversity of stream algae. *Ecology*, **71**, 905–20.

Pringle, C. M. & Bowers, J. A. (1984). An *in situ* substratum fertilization technique: Diatom colonization on nutrient-enriched, sand substrata. *Canadian Journal of Fisheries and Aquatic Sciences*, **41**, 1247–51.

Prygiel, J. (1991). Use of benthic diatoms in surveillance of the Artois–Picardie basin hydrobiological survey. In *Use of Algae for Monitoring Rivers,* ed. B. A. Whitton, E. Rott & G. Friedrich, pp. 89–96. Universität Innsbruck, Austria: Institut für Botanik.

Prygiel, J. & Coste, M. (1993). The assessment of water quality in the Artois–Picardie water basin (France) by the use of diatom indices. *Hydrobiologia*, **269/270**, 343–9.

Raschke, R. L. (1993). Diatom (Bacillariophyta) community response to phosphorus in the Everglades National Park, USA. *Phycologia*, **32**, 48–58.

Reavie, E. D., Hall, R. I. & Smol, J. P. (1995). An expanded weighted-averaging model for inferring past total phosphorus concentrations from diatom assemblages in eutrophic British Columbia (Canada) lakes. *Journal of Paleolimnology*, **14**, 49–67.

Richards, C., Johnson, L. B. & Host, G. E. (1996). Landscape-scale influences on stream habitats and biota. *Canadian Journal of Fisheries and Aquatic Sciences*, **53** (Supplement 1): 295–311.

Robinson, C. T., Rushforth, S. R. & Liepa, R. A. (1995). Relationship of land use to diatom assemblages of small streams in Latvia. In *A Century of Diatom Research in North America: A Tribute to the Distinguished Careers of Charles W. Reimer and Ruth Patrick*, ed. J. P. Kociolek & M. J. Sullivan, pp. 47–59, Champaign, IL: Koeltz Scientific Books.

Rosemond, A. D. (1994). Multiple factors limit seasonal variation in periphyton in a forest stream. *Journal of the North American Benthological Society*, **13**, 333–44.

Rosen, B. H. (1995). Use of periphyton in the development of biocriteria. In *Biological Assessment and Criteria: Tools for Water Resource Planning and Decision Making*, ed. W. S. Davis & T. P. Simon, pp. 209–15. Boca Raton, FL: Lewis Publishers.

Rosen, B. H. & Lowe, R. L. (1984). Physiological and ultrastructural responses of *Cyclotella meneghiniana* (Bacillariophyta) to light intensity and nutrient limitation. *Journal of Phycology*, **20**, 173–93.

Rott, E. (1991). Methodological aspects and perspectives in the use of periphyton for monitoring and protecting rivers. In *Use of Algae for Monitoring Rivers*, ed. B. A. Whitton, E. Rott & G. Friedrich, pp. 9–16. Institut für Botanik, Universität Innsbruck, Innsbruck, Austria: E. Rott, Publisher.

Rott, E. & Pfister, P. (1988). Natural epilithic algal communities in fast-flowing mountain streams and rivers and some man-induced changes. *Verhandlungen Internationale Vereinigung für Theoretische und angewandte Limnologie*, **23**, 1320–4.

Round, F. E. (1991). Diatoms in river water-monitoring studies. *Journal of Applied Phycology*, **3**, 129–45.

(1993). *A Review and Methods for the Use of Epilithic Diatoms for Detecting and Monitoring Changes in River Water Quality 1993*. Methods for the Examination of Waters and Associated Materials. London: HMSO.

Rumeau, A. & Coste, M. (1988). Initiation à la systématique des diatomées d'eau douce pour l'utilisation pratique d'un indice diatomique générique. *Bulletin Français Pêche Piscuiculture*, **309**, 1–69.

Schiefele, S. & Schreiner, C. (1991). Use of diatoms for monitoring nutrient enrichment, acidification, and impact of salt in rivers in Germany and Austria. In *Use of Algae for Monitoring Rivers*, ed. B. A. Whitton, E. Rott & G. Friedrich, pp. 103–10. Universität Innsbruck, Austria: Institut für Botanik.

Schindler, D. W. (1990). Experimental perturbations of whole lakes as tests of hypotheses concerning ecosystem structure and function. *Oikos*, **57**, 25–41.

Schoeman, F. R. (1976). Diatom indicator groups in the assessment of water quality in the Jukskei-Crocodile River System (Transvaal, Republic of South Africa). *Journal of the Limnological Society of South Africa*, **2**, 21–4.

Shannon, C. F. (1948). A mathematical theory of communication. *Bell Systems Technical Journal*, **27**, 37–42.

Sicko-Goad, L., Stoermer, E. F. & Ladewski, B. G. (1977). A morphometric method for correcting phytoplankton cell volume estimates. *Protoplasma*, **93**, 147–63.

Simpson, E. H. (1949). Measurement of diversity. *Nature*, **163**, 688.

Slàdecek, V. (1973). System of water quality from the biological point of view. *Archiv für Hydrobiologie und Ergebnisse Limnologie*, **7**, 1–218.

Squires, L. E., Rushforth, S. R. & Brotherson, J. D. (1979). Algal response to a thermal effluent: Study of a power station on the Provo River, Utah, USA. *Hydrobiologia*, **63**; 17–32.

Steinberg, C. & Schiefele, S. 1988. Indication of trophy and pollution in running waters. *Zeitschrift für Wasser-Abwasser Forschung*, **21**, 227–34.

Steinman, A. D., McIntire, C. D. Gregory, S. V. Lamberti, G. V. & Ashkenas, L. (1987*a*). Effect of herbivore type and density on taxonomic structure and physiognomy of algal assemblages in laboratory streams. *Canadian Journal of Fisheries and Aquatic Sciences*, **44**, 1640–8.

Steinman, A. D., McIntire, C. D. & Lowry, R. R. (1987*b*). Effect of herbivore type and density on chemical composition of algal assemblages in laboratory streams. *Journal of the North American Benthological Society*, **6**, 189–97.

Stevenson, R. J. (1983). Effects of current and conditions simulating autogenically changing microhabitats on benthic algal immigration. *Ecology*, **64**, 1514–24.

(1984*a*). How currents on different sides of substrates in streams affect mechanisms of benthic algal accumulation. *Internationale Revue für gesamten Hydrobiologie*, **69**, 241–62.

(1984*b*). Procedures for mounting algae in a syrup medium. *Transactions of the American Microscopical Society*, **103**, 320–1.

(1984c). Epilithic and epipelic diatoms in the Sandusky River, with emphasis on species diversity and water quality. *Hydrobiologia*, **114**, 161–75.

(1986*a*). Mathematical model of epilithic diatom accumulation. In *Proceedings of the Eighth International Diatom Symposium*, ed. M. Ricard, pp. 209–31. Koenigstein, Germany: Koeltz Scientific Books.

(1986*b*). Importances of variation in algal immigration and growth rates estimated by modelling benthic algal colonization. In *Algal Biofouling: Studies in Environmental Sciences 28*, ed. L. Evans & K. Hoagland, pp. 193–210. Amsterdam: Elsevier Press.

(1990). Benthic algal community dynamics in a stream during and after a spate. *Journal of the North American Benthological Society*, **9**, 277–88.

(1995). Community dynamics and differential species performance of benthic diatoms along a nitrate gradient. In *A Century of Diatom Research in North America. A Tribute to the Distinguished Careers of Charles W. Reimer and Ruth Patrick*. ed. J. P. Kociolek & M. J. Sullivan, pp. 29–46. Champaign, IL: Koeltz Scientific Books USA.

(1996). An introduction to algal ecology in freshwater benthic habitats. In *Algal Ecology: Freshwater Benthic Ecosystems*. ed. R. J. Stevenson, M. Bothwell & R. L. Lowe. pp. 3–30. San Diego, CA: Academic Press.

(1997). Scale-dependent causal frameworks and the consequences of benthic algal heterogeneity. *Journal of the North American Benthological Society*, **16**, 248–62.

(1998). Diatom indicators of stream and wetland stressors in a risk management framework. *Environmental Monitoring and Assessment*, **51**, 107–118.

Stevenson, R. J., & Lowe, R. L. (1986). Sampling and interpretation of algal patterns for water quality assessment. In *Rationale for Sampling and Interpretation of Ecological Data in the Assessment of Freshwater Ecosystems*, ed. B. G. Isom, ASTM STP 894. pp. 118–149. Philadelphia, PA: American Society for Testing and Materials Publication.

Stevenson, R. J. & Pan, Y. (1995). Are evolutionary tradeoffs evident in responses of benthic diatoms to nutrients? In *Proceedings of the 13th International Diatom Symposium 1994*. ed. D. Marino, pp. 71–81.

Stevenson, R. J., & Peterson, C. G. (1991). Emigration and immigration can be important determinants of benthic diatom assemblages in streams. *Freshwater Biology*, **6**, 295–306.

Stevenson, R. J., & Stoermer, E. F. (1981). Quantitative differences between benthic algal communities along a depth gradient in Lake Michigan. *Journal of Phycology*, **17**, 29–36.

Stevenson, R. J., & White, K. D. (1995). A comparison of natural and human determinants of phytoplankton communities in the Kentucky River basin, USA. *Hydrobiologia*, **297**, 201–16.

Stevenson, R. J., Bothwell, M., & Lowe, R. L. (1996). *Algal Ecology: Freshwater Benthic Ecosystems*. San Diego, CA: Academic Press.

Stevenson, R. J., Peterson, C. G., Kirschtel, D. B., King, C. C., & Tuchman, N. C. (1991). Density-dependent growth, ecological strategies, and effects of nutrients and shading on benthic diatom succession in streams. *Journal of Phycology*, **27**, 59–69.

Stevenson, R. J., Sweets, P. R., Pan, Y., & Schulz, R. E. (in press). Algal community patterns in wetlands and their use as indicators of ecological conditions. *Proceedings of INTECOL's Vth International Wetland Conference*, Gleaneagles Press, Adelaide. ed. A. J. McComb & J. A. Davis.

Suter, G. W. II. (1984). A critique of ecosystem health concepts and indexes. *Environmental Toxicology and Chemistry*, **12**, 1533–9.

Swanson, C. D. & Bachmann, R. W. (1976). A model of algal exports in some Iowa stream. *Ecology*, **57**, 1076–80.

Sweets, P. R. (1992). Diatom paleolimnological evidence for lake acidification in the Trial Ridge region of Florida. *Water, Air and Soil Pollution*, **65**, 43–57.

ter Braak, C. J. F. (1986). Interpreting a hierarchical classification with simple discriminate functions: An ecological example. In *Data Anlaysis and Informatics*. ed. E. Diday, pp. 11–21. Amsterdam: North Holland.

(1987–1992). *CANOCO – a FORTRAN program for Canonical Community Ordination*. Ithaca, New York: Microcomputer Power,

ter Braak, C. J. F. & van Dam, H. 1989. Inferring pH from diatoms: A comparison of old and new calibration methods. *Hydrobiologia*, **178**; 209–23.

Tett, P., Gallegos, C., Kelly, M. G., Hornberger, G. M. & Cosby, B. J. (1978). Relationships among substrate, flow, and benthic microalgal pigment density in the Mechums River, Virginia. *Limnology and Oceanography*, **23**, 785–97.

Thorp, J. H., Black, A. R., Jack, J. D. & Casper, A. F. 1996. Pelagic enclosures – modification and use for experimental study of riverine plankton. *Archiv für Hydrobiologie* (Supplement) **113**, 583–9.

Tuchman, M. & Stevenson, R. J. (1980). Comparison of clay tile, sterilized rock, and natural substrate diatom communities in a small stream in southeastern Michigan, USA *Hydrobiologia*, **75**, 73–9.

US Environmental Protection Agency. (1992). *Framework for Ecological Risk Assessment*. EPA 630/R-92/001, Washington, DC.

   (1993). *A Guidebook to Comparing Risks and Setting Environmental Priorities*. EPA 230–B–93–003, Washington, DC.

van Dam, H. (1982). On the use of measures of structure and diversity in applied diatom ecology. *Nova Hedwigia*, **73**, 97–115.

Van Dam, H., Mertenes, A. & Sinkeldam, J. (1994). A coded checklist and ecological indicator values of freshwater diatoms from the Netherlands. *Netherlands Journal of Aquatic Ecology*, **28**, 117–33.

Vannote, R. L., Minshall, G. W., Cummins, K. W., Sedell, J. R. & Cushing, C. E. 1980. The river continuum concept. *Canadian Journal of Fisheries and Aquatic Sciences*, **37**, 130–137.

Vollenweider, R. A. & Kerekes, J. J. (1981). Background and summary results of the OECD cooperative program on eutrophication. In *Restoration of Inland Lakes and Waters*, pp. 25–36. Washington, DC: US Environmental Protection Agency.

Watanabe, T., Asai, K., Houki, A. Tanaka, S. & Hizuka, T. (1986). Saprophilous and eurysaprobic diatom taxa to organic water pollution and diatom assemblage index (DAIpo). *Diatom*, **2**; 23–73.

Weber, C. I. (1973). Recent developments in the measurement of the response of plankton and periphyton to changes in their environment. In *Bioassay Techniques and Environmental Chemistry*, ed. G. Glass, pp. 119–38. Ann Arbor, MI: Ann Arbor Science Publishers.

Whitton, B. A. & Kelly, M. G. (1995). Use of algae and other plants for monitoring rivers. *Australian Journal of Ecology*, **20**, 45–56.

Whitton, B. A., Rott, E. & Friedrich, G., (ed.) (1991). *Use of Algae for Monitoring Rivers*. Institut für Botanik, Universität Innsbruck, Innsbruck, Austria: E. Rott, Publisher.

Wright, J. F., Furse, M. T. & Armitage, P. D. (1993). RIVPACS: A technique for evaluating the biological quality of rivers in the UK. *European Water Pollution Control*, **3**, 15–25.

Yoder, C. O. & Rankin, E. T. (1995). Biological response signatures and the area of degradation value: New tools for interpreting multimetric data. In *Biological Assessment and Criteria: Tools for Water Resource Planning and Decision Making*, ed. W. S. Davis & T. P. Simon, pp. 263–86. Boca Raton, FL: Lewis Publishers.

Zelinka, M. & Marvan, P. (1961). Zur Prazisierung der biologischen Klassifikation des Reinheit fliessender Gewässer. *Archiv für Hydrobiologie*, **57**, 389–407.

3 Diatoms as indicators of hydrologic and climatic change in saline lakes

SHERILYN C. FRITZ, BRIAN F. CUMMING, FRANÇOISE GASSE
AND KATHLEEN R. LAIRD

Introduction

Lakes are intricately tied to the climate system in that their water level and chemistry are a manifestation of the balance between inputs (precipitation, stream inflow, surface runoff, groundwater inflow) and outputs (evaporation, stream outflow, groundwater recharge) (Mason *et al.*, 1994). Hence, changes in the hydrologic budget, caused by either climatic change or human activity, have the potential to alter lake level and lake chemistry. These, in turn, may affect the physiological responses and species composition of the lake's biota, including those of diatoms. Here, we review the use of diatoms as indicators of hydrologic and climatic change, with an emphasis on environmental reconstruction in arid and semi-arid regions. First we discuss linkages among climate, hydrology, lake hydrochemistry, and diatoms that form the foundation for environmental reconstruction and then review selected examples of diatom-based studies.

LAKE HYDROLOGY AND HYDROCHEMISTRY

Lakes vary in their hydrologic sensitivity to climatic change (Winter, 1990). In basins with a surface outlet, lake-level increase is constrained by topography, and any change in input is usually balanced by outflow. Thus, in open basins, lake level fluctuates relatively little, unless hydrologic change is sufficiently large to drop water level below the outlet level. In contrast, closed-basin lakes, that is lakes without surface outflow, often show changes in level associated with changes in the balance between precipitation and evaporation $(P - E)$. The magnitude of response to fluctuations in $P - E$ depends on the relative contribution of groundwater inflow and outflow to the hydrologic budget; lake-level change is greatest in terminal basins, which have neither surface nor groundwater outflow. In closed-basins, seasonal and interannual changes in $P - E$ result in concentration or dilution of lakewater and hence changes in ionic concentration (salinity) and ionic composition (Fig. 3.1). Although lake-level change can affect diatom communities directly, here we consider primarily linkages between diatoms and changes in salinity and brine composition. The response of diatoms to lake-level change is described in detail by Wolin and Duthie (this volume).

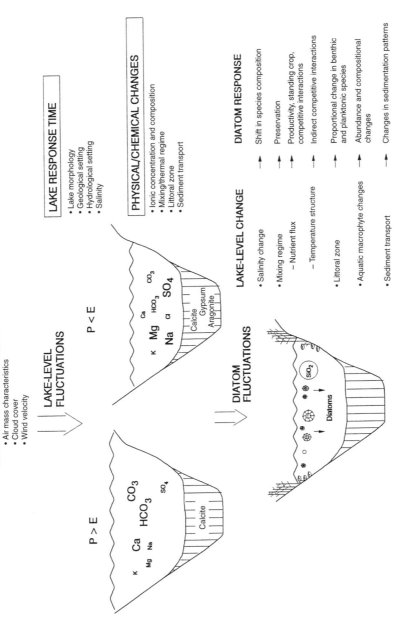

Fig. 3.1. Cartoon illustrating the potential influences of shifts in the balance between precipitation and evaporation on lakewater hydrochemistry and on diatom assemblages. Modified from Cumming *et al.* (1995).

Closed-basin lakes are common in semi-arid and arid regions, where moisture balance $(P - E)$ is negative. Although lakes in arid/semi-arid regions commonly receive far less attention than their temperate counterparts, on a global basis they contain a nearly equal volume of water (Meybeck, 1995) and occur on every continent. Furthermore, because water resources are often marginal in regions of net negative water balance, the status of these lakes is intricately tied to human culture.

LAKE SALINITY AND BRINE COMPOSITION

Lake salinity is a product of the source waters, which determine the initial conditions, and climatic setting. Most lakes of moderate to high salinity occur in areas of negative water balance, where evaporation concentrates dissolved salts. Occasionally, saline basins occur in mesic regions where source waters flow through salt deposits. A variety of classification schemes have been used to categorize lakes based on salinity. The limit separating 'freshwater' from 'saline' water is commonly set at 3 g/l (Williams, 1981). Although this limit is somewhat arbitrary, it also has biological validity, in that a number of organisms, including species of diatoms, have distribution limits at about 3 g/l. This is illustrated with diatom data from the North American Great Plains (Fig. 3.2), where a few species, such as *Stephanodiscus minutulus* Kütz. Cleve & Möller, *Stephanodiscus niagarae* Ehr., and *Fragilaria capucina* var. *mesolepta* (Rabenh.) Grun. in Van Heurck are rare or absent above 3 g/l and others, including *Cyclotella quillensis* Bailey, *Chaetoceros elmorei* Boyer, and *Navicula* cf. *fonticola* only occur above this limit.

In contrast to marine waters, lakes are highly variable in ionic composition and may be dominated by carbonates, sulfates, or chlorides in combination with major cations (Ca, Mg, Na). The initial solute composition is acquired via chemical weathering reactions and the interaction of inflowing ground and surface waters with bedrock and surface deposits. However, significant alteration of ionic composition in closed-basin lakes can occur via the precipitation of minerals during evaporative concentration (Eugster & Jones, 1979).

Within any geographic region, lake systems may vary spatially in ionic concentration and composition as a result of variability in groundwater sources and flow. A lake fed by a deep saline aquifer may lie adjacent to a relatively dilute basin fed by surficial flow, and the two may differ in brine composition, as well. Similarly, a chain of lakes connected by groundwater flow are often progressively more saline, because of progressive concentration of groundwater recharge (Donovan, 1994) from up-gradient to down. Thus ionic concentration and composition of closed-basin lakes are a result of the interplay between the chemistry of source waters, groundwater flow, and climate.

DIATOMS AND LAKEWATER CHEMISTRY

Diatoms show distributional patterns based on both salinity and brine composition. The earliest attempt to classify diatoms based on salinity

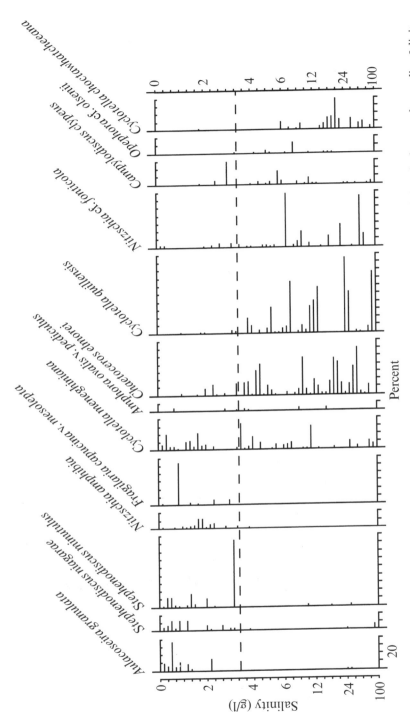

Fig. 3.2. Distribution of diatoms in a series of lakes in the North American Great Plains. Lakes are ranked sequentially from freshwater to hypersaline. Salinity is indicated on the y axis. The dotted line is at 3 g/l, which divides fresh from saline waters (Williams, 1981). Modified from Fritz (1990).

preference was a scheme developed by Kolbe (1927) and subsequently modified by Hustedt (Hustedt, 1953). This 'halobion' system of classification, however, was developed primarily for marine water. It is not really appropriate for inland waters, where lakes are often characterized by brines other than the NaCl type and span a far greater range of salinity, from dilute freshwater to hypersaline brines, often greater than 100 g/l. It is the series of species replacements along the salinity gradient (Fig. 3.5(b)), that makes diatoms powerful indicators of chemical change driven by changes in hydrology and climate. Although many diatom species occur in lakes of varied brine types, others show clear specificity and may be characteristic of carbonate, chloride, or sulfate.

The mechanisms that relate diatom distribution and ionic concentration and composition have not been well studied. Clearly, increased salinity represents an osmotic stress, but it is not clear whether distributional patterns are a result of differential responses to salinity directly or to other physiochemical processes correlated with salinity. Accumulation of proline or glycerol for osmoregulation has been investigated in only a very few diatom taxa (Fisher, 1977; Shobert, 1974). In the eurytopic species, *Cyclotella meneghiniana* Kütz., experimental studies suggest that high salinity affects the thickness of the silica cell wall (Tuchman *et al.*, 1984), and studies by Bhattacharyya and Volcani (1980) suggest that, at least for marine diatoms, high salinity may affect nutrient transport across the cell membrane. Thus, the impact of salinity on diatom distributions may only indirectly reflect salinity and may be driven by salinity effects on nutrient uptake or some other physiological process.

Tools for environmental reconstruction

The diatom species composition of a lake is related to chemical, physical, and biological variables. Relationships between diatom taxa and their preferred environmental conditions can be estimated by a surface-sediment calibration set or 'training set' (Birks, 1995; Charles & Smol, 1994). The basic approach, illustrated in a hypothetical example (Fig. 3.3(a)), is to choose a suite of lakes, commonly referred to as the training set, that span the limnological gradients of interest, in this case lakewater salinity. Given unlimited resources and time, the ideal training set of lakes would span all biologically relevant limnological gradients (e.g., pH, nutrients, salinity, ionic composition, light, etc.).

Limnological variables are measured from each lake in the training set and related to the diatom composition of surface-sediment samples (the uppermost 0.5 or 1 cm of sediment, representing the last few years of sediment accumulation). Ideally, multiple estimates of limnological conditions should be taken throughout the year, or sampling should be restricted to a specific period (e.g., spring turnover or late summer) to minimize noise introduced due to seasonal changes in water chemistry. Based on the distribution of taxa preserved in the surface sediments of training-set lakes, regression techniques

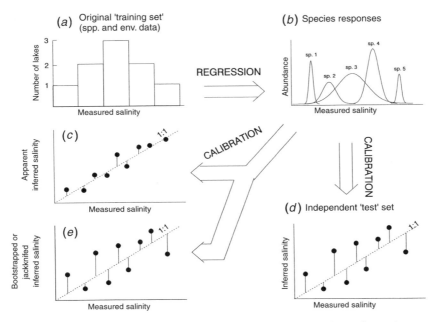

Fig. 3.3. Simplified illustration of the typical steps involved in developing and assessing the predictive ability of an inference model including: (*a*) selection of a modern set of lakes that span the limnological gradient of interest from which both environmental variables and species assemblages are collected; (*b*) the regression step, where species responses are estimated based on their distribution in the training set; and the calibration step where (i) the same set of lakes from which the species responses were estimated are used to generate an inference model (*c*), (ii) an independent set of lakes are chosen to assess the inference model (*d*), or (iii) computer resampling techniques, such as bootstrapping or jackknifing (*e*), are used to assess the inference model.

can be used to quantify the response of each diatom taxon to a given environmental variable (Fig. 3.3(*b*)). For example, the response of species 1 indicates that it prefers freshwater conditions in comparison to species 5. Similarly, species 3 would be seen as a generalist with respect to lakewater salinity. A number of different regression techniques can be used, ranging from intuitively simple methods, such as weighted-averaging, to mathematically more complex techniques, such as Gaussian logit regression (Birks *et al.*, 1990).

Subsequently, environmental conditions may be calculated from the species composition of fossil samples, based on quantitative estimates of species responses derived from the training set (Fig. 3.3(*b*)). This calibration step assumes that the environmental variable to be reconstructed is highly correlated with diatom distribution, that the ecological preferences of taxa can be assessed accurately based on their occurrences in the training set, that the ecological preferences of taxa have not changed over time, and that taxa present in the fossil samples are well represented in the training set. There are a number of ways to assess whether downcore assemblages are represented by the present-day assemblages, including analogue matching techniques and a number of multivariate approaches (Birks *et al.*, 1990).

S. C. FRITZ *ET AL.*

The predictive ability of a training set can be evaluated in several ways. One can calculate the target environmental variable (salinity) from the species composition of the same set of samples used to derive the species response model. Statistics determined using this approach are termed the 'apparent' error (Birks et al., 1990). However, this may lead to overly optimistic, and in some cases totally misleading, estimates of the predictive ability of the transfer function (Fig. 3.3(c)). Ideally, independent datasets that are similar in size and span gradients similar to those in the original training set would be used to evaluate the predictive ability of inference models. However, independent datasets are rarely collected because of the cost and time involved in collecting and analyzing additional samples and the loss of ecological information that results from failing to use these additional samples to form a larger training set. Thus, there is a trade-off between having an independent 'test' dataset vs. a larger training-set that will produce more realistic estimates of taxon responses. This is especially true in species-rich biological groups, such as diatoms. Fortunately, computer-intensive approaches, such as jackknifing and bootstrapping, can be used to provide realistic error estimates similar to those based on an independent 'test' dataset (Fig. 3.3(d)), while at the same time utilizing the ecological information inherent in the entire set of samples (Fig. 3.3(e)).

In both jackknifing and bootstrapping, estimates of species responses are independent of the samples used for evaluation of predictive power. This is possible because these techniques involve creating a series of new training sets that are subsets of the original datasets. For example, in 'jackknifing' (or leave-one-out substitution), new training sets are formed, each of which excludes one of the original training-set lakes. In the hypothetical example presented in Fig. 3.4(a), nine 'new' training sets would be formed, and each lake would appear in the test set only once. For each of the jackknife training sets, species responses are estimated, and the limnological condition of the lake in the test set is estimated (e.g., lake 1 in Fig. 3.4(a)). Overall estimates of the predictive ability of the inference model are based on estimates from each test set.

Bootstrapping is more complex (Fig. 3.4(b)). From the original training set, a 'new' set of lakes is randomly selected with replacement, so that the 'new' training set is the same size as the original training set (i.e., some samples will be chosen twice). Because sampling is with replacement, the samples that are not selected form an independent test-set. This procedure is repeated many times (e.g., 1000), so that many combinations of 'training' and 'test' sets are selected. The overall estimate of the predictive ability of the transfer function is based on estimates of the environmental variable of interest for lakes in each test set. Consequently, the predictive abilities of bootstrapping and jackknifing are similar to those from independent test sets, but contain better estimates of species responses because these estimates are based on information from all lakes.

From the regional calibration studies completed to date, it is clear that the distribution of diatom taxa is highly correlated with lakewater salinity and

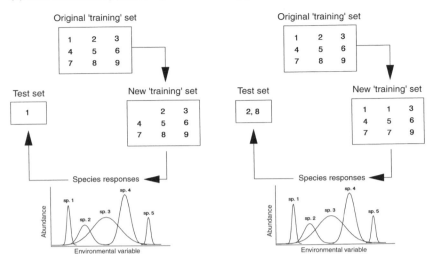

(a) Error estimation by jackknifing

Original 'training' set

1	2	3
4	5	6
7	8	9

Test set

| 1 |

New 'training' set

	2	3
4	5	6
7	8	9

Species responses

Abundance

sp. 1
sp. 2
sp. 3
sp. 4
sp. 5

Environmental variable

(b) Error estimation by bootstrapping

Original 'training' set

1	2	3
4	5	6
7	8	9

Test set

| 2, 8 |

New 'training' set

1	1	3
4	5	6
7	7	9

Species responses

Abundance

sp. 1
sp. 2
sp. 3
sp. 4
sp. 5

Environmental variable

Fig. 3.4. (a) Schematic representation of one iteration of a jackknife cycle in a hypothetical dataset of nine lakes. In each of the nine jackknife cycles, one sample will be left out (e.g., lake 1), and eight samples are left to form a training set of eight lakes, from which species responses (e.g., Gaussian curves in this example) are estimated. Following all nine jackknife cycles, estimates of the predictive ability of the inference models are based solely on the estimates when the lakes were in the test set. (b) Schematic representation of one iteration of a bootstrap cycle. From the nine-lake training set, a 'new' set of nine lakes is randomly selected with replacement. Because sampling is with replacement, the samples that are not selected will form an independent test-set (e.g., lake 2 and 8). After, for example, 1000 iterations, the predictive ability of a transfer function can be calculated based on estimates of the environmental variable only when lakes were in the test set.

that quantitative estimates of historical changes in salinity can be inferred from the species composition of sediment cores. There are presently three large calibration datasets (> 50 lakes) that were developed to assess the strength of the relationship between diatom species composition and lakewater ionic concentration and composition. These larger datasets include a 219–lake data-set mainly from British Columbia (Wilson et al., 1996), a 55–lake dataset from the northern Great Plains (Fritz, 1990; Fritz et al., 1991, 1993), which has recently been expanded to include over 100 lakes (S. C. Fritz, unpublished observations), and a dataset from Africa (Gasse et al., 1995), which includes 282 diatom samples from 164 lakes.

The dataset from British Columbia includes lakes from the Cariboo/Chilcotin region, Kamloops region, southern Interior plateau, and southern Rocky Mountain Trench. The lakes range from subsaline through hypersaline (Fig. 3.5(a)) and include both bicarbonate/carbonate dominated systems in combination with magnesium and sodium, as well as sodium sulfate dominated lakes (Cumming & Smol, 1993; Cumming et al., 1995; Wilson et al., 1994, 1996). In general, the freshwater lakes are dominated by bicarbonate/carbonate in combination with calcium, although in a few cases they also

S. C. FRITZ ET AL.

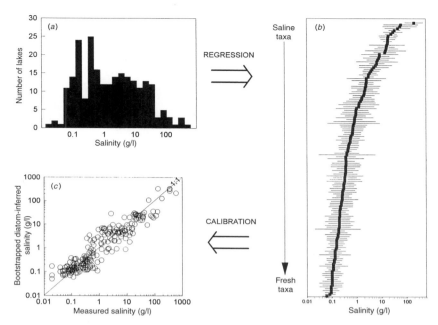

Fig. 3.5. Illustration of the steps involved in the development of a diatom-based salinity inference model from the 219-lake modern BC dataset (Wilson *et al*. 1996) showing: (*a*) the frequency distribution of the 219 lakes along the salinity gradient; (*b*) the estimated salinity optima (solid squares) and tolerances (bars) of diatom taxa arranged according to salinity optima; and (*c*) the relationship between measured and diatom-inferred salinity after bootstrapping. Modified from Wilson *et al*. (1996).

have a high proportion of sulfate. From these data the optima and tolerance of the dominant diatom taxa (those that occurred in at least 1% relative abundance in at least two lakes, Fig. 3.5(*b*)) were estimated using a weighted-averaging (WA) approach, and a salinity inference model was developed using WA calibration (Fig. 3.5(*c*), r^2 boot = 0.87).

The modern northern Great Plains diatom dataset consists of 55 lakes, which range in salinity from just under 1 to 270 g/l (Fritz, 1990; Fritz *et al*., 1991, 1993). The majority of these lakes are dominated by magnesium and sodium sulfates or sodium carbonate, but a few are sodium chloride systems. Ordination techniques showed that lakewater ionic concentration and composition (sulfate vs. carbonate) account for major directions of variation in the diatom species composition in these lakes. Based on the strong relationship between diatom species composition and lakewater salinity, a weighted-averaging model was developed that had an apparent coefficient of determination of 0.83.

The modern African dataset was created by combining regional datasets from Northwest Africa (125 samples; Khelifa, 1989; F. Gasse, unpublished observations), East Africa (167 samples; Gasse, 1986), and Niger (20 samples; Gasse, 1987). The samples include benthic, epiphytic, and epipelic diatom assemblages and are from a diverse array of aquatic environments, including lakes, swamps, bogs, and springs, that cover vast gradients in ionic concentration (40 to approximately 100 000 μS/cm) and exhibit virtually every possible combination of ionic composition. Environmental variables that account for significant variation in the diatom assemblage data include conductivity, pH, and cation and anion composition. Based on this diverse dataset, diatom models were developed to infer conductivity (bootstrapped coefficient of determination of 0.81), as well as pH, alkali and alkaline earth metals, and ion ratios of carbonate + bicarbonate: sulfate + chloride (Gasse *et al.*, 1995).

The strong association between diatom species distributions and lakewater ionic concentration and composition has been demonstrated in a number of smaller regional datasets from around the world. In Central Mexico, Metcalfe (1988) investigated the diatom assemblages collected from plankton, periphyton, and bottom sediments from 47 aquatic environments including lakes, springs, marshes, and rivers. The majority of these sites are dominated by carbonate/bicarbonate. Servant and Roux (Roux *et al.*, 1991; Servant-Vildary & Roux, 1990) sampled 14 saline lakes in the Bolivian Altiplano, variously dominated by sodium carbonate, sulfate, and chloride salts, and related diatom distribution to chemical variables, including salinity and ion composition. In a study of 32 lakes from Southeastern Australia that span a conductivity range from 1000 to 195 000 μS/cm, Gell and Gasse (1990) found that diatom species collected from the plankton and scrapings from floating and submerged objects were clearly related to lakewater conductivity. Similarly, Blinn (1995) showed that specific conductance was one of the most important predictors of epipelic and epilithic diatom communities in a set of 19 lakes in Western Victoria, Australia that span a specific conductance gradient of 1500 to 262 000 μS/cm. Kashima (1994) quantified the ecological preferences of diatom taxa collected from periphyton scrapes and water samples from 51 samples collected from 23 lakes, ponds, and rivers that range in salinity from 8 to 100 g/l from the Anatolia Plateau of Turkey. Based on these results, a diatom transfer function was developed using weighted-averaging techniques with an apparent coefficient of determination of 0.9. Additionally, Roberts & McMinn (1996) have recently shown that diatom assemblages are significantly correlated with lakewater salinity in a dataset of 33 lakes from the Vestfold Hills in the Antarctic.

Environmental reconstructions

One of the primary uses of stratigraphic diatom studies in arid and semi-arid regions is in reconstruction of past climate and hydrology. These reconstructions can be undertaken at a variety of temporal scales dependent on the nature of the stratigraphic record. To date, most of the diatom-based studies of climatic change have been of long-term patterns of climatic variability at scales of centuries or millennia. Case studies from Africa, Asia (Tibet), and western North America illustrate how diatoms record long-term climatic changes induced primarily by changes in solar insolation, as well as large-scale yet abrupt climatic events, which cannot be explained solely by changes in orbital parameters. More recently, a few high-resolution diatom records have been used to examine decadal scale climatic variability, rather than long-term trends (see below).

AFRICAN AND ASIAN RECORDS OF QUATERNARY CLIMATE AND HYDROLOGY

Changes in the monsoonal circulation in East Africa over the past glacial/interglacial cycle
Climate of intertropical Africa is mainly controlled by monsoonal circulation. The latter follows the seasonal latitudinal migration of the Inter-Tropical Convergence Zone (ITCZ) and is related to changes in the seasonal distribution of solar radiation. In East Africa, several closed rift lakes have undergone spectacular water-level fluctuations and drastic changes in salinity reflecting monsoon variability, as documented by numerous diatom studies (Barker, 1990; Gasse, 1977; Gasse & Fontes, 1989; Richardson & Dussinger, 1986; Stager, 1982, 1988; Taieb *et al.*, 1991). As an example, we consider here the ^{14}C-dated section of a diatom record from Lake Abhé (Ethiopia and Djibouti).

Lake Abhé (11°5′ N, 41°50′ E, 240 m.a.s.l.) is the terminal lake of the Awash River, which flows from the Ethiopian Plateau to the Afar desert. Presently, Lake Abhé is nearly dry, hypersaline, and hyperalkaline (S = 160 g/l, CO_3^{2-} > Cl^-; pH = 10), with water output due primarily to evaporation from the lake surface. Reconstruction of lake-level fluctuations (Fig. 3.6) is based on the study of well-preserved Holocene shorelines and two cores taken off the lake margin . Reconstruction of conductivity (C), water pH, and anion ratio ([carbonate/bicarbonate]/[chloride + sulfate]) (Fig. 3.6) is based on transfer functions established by Gasse *et al.* (1995).

A short-term low-level stage around 37 cal. ka BP is characterized by the mesosaline alkaline species *Thalassiosira faurii* (Gasse) Hasle. Subsequently, diatom-inferred conductivity dropped (35–27 cal. ka BP), as the lake rose 170 meters above its present-day level, and *Aulacoseira granulata* (Ehr.) Simonsen associated with *Cyclotella ocellata* Pant. predominated. Arid conditions prevailed during the Last Glacial Maximum (LGM) and culminated in a lake level lower than that of today from about 19 to 11.5 cal. ka BP, as indicated by a paleosoil with Gramineae remains. The absence of diatoms during these very low stands is attributed to excessive salinity and/or dissolution of biogenic silica in strongly alkaline solutions. The change from carbonate–bicarbonate facies to chloride–sulfate type may be due either to brine evolution under hyper-arid conditions or the local influence of hydrothermal springs. Two diatom-bearing levels interbedded in the paleosoil were tentatively correlated with short-term humid pulses that are documented in neighboring basins around 15 and 14 cal. ka BP (Gasse & Van Campo, 1994). Lake Abhé filled up rapidly around 11.5 cal. ka BP. The diatomaceous carbonates, dominated by *Stephanodiscus minutulus*, reflect a deep clear lake with oligosaline, slightly

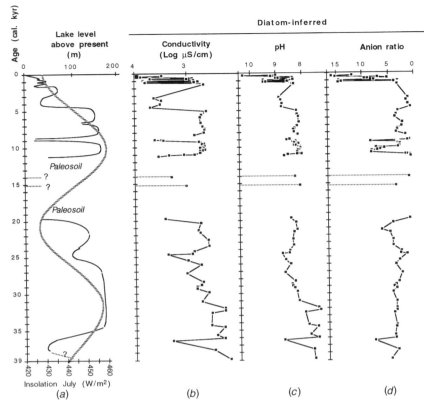

Fig. 3.6. Lake Abhé (Ethiopia–Djibouti). Radiocarbon ages are converted in calendar ages (after Stuiver *et al.*, 1993, for the past 18 ^{14}C yr BP, and E. Bard, pers. comm. prior to 18 ^{14}C yr BP). (*a*) Lake-level fluctuations (after Gasse, 1977; and Fontes *et al.*, 1985) and summer solar radiation (July) at 10° N (Berger, 1978); (*b*)–(*d*) diatom-inferred conductivity (*b*), pH (*c*) and (*d*) anion ratio ($HCO_3^- + CO_3^{2-}$)/($Cl^- + SO_4^{2-}$).

alkaline water and low-Si, high-P supply rates. The late-Holocene period shows several positive oscillations of minor amplitude; the largest one is recorded around 2 cal. ka BP. These are imposed on a general regressive trend, recorded by the development of *Thalassiosira faurii* and *Cyclotella meneghiniana*, which reflect sodium-carbonate rich, alkaline water.

The general pattern of lake-level and salinity change can be accounted for by changes in summer insolation in the northern tropics (Fig. 3.6). The 35–27 cal. ka BP and early- to mid-Holocene highstands occurred when seasonal contrasts in solar radiation were at their maximum. Conversely, radiation input was close to that of today during the Last Glacial Maximum and the late-Holocene period. However, solar radiation does not explain the differences observed in diatom assemblages and inferred salinity between the late Pleistocene and Holocene highstands. Additional causes, e.g., lower temperature during the Glacial period or different vegetative cover, weathering, and runoff coefficients must be considered. Moreover, abrupt changes are not consistent with purely orbital effects. Other global mechanisms, e.g., changes in sea surface conditions or atmospheric greenhouse gas concentration, may explain events, such as the dry spells around 9–8 and 4–3 cal. ka BP, which are recorded at large geographical scales in the Indian and African monsoon domains (Gasse & Van Campo, 1994).

Holocene lakes of the Sahara and Sahel From about nine to five thousand years ago, Neolithic civilizations flourished in a green Sahara and woody Sahel, where a multitude of lakes occupied presently dry eolian and fluvio-eolian depressions. Many paleo-lakes were the surface expression of aquifers that now lie several meters or tens of meters below ground surface and which outcropped in response to increased precipitation. Several sedimentary profiles have been analyzed for their diatom content (for a review see Gasse, 1988).

In the Sahelian zone, Bougdouma (13°19′ N, 11°40′ E, 337 m a.s.l.) is an interdunal depression that is occupied by an hyperalkaline playa with a trona crust. It is supplied by an unconfined aquifer, which is recharged by local summer rainfalls. A lake/swamp sequence from 12 ka BP shows large fluctuations in $P - E$ (Gasse, 1994; Gasse *et al.*, 1990). A freshwater to oligosaline swamp existed in response to sudden groundwater influxes at >12, ≈12 and ≈10.8 ka BP, followed by increased aridity at ≈10.3–10 ka BP and then a freshwater lake from 9.7–6.4 ka BP. The upper part of the section reflects unstable hydroclimatic conditions, with dry episodes at 6.3–6, ≈4.5 ka BP, and after 2 ka BP. Freshwater episodes are characterized by *Aulacoseira granulata*, *A. ambigua* (Grun.) Simonsen, and *Cyclotella stelligera* var. *glomerata* (Bachmann) Haworth & Hurley living in very dilute water of the calcium-bicarbonate type. Hydrologic deficit is recorded by the development of saline/alkaline species of the sodium-chloride type, with *Navicula elkab* O. Müller, *Anomoeoneis sphaerophora* (Kütz.) Pfitzer, *Cyclotella meneghiniana*, *Rhopalodia gibberula* (Ehr.) O. Müller, and *Campylodiscus clypeus* (Ehr.) Ehr. *ex* Kütz.

Adrar Bous (20°20′ N, 9° E) (Fig. 3.7) is a dry depression in the central Sahara lying at the western edge of the Tenere Desert (Dubar, 1988; Fontes & Gasse, 1991), which is mainly supplied by surface runoff and infiltration of water precipitated on the massif. A humid pulse is recorded by a layer of calcareous diatomite interbedded between eolian sand, dated at ≈12.9 ka BP. It is rich in the freshwater planktonic species *Aulacoseira ambigua* and *A. granulata*. The filling of the Holocene lake occurred rapidly prior to 9.7 ka BP and led to a shallow freshwater lake with *Fragilaria* and *Aulacoseira* dominant in bicarbonate water. Water then concentrated through evaporation and/or weathering of the surrounding salt crust, as indicated by a diatom flora of the mesosaline type, with abundant *Cyclotella chochtawhatcheana* Prasad and *Campylodiscus clypeus*, which indicate a change towards the chloride–sulfate chemical facies. The subsequent lacustrine optimum (≈8–7 ka BP) is recorded by a pure freshwater diatomite, where *Aulacoseira granulata* and *A. ambigua* represent up to 98% of the assemblages. The lake dried up after 6.5 ka BP, with a subsequent return to moist conditions at 4.9 ka BP.

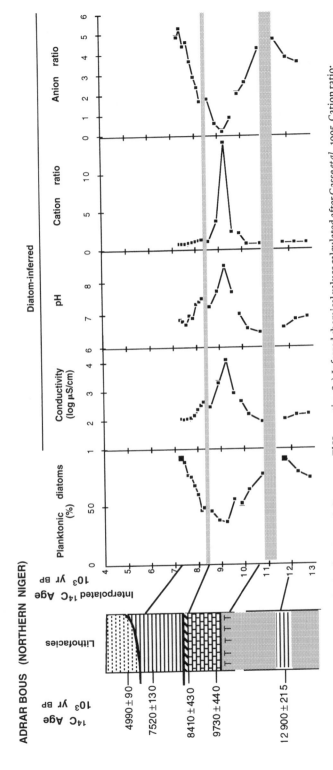

Fig. 3.7. Adrar Bous (northern Niger). After Fontes & Gasse, 1991; El Hamouti, 1989). Inferred chemical values calculated after Gasse *et al.*, 1995. Cation ratio: $(Na^+ + K^+)/(Ca^{2+} + Mg^{2+})$. Anion ratio: $(HCO_3^- + CO_3^{2-})/(Cl^- + SO_4^{2-})$.

Climatic changes in the northern Sahara are documented by Holocene swamp and lacustrine diatom-bearing sediments lying in depressions along the northern margin of the Great Western Erg (Fontes *et al.*, 1985*a*; Gasse *et al.*, 1996; Gasse & Seyve, 1987), where present-day mean annual rainfall is <50 mm/yr. Precipitation is brought through the westerlies from the Atlantic Ocean and/or from Mediterranean depressions. Hassi el Mejnah (31°30' N, 2°15' E) is presently a closed dry depression that was occupied from 9.3 to 4 ka by a saline lake, supplied through the aquifer of the Great Western Erg, wadi underflows, and/or local rainfall. Water losses were due to evaporation. Two successive lacustrine episodes are dated from >9.3 to 7.2, and >6.2 to ≈4.0 ka BP, respectively, and are followed by the deposition of a gypsum crust. An assemblage dominated by abundant *Cyclotella chochtawhatcheana* indicates a mesosaline lake, with conductivity from 3 to 50 mS/cm and chloride–sulfate waters. The occurrence of several species that usually live in marine environments is remarkable (e.g., *Mastogloia aquilegia* Grun. in Möller, *M. baltica* Grun., *N. zosteretii* Grun.).

All diatom records from the Sahara and the Sahel indicate climatic conditions much wetter than those of today during the early-mid Holocene period. Maximal $P - E$ occurred everywhere between 10 and 4 ka BP. After the arid LGM period, humid conditions reestablished in two steps, at ≥12 and 10 ka BP, respectively. The latter oscillation, which was very large in amplitude, is observed at all sites. These two steps are separated by a dry episode synchronous with the Younger Dryas event. In the Sahel and the central Sahara, long-term climatic trends may be accounted for by northern migration of the ITCZ inducing increased squalls and/or monsoon precipitation up to 20° N, in response to changes in solar radiation. However, the dry spells that punctuate the interval 10–4 ka BP remain unexplained. Causes other than monsoon penetration over the continent, e.g., increase of the North Atlantic sea-surface temperature or changes in land surface conditions, also have to be considered in order to explain the synchronous occurrence of Holocene lakes in the northern Sahara (Gasse *et al.*, 1990). Despite different atmospheric mechanisms, the spatial coherence over Northwest Africa in the major arid–humid transition and reversal events is remarkable.

Post-glacial climatic changes in western Tibet Although no transfer functions are available to date for western Tibet, detailed diatom records have been established from two lake systems, the Sumxi–Longmu lake system (34°30' N, 80°23' E) (Van Campo & Gasse, 1993) and Bangong Lake (33°40' N, 79° E) (Hui *et al.*, 1996), which have provided significant insights on climatic changes over the post-glacial period. Western Tibet is presently outside the monsoon influence; although it penetrates the Tibetan Plateau from the southeast. It is a cold desert, where rare precipitation (≤60 mm/a) falls in summer as convective rain or snow.

The Sumxi–Longmu and the Bangong records differ in several respects. This is attributed to the different hydrological regime of individual systems induced by topographical factors. However, the two records lead to similar conclusions about regional climatic changes (Fig. 3.8). Both records show rapid

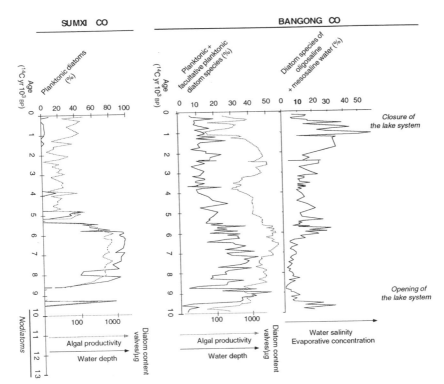

Fig. 3.8. Summary diagram of the Holocene diatom stratigraphy of Western Tibet. Lake Sumxi-Longmu (after Van Campo & Gasse, 1993) and Lake Bangong (after Fan Hui *et al.*, 1996).

establishment of conditions much wetter and warmer than those of today from 9.9–9.6 ka BP. The two lakes Sumxi and Longmu connected, and Lake Bangong opened in response to increased meltwater and/or rainfall supplies. This is documented by the predominance of freshwater planktonic forms (*Cyclotella* sp. aff. *comensis*). High diatom concentrations reflect intense algal productivity, suggesting relatively warm conditions as confirmed by pollen records. A general trend toward cold/arid conditions is then observed from ≈6.0 ka BP. This is recorded by decreasing percentages of planktonic diatoms, decreasing diatom concentrations, and increasing proportions of saline species. The abrupt increase in salinity recorded in Lake Bangong at ≈2 ka BP reflects the closure of the lake due to a climate-induced hydrological deficit. The wet/warm conditions, which persisted from 10 up to 6 ka BP, are attributed to an enhanced monsoon circulation and its penetration across the entire Tibetan Plateau.

NORTH AMERICAN RECORDS OF QUATERNARY CLIMATE AND HYDROLOGY

On the North American continent Quaternary hydrology and climate have been inferred from the diatom stratigraphy of closed-basin lakes. In glaciated

S. C. FRITZ ET AL.

regions, lake sediments record the pattern of limnological change following retreat of the Laurentide ice sheet *ca.* 12 ka BP. In the northern Great Plains, freshwater diatoms, such *as Cyclotella bodanica* Grun., *C. michiganiana* Skvortzow, *Aulacoseira granulata, Stephanodiscus* spp. and benthic *Fragilaria,* dominate in the centuries following lake formation and reflect the cool moist climate of the early Holocene. Subsequent replacement of freshwater species by taxa indicative of saline waters (*Cyclotella quillensis, Cyclotella chochtawhatcheana, Chaetoceros elmorei/muelleri*) suggests hydrologic closure of lakes in response to climatic warming/drying and occurs coincident with a shift in vegetation from forest to prairie. The timing of the transition from freshwater to saline and from forest to prairie varies among sites and occurs as early as 9.5 ka BP in southeastern South Dakota (Radle *et al.*, 1989)but not until after *ca.* 8 ka BP at sites further north (Laird *et al.*, 1996a). At some sites the transition is apparently quite rapid and occurs over a period of several hundred years, whereas in other cases changes in hydrochemistry occur gradually over one to two millennia . The abruptness of response undoubtedly is, in part, a function of lake morphometry and the amount of change in the lake's water budget required to drop lake level below the threshold of the outlet (Kennedy, 1994). In any case, the timing of hydrologic change in these parts of the northern Great Plains suggests interaction between solar insolation and a retreating ice sheet in regional climates during the late-Glacial and early-Holocene periods. Further to the west, near the eastern edge of the Rocky Mountains, the timing of maximum aridity may have occurred earlier, closer to the summer solar insolation maximum at 9 ka BP, because of the reduced influence of the Laurentide ice sheet (Schweger & Hickman, 1989).

Most diatom records suggest that the mid-Holocene in the continental interior was quite arid, as reflected by high diatom-inferred lakewater salinity. Some sites, such as Devils Lake, North Dakota (Fritz *et al.*, 1991) show considerable century-scale variability in salinity and hence in moisture, whereas at other sites, salinity appears less variable (Laird *et al.*, 1996a). Again, the differences in pattern are probably a result of hydrochemical differences rather than differences in local climate. The more saline lakes are often dominated by taxa with very broad salinity tolerances, particularly *Cyclotella quillensis* and *Chaetoceros elmorei/muelleri,* and in these systems large shifts in hydrochemistry are required to cross the salinity tolerance thresholds of the common taxa. In contrast, a number of species have thresholds around the freshwater/saline water boundary of 3 g/l and within the hyposaline range (Fig. 3.2), and fluctuations across this gradient result in shifts in species composition. Clearly arid climate dominated the continental interior throughout much of the Holocene, and it is only within the last *ca..* three to four millennia that freshwater conditions, indicative of increased moisture, have recurred with greater frequency (Fritz *et al.*, 1991; Laird *et al.*, 1996a).

Closed-basin lakes are also common in non-glaciated regions of western North America, particularly in intermontane basins west of the Rocky Mountains. These basins often contain a much longer record of climatic change, in some cases extending back as far as the late Tertiary. The long-term records from the North American Great Basin are reviewed by Bradbury (this volume). Pluvial lakes, such as Walker Lake, Nevada, were part of the extensive Lake Lahontan system, which covered much of the Great Basin in the late Pleistocene. In Walker Lake (Bradbury *et al.*, 1989) diatom assemblages dominated by *Stephanodiscus excentricus* Hust. and *Surirella nevadensis* Hanna & Grant indicate a large hyposaline lake in the mid Wisconsin, maintained by a climate much moister than today. Lake-level dropped at about 25 ka BP as the climate dried, to form a shallow saline lake with *Anomoeoneis costata* (Kütz.) Hust. and *Navicula subinflatoides*. Hust. This shallow lake persisted until *ca.* 16 ka BP and then desiccated either permanently or intermittently, as indicated by sediment texture and the absence of microfossils. Diatom abundance remains low until the mid-Holocene (5 ka BP), when the Walker River rapidly refilled the lake basin. The lake sediments during the filling event show a progression from saline species (*N. subinflatoides*) to species with progressively fresher salinity optima (*Chaetoceros elmorei, Cyclotella meneghiniana, S. excentricus, S. niagarae, C. ocellata*) as the fresh river water mixed with saline lake waters. Subsequent oscillations between freshwater and saline taxa document short-term variations in climate during the late Holocene.

Saline lakes are also present on the Colorado Plateau, formed by dissolution and collapse in a karstic terrain. Diatom-based studies of one site in north-central Arizona have been used to infer fluctuations in hydrology and climate during the Holocene (Blinn *et al.*, 1994). Attached diatoms characteristic of alkaline freshwater dominate the late-Glacial and early-Holocene, whereas pulses of *Anomoeoneis sphaerophora*, a taxon characteristic of carbonate-dominated hyposaline systems, suggest periods of reduced water level and evaporative concentration during the mid-Holocene. Prehistoric occupation of the lake's catchment coincides with a late-Holocene dry period, as indicated by increases in *A. sphaerophora* between *ca.* 0.6 to 1.2 ka BP, and it may be that native peoples were driven to settle near the few open water sources during this interval of drought.

HIGH-RESOLUTION STUDIES OF SHORT-TERM CLIMATIC VARIABILITY

High-resolution paleoclimatic studies can provide information on patterns of climatic variability over decades to centuries. Several explanations of decadal-to centennial-scale climatic variability have been hypothesized, including inherent variability of the linked atmospheric–oceanic system, solar activity, volcanic eruptions, and variability of greenhouse gases (Rind & 1993). The spatial patterns of high-resolution records can help discern the climatic processes that occur on these time scales and the sensitivity of the climatic system to these various forcings. At present, few diatom records are of sufficiently high resolution or of long enough duration to understand natural climatic cycles on decadal and centennial scales (but see Bradbury & Dietrich-Rurup, 1993).

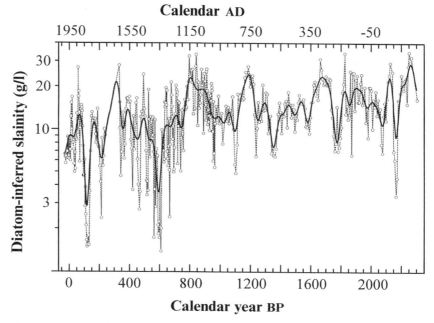

Fig. 3.9. Reconstruction of diatom-inferred salinity for Moon Lake, ND over the last 2300 years (dotted line). The solid line is a fast Fourier transformation of interpolated 5-year equal-interval diatom-inferred salinity values using a 10-point smoothing window. Chronology is based on 4 AMS radiocarbon measurements calibrated to calendar years (Stuiver & Reimer, 1993). Modified from Laird *et al.* (1996*b*).

In many arid regions, closed-basin lakes are important repositories that can reveal long-term patterns in the intensity, duration, and frequency of droughts. Because drought is a natural recurring feature of climate that has had dramatic environmental, economic, and social impacts on modern (Borchert, 1971; Riebsame *et al.*, 1991; Rosenberg, 1978) and ancient civilizations (Hodell *et al.*, 1995), it is important to understand patterns of natural variability. However, the longest instrumental climate records only extend over the last two centuries, and in North America only the last 100 years at best, and tree-ring reconstructions of climate are not possible in regions such as the Great Plains, which are mostly devoid of long-lived trees (Meko, 1992).

A sub-decadal record (ave. resolution 5.3 years) of diatom-inferred salinity for the past 2300 years from Moon Lake in eastern North Dakota (Laird *et al.*, 1996*b*) provides a long-term record of drought conditions (Fig. 3.9). High-salinity intervals are dominated by taxa, such as *Cyclotella quillensis*, *Chaetoceros elmorei/muelleri*, and *Nitzschia* cf. *fonticola*, whereas fresher intervals are dominated by taxa, such as *Stephanodiscus minutulus* and *S. parvus*. The Moon Lake record indicates that, prior to AD 1200, recurring severe droughts of greater intensity than those during the 1930s 'Dust Bowl' were the norm. This high frequency of extreme and persistent droughts, for which we have no modern equivalents, was most pronounced from AD 200–370, AD 700–850, and

AD 1000–1200. This latter drought period is coeval with the 'Medieval Warm Period'. A pronounced shift to generally wetter conditions at AD 1200 coincided with the end of the 'Medieval Warm Period' (AD 1000–1200; Lamb, 1982) and the onset of the 'Little Ice Age' (LIA, AD 1300–1850; Porter, 1986). The Moon Lake record provides support for a hydrologically complex 'LIA' interval (Bradley & Jones, 1992), with periods of wet conditions interspersed with short episodes of drier conditions, at times comparable to the droughts of the 1930s. Although the 'LIA' was not a continuously cold or wet period, at Moon Lake this interval was generally wetter in comparison to the previous 1500 years, with salinities reaching lows not recorded in the diatom record since the early Holocene (Laird *et al.*, 1996*b*).

The Moon Lake diatom record suggests that the recorded climate of the last 100 years in the Northern Great Plains is not representative of natural variation in drought intensity and frequency. Tree-ring records from western North America (Graumlich, 1993; Hughes & Brown, 1992; Stine, 1994) suggest a similar pattern. In comparison to fluctuations during glacial periods, the Holocene has typically been thought to be relatively stable (Dansgaard *et al.*, 1993, Meese *et al.*, 1994). However, if drought of similar magnitude to that inferred from the Moon Lake record prior to AD 1200 occurred today, it would have severe political, social, and economic consequences.

Problems in interpretation of climate and hydrology from fossil diatom records

There are a number of problems associated with unambiguous reconstruction of climate from fossil diatom records. These problems relate to each of the logical steps in the reconstruction process: preservation and taphonomy of fossil assemblages, the adequacy of salinity reconstruction, and the degree of coupling of lakewater salinity and climate. These issues are discussed extensively in a recent review by Gasse *et al.* (1996) and are reviewed more briefly below.

TAPHONOMY

Mixing and transport before burial may cause distorted or inaccurate environmental reconstructions. Mixture of different source communities in relatively large, permanent saline lakes has been thoroughly analyzed in Southeastern Australia (Gasse *et al.*, 1996; Gell, 1995). Diatoms were collected at regular time intervals over more than one hydrological cycle from sediment traps, phytoplankton tows, and artificial substrates set at varying distances from the lake margin. The occurrence of littoral forms in plankton samples and the differences in assemblage composition between phytoplankton tows and sediment traps indicate spatial heterogeneity, transport of diatoms from their source, and/or resuspension from sediment. These taphonomic processes may result in distortions and biases to the subfossil assemblages (Anderson &

S. C. FRITZ ET AL.

Battarbee, 1994). Temporal and spatial homogenization of diatom assemblages may have benefits in that surface sediments include the whole hydrological cycle and the entire basin. However, these mixing processes limit the accuracy of predictive models if calibration data sets integrate assemblages from different biotopes with vastly different salinity. In addition, inputs of diatoms to the sediment represent a mixture of successional communities, whose proportion in the sediment depends primarily on productivity. Unless seasonal fluxes are fully measured, assemblages from surficial bulk sediment cannot be regarded as reflecting mean annual lake surface salinity accurately.

An extreme situation is the inflow of allochthonous diatoms from rivers. Such a contamination explains the occurrence of *Cyclotella ocellata*, a taxon well known for living in freshwater but which occurs in samples of the hyperalkaline Lake Bogoria, Kenya (C = 50 mS/cm) and is integrated in the modern African calibration data set (Gasse *et al.*, 1995). Although the species occurs in 28 other waterbodies with lower salinities, the Bogoria samples introduce a strong bias in the calibration because of the very high conductivity, alkalinity, and pH values of host waters.

Detection of allochthonous inputs in fossil records is sometimes possible, when assemblages contain a mixture of diatom species with non-compatible ecological requirements. This is the case in the sedimentary sequence collected from the hypersaline–alkaline Lake Magadi, at the Kenya–Tanzania border (Barker, 1990; Gasse *et al.*, 1996; Taieb *et al.*, 1991). Paleoenvironmental reconstruction of the Late Pleistocene portion of the Lake Magadi sequence is problematic as it contains diatoms representative of environments from freshwater (*Fragilaria* spp., *Cymbella* spp., *Gomphonema* spp.) to hyperalkaline (*Anomoeoneis sphaerophora*, *Navicula elkab*, *Thalassiosira faurii*) waters. Taphonomy may have incorporated spatially separate and yet contemporaneous diatom assemblages, as would occur if a stream carrying freshwater diatoms were to discharge into a saline lake. Several short paleochannels of unknown age in the vicinity of the core sites could have been the source of this freshwater. Proceeding under this hypothesis, the total diatom assemblage was split into its fresh and saline components to estimate the pH and conductivity of the hypothesized lake and river groups separately. The difference between these extreme ranges (2 pH units, about 15 000 µS/cm) demonstrates the large error that can be introduced into environmental reconstructions if the taphonomy of the assemblage had not been considered and only mean values were used.

DISSOLUTION

In saline waters, a severe problem in interpretation of some diatom records is dissolution and diagenesis of diatom silica, which selectively removes certain taxa from the record (Barker *et al.*, 1994; Barker, 1990; Gell, 1995; Ryves, 1994). These phenomena occur during sedimentation and from contact between fossil diatoms and interstitial brines.

Experimental dissolution of lacustrine diatom silica in concentrated salt solutions (Barker, 1990) was performed to better understand the roles of the solution in which the silica is immersed and of the shape and structure of the diatom valves themselves. Dissolution of diatom silica from sediments and live diatoms was measured over two or three months at room temperature in 3M salt solutions (NaCl, Na_2CO_3, KNO_3, $CaCl_2$, $LiNO_3$, $MgCl_2$) and in distilled water. One major result is that composition of salt solutions affects not only rates of dissolution but also the final silica concentration at which saturation and molecular diffusion are achieved. As expected, dissolution rate and level of saturation is maximal in Na_2CO_3 solutions where pH > 9 enhances the dissociation of silicic acid. The diminishing levels of dissolution in other solutions, from NaCl, strong bases and weak acids (KNO_3 and $LiNO_3$), and then strong acid and weak bases with divalent cations ($CaCl_2$, $MgCl_2$), are more difficult to explain as solubility is independent of pH within the range of pH values in these solutions (4–7.5). Whatever the cause (hydration number of cations, ionic dissociation even for pH values close to neutrality, influence of the ionic strength of the solution on the activity coefficient of silicic acid), these results suggest that the final level of dissolution in saline lakes depends on brine type.

Another interesting observation from these experiments is the impact of differential dissolution on diatom assemblages. Clear interspecific differences occurred in diatom dissolution rates. The gross valve surface area/volume ratio was found to be a useful index of a particular diatom species' propensity to dissolve. The assemblage resulting from 46 days of immersion in Na_2CO_3 solution is composed mainly of partially dissolved *Stephanodiscus neoastraea* Håk. & Hickel (80%), which represented only 12% of the initial community. Conversely, the predominant taxa at the starting point, *Fragilaria capucina* Desm. and *Fragilaria crotonensis* Kitton, had totally disappeared. The outcome of these dissolution experiments can be used to interpret the integrity of fossil sequences.

Experimental results also suggest varying capacities of brines to induce formation of diagenetic silicates from diatom silica in divalent cation solutions (e.g., Mg^{2+} clays or zeolites), which decrease the stability of this amorphous silica phase. Neogenesis of smectites from diatom silica has been demonstrated from the Bolivian Altiplano (Badaut & Risacher, 1983). At Bougdouma (Niger), observations in scanning electron microscopy show that the original internal structure of *Aulacoseira* is replaced *in situ* by clay-like particles. This alteration is suspected to have occurred during early diagenesis, since similar changes in morphology have been observed in floating frustules of dead *Navicula elkab* in the neighboring alkaline pond of Guidimouni (Gasse, 1987). In several levels of the Late Quaternary record of the hyperalkaline Lake Bogoria (Gasse & Seyve, 1987), dissolution and neogenesis from *Thalassiosira rudolfii* towards zeolite mineral species (analcime) leads to the total disappearance of diatoms. In saline environments of the chloride–sulfate type, fossil assemblages are commonly reduced to the heavily silicified parts of

S. C. FRITZ ET AL.

Campylodiscus clypeus, Mastogloia spp., or septae of *Rhopalodia gibberula* or *Epithemia argus* (Ehr.) Kütz. Delicate taxa have obviously disappeared.

Quantification of biogenic silica dissolution and diagenesis for the reconstruction of initial assemblages must be investigated, as is tentatively done in the marine (Pichot *et al.*, 1992) and lacustrine (Ryves, 1994) contexts. In cases of poorly preserved diatoms in the fossil records, salinity reconstruction may be improved by using a dissolution index of heavily silicified taxa, as attempted for the Prairie region of North America (Ryves, 1994).

SALINITY RECONSTRUCTION

Modern calibration data sets show a high correlation between diatom-inferred salinity and measured lakewater salinity (see above), which suggests that these quantitative inference techniques can be used with some confidence to estimate past salinity. However, despite the statistical robustness of these methods, ambiguities may still arise. Calibration data sets are often composed of a single surface-sediment sample from each site, which integrates diatom production over one to several years, dependent on rates of sediment accumulation. Estimates of the salinity 'optimum' of individual taxa from surface sediments can be made with confidence in lakes where salinity fluctuates within a modest range or where the full range of seasonal variation in salinity is known. However, in some cases, calibration data sets are based on limited water-chemistry sampling, and in these cases a single measurement of salinity may not be representative of the water chemistry in which the diatom lived. Problems associated with the estimation of salinity optima, however, can be minimized by sampling a large number of sites (Wilson *et al.*, 1996) and by modest efforts to characterize the range of chemical variability.

A more significant limitation of the method occurs at the high end of the salinity gradient, where species diversity is low and many of the taxa have large salinity ranges. In these cases, the error surrounding any salinity estimate may be quite large. This is probably the cause of an incoherence between measured and diatom-inferred salinity in Devil's Lake, North Dakota, during intervals of higher salinity (Fig. 3.10). Devil's Lake represents an ideal situation in that the accuracy of diatom-based salinity reconstruction could be tested by comparing a diatom-inferred history of salinity change from a sediment core encompassing the last 100 years with measured salinity values from the same time interval (Fritz, 1990).

The ability of a calibration data set to reconstruct past salinity depends on having adequate modern analogs for fossil assemblages. The adequacy of a calibration data set in this regard can be evaluated by multivariate statistical techniques (Fig. 3.11). Poor analog situations exist where species occur in stratigraphic records in percentages greater than those in modern samples or in some cases where individual fossil species are not extant in the modern landscape. This is the case *for Cyclotella choctawhatcheeana*, which dominates some African sections of Holocene age, as well as core sequences in British

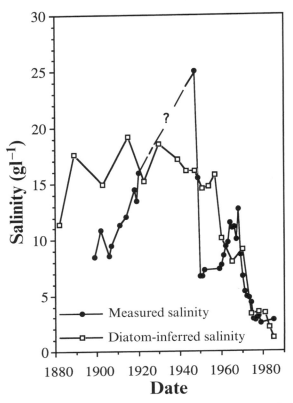

Fig. 3.10. Comparison of diatom-inferred salinity with measured salinity in Devils Lake, N. D. No salinity measurements were recorded between 1923 and 1948. Core dates are based on ^{210}Pb and are calculated with the constant rate of supply (c.r.s.) model. Modified from Fritz (1990).

Columbia (Wilson *et al.*, 1996), but has no modern analog in over several hundred modern samples from Africa (Gasse *et al.*, 1987) or British Columbia. However, this taxon occurs in modern sulfate-dominated saline lakes in the North American Great Plains (Fritz *et al.*, 1993) and in Australia (Gell & Gasse, 1990), and thus combining regional data sets from widespread geographic areas may enhance our ability to interpret fossil environments (Juggins *et al.*, 1994).

CLIMATIC RECONSTRUCTION

The problems associated with reconstruction of climate from salinity records are more difficult to address than those associated with salinity reconstruction itself and require evaluation of the nature of the hydrologic response to fluctuations in moisture. For example, salinity may not be linearly coupled to lake level or to the balance between $P - E$. In Devils Lake, lakewater salinity for any given lake stage was higher when lake level was falling than during

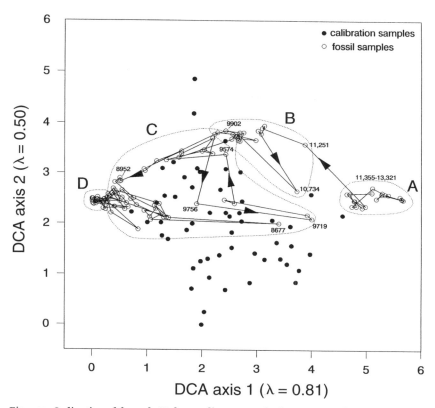

Fig. 3.11. Ordination of the early-Holocene diatom samples from Moon Lake, ND with the 55-lake calibration data set of Fritz *et al.* (1993) using detrended correspondence analysis (DCA). Circled areas with lettered designations correspond to diatom zones. See Laird *et al.* (1997) for further detail.

refilling. Presumably salts are lost to the system either by deflation or by precipitation and do not redissolve when the lake refills. Alternatively, in some lakes dissolution of evaporites during wet intervals can lead to elevated salinity (Gasse *et al.*, 1996), contrary to what one would predict if salinity were controlled by $P - E$. A related situation is hypothesized for areas within the Cuenca de Mexico (Bradbury, 1989), a complex of lakes, small pools, and marshes, where rising levels of saline lakes during periods of increased precipitation may have inundated marginal springs and pools with saline waters and elevated their salinity, whereas during arid periods the pools and springs were isolated and fed by fresh groundwater discharge.

Lakes also vary greatly in their sensitivity and response times to changes in moisture balance. Devils Lake, for example, is connected to several upstream basins during times of high precipitation and isolated during intervals of extreme drought. Thus its water budget and the magnitude of hydrological response to changes in moisture balance will differ during wet and dry periods. Likewise, a lake that fluctuates from open to closed hydrology may exhibit large shifts in salinity that are not proportionate to climatic forcing (Radle *et al.*, 1989; Van Campo & Gasse, 1993). Lakes with a large groundwater component to their hydrologic budget may show very small responses to changes in $P - E$, whereas the same climatic forcing may more strongly affect a system with limited groundwater inflow. Similarly a meromictic system may respond differently to changes in precipitation than a holomictic system. Thus periods of limnological stability do not necessarily reflect a stable climate (Laird *et al.*, 1996*a*). One cannot simply equate the magnitude of salinity change and the magnitude of climatic forcing; some understanding of basin hydrology is required in climatic interpretation of paleosalinity records.

The coupling between climate and salinity can be assessed, to some extent, by comparison of measured climate with diatom-inferred salinity fluctuations from cores that span the period of historic record. In the northern Great Plains, for example, meteorological measurements of precipitation and temperature are available for approximately the last 100 years. Comparison of diatom-inferred salinity fluctuations with a measure of moisture availability, such as $P - ET$ (Laird *et al.*, 1996*a*) or the Bhalme and Mooley Drought Index (Fig. 3.12), can be used to assess the relationship between moisture balance and salinity. In lakes where the correspondence is strong, one can infer climatic patterns from diatom-inferred salinity with some confidence.

Additional problems of interpretation can occur in settings where hydrologic change may be related to non-climatic factors. In the late-Quaternary record from Walker Lake, Nevada (Bradbury *et al.*, 1989), changes in lakewater salinity are driven by both changes in $P - E$ and by inflow or diversion of the freshwater Walker River into the lake. Although river flow patterns may, at times, be driven by climatic change, they may also be under geomorphic or tectonic control, and thus geologic setting must be carefully evaluated prior to interpretation of paleosalinity records.

Summary

The distribution of diatom species is clearly related to ionic concentration and composition. Presently, the mechanisms that relate diatoms and ions are not well studied, and the extent to which salinity exerts a direct vs. an indirect impact on individual taxa or species assemblages is not known. None the less, the strength of correlations between diatoms and ion strength and balance enables diatoms to be used to reconstruct past changes in lakewater salinity and brine composition driven by hydrologic and climatic change. The short generation times and rapid response of diatoms to hydrochemical change make them a particularly useful indicator of high frequency environmental variation.

The primary value of paleolimnologically based reconstructions of past hydrology and climate is to establish natural patterns of climatic variability at a variety of spatial and temporal scales. At millennial time scales, diatom-based studies have been important in establishing shifts in moisture balance

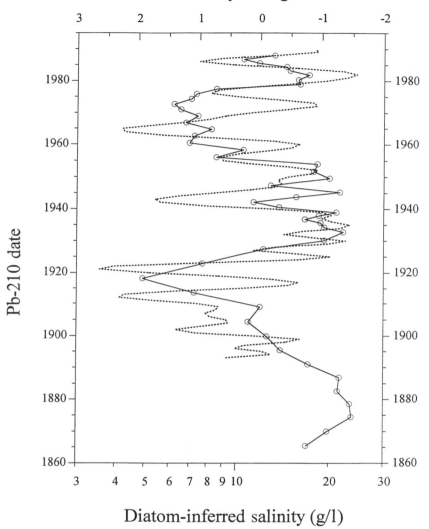

Fig. 3.12. Comparison of the diatom-inferred salinity (solid line) of Moon Lake, N. D. based on a short core covering the period of historic record with the Bhalme-Mooley Drought Index (BMDI) (broken line) calculated from precipitation measurements at a climate station near the lake. Chronology is based on ^{210}Pb dates calculated with the c.r.s. model.

associated with glacial/interglacial cycles across broad geographic regions. These data have been used to evaluate the relative importance of various climatic forcings and drivers, particularly the extent to which the influence of insolation is modulated by local or regional variables, such as oceanic circulation, topography, land-surface characteristics, or aerosols. Efforts to discern decadal to centennial-scale patterns of climatic variability and their causal mechanisms from high-resolution studies have intensified only recently

(Overpeck, 1990). The frequency of change in these high-resolution records, as well as the spatial distribution of coherent patterns of change, may provide clues regarding the relative roles of processes, such as variability of the linked atmospheric–oceanic system, solar and lunar activity, volcanic eruptions, and greenhouse gas concentration in causing decadal to centennial variation.

Diatom-based reconstruction of lakewater hydrochemistry can also be used to examine responses of lake systems to independently documented climatic changes. Thus comparisons of diatom-inferred salinity with instrumental climate data may be useful not only to choose suitable sites for climatic reconstruction (Laird *et al.*, 1996*a*) but also as a tool to better understand the relative importance of various components of the hydrologic budget in different lake types and geographic regions, particularly the importance of groundwater in controlling lake hydrochemistry (Crowe, 1993).

Stratigraphic records are critical to our understanding of modern climate and the extent of human modification of the climate system. Diatom-based climatic reconstructions provide a baseline for evaluating current climatic patterns, enabling us to determine whether the modern behavior of the climate system falls outside the range of natural variability and whether or not recent directional changes are a manifestation of natural cyclic behavior or alternatively may be a result of anthropogenic perturbations. Stratigraphic records are also a tool for testing the accuracy of climate models under boundary conditions that fall outside the modern norm, such as may occur with double CO_2. At present, there are relatively few paleolimnological studies from arid and semi-arid regions, and synchroneity of behavior among regions, which might suggest climatic teleconnections, can not be clearly defined. Nor do we have a sufficient number of records to understand the extent of climatic variability in the past, and to assess whether the modern climate is more or less variable either spatially or temporally. As is true of many types of diatom-based environmental reconstruction, our ability to resolve patterns of environmental change is often limited by our knowledge of the mechanisms that link diatoms, lake hydrochemistry, and climate. Thus not only do we need more stratigraphic records to discern patterns and mechanisms of climatic change, but we also need more modern process-based studies, particularly in the areas of lake hydrology and diatom physiology.

References

Anderson, N. J. & Battarbee, R. W. (1994). Aquatic community persistence and variability: A palaeolimnological perspective. In *Aquatic Ecology: Scale, Pattern, and Process,* ed. P. S. Giller, A. G. Hildrew & D. G. Raffaelli, pp. 233–59. Oxford: Blackwell Scientific Publications.

Badaut, D. & Risacher, F. (1983). Authigenic smectite on diatom frustules in Bolivian saline lakes. *Geochimica et Cosmochimica Acta,* 47, 363–75.

Barker, P. A. (1990). Diatoms as palaeolimnological indicators: A reconstruction of Late Quaternary environments in two East Africa salt lakes. PhD thesis, Loughborough University, 267 pp.

Barker, P. A., Fontes, J-C., Gasse, F., & Druart, J-C. (1994). Experimental dissolution of diatom silica in concentrated salt solutions and implications for palaoenvironmental reconstruction. *Limnology and Oceanography*, **39**, 99–110.

Bhattacharyya, P. & Volcani, B. E. (1980). Sodium – dependent silicate transport in the apochlorotic marine diatom *Nitzschia alba*. *Proceedings, Academy of Natural Sciences of Philadelphia*, **77**, 6386–90.

Birks, H. J. B. (1995). Quantitative paleoenvironmnetal reconstructions. In *Statistical Modelling of Quaternary Science Data*, ed. D. Maddy & J. S. Brew, pp. 161–254. Cambridge: Quaternary Research Association.

Birks, H. J. B., Line, J. M., Juggins, S., Stevenson, A. C., & ter Braak, C. J. F. (1990). Diatoms and pH reconstruction. *Philosophical Transactions, Royal Society of London, Series B*, **327**, 263–78.

Blinn, D. (1995). Diatom community structure along salinity gradients in Australian saline lakes: Biogeographic comparisons with other continents. In *A Century of Diatom Research in North America: a Tribute to the Distinguished Careers of C. W. Reimer and R. Patrick*, ed. J. P. Kociolek & M. J. Sullivan, pp. 156–67. Champaign, IL: Koeltz Scientific.

Blinn, D. W., Hevly, R. H., & Davis, O. K. (1994). Continuous Holocene record of diatom stratigraphy, palaeohydrology, and anthropogenic activity in a spring-mound in southwestern United States. *Quaternary Research*, **42**, 197–205.

Borchert, J. R. (1971). The dust bowl in the 1970s. *Annals of the Association of American Geographers*, **40**, 1–39.

Bradbury, J. P. (1989). Late Quaternary lacustrine paleoenvironments in the Cuenca de Mexico. *Quaternary Science Reviews*, **8**, 75–100.

Bradbury, J. P., & Dieterich-Rurup, K. (1993). Holocene diatom paleolimnology of Elk Lake, Minnesota. In *Elk Lake Minnesota: Evidence for Rapid Climate Change in the North-Central United States,* ed. J. P. Bradbury & W. E. Dean, pp. 215–37. Boulder: Geological Society of America.

Bradbury, J. P., Forester, R. M., & Thompson, R. S. (1989). Late Quaternary paleolimnology of Walker Lake, Nevada. *Journal of Paleolimnology*, **1**, 249–67.

Bradley, R. S., & Jones, P. D. (ed.) 1992. *Climate Since A. D. 1500*. New York: Routledge .

Charles, D. F., & Smol, J. P. (1994). Long-term chemical changes in lakes: quantitative inferences from biotic remains in the sediment record. In *Environmental Chemistry of Lakes and Reservoirs,* ed. L. Baker, pp. 3–31. Washington, DC: American Chemical Society.

Crowe, A. S. (1993). The application of a coupled water-balance–salinity model to evaluate the sensitivity of a lake dominated by groundwater to climate variability. *Journal of Hydrology*, **141**, 33–73.

Cumming, B. F., & Smol, J. P. (1993). Development of diatom-based salinity models for paleoclimatic research from lakes in British Columbia (Canada). *Hydrobiologia*, **269/270**, 179–96.

Cumming, B. F., Wilson, S. E., Hall, R. I., & Smol, J. P. (1995). *Diatoms from British Columbia (Canada) Lakes and Their Relationship to Salinity, Nutrients, and Other Limnological Variables*. Stuttgart, Germany: Koeltz Scientific Publishers. 207 pp.

Dansgaard, W., Johnsen, S. J., Clausen, H. B., Dahl-Jensen, D., Gunderstrup, N. S., Hammer, C. U., Hvidberg, C. S., Steffensen, J. P., Sveinbjirnsdottir, A. E., Jouzel, J., & Bond, G. (1993). Evidence for general instability of past climate from a 250-year ice core record. *Nature*, **364**, 218–20.

Donovan, J. J. (1994). Measurement of reactive mass fluxes in evaporative groundwater-source lakes. In *Sedimentology and Geochemistry of Modern and Ancient Saline Lakes,* ed. R. Renaut & W. M. Last, pp. 33–50. SEPM Society for Sedimentology Geology.

Dubar, C. (1988). Elements de paleohydrologie de l'Afrique Saharienne: les depots quaternaries d'origine du Nord-Est de l'Air (Niger). PhD thesis, *Université de Paris-Sud*.

Eugster, H. P., & Jones, B. F. (1979). Behavior of major solutes during closed-basin brine evolution. *American Journal of Science*, **279**, 609–31.

Fisher, N. S. (1977). On the differential sensitivity of estuarine and open-ocean diatoms to exotic chemical stress. *American Naturalist*, **111**, 871–95.

Fontes, J. C., & Gasse, F. (1991). PALHYDAF Programme. Objectives, methods, major results. *Palaeogeography, Palaeoclimatology, Palaeoecology*, **84**, 191–215.

Fontes, J. C., Gasse, F., Callot, Y., Plaziat, J-C., Carbonel, P., Dupeuble, P. A., & Kaczmarska, I. (1985*a*). Freshwater to marine-like environments from Holocene lakes in northern Sahara. *Nature*, **317**, 608–10.

Fontes, J. C., Gasse, F., Camara, E., Millet, B., Saliege, J. F., & Steinberg, M. (1985*b*). Late Holocene changes in Lake Abhe hydrology. *Zeischrift für Gletscherkunde und Glazial Geologie*, **21**, 89–96.

Fritz, S. C. (1990). Twentieth-century salinity and water-level fluctuations in Devils Lake, N. Dakota: A test of a diatom-based transfer function. *Limnology and Oceanography*, **35**, 1771–81.

Fritz, S. C., Juggins, S., & Battarbee, R. W. (1993). Diatom assemblages and ionic characterization of lakes of the Northern Great Plains, North America: A tool for reconstructing past salinity and climate fluctuations. *Canadian Journal of Fisheries and Aquatic Sciences*, **50**, 1844–56.

Fritz, S. C., Juggins, S., Battarbee, R. W., & Engstrom, D. R. (1991). Reconstruction of past changes in salinity and climate using a diatom-based transfer function. *Nature*, **352**, 706–8.

Gasse, F. (1977). Evolution of Lake Abhe (Ethiopia and T. F. A. I.) from 70,000 B. P. *Nature*, **2**, 42–5.

(1986). *East African Diatoms: Taxonomy, Ecological Distribution*. Stuttgart: Cramer. 201 pp.

(1987). Diatoms for reconstructing palaeoenvironments and palaeohydrology in tropical semi-arid zones: Examples of some lakes in Niger since 12000 BP. *Hydrobiologia*, **154**, 127–63.

(1988). Diatoms, palaoenvironments and palaeohydrology in the western Sahara and the Sahel. *Würzburg Geographische Arbeite*, **69**, 233–54.

(1994). Lacustrine diatoms for reconstructing past hydrology and climate. In *Long-term Climatic Variations: Data and Modelling*, ed. J. C. Duplessy & M. T. Spyridakis, pp. 335–69. NATO ASI series.

Gasse, F., & Fontes, J. C. (1989). Paleoenvironments and paleohydrology of a tropical closed lake (L. Asal, Djibouti), since 10,000 yr. B. P. *Palaeogeography, Palaeoclimatology, Palaeoecology*, **69**, 67–102.

Gasse, F., & Seyve, C. (1987). Sondages du Lac Bogoria: Diatomees. In *Le demi-graben de Baringo-Bogoria, Rift Gregory, Kenya. 30.000 ans d'histoire hydrologique et sedimentaire*, ed. J. J. Tiercelin & A. Vincens, pp. 414–37. Bulletin du Centre de Recherche et d'Exploration-Production Elf-Aquitaine Boussens.

Gasse, F., & Street, F. A. (1978). Late Quaternary lake level fluctuations and environments of the northern Rift Valley and Afar region (Ethiopia and Djibouti). *Palaeogeography, Palaeoclimatology, Palaeoecology*, **25**, 145–50.

Gasse, F., & Van Campo, E. (1994). Abrupt post-glacial climate events in west Asia and north Africa monsoon domains. *Earth and Planet Science Letters*, **126**, 435–65.

Gasse, F., Fontes, J. C., Plaziat, J. C., Carbonel, P., Kaczmarska, I., De Deckker, P., Soulié-Marsche, I., Callot, Y., & Dupeuble, P. A. (1987). Biological remains, geochemistry and stable isotopes for the reconstruction of environmental and hydrological changes in the Holocene lakes from North Sahara. *Palaeogeography, Palaeoclimatology, Palaeoecology*, **60**, 1–46.

Gasse, F., Gell, P., Barker, P., Fritz, S. C., & Cahlie, F. (1996). Diatom-inferred salinity of palaeolakes: An indirect tracer of climate change. *Quaternary Science Reviews*, **15**, 1–19.

Gasse, F., Juggins, S., & Ben Khelifa, L. (1995). Diatom-based transfer functions for inferring hydrochemical characteristics of African palaeolakes. *Palaeogeography, Palaeoclimatology, Palaeoecology*, **117**, 31–54.

Gasse, F., Tehet, R., Durand, A., Gibert, E., & Fontes, J. C. (1990). The arid-humid transition in the Sahara and the Sahel during the last deglaciation. *Nature*, **346**, 141–6.

Gell, P. A. (1995). The development and application of a diatom calibration set for lake salinity, Western Victoria, Australia. PhD thesis, Monash University.

Gell, P. A., & Gasse, F. (1990). Relationships between salinity and diatom flora from some

Australian saline lakes. In *Proceedings of the 11th International Diatom Symposium,* ed. J. P. Kociolek, pp. 631–47. San Francisco: California Academy of Sciences.

Graumlich, L. J. (1993). A 1000 year record of temperature and precipitation in the Sierra Nevada. *Quaternary Research,* **39,** 249–55.

Hodell, D. A., Curtis, J. H., & Brenner, M. (1995). Possible role of climate in the collapse of classic Maya civilization. *Nature,* **375,** 391–5.

Hughes, M., K., & Brown, P. M. (1992). Drought frequency in central California since 101 BC recorded in giant sequoia tree rings. *Climate Dynamics,* **6,** 161–7.

Hui, F., Gasse, F., Huc, A., Yuanfang, L., Sifeddine, A., & Soulie-Marsche, I. (1996). Holocene environmental changes in Bangong Co Basin (Western Tibet). Part 3: Biogenic remains. *Palaeogeography, Palaeoclimatology, Palaeoecology,* **120,** 65–78.

Hustedt, F. (1953). Die Systematik der Diatomeen in ihren Beziehungen zur Geologie und Ökologie nebst einer Revision des Halobien-systems. *Botanisk Tidskkrift,* **47,** 509–19.

Juggins, S., Battarbee, R. W., Fritz, S. C, & Gasse, F. (1994). The CASPIA project: Diatoms, salt lakes, and environmental change. *Journal of Paleolimnology,* **2,** 191–6.

Kashima, K. (1994). Sedimentary diatom assemblages in freshwater and saline lakes of the Anatolia Plateua, central part of Turkey: An application of reconstruction of palaeosalinity change during the Late Quaternary. In *Proceedings of the 13th International Diatom Symposium,* ed. D. Marino & M. Montresor, pp. 93–100. Bristol: Biopress Ltd.

Kennedy, K. A. (1994). Early-Holocene geochemical evolution of saline Medicine Lake, South Dakota. *Journal of Paleolimnology,* **10,** 69–84.

Khelifa, L. B. (1989). Diatomees continentales et paleomillieux du Sud-Tunisien aux Quaternaire superieur. PhD thesis, Université de Paris-Sud.

Kolbe, R. W. (1927). Zur Okologie, Morphologie und Systematik der Brackwasser-Diatomeen. *Pflanzenforschung,* **7,** 1–146.

Laird, K., Fritz, S. C., Grimm, E. C., & Mueller, P. G. (1996a). The paleoclimatic record of a closed-basin lake in the northern Great Plains: Moon Lake, Barnes Co., N. D. *Limnology and Oceanography,* **40,** 890–902.

Laird, K. R., Fritz, S. C., Maasch, K. A., & Cumming, B. F. (1996b). Greater drought intensity and frequency before AD 1200 in the Northern Great Plains, USA. *Nature,* **384,** 552–5.

Laird, K. R., Fritz, S. C., Grimm, E. C., & Cumming, B. F. (1997). Early Holocene limnologic and climatic variability in the northern Great Plains. *Holocene,* **8,** 275–86.

Lamb, H. H. (1982). *Climate, History, and the Modern World.* London: Methuen. 387 pp.

Mason, I. M., Guzkowska, M. A. J., & Rapley, C. G. (1994). The response of lake levels and areas to climatic change. *Climatic Change,* **27,** 161–97.

Meese, D. A., Gow, A. J., Grootes, P., Mayewski, P. A., Ram, M., Stuiver, M., Taylor, K. C., Waddington, E. D., & Zielinski, G. A. (1994). The accumulation record from the GISP2 core as an indicator of climate change through the Holocene. *Science,* **266,** 1680–2.

Meko, D. M. (1992). Dendroclimatic evidence from the Great Plains of the United States. In *Climate Since A. D. 1500,* ed. R. S. Bradley & P. D. Jones, pp. 312–30. New York: Routledge.

Metcalfe, S. E. (1988). Modern diatom assemblages in Central Mexico: The role of water chemistry and other environmental factors as indicated by TWINSPAN and DECO-RANA. *Freshwater Biology,* **19,** 217–33.

Meybeck, M. (1995). Global distribution of lakes. In *Physics and Chemistry of Lakes,* ed. A. Lerman, D. Imboden, & J. Gat, pp. 1–35. Berlin: Springer Verlag.

Overpeck, J. (1996). Warm climate surprises. *Science,* **271,** 1820–1.

Pichon, J. J., Bareille, G., Labracherie, M., Labeyrie, L. D., Baudrimont, A., & Turon, J. L. (1992). Quantification of the biogenic silica dissolution in southern ocean sediments. *Quaternary Research,* **37,** 361–78.

Porter, S. C. (1986). Pattern and forcing of Northern Hemisphere glacier variations during the last millennium. *Quaternary Research,* **26,** 27–48.

Radle, N. J., Keister, C. M., & Battarbee, R. W. (1989). Diatom, pollen, and geochemical evidence for the paleosalinity of Medicine Lake, S. Dakota, during the Late Wisconsin and early Holocene. *Journal of Paleolimnology*, 2, 159–72.

Richardson, J. L., & Dussinger, R. A. (1986). Paleolimnology of mid-elevation lakes in Kenya Rift Valley. *Hydrobiologia*, 143, 167–74.

Riebsame, W. E., Changon, S. A., & Karl, T. R. (1991). *Drought and Natural Resources Managment in the United States: Impacts and Implications of the 1987–1989 Drought.* Boulder: Westview Press. 174 pp.

Rind, D., & Overpeck, J. (1993). Hypothesized causes of decade to century scale climate variability: Climate model results. *Quaternary Science Reviews*, 12, 357–74.

Roberts, D., & McMinn, A. (1996). Relationships between surface sediment diatom assemblages and water chemistry gradients of the Vestfold Hills, Antarctic. *Antarctic Science*, 8, 331–41.

Rosenberg, N. J. (1978). *North American Droughts.* Boulder: Westview Press. 177 pp.

Roux, M., Servant-Vildary, S., & Servant, M. (1991). Inferred ionic composition and salinity of a Bolivian Quaternary lake, as estimated from fossil diatoms in the sediments. *Hydrobiologia*, 210, 3–18.

Ryves, D. B. (1994). Diatom dissolution in saline lake sediments. An experimental study in the great Plains of Northern America. PhD thesis, University College London, London, UK.

Schweger, C. E., & Hickman, M. (1989). Holocene paleohydrology of central Alberta: testing the general-circulation-model climate simulations. *Canadian Journal of Earth Sciences*, 26, 1826–33.

Servant-Vildary, S., & Roux, M. (1990). Multivariate analysis of diatoms and water chemistry in Bolivian saline lakes. *Hydrobiologica*, 197, 267–90.

Shobert, B. (1974). The influence of water stress on the metabolism of diatoms. I. Osmotic resistance and proline accumulation in *Cyclotella meneghiniana. Zeitschrift für Pflanzenphysiologie*, 74, 106–20.

Stager, J. C. (1982). The diatom record of Lake Victoria (East Africa): the last 17,000 years. In *Proceedings of the 7th International Diatom Symposium,* ed. D. G. Mann, pp. 455–76. Philadelphia: O. Koeltz Science Publishers.

(1988). Environmental changes at Lake Cheshi, Zambia since 40,000 years B. P. *Quaternary Research*, 29, 54–65.

Stine, S. (1994). Extreme and persistent drought in California and Patagonia during mediaeval time. *Nature*, 369, 546–9.

Taieb, M., Barker, P., Bonnefille, R., Damnati, B., Gasse, F., Goetz, C., & Hillaire (1991). Histoire paleohydrologique du lac Magadi (Kenya) au Pleistocene superieur. *Comptes Rendus de l'Academie des Sciences, Paris series II*, 313, 339–46.

Tuchman, M. L., Theriot, E., & Stoermer, E. F. (1984). Effects of low level salinity concentrations on the growth of *Cyclotella meneghiniana* Kütz. (Bacillariophyta). *Archiv für Protistenkunde* 128, 319–26.

Van Campo, F., & Gasse, F. (1993). Pollen and diatom-inferred climatic and hydrological changes in Sumxi Co. Basin (western Tibet) since 13,000 yr B. P. *Quaternary Research*, 39, 300–13.

Williams, W. D. (1981). Inland salt lakes: An introduction. *Hydrobiologia*, 81, 1–14.

Wilson, S. E., Cumming, B. F., & Smol, J. P. (1994). Diatom-based salinity relationships in 111 lakes from the Interior Plateau of British Columbia, Canada: The development of diatom-based modesl for paleosalinity and paleoclimatic reconstructions. *Journal of Paleolimnology*, 12, 197–221.

(1996). Assessing the reliability of salinity inference models from diatom assemblages: An examination of a 219 lake data set from Western North America. *Canadian Journal of Fisheries and Aquatic Sciences*, 53, 1580–94.

Winter, T. C. (1990). Hydrology of lakes and wetlands. In *Surface Water Hydrology.*, ed. M. G. Wolman & H. C. Riggs, pp. 159–88. Boulder: Geological Society of America.

4 Diatoms as mediators of biogeochemical silica depletion in the Laurentian Great Lakes

CLAIRE L. SCHELSKE

Introduction

The silica depletion hypothesis (Schelske & Stoermer, 1971, 1972), advanced in the early 1970s, stated that increases in diatom production driven by increases in phosphorus loading increased permanent sedimentation of diatoms, and that the associated progressive utilization of silica in the water column reduced silica reserves and eventually induced epilimnetic silica depletion. This hypothesis, in fact, encompassed several related hypotheses: (i) increased phosphorus loading increases the production of diatoms and other algae with, or without, an obligate requirement for silica, and concomitantly increases silica utilization and production of biogenic silica; (ii) permanent sedimentation of siliceous organisms increases concomitantly with increased production, either reducing or depleting silica reserves in the water mass; (iii) increased production and sedimentation of diatoms, coupled with the imbalance in external loading and supplies of phosphorus and silica relative to the requirements for diatom production, ultimately causes silica-limited diatom growth; and (iv) silica depletion or limitation progressively shifts species composition as assemblages with smaller proportions of diatoms and other siliceous algae and increasing proportions of green and blue-green (cyanobacteria) algae replace assemblages dominated by diatoms. Such changes in composition of phytoplankton affect ecosystem processes by changing trophic structure and trophic efficiency (Rabalais *et al.*, 1996). Subsequent studies refined the hypothesis by finding that epilimnetic silica depletion occurred more rapidly than initially expected (Schelske *et al.*, 1983, Schelske, 1988) and that silica depletion developed not only in the epilimnion but also over the entire water column in the lower Great Lakes (Schelske *et al.*, 1986). This summary outlines how a new paradigm on the role of diatoms as biogeochemical mediators of silica depletion in the Great Lakes evolved from the initial studies on Lake Michigan.

Historical phosphorus concentrations in the Great Lakes have been measured in the last few decades (Johengen *et al.*, 1994, Schelske, 1991) and were calculated from computer simulations for earlier periods (Chapra, 1977). Two periods of increased nutrient enrichment modified the silica cycle in these phosphorus-limited systems by stimulating the production and sedimentation of diatoms. The first increase in nutrient loading occurred in the late 1800s after forest clearance and settlement by Europeans and an exponential increase began after 1940 when the use of phosphate detergents and the population served by municipal sewerage systems increased. Concerns about water quality led to an international agreement between Canada and the United States in 1972 to decrease phosphorus loading to the Great Lakes. The success of this effort is now evident from concentrations of total phosphorus (*TP*) in Lake Ontario and the eastern basin of Lake Erie that decreased from 25 μg TP/l in the early 1970s to well below 10 μg TP/l in the mid-1990s.

Some of the terms that are important in discussing the role of diatom-mediated silica depletion in the Laurentian Great Lakes are listed below:

- Biogenic silica – amorphous silica (opal) incorporated into diatom frustules and chrysophyte cysts and scales.
- Silica reservoir – quantity of soluble silica in the water mass. Under steady-state conditions, the reservoir is relatively constant from year to year.
- Silica utilization – amount of biogenic silica incorporated into new cells as frustules or other siliceous structures. As a result, production of biogenic silica is quantitatively equivalent to the loss in soluble silica from the water column.
- Silica depletion – long-term decrease in the silica reservoir. Silica is depleted when silica utilization increases due to nutrient enrichment and is limiting if the availability of silica limits the production of bio-genic silica. In the geochemical sense, silica-limited production occurs when supplies of silica are so low that no additional diatom biomass is produced.
- Primary limiting nutrient – nutrient that first limits growth rate or biomass production.
- Secondary nutrient limitation – silica limitation, for example, induced by increased enrichment of the primary limiting nutrient; or limitation by the second least available nutrient.

Mass balance considerations

Silica depletion on a lake-by-lake basis is quantifiable using a mass balance analysis of annual inputs and outputs (eqns. 1 and 2).

$$\Delta R = \text{inputs} - \text{losses} \qquad (1)$$
$$\Delta R = TI + AI - (SL + OL) \qquad (2)$$
$$SL = DP - DD \qquad (3)$$

where R is the annual maximum quantity of silica in the water mass, ΔR is the annual change in this silica reservoir, TI is tributary input, AI is atmospheric input, SL is sediment loss, OL is output loss, DP is diatom production and DD is diatom dissolution (Schelske, 1985).

Nutrient enrichment of waters increases the production and sedimenta-

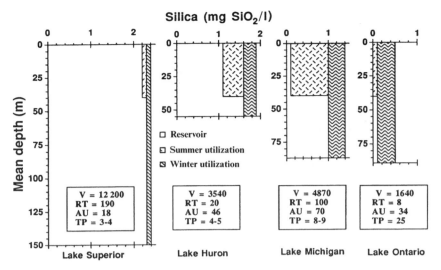

Fig. 4.1. Annual silica dynamics in the Laurentian Great Lakes in the early 1970s. Silica concentrations (mg SiO2/l) as annual maximum and minimum for each lake are plotted vs. mean depth. The hatched areas define the annual minimum and maximum silica concentration and represent areal utilization (AU) during winter–spring mixing and summer stratification and the remaining area represents the silica reservoir remaining after seasonal uptake and utilization by diatoms. Also shown for each lake are volume (V) in km³, hydraulic residence time (RT) in years, AU in g SiO2/m², and total phosphorus (TP) in μg/l. Lake Erie is not shown because silica dynamics differ among its three morphometric basins; characteristics of the eastern basin are similar to those in Lake Ontario.

tion of diatoms and, consequently, affects the annual mass balance of silica in an aquatic ecosystem. Silica loss (SL) to the sediment sink is equal to diatom production (DP) less diatom dissolution (DD) (see eqn. 3). Because dissolution is >90% of diatom production in the Great Lakes, its rate must be estimated indirectly (Schelske, 1985). Calculating SL directly from eqn. 3, therefore, yields large errors because sediment loss is <10% of diatom production. When diatom production increases as the result of nutrient enrichment, the resulting increased loss of biogenic silica to the sediments causes a compensating decrease in the silica reservoir (see eqn. 2). If inputs are constant, eqns. 2 and 3 show that maximum historical diatom production is a function of the silica reservoir and that production will first increase and then decrease with depletion of R leading to a peak in biogenic silica accumulation in the sediment record (Schelske et al., 1983). Two different steady states of nutrient limitation occur during this transition. Initially, with an adequate reservoir of silica, diatom growth is phosphorus limited whereas diatom growth becomes silica limited, as the reservoir decreases.

Hydrologic characteristics make the Great Lakes ideal environments to study silica depletion on an ecosystem basis and to quantify silica depletion using mass balances. Large silica reservoirs characterized the upper Great Lakes historically (Fig. 4.1), and losses of biogenic silica to the sediments were small compared to the silica reservoir in these lakes (Schelske, 1985). The reservoir under steady-state conditions was not affected greatly by annual inputs or losses in Lake Superior and Lake Michigan because residence times are long (190 and 100 years, respectively). Silica depletion that developed over periods of one or two decades, therefore, clearly resulted from increased sediment losses of biogenic silica, a proxy for a lake's response to increased diatom production that resulted from increased nutrient loading. By contrast, the residence time of Lake Erie is so short (3 years) that effects of phosphorus enrichment on diatom production are evident in real time by comparing inputs and outputs of silica. Finally, a change in outflow silica concentration (OL in eqn. 2) over time reflects time-integrated responses of the lakes to nutrient enrichment.

Role of nutrient enrichment

Comparing nutrient data among lakes shows the relationship between nutrient enrichment and silica depletion in the Great Lakes. Because phosphorus is the limiting nutrient, TP (the most conservative form of phosphorus) is an index of nutrient enrichment. Chemical data from the 1970s shows a fivefold difference in TP concentration among the five lakes (Fig. 4.1). Lake Superior and Lake Huron with TP concentrations ranging from 4–5 μg TP/l are the most oligotrophic; Lake Erie and Lake Ontario with a TP concentration of approximately 25 μg TP/l are the most nutrient enriched. Diatom production is not silica limited in Lake Superior or Lake Huron, the lakes with the lowest TP concentration, and is seasonally limited in epilimnetic waters during summer stratification in Lake Michigan. Lake Michigan with a TP concentration of 8 μg TP/l is only slightly enriched compared to Lake Superior and Lake Huron, but this small degree of enrichment was adequate to induce epilimnetic silica depletion. Water column silica limitation characterizes Lake Erie and Lake Ontario, the lakes with the highest TP concentration. With this degree of phosphorus enrichment, silica in the entire water column is depleted to the extent that diatom production is limited before the lake stratifies thermally. The degree of silica limitation therefore, is related to TP enrichment among the Great Lakes.

Limiting nutrients added experimentally to natural lakewater with naturally occurring phytoplankton assemblages increase phytoplankton growth. Experiments conducted *in situ* using large polyethylene spheres simulate natural light and temperature conditions, but laboratory experiments provide essentially the same information and employ more treatments required for statistical tests. Diatom production in these nutrient perturbation experiments depleted silica supplies in 5–10 days with phosphorus enrichments as small as 3–5 μg TP/l, relatively small enrichments compared to most experimental studies (Schelske et al., 1986). These results demonstrate that a relatively small nutrient enrichment of natural lakewater profoundly affects silica dynamics in the Great Lakes during an annual cycle of phosphorus-limited diatom production.

Several findings show that effects of phosphorus enrichment on diatom

C. L. SCHELSKE

production and silica dynamics are even more pronounced than expected from first principles. First, phosphorus is recycled more efficiently than silica (Conley *et al.*, 1988). During the annual production cycle, phosphorus is recycled several times, and must be recycled efficiently to maintain nutrients to sustain phytoplankton production in the mixed layer after thermal stratification. Biological recycling of phosphorus is efficient because it depends mainly on biological processes of uptake, excretion, and mineralization. By contrast, the silica cycle involves biological uptake (frustule formation) and chemical dissolution of biogenic silica as the major factor in mineralization. Dissolution of biogenic silica is a relatively slow process compared to rapid cycling of phosphorus, involving a time constant of approximately 1 year in Lake Michigan (Schelske, 1985). Second, accessory growth substances added in low concentrations in combination with phosphorus enrichments increase diatom production more than comparable phosphorus enrichments without these substances. These accessory growth substances which include vitamins (thiamin, biotin and cobalamin), trace metals (iron, manganese, cobalt, zinc and molybdenum), and EDTA (a chelating agent) may be present in waste water and thus enhance effects of nutrient enrichment. Effects of individual components in these mixtures have not been tested, primarily because possible combinations of treatments are so large. Third, light–nutrient interactions at low light compensate to increase responses of shade-adapted populations to nutrient enrichment. Adding nutrients, including accessory growth substances, stimulates diatom production at low irradiances (5–50 μEin m^{-2} s^{-1}) relative to responses without nutrient enrichment (Fahnenstiel *et al.*, 1984). This light/nutrient interaction compensates for low light intensities generally considered to be growth limiting. Synergistic effects among phosphorus, other major nutrients, light, and accessory growth substances enhance the response of phytoplankton to small nutrient enrichments and interact to accelerate silica utilization and depletion as phosphorus or nutrient loading increases.

Results of nutrient enrichment experiments demonstrate that phosphorus is the primary nutrient limiting production of phytoplankton biomass or standing crop in the Great Lakes. In addition, both comparative limnology (Fig. 4.1) and nutrient enrichment experiments show that phosphorus enrichment can induce secondary silica limitation of diatom production. Silica depletion and its associated biological consequences also result from phosphorus and other coincident nutrient enrichment in the Mississippi River and adjacent continental shelf (Rabalais *et al.*, 1996) and in other marine, estuarine and freshwater systems (Conley *et al.*, 1993).

Development and degree of silica limitation

Several examples are presented to show that silica limitation develops rapidly and that phosphorus enrichment affects the degree of silica limitation (see Fig. 4.1).

Several studies of silica in the water column provide evidence of nutrient-driven silica depletion in Lake Michigan (see Schelske, 1988). Silica data collected at the municipal water intakes for Chicago show a long-term decrease in concentration from 1926 to 1964. Surface silica concentrations in August 1955 were <1.0 mg SiO_2/l in the southern end of the lake, much lower than the remainder of the basin. Finally, comparing open-lake data collected in 1954–1955 with those from 1970 show that epilimnetic silica decreased approximately 2.0 mg SiO_2/l in this 15-year period. These decreases in concentration combined with lake-wide epilimnetic concentrations in 1970 as low as 0.1 mg SiO_2/l strongly support the case for epilimnetic silica depletion in Lake Michigan.

RAPID EPILIMNETIC SILICA DEPLETION

A comprehensive analysis of historic water column data from Lake Michigan showed that silica decreased rapidly from 1955–1970 (Schelske, 1988). The winter maximum concentration decreased from 4.4 to 1.4 mg SiO_2/l and the summer epilimnetic concentration decreased from 2.2 to 0.1 mg SiO_2/l in this 15-year period. Rapid silica depletion inferred from paleolimnological data (Fig. 4.2) occurred in Lake Ontario, but in the late 1800s after early settlement and forest clearance by Europeans (Schelske, 1991). Such a drastic change in system biogeochemistry mediated by anthropogenic factors in the late 1800s is surprising in such a large lake, but studies of siliceous microfossils also confirm this drastic perturbation (Stoermer *et al.*, 1985*a*). A relatively small phosphorus enrichment induced rapid silica depletion; epilimnetic silica depletion developed in Lake Michigan while the concentration of TP was <10 mg TP/l and at a comparably low level in Lake Ontario (Schelske *et al.*, 1983). Synergisms discussed above probably enhanced effects that might be predicted only from phosphorus enrichment of lake waters.

WATER COLUMN SILICA DEPLETION

After epilimnetic silica depletion develops, additional nutrient loading can produce water column silica depletion (Schelske *et al.*, 1986), i.e. silica concentrations in the water column <0.50 mg SiO_2/l throughout the year. Such severe silica depletion that characterizes Lake Erie and Lake Ontario was induced by the exponential increase in phosphorus loading that began in the 1940s. Under these conditions, diatom production is silica-limited throughout the year because silica is depleted to limiting levels for maximum diatom growth before the lake is stratified thermally (see Fig. 4.1).

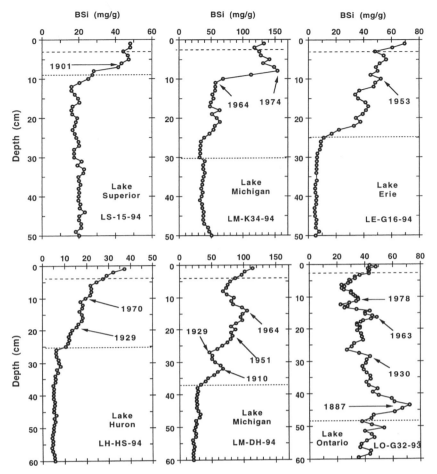

Fig. 4.2. Biogenic silica concentration (mg SiO$_2$/g dry sediment) plotted vs. depth for high resolution sediment cores from the Laurentian Great Lakes. Cores were sectioned at 1.0-cm intervals except the upper 20 cm of LO-G32-93 were sectioned at 0.5-cm intervals. The dashed line near the top of the core is drawn to roughly define the zone of dissolution; the dotted line at depth depicts the settlement horizon which varied from approximately 1865 in the lower lakes to 1890 in the upper lakes. Dates are ^{210}Pb ages. Data are from Schelske and Hodell (in preparation).

LAKE ERIE: A NATURAL CHEMOSTAT

Lake Erie can be viewed as a natural chemostat in which the change in silica concentration between input and output provides a measure of phosphorus enrichment effects on silica utilization by diatoms and a real-time demonstration of the silica depletion hypothesis (Schelske *et al.*, 1986). Water with a low TP concentration and a relatively large silica concentration flowing from Lake Huron (Fig. 4.1) accounts for 80% of the inflow to Lake Erie. In Lake Erie, anthropogenic phosphorus loading enriches water five-fold compared to the Lake Huron outflow and augments production and sedimentation of biogenic silica. Sedimentation of biogenic silica can be bounded crudely from estimates of silica inputs and outputs. The input from Lake Huron is 20×10^7 kg SiO_2/yr compared to an output of 4×10^7 kg SiO_2/yr. This large retention of silica in the lake, 16×10^7 kg SiO_2/yr, results specifically from diatom sedimentation in Lake Erie. The calculated retention is conservative because silica inputs other than those from Lake Huron are unmeasured losses in the analysis. Production and sedimentation of diatoms, therefore, must be considerably larger than inferred from the crude calculation. Finally, during the three-year hydraulic residence time, phosphorus enrichment of Lake Erie waters induces water column silica limitation and establishes water column silica limitation in Lake Ontario, the downstream lake (Fig. 4.1).

Paleolimnological evidence: comparison among lakes

Different patterns of biogenic silica accumulation in sediment cores are presented to illustrate the historical development and different degrees of silica depletion in the Great Lakes (Fig. 4.2). Historical changes in diatom production and sedimentation are inferred from measurements of biogenic silica which are used as proxy for ecosystem responses to phosphorus loading. In this model, phosphorus-limited diatom production increases in response to increased phosphorus loading until silica becomes limiting and then decreases because diatom production is silica limited (Schelske *et al.*, 1983). A peak in diatom production (biogenic silica) is found in the sediment record because increases in diatom production are initially sustained by the silica reservoir and decrease only after the silica reservoir is reduced to levels that limit diatom production.

LAKE SUPERIOR AND LAKE HURON – NO SILICA DEPLETION OR SEASONAL SILICA LIMITATION (FIG. 4.1)

Accumulation of biogenic silica increased after settlement in response to phosphorus enrichment, but a peak is not found because phosphorus loading was low historically and diatom production was never silica limited (Fig. 4.2).

LAKE MICHIGAN – RECENT SEASONAL (EPILIMNETIC) SILICA LIMITATION.

Accumulation of biogenic silica increased markedly in recent sediments, signaling increased production and sedimentation of diatoms since 1950 (Fig. 4.2). The peak in accumulation, signaling epilimnetic silica depletion, may be obscured because the decrease in diatom production is too recent to be recorded in the sediment record of the open lake (LM-K34–94). Temporal resolution of the peak is improved by higher sedimentation rates in Grand Traverse Bay (LM-DH-94).

C. L. SCHELSKE

A peak in biogenic silica accumulation in the late 1800s signals depletion of the epilimnetic silica reservoir as the result of increased production and sedimentation of diatoms. This peak is resolved in pre-1900 sediments in Lake Ontario in contrast to the more recent event in Lake Michigan (LM-K34–94).

LAKE ONTARIO AND LAKE ERIE – WATER COLUMN SILICA LIMITATION

A secondary peak in biogenic silica accumulation beginning in the 1950s (Fig. 4.2) signals water column silica depletion in Lake Ontario as the result of increases in diatom production during the winter–spring isothermal period that deplete the silica reservoir (see Fig. 4.1). Increased biogenic silica accumulation in the 1880s in Lake Erie reflects epilimnetic silica depletion after early settlement and water column silica depletion induced by increased diatom production in the 1950s. Biogenic silica accumulation upcore is maintained at high levels by inputs of soluble silica from Lake Huron and increases in response to historic increases in phosphorus loading. Sediment records in Lake Erie differ among the three hydrographic basins because high silica loading from Lake Huron sustains diatom production in the shallower central and western basins and severe silica limitation develops downstream in the eastern basin where biogenic silica accumulation is similar to Lake Ontario.

Patterns of biogenic silica accumulation in the five Great Lakes (Fig. 4.2) are clearly related to the degree of phosphorus loading among lakes that affects silica utilization by diatoms and, if adequate, depletes the silica reservoir causing silica limitation (see Fig. 4.1). Analysis of siliceous microfossils in sediment cores provides supporting evidence (Stoermer et al., 1993). The microfossil evidence of silica depletion includes changes in deposition rate of frustules, shifts in the trophic preference of species, reduction in silicification of frustules (Stoermer et al., 1985b), and changes in growth habitat (Stoermer et al., 1985a). Responses to decreased phosphorus loading beginning in the mid-1970s have now been inferred from the paleolimnological record, both from studies of microfossils (Wolin et al., 1991; Stoermer et al., 1996) and from chemical proxies including stable isotopes (Schelske & Hodell, 1991, 1995).

Future Research

Phosphorus abatement initiated in the early 1970s by an international agreement between Canada and the United States reduced phosphorus loading to the Great Lakes. Actions taken were successful in that the TP concentration in Lake Ontario and the eastern basin of Lake Erie decreased in 20 years from highs 25 µg TP/l to <10 µg TP/l in the mid-1990s. This raises interesting questions related to silica dynamics in the presently silica-limited lakes: Lake

Michigan, Lake Erie and Lake Ontario. What will be the long-term effect of reduced phosphorus loading on silica dynamics? Will production of biogenic silica decrease to the extent that inputs exceed losses thereby increasing the silica reservoir (eqn. 2)? I predict such an effect will be evident first in Lake Erie because the large, upstream input of silica is a major component in the annual mass balance in this lake with a short residence time. With the unexpected reduction in phosphorus concentration resulting from establishment of zebra mussels in 1989, an increase in the silica reservoir may be measurable in this century (Stoermer *et al.*, 1996; Schelske & Hodell 1995). If the output of silica from Lake Erie increases from the combination of increased input and reduced demand for diatom production, the resulting increase in silica loading to Lake Ontario will also relax silica limitation fairly rapidly in this lake with an 8-year residence time. Relaxation of silica limitation will increase diatom production until the silica reservoir is increased to the extent that diatom production is phosphorus limited. A recent study of Lake Ontario suggests this initial effect of reduced phosphorus loading on silica utilization and diatom production (Johengen *et al.*, 1994). An increase in the silica reservoir of Lake Michigan will be much slower than in the lower lakes because of its larger volume and 100-year residence time. Approximately half of the water column silica reservoir, 14 million tonnes, was depleted during the rapid development of epilimnetic silica limitation from 1955–1970 (Schelske, 1985). Annual inputs of silica, approximately 3% of the 1970 silica reservoir; therefore, provide a small input compared to the reservoir in Lake Michigan. Any future increase in the silica reservoir of individual lakes will be reflected by an increase in silica concentration of outflowing waters.

Summary

Several types of investigations show that diatom production in the Laurentian Great Lakes was stimulated historically by increased anthropogenic phosphorus loading. Increased diatom sedimentation that was associated with increased diatom production was so large in Lake Michigan, Lake Erie and Lake Ontario that silica reserves were depleted seasonally in the epilimnion to limiting levels for diatom growth. Diatom production and sedimentation also increased historically in Lake Huron and Lake Superior, but silica limitation was not induced by the small phosphorus enrichment in these lakes. In Lake Michigan, epilimnetic silica limitation developed about 1970 after a 15-year period of rapid decrease in silica concentration in the water mass. In Lake Erie and Lake Ontario, epilimnetic silica depletion that developed in the mid to late 1800s was inferred from paleolimnological data to be the result of increased phosphorus loading associated with European settlement and forest clearance. Epilimnetic silica depletion in these lakes developed rapidly with relatively small phosphorus enrichment, i.e., when TP concentration in the water mass was <10 μg TP/l. At higher levels of TP, silica reserves in the

C. L. SCHELSKE

entire mass were depleted to limiting levels for diatom growth in Lake Erie and Lake Ontario; this water column silica depletion that developed in the 1950s and 1960s resulted from the exponential increase in anthropogenic phosphorus loading associated with expansion of domestic sewerage systems and introduction of phosphate detergents. Diatom-mediated silica depletion that is driven by phosphorus enrichment has biological consequences. The development of silica limitation progressively alters phytoplankton species composition as assemblages with smaller proportions of diatoms and increasing proportions of green and blue-green (cyanobacteria) algae replace assemblages dominated by diatoms and affects trophic structure and trophic efficiency. Phosphorus loading to the Great Lakes has decreased dramatically since the mid-1970s, reducing TP concentrations from 25 μg TP/l to <10 μg TP/l in Lake Ontario. This reduction in phosphorus loading may eventually reduce diatom production so that it is again phosphorus limited in all five Great Lakes and no longer silica limited.

Acknowledgments

This chapter summarizes more than 25 years of research on the Laurentian Great Lakes. Many colleagues and agencies have contributed to this work. I would like to acknowledge E. F. Stoermer for the original suggestion and insistence that we investigate diatoms and silica in the Great Lakes, D. J. Conley for work on biogenic silica, J. A. Robbins for willingly sharing dated core material and discussing paleolimnological problems, G. L. Fahnenstiel for work on nutrient–light interactions, D. A. Hodell for collaboration on recent paleolimnological investigations, M. Haibach for enrichment experiments with natural phytoplankton assemblages, L. E. Felt and M. S. Simmons for early studies on the Great Lakes and many other associates at the University of Michigan and University of Florida who cannot be acknowledged individually. Grants and contracts from the National Science Foundation, Environmental Protection Agency and its forerunners, and Department of Energy and its forerunners were essential in this research as was recent support from the University of Florida Foundation, Carl S. Swisher Endowment. Finally, I would like to acknowledge the support and encouragement from David C. Chandler throughout my career. He, and other dedicated educators and supervisors, provided important professional guidance over the years.

References

Chapra, S. C. (1977). Total phosphorus model for the Great Lakes. *Journal of Environmental Engineering Division, American Society of Chemical Engineering*, **103**, 147–61.

Conley, D. J., Quigley, M. A., and Schelske, C. L. (1988). Silica and phosphorus flux from sediments: Importance of internal recycling in Lake Michigan. *Canadian Journal of Fisheries and Aquatic Sciences*, **45**, 1030–5.

Conley, D. J., Schelske, C. L. and Stoermer, E. F. (1993). Modification of the biogeochemical cycle of silica with eutrophication. *Marine Ecology Progress Series*, **101**, 179–92.

Fahnenstiel, G. L., Scavia, D., & Schelske, C. L. (1984). Nutrient–light interactions in the Lake Michigan subsurface chlorophyll layer. *Verhandlungen Internationale Vereinigung für Theoretische und Angewandte Limnologie*, **22**, 440–4.

Johengen, T. H., Johannsson, O. E., Pernie, G. L., & Millard, E. S. (1994). Temporal and seasonal trends in nutrient dynamics and biomass measures in Lakes Michigan and Ontario in response to phosphorus control. *Canadian Journal of Fisheries and Aquatic Sciences*, **51**, 2570–8.

Rabalais, N. N., Turner, R. E., Justic, D., & Dortch, Q. (1996). Nutrient changes in the Mississippi River and system responses on the adjacent continental shelf. *Estuaries*, **19**; 386–407.

Schelske, C. L. (1985). Biogeochemical silica mass balances in Lake Michigan and Lake Superior. *Biogeochemistry*, **1**, 197–218.

(1988). Historic trends in Lake Michigan silica concentrations. *International Revue gesamten Hydrobiologie*, **73**, 559–91.

(1991). Historical nutrient enrichment of Lake Ontario: Paleolimnological evidence. *Canadian Journal of Fisheries and Aquatic Sciences*, **48**, 1529–1538.

Schelske, C. L., & Hodell, D. A. (1991). Recent changes in productivity and climate of Lake Ontario detected by isotopic analysis of sediments. *Limnology and Oceanography*, **36**, 961–975.

(1995). Using carbon isotopes of bulk sedimentary organic matter to reconstruct the history of nutrient loading and eutrophication in Lake Erie. *Limnology and Oceanography*, **40**, 918–929.

Schelske, C. L., & Stoermer, E. F. (1971). Eutrophication, silica and predicted changes in algal quality in Lake Michigan. *Science*, **173**, 423–4.

(1972). Phosphorus, silica and eutrophication of Lake Michigan, In *Nutrients and Eutrophication, Special Symposia*, ed. G. E. Likens, vol. 1, pp. 157–71. American Society of Limnology and Oceanography. Lawrence, Kansas: Allen Press.

Schelske, C. L., Stoermer, E. F., Conley, D. J., Robbins, J. A., & Glover, R. M. (1983). Early eutrophication of the lower Great Lakes: New evidence from biogenic silica in sediments. *Science*, **222**, 320–2.

Schelske, C. L., Stoermer, E. F., Fahnenstiel, G. L., & Haibach, M. (1986). Phosphorus enrichment, silica utilization, and biogeochemical silica depletion in the Great Lakes. *Canadian Journal of Fisheries and Aquatic Sciences*, **43**, 407–15.

Stoermer, E. F., Emmert, G., Julius, M. L., & Schelske, C. L. (1996). Paleolimnologic evidence of rapid recent change in Lake Erie's trophic status. *Canadian Journal of Fisheries and Aquatic Sciences*, **53**, 1451–8.

Stoermer, E. F., Wolin, J. A., & Schelske, C. L. (1993). Paleolimnological comparison of the Laurentian Great Lakes based on diatoms. *Limnology and Oceanography*, **38**, 1311–16.

Stoermer, E. F., Wolin, J. A., Schelske, C. L., & Conley, D. J. (1985a). An assessment of ecological changes during the recent history of Lake Ontario based on siliceous algal microfossils preserved in the sediments. *Journal of Phycology*, **21**, 257–76.

(1985b). Variations in *Melosira islandica* valve morphology in Lake Ontario sediments related to eutrophication and silica depletion. *Limnology and Oceanography*, **30**, 414–18.

Wolin, J. A., Stoermer, E. F., & Schelske, C. L. (1991). Recent changes in Lake Ontario 1981–1987: Microfossil evidence of phosphorus reduction. *Journal of Great Lakes Research*, **17**, 229–40.

5 Diatoms as indicators of surface water acidity

RICHARD W. BATTARBEE, DONALD F. CHARLES, SUSHIL S. DIXIT,
AND INGEMAR RENBERG

Introduction

Lake acidification became an environmental issue of international significance in the late 1960s and early 1970s when Scandinavian scientists claimed
that 'acid rain' was the principal reason why fish populations had declined
dramatically in Swedish and Norwegian lakes (Odén, 1968; Jensen &
Snekvik, 1972; Almer *et al.*, 1974). Similar claims were being made at about
the same time in Canada (Beamish & Harvey, 1972). However, these claims
were not immediately accepted by all scientists. It was argued instead that
acidification was due to natural factors or to changes in catchment land-use
and management (Rosenqvist 1977, 1978; Pennington 1984; Krug & Frink,
1983).

In the scientific debate that followed, diatom analysis played a pivotal
role. It enabled the timing and extent of lake acidification to be reconstructed
(Charles *et al.*, 1989; Battarbee *et al.*, 1990; Dixit *et al.*, 1992*a*) and allowed the
various competing hypotheses concerning the causes of lake acidification to
be evaluated (Battarbee *et al.*, 1985; Battarbee & Charles 1994; Emmett *et al.*,
1994). However, diatoms had been recognized and used as indicators of water
pH well before the beginning of this controversy. The 'acid rain' issue served
to highlight the importance of diatoms and stimulated the advance of more
robust and sophisticated techniques, especially the development of transfer
functions for reconstructing lakewater pH and related hydrochemical
variables.

This chapter outlines the history of diatoms as pH indicators, and describes
how diatoms are currently used in studies of acid and acidified waters. It then
describes how diatom-based paleolimnological methods have been used to
trace the pH and acidification history of lakes and how diatoms are being used
in the management of surface-water acidification problems. It draws on a wide
literature base and on a range of previous reviews (e.g., Charles & Norton, 1986;
Battarbee & Charles, 1986; Battarbee, 1991; Charles, *et al.*, 1989, 1994*a,b*).
Although diatoms are being used increasingly as indicators of acidity in
running waters (Round, 1991, Steinberg & Putz, *et al.*, 1991; van Dam & Mertens,
1995; Coring 1996; Lancaster *et al.*, 1996) the principal focus of the chapter is on
standing waters.

Diatoms and pH in acid waters

Modern measurements of stream and lakewater chemistry show that acidic waters occur in many parts of the world, but almost always in regions where catchment soils have low concentrations of base cations. Some acidic waters are very clear, others with peaty catchments can be colored containing high concentrations (>10 mg/l) of dissolved organic carbon (DOC).

Almost all these waters are unproductive, but, except in the most extreme of cases, they are capable of supporting highly diverse and characteristic diatom floras. In streams, the epilithic community is usually the most common and diverse. In lakes there are more habitats for diatoms. The epilithon and epiphyton have many taxa in common and have floras that can be similar to stream floras. For those lakes with sandy shorelines, the most distinctive assemblage, however, is the epipsammon. In contrast to the benthos, diatom plankton in acid lakes is often poorly developed and very acid and acidified lakes often completely lack a planktonic diatom component.

PH CLASSIFICATION: THE HUSTEDT SYSTEM

The strong relationship between diatom distribution and pH has been recognized for many decades. The earliest classification of diatoms according to pH was presented by Hustedt in his monograph on the diatoms of Java, Bali and Sumatra (Hustedt, 1937–1939). He studied over 650 samples from a wide variety of habitats and from his data argued that diatoms have different pH 'preferences', and could be classified into the following groups:

(i) alkalibiontic: Occurring at pH values > 7
(ii) alkaliphilous: Occurring at pH about 7 with widest distribution at pH > 7
(iii) indifferent: Equal occurrences on each side of pH 7
(iv) acidophilous: Occurring at pH about 7 with widest distribution at pH < 7
(v) acidobiontic: Occurring at pH values < 7, with the optimum distribution at pH = 5.5 or below.

This classification has been immensely influential. It was adopted by most following diatomists and it became common practice for diatom floras to use the Hustedt terminology and concepts (Foged, 1977; Cleve-Euler, 1951–1955). However, the use of the term 'indifferent' has been questioned. Although some taxa may have very broad pH tolerances, true indifference to pH is very unlikely and has never been demonstrated. Instead the term 'circumneutral' is preferred (Renberg, 1976) to apply to those diatoms that 'occur at around pH 7'.

PH RECONSTRUCTION USING THE HUSTEDT PH CLASSIFICATION

The Hustedt classification became the basis of many attempts to reconstruct pH from diatom assemblages in lake sediment cores. The first was by Nygaard

(1956). He developed a number of indices for deriving pH, the most useful of which, index α, uses the ratio of the abundance of acidophilous and acidobiontic taxa to the abundance of alkaliphilous and alkalibiontic taxa in a sample, with the acidobiontic and alkalibiontic taxa being weighted by a factor of 5:

$$\text{Index } \alpha = \frac{\% \text{ acidophilous} + 5 \times \% \text{ acidobiontic taxa}}{\% \text{ alkaliphilous} + 5 \times \% \text{ alkalibiontic taxa}}$$

This basic approach was further developed by Meriläinen (1967), Renberg & Hellberg (1982), Charles (1985), Davis & Anderson (1985), Flower (1986), Baron *et al.* (1986), Arzet *et al.* (1986*a,b*), Charles & Smol 1988, Dixit *et al.* (1988), Steinberg *et al.* (1988) and Whiting *et al.* (1989) using a variety of simple and multiple linear regression statistics. These methods have been extensively reviewed (Battarbee, 1984, Battarbee & Charles 1987; Davis 1987).

PH RECONSTRUCTION USING NON-LINEAR MODELS AND A SPECIES-BASED APPROACH

Despite the demonstrable success of using the Hustedt classification system coupled with linear regression models for pH reconstruction, this approach has many weaknesses and several assumptions that are not fully justified. The main problem, as with any similar ecological classification, is the difficulty of assigning a taxon unambiguously to an individual class. The literature has many examples of authors placing the same taxon into different classes, and the pH reconstruction models can be shown to be extremely sensitive to misallocation of this kind (Oehlert, 1988).

Moreover, a number of invalid ecological and statistical assumptions associated with the linear regression approach to pH reconstruction have been pointed out by Birks (1987). The main issue concerns the requirement in linear regression that species have a linear or monotonic relationship to pH. Whilst this is sometimes the case along short gradients, examination of species' distribution along longer gradients shows, as would be predicted from ecological theory, that species' responses are non-linear and predominantly unimodal, with a species abundance rising to a maximum at the centre of its pH range (Fig. 5.1).

Training sets On the basis of the unimodal response function, a pH reconstruction methodology can be developed if the pH optima of individual taxa in a diatom assemblage can be derived. Although it is possible to use data reported in the literature, pH optima for diatoms are best derived from purposely compiled modern data sets (usually called 'training sets'). These combine modern diatom and water chemistry data from lakes spanning the pH gradient of interest, and are collected from a large number of sites located in the geographic regions that contain the study sites. Charles (1990*a*) provides a useful guide to the compilation of training sets.

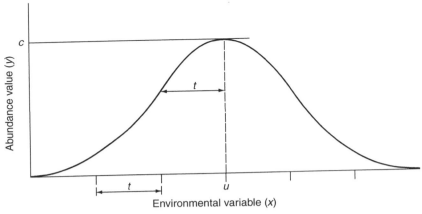

Fig. 5.1. Gaussian response curve (from ter Braak, 1987*b*). *t* = tolerance (1 standard deviation); *u* = optimum; *c* = maximum.

Whilst it would seem preferable, on ecological grounds, to use samples of living diatoms collected from individual benthic and planktonic habitats, paleolimnologists have preferred to use diatoms from surface sediment samples (usually the top 0.5 or 1 cm levels) in constructing diatom–pH training sets. Surface sediment samples have the advantage that they contain, or are assumed to contain, diatoms representative of all habitats in the lake (cf. DeNicola, 1986; Cameron, 1995) and to provide an integrated (in time and space) flora. They are also most analogous, in terms of their sedimentary environment, to the core material to be used in pH reconstruction. On the other hand, surface sediment samples have the disadvantage that they may contain contaminants from older reworked sediment and the age span of the uppermost sample is usually unknown, and varies both within and between lakes.

Whilst the collection of surface sediment samples usually only requires a single visit to each lake, samples for water chemistry need to be taken more frequently. pH can vary through the year by well over 1 pH unit in low alkalinity waters and it is advisable to collect samples on at least four occasions on a seasonal basis to generate a mean or median value for each site. If only a single sampling visit is possible, then this should take place in autumn after overturn (e.g., Huttunen & Meriläinen, 1983) or in spring prior to stratification.

The earliest pH training sets were relatively small, often between 30 and 40 sites (e.g., Charles, 1985; Flower, 1986). Whilst such datasets perform quite well with older methods of pH reconstruction (e.g., Renberg & Hellberg, 1982), where pH classes based on literature reports are used, they are too small for current methods that require pH optima to be calculated from field measurements and from new diatom collections. In particular, more sites are needed across longer pH gradients both to avoid artificial truncation of species' ranges and to provide more reliable estimates of pH optima, especially for uncommon taxa.

Combining training sets and taxonomic quality control Large training sets can be built most rapidly by combining smaller ones. However, where this involves more than one laboratory, problems of taxonomic consistency between laboratories become important. This is a concern, not only in ensuring correct estimates of optima, but also in allowing the correct application of transfer functions to sediment core assemblages (Birks, 1994). This issue has been tackled in both the Surface Water Acidification Project (SWAP) (Battarbee *et al.*, 1990) and the Paleoecological Investigation of Recent Lake Acidification (PIRLA) project (Charles & Whitehead, 1986; Kingston *et al.*, 1992*a*), where differences in conventions between laboratories were resolved following taxonomic workshops, slide exchanges, the circulation of agreed nomenclature (e.g., Williams *et al.*, 1988), taxonomic protocols (Stevenson *et al.*, 1991) and taxonomic revisions (e.g., Flower & Battarbee 1985; Camburn & Kingston 1986). The PIRLA project also produced an iconograph (Camburn *et al.*, 1986). In the case of SWAP, the effectiveness of this approach was tested by an interlaboratory comparison of counts from test slides before and after the harmonization workshops (Kreiser & Battarbee 1988; Munro *et al.*, 1990). An example of this is shown in Fig. 5.2.

The results of this taxonomic quality control exercise demonstrated the three main problems encountered: differences in nomenclature, splitting versus amalgamation of taxa, and differing criteria used in the identification of taxa. For SWAP, the harmonization process allowed the merging of datasets from the UK, Sweden and Norway to take place, generating a dataset of 167 sites (Stevenson *et al.*, 1991). For PIRLA, a similar exercise has produced datasets with a consistent taxonomy for over 200 lakes from the Adirondack Mountains in northern New York State, Northern New England, the Upper Mid-West and Northern Florida.

Full documentation of taxonomic decisions made in merging datasets in this way is essential, especially in cases where unknown taxa or taxa not previously described occur in the samples, as the training set can only be used for reliable pH reconstruction where the taxonomy of diatoms in core assemblages are subsequently harmonised with those in the training set.

Exploring diatom–environment relationships from surface sediment training sets Whilst the main patterns of water chemistry and diatom variation in a training set can be explored independently using techniques such as principal components analysis (PCA) and detrended correspondence analysis (DCA), respectively, canonical correspondence analysis (CCA) implemented by the computer program CANOCO (ter Braak, 1987*a*) allows both datasets to be analysed together. Diagrams generated by CCA show ordination axes scores for samples and for species, and vectors for the environmental variables. The vectors indicate the direction of maximum variation of each environmental variable, and their length is directly proportional to their importance in explaining variation in the dataset.

A CCA diagram for a 138 lake data subset of the SWAP training set (Fig. 5.3,

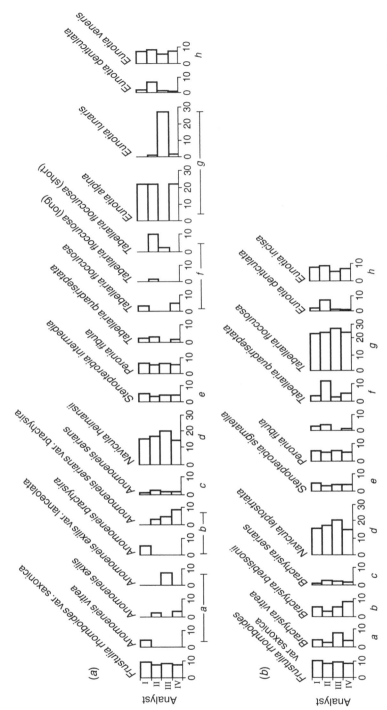

Fig. 5.2. (*a*) Dominant taxa in a diatom slide from Lingmoor Tarn (Cumbria, England) illustrating problems of nomenclature in group *a, b, c, d, e, g, h*, problems of splitting vs. amalgamation in groups *a, f* and the use of differing identification criteria in group *g*. (*b*) Dominant taxa in the Lingmoor Tarn slide after full taxonomic and nomenclatural revision. Horizontal scale in percentage occurrence, Vertical scale indicates participating analysts in the quality control exercise (from Munro *et al.*, 1990).

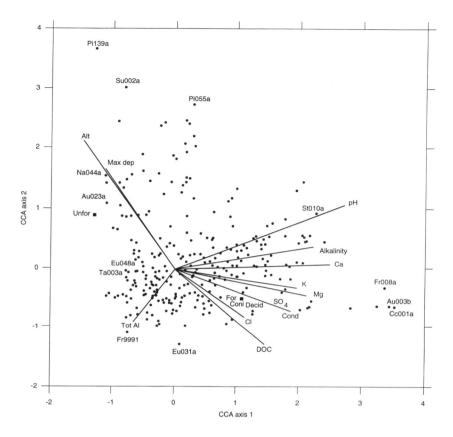

Fig. 5.3. Canonical correspondence analysis (CCA) plot of diatom taxa included in the CCA of the 138 samples in the Surface Water Acidification (SWAP) dataset of lakes from the UK, Sweden and Norway. The diagram shows the 16 variables included in the analysis. Only selected taxa are labelled with taxon codes (from Stevenson *et al.*, 1991). Summary statistics of the analysis are shown in Table 5.1.

Table 5.1) shows an example of this analysis. The data indicate that axis 1 is strongly related to pH ($r = 0.84$), along with conductivity (0.58), calcium (0.78) and alkalinity (0.69), and negatively correlated with altitude (-0.45), maximum depth (-0.35), and total aluminium (-0.21). Axis 2 is primarily a DOC gradient.

Analysis such as this has been carried out on many training sets of softwater lakes (Stevenson *et al.*, 1989; Kingston *et al.*, 1992*b*; Dixit *et al.*, 1993) and all show the prime importance of pH as an explanatory variable, supporting conclusions from the early literature from Hustedt onwards. They underpin the use of diatoms as indicators of water pH and provide a sound empirical basis for the development of diatom–pH transfer functions.

pH reconstruction by weighted-averaging regression and calibration After assessing various statistical approaches in the SWAP and the PIRLA projects, it was

Table 5.1. *Summary statistics of the environmental variables included in the CCA of the 138 diatom–environmental data set used in SWAP*

Environmental variable	Weighted mean	Standard deviation
pH	5.5598	0.7050
Conductivity	43.6492	26.2270
DOC	3.8761	3.2846
Ca	101.5286	107.7411
Mg	64.9368	49.5752
K	9.9082	6.9200
SO$_4$	105.1257	64.1189
Cl	182.5068	141.3208
Alkalinity	46.2302	98.4989
Total Al	125.3340	94.9714
Altitude	352.1492	262.1623
Unafforested	0.3984	0.4896
Forested	0.5938	0.4911
Conifer	0.5709	0.4949
Deciduous	0.0228	0.1494
Maximum depth	15.7617	11.2323

CCA axes	1	2	3	4
Eigenvalues	0.483	0.248	0.179	0.141
Species–environment correlations	0.932	0.847	0.783	0.788
Cumulative percentage variance				
(i) of species data	7.8	11.8	14.6	17.0
(ii) of species–environment relationship	30.4	46.1	57.4	66.7

Source: From Stevenson *et al.*, 1991.

concluded that weighted averaging regression and calibration, implemented by the computer program WACALIB (Line & Birks, 1990; Line *et al.*, 1994) is the most effective statistical method for pH reconstruction. Since 1989, this approach has been widely used in diatom inferred pH calibration studies in both Europe (e.g., Birks *et al.*, 1990a; Korsman & Birks, 1996) and North America (e.g., Kingston & Birks, 1990; Dixit *et al.*, 1991, 1993).

The weighted-averaging regression and calibration approach does not assume a linear relationship between the inferred variable (i.e., pH) and diatoms, but assumes unimodal relationships. Based on the autecological data available for diatoms, this is a sound ecological assumption. Moreover, as for direct gradient analysis, WA maximizes the covariance between the diatom data and the measured environmental variable (Korsman & Birks, 1996). In WA the main assumption is that, at a given pH value, taxa that have optima nearest to that pH will be most abundant. The pH optimum (u_k) for a taxon would then be an average of all the pH values of lakes in the training set in which the taxon occurred, weighted by its relative abundance. This is the regression step:

$$\hat{u}_k = \sum_{i=1}^{n} y_{ik} x_i \bigg/ \sum_{i=1}^{n} y_{ik}$$

where x_i is the value of the environmental variable at site i, and y_{ik} is the abundance of species k at site i.

Computed pH optima of various taxa are then used to infer pH by taking an average of the taxa abundances in a sample, each weighted by its pH optimum. This is the calibration step:

$$\hat{x}_i = \sum_{k=1}^{m} y_{ik} \hat{u}_k \bigg/ \sum_{k=1}^{m} y_{ik}$$

Tolerance values of individual diatom taxa can also be estimated as the weighted-average standard deviation. When tolerance values of taxa for any given environmental variable vary substantially, taxa can be weighted by their squared tolerance in the weighted averaging equation. Taxa that have a narrow tolerance can be given more weight in WA than taxa with a wide pH tolerance (Birks *et al.*, 1990a). Ideally, tolerance downweighting should provide more accurate pH inferences. However, in large data sets (e.g., SWAP) it has been shown that, although tolerance downweighting gives a lower apparent root mean squared error (RMSE) than WA, it does not improve RMSE in cross-validation exercises or bootstrapping. Thus, WA without tolerance correction has been generally adopted for reconstruction purposes.

In WA, averages are taken twice, once in the regression step and once in the calibration step. This results in shrinkage of the range of inferred pH values, which is normally corrected by a linear deshrinking regression ('classical regression'). An 'inverse regression' approach has been also used to deshrink the inferred pH data because this reduces the RMSE in the training set (ter Braak & van Dam, 1989). However, in the SWAP and PIRLA projects, classical regression was preferred (Birks *et al.*, 1990a, Dixit *et al.*, 1993) as this performs better for reconstructions at the ends of gradients (cf. Birks, 1995).

The predictive ability of inferred pH models can be assessed by examining the correlation coefficient between the measured and diatom-inferred pH. For example, the regression plot of measured vs. inferred pH for lakes in the Adirondack Mountains, shows that the inferred pH model is strong (Fig. 5.4). The r^2 value (0.91) of this WA model is close to the value reported for the SWAP data set (Birks *et al.*, 1990a).

The greatest advances in paleolimnological error estimation have come from the SWAP (Birks *et al.*, 1990a) and PIRLA II (Kingston *et al.*, 1990; Dixit *et al.*, 1993) projects. Initially, the root mean squared error (RMSE) was generally used to assess the predictive abilities of various pH inference models. The RMSE based on the training set alone gives the 'apparent error' in the calibration

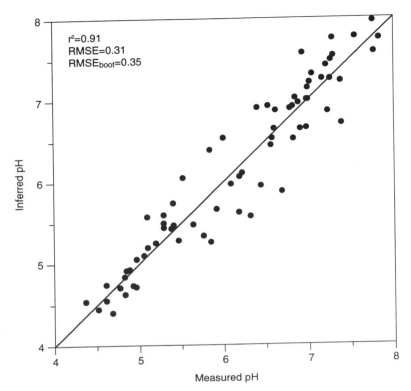

Fig. 5.4. The relationship between measured and WA diatom inferred pH for the 71 Adirondack drainage lakes (modified from Dixit *et al.*, 1993).

model, and always underestimates the 'true error' (ter Braak & van Dam 1989; Birks *et al.*, 1990*a*). Better estimates of RMSE can be obtained by cross-validation and splitting the data into a training set and a test set. However, this approach would have the undesired effect of lowering the number of lakes in the training set. To have more realistic estimates of taxa responses, it is desirable to have large training sets. This also assures the maximum probability of surface sediment diatom analogues for core (passive) samples.

To assess realistic error estimates in pH reconstruction, computer intensive resampling approaches, such as jackknifing and bootstrapping have been used (Birks *et al.*, 1990*a*, ter Braak & Juggins, 1993). These approaches maintain independence of the calibration data set, while using the ecological information of all the samples. Jackknifing, also known as a 'leave one out' approach, is less computer intensive than bootstrapping. In each jackknife, only one sample is excluded from the original training set, and the inferred pH of that sample is based on the optima and tolerance of the taxa in the remaining training set samples. It has been observed that errors generated by jackknifing are generally slightly lower than errors estimated by bootstrapping (H. J. B. Birks, personal communication).

Bootstrapping is a computer-intensive resampling procedure where a sub-

R. W. BATTARBEE *ET AL.*

set of training samples that is the same size as the original training set is selected at random with replacement. The remaining unselected samples form an independent test set. After a large number of bootstrap cycles (usually 1000), a new $RMSE_{boot}$ is obtained which is less subject to bias because it uses a larger training set. Moreover, besides calibration samples, $RMSE_{boot}$ can be also computed for passive samples. Bootstrap error estimates can be readily made using the WACALIB (version 3.3) program (Line et al., 1994).

In the Adirondack data set (Dixit et al., 1993) the $RMSE_{boot}$ of 0.35 is only slightly higher than the apparent RMSE of 0.31 (Fig. 5.4), and is comparable to the WA $RMSE_{boot}$ of 0.31 for the SWAP (Birks et al., 1990a) and 0.40 for the Swedish data sets (Korsman & Birks, 1996). The bootstrap error can be split into an estimate of predictive error unique for each sample (s_{i1}) and a constant error (s_2) for the entire data set that may largely represent inherent natural variation (Birks et al., 1990b). The s_{i1} becomes smaller as the training set gets larger. In the Adirondack calibration the low prediction error $(s_{i1} = 0.09$ pH) suggests that most of the predictive error $(s_2 = 0.34$ pH) was associated with the inherent variation in the data set (Dixit et al., 1993). These errors are also greatly affected by data screening, and the objective removal of rogue samples from the calibration. By data screening and removing the rogue samples, Birks et al. (1990a) improved the predictive ability of their inferred pH model. Similarly, for the Adirondack lakes, separate pH calibration models for drainage (Dixit et al., 1993) and seepage lakes (Dixit & Smol, 1995) provided better estimates of inferred pH.

In the calibration step, it is generally assumed that the diatom taxa present in a sediment core sample are well represented in the training set. However, when this is not the case, inferred pH values will have a higher error than the error in the training set. This is a problem of analogues. However a number of techniques are available that allow the representativeness of the core diatom assemblages in the training set to be assessed (Birks et al., 1990b). For example, Birks et al. (1990b) assessed the reliability of their inferred pH values by (i) measures of lack-of-fit of diatom assemblage to pH, and (ii) estimated mean-squared errors for each inferred pH.

There are some weaknesses in the weighted averaging approach. The main problem is its sensitivity to the distribution of the environmental variable in the training set, as each environmental variable is considered separately, disregarding the residual correlations in species data (ter Braak & Juggins, 1993). An alternative method is 'weighted averaging-partial least squares' (WA-PLS), which uses the residual structure in the species data (ter Braak & Juggins, 1993). However, for pH reconstruction this approach has not yet been shown to provide an improvement over the simple WA predictions (ter Braak & Juggins, 1993; Birks, 1995; Korsman & Birks, 1996). For example, predictive pH errors for the WA and a two component WA-PLS model for the Adirondack lakes were 0.35 and 0.36 of a pH unit, respectively (S. S. Dixit, unpublished observations). It has been suggested that greatest improvement by WA-PLS may occur: (i) in lakes with diatom taxa that have broad tolerances; (ii) in lakes that have high percentage abundances of dominant taxa and thus low sample heterogeneity (Korsman & Birks, 1996); and/or (iii) in data sets where species response is influenced by some structured secondary or 'nuisance' variable (ter Braak & Juggins, 1993). Some of these features characterize diatom responses to nutrient gradients and probably explain why WA-PLS performs better than WA in diatom-TP transfer functions rather than in diatom-pH transfer functions (Bennion *et al.*, 1996).

Errors and assumptions in the application of transfer functions for pH Transfer functions, especially using the WA approach, have been widely used to reconstruct lakewater pH and associated variables for many lake regions (Battarbee *et al.*, 1990; Dixit *et al.*, 1991, 1993; Huttunen & Turkia, 1990). However, the errors described above on pp. 94–95 relate mainly to the statistical precision of pH reconstruction and to relationships between the composition of the training set and the sediment core assemblages. These are not the only potential errors as the accuracy of pH reconstruction also depends on the validity of other assumptions that are made when carrying out diatom analysis of sediment cores.

The most important of these are:
   (i) that the deep water sediment assemblages faithfully reflect the composition of the live communities in the lake (cf. DeNicola, 1986; Jones & Flower, 1986; Cameron, 1995);
   (ii) that the core assemblages are not significantly biased by problems of diatom preservation;
   (iii) that the core assemblages are not contaminated by the reworking of older diatoms (Anderson, 1990); and
   (iv) that the true pH history of a lake can be reconstructed from one or a very small number of cores (e.g., Davis *et al.*, 1990; Charles *et al.*, 1991).

These assumptions have been evaluated more thoroughly for acid lakes than for other lake types. In acid waters diatom preservation is usually very good, with valve breakage being a more serious issue than dissolution. This is not the case in some eutrophic and many saline lakes where diatom dissolution can be a serious concern (Barker 1992; Ryves 1994). In a detailed taphonomic study of diatoms in a small Scottish loch, Cameron (1995) not only showed that preservation in sediments was excellent but there was also a very close agreement between the relative abundance of benthic and planktonic diatoms growing in the lake with the composition of diatoms in sediment trap samples. Some surface sediment samples, on the other hand, contained small numbers of diatoms not currently growing in the lake, suggesting that reworking of older sediments can introduce contaminants into the record that could potentially bias pH reconstructions. A similar finding was also reported by Allott (1992) for the Round Loch of Glenhead, where taxa characteristic of early Holocene sediments were found to occur in recent sediments.

There have been a number of studies to assess the amount of variation between diatom records and pH reconstruction within a lake basin (Davis *et al.*,

1990; Charles *et al.*, 1991; Allott 1992). Whilst there are considerable variations in sediment accumulation rates between cores that affect the resolution of pH reconstructions, all studies show that the between-core variations in the diatom composition of sediment cores within a lake-basin is relatively small. Charles *et al.* (1991) concluded that between core error in pH reconstruction was less than the statistical error of the transfer function.

In summary, whilst care is needed at every step, the evaluations carried out so far suggest that diatom–pH reconstructions from sediment records can be robust and reliable.

Diatoms, dissolved organic carbon (DOC) and aluminium in acid waters

Although pH is the primary variable associated with lake acidification, the biological impact of acidification is also influenced by dissolved organic carbon (DOC) and by aluminium (Muniz & Leivestad, 1980; Driscoll *et al.*, 1980).

In brown water lakes the high concentrations of organic acids can substantially lower pH. However, low pH associated with acid humic waters are less damaging than very acidic clear waters as DOC can reduce the concentration of toxic aluminium by forming organic complexes (Driscoll *et al.*, 1987). On the other hand, it has been hypothesized that deposition of strong acids can cause a lowering of DOC in lake waters (Almer *et al.*, 1974, Driscoll *et al.*, 1987), exposing fish and other biota to higher concentrations of more toxic aluminium fractions, especially the labile monomeric form (Muniz & Leivestad, 1980).

Because of these relationships, there has been significant interest in the extent to which the diatom record can be used to reconstruct DOC and Al in addition to pH (Davis *et al.*, 1985; Anderson *et al.*, 1986; Huttunen *et al.*, 1988; Kingston & Birks 1990; Birks *et al.*, 1990*b*, Kingston *et al.*, 1992*b*). Anderson *et al.* (1986) analyzed a training set of surface sediments from 35 acid lakes from Southern Norway, and showed from an ordination of the data using PCA, that 14% of the variance in the dataset was related to total organic carbon (TOC). *Frustulia rhomboides* (Ehrenb.) De Toni and *Anomoeoneis serians* v. *brachysira* (Breb. ex Rabenh.) (=*Brachysira brebissonii* R. Ross) were the most indicative taxa. Using this dataset Davis *et al.* (1985) derived a multiple regression equation to reconstruct TOC separately from pH, and applied the equations to two lakes, Hovvatn and Holmvatn, in Southern Norway. The results showed recent declines in both inferred pH and inferred TOC at both sites. As the preacidification inferred pH for both sites was about 5.0, the authors argued that fish loss was caused by an increase in the concentration of toxic inorganic aluminium following the reduction in TOC.

Kingston and Birks (1990) explored a range of statistical techniques, including weighted averaging and maximum likelihood regression, to estimate DOC optima from North American datasets created during the PIRLA project (Charles & Whitehead, 1986). From the 117 taxa considered, only 35 showed a significant Gaussian unimodal response to DOC, and taxa with the strongest unimodal response did not include *Frustulia rhomboides* and *Anomoeoneis serians* v. *brachysira* (= *Brachysira brebissonii*) that were most significant in the Norwegian study. A WA reconstruction, based on these data, from two sites in North America, Big Moose Lake in the Adirondacks and Brown Lake in Wisconsin, showed DOC declines coincident with pH declines, consistent with the hypothesis that the addition of strong mineral acids to lakes can cause a decrease in DOC.

Analysis of the Northwest European SWAP training set (Birks *et al.*, 1990*b*, Stevenson *et al.*, 1991) showed from the CCA of the 138 sites for which pH, DOC and total aluminium values were available (Fig. 5.3) that there was (i) a significant diatom gradient strongly and positively related to alkalinity, calcium and pH and negatively, but less strongly correlated to total Al (axis 1), and (ii) a significant gradient strongly correlated with DOC (axis 2).

This analysis also allowed potential diatom indicator taxa for pH, DOC and total Al to be identified. Interestingly, as for the North American dataset, *Frustulia rhomboides* and *Anomoeoneis serians* var. *brachysira* (= *Brachysira brebissonii*), the two species identified by Davis *et al.* (1985) as most indicative of humic waters do not emerge as good DOC indicators. This suggests either that there are regional differences in the composition of dissolved organic carbon or that the importance of these taxa were overestimated by Davis *et al.* (1985).

The results of the CCA analysis of the SWAP dataset indicate that pH, DOC and Al could all be potentially reconstructed from fossil diatom assemblages. The application of WA reconstruction for pH, DOC and Al for a core from the Round Loch of Glenhead, Galloway, showed values and trends close to expectation with a decrease in pH, decrease in DOC, and an increase in total aluminium, occurring during the twentieth century (Fig. 5.5).

Although aluminium was known to reduce fish survival in acid waters (e.g., Baker & Schofield, 1982), early analytical techniques did not allow separate measurement of the most toxic inorganic monomeric fraction. Consequently, many of the early diatom training sets, such as the SWAP dataset, included Al only as 'total Al'. However, the later development of methods for fractionating aluminium (Driscoll, 1984) allowed Kingston *et al.* (1992*b*) to explore the relationships between diatoms and labile monomeric Al in a dataset from the Adirondacks. They showed that pH, labile monomeric Al, Secchi disc transparency, and DOC were the four most important variables explaining the diatom gradients, and that Al contributed additional variance not attributable to pH and DOC. On this basis, they used diatom assemblages of sediment cores from four contrasting Adirondack lakes to reconstruct labile monomeric Al, and compared the results with the fish history of the lakes inferred from documentary sources and from an analysis of *Chaoborus* (phantom midge) remains in the cores (cf. Uutala 1990) (Fig. 5.6). The results demonstrated the importance of the paleolimnological record in explaining the present status of fish populations in acid lakes. Although some fish were absent as a result of recent acidification processes, the data from Upper Wallface Pond showed that

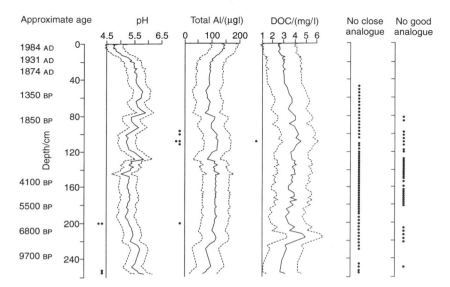

Fig. 5.5. Reconstructed pH, total Al, and DOC values for the Round Loch of Glenhead plotted against depth (solid lines) based on weighted averaging. The RMSE of prediction for these estimates are shown as broken lines. Samples with poor (.) or very poor(..) fit to particular chemical variables and those lacking good or close modern analogues in the SWAP training set are indicated (from Birks *et al.*, 1990*b*).

the lake had naturally high aluminium and low pH levels and had been fishless, even in preindustrial times.

Application of diatom analysis to questions of changing surface water acidity

Surface water acidification can be caused in many ways relating either to changes in the characteristics of stream and lake catchments or to the incidence of acidic pollution. The main processes are long-term soil acidification over the postglacial period (*ca.* 11 000 yrs), alterations in the base cation status of soils associated with changes in catchment characteristics and use, and an increase in S deposition from the combustion of fossil fuels. More locally, acidification can also be caused by acid-mine drainage and by the direct discharge of industrial effluents by the disturbance and drainage of sulfide-rich soils.

NATURAL ACIDIFICATION PROCESSES

Long-term soil acidification The tendency for certain lakes with catchments on base-poor or slow-weathering bedrock to become gradually more acidic during the postglacial time period has been recognized for many decades (Lundqvist, 1924; Iversen 1958). Long-term acidification has been reported

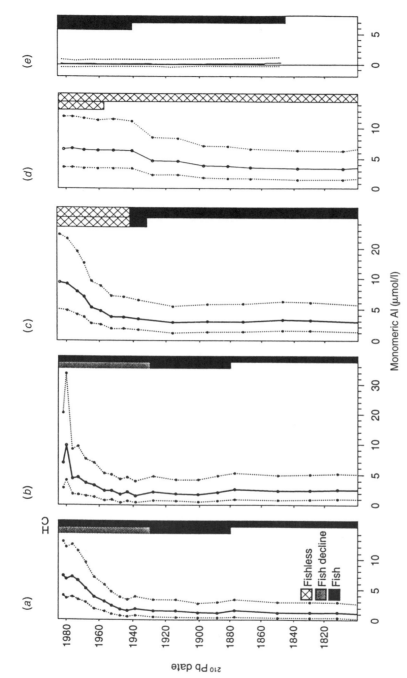

Fig. 5.6. Reconstructions of (a) diatom-based monomeric Al; (b) chrysophyte-based monomeric Al for Big Moose Lake; and (c) diatom-based monomeric Al for Deep Lake; (d) Upper Wallface Pond; and (e) Windfall Pond. Bars to the right of each reconstruction indicate historical (H) and *Chaoborus*-based (C) reconstructions of fish populations (from Kingston *et al.*, 1992).

from many countries (Charles *et al.*, 1989), and has been ascribed to both base-cation leaching and the paludification of catchment soils (Pennington, 1984). Round's study of Kentmere in the English Lake District (Round, 1957) showed, from diatom analysis, that the beginning of acidification was coincident with the Boreal–Atlantic transition (about 7500 BP), a time when upland blanket peat was beginning to form in the surrounding hills. The change was characterized by a replacement of a predominantly alkaliphilous *Epithemia–Fragilaria–Cyclotella* assemblage by a more acidophilous *Eunotia–Achnanthes–Cymbella–Anomoeoneis (= Brachysira) –Gomphonema* assemblage. Further early studies of this kind, using diatoms and identifying long-term acidifying trends in lakes, have been carried out, especially in northwest Europe, including Scotland (Alhonen, 1968; Pennington *et al.*, 1972), northwest England (Round, 1961; Haworth, 1969; Evans, 1970), North Wales (Crabtree, 1969; Evans & Walker, 1977), Sweden (Renberg, 1976, 1978) and Finland (Huttunen *et al.*, 1978).

Following the emergence of interest in acid deposition and its effects, there has been an increase in long-term acidification studies (Renberg & Hellberg, 1982; Whitehead *et al.*, 1986, 1989; Jones *et al.*, 1986, 1989; Winkler, 1988, Renberg, 1990; Ford, 1990; Korhola & Tikkanen, 1991). There were a number of reasons for this interest: (i) to test the hypothesis that the present low values of pH in many uplands waters were not due to acid deposition but to long-term natural processes (cf. Pennington, 1984; Jones *et al.*, 1986; Renberg, 1990); (ii) to show that long-term acidification could reduce catchment alkalinity and make lakes more vulnerable to the impact of acid deposition (Battarbee, 1990); and (iii) to demonstrate the difference between the nature of recent change in comparison to the rate and amplitude of earlier natural changes (Jones *et al.*, 1986; Winkler, 1988; Ford, 1990; Birks *et al.*, 1990*b*).

Although some sites have been acidic (pH, 5.6) for the entire Holocene (Jones *et al.*, 1989; Winkler, 1988), most acid lakes studied so far were less acidic in their initial stages. In the case of Lilla Öresjön (Fig. 5.7), and other lakes in southern Sweden, Renberg (1990) and Renberg *et al.* (1993*a,b*) recognized a natural long-term acidification period following deglaciation around 12 000 BP to about 2000 BP, during which period pH decreased from about 7 to 5.5 resulting from soil acidification. At Loch Sionascaig in Scotland, and Devoke Water, in England (Atkinson & Haworth, 1990), the basal diatom assemblages indicate pH values of 7 and above. At both sites, pH declined during the early postglacial period, stabilising at about pH 6.5 before 8500 BP in L. Sionascaig, and by about 5000 BP in Devoke Water. Unlike Lilla Öresjön, neither of these two lakes continued to acidify and pH at both sites remains constant at about 6.2–6.6 until the twentieth century. A further example of early acidification is Pieni Majaslampi in Finland (Korhola & Tikkanen, 1991). Here conditions are more extreme, with the earliest lake phase being acidic (pH about 5.5 to 6.0) and falling rapidly within the first 2000 years of lake history to pH, 5.0. This is a rare example of acidification to such extremely low pH values as a result of natural processes.

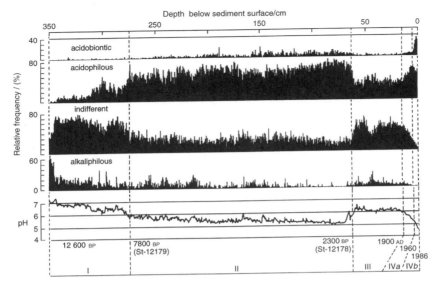

Fig. 5.7. Hustedt pH categories, diatom-inferred pH values (from weighted averaging), calibrated radiocarbon dates, ^{210}Pb dates and pH periods in the history of Lilla Öresjön, southwest Sweden (from Renberg, 1990).

In North America, Whitehead *et al.* (1986, 1989) studied the postglacial acid-ification history of three sites along an elevational gradient in the Adirondack Mountains (NY). They argued that the rate of acidification was most rapid at the highest site and slowest at the lowest site, corresponding to catchment soil depth, slope and amount of rainfall. Ford (1990) demonstrated that the pH of Cone Pond in New Hampshire had dropped to 5.5 by 6500 BP and to around pH 5.0 by as early as 2000 BP, whereas nearby S. King Pond, which has thicker glacial tills and some limestone in its catchment, did not undergo acidifica-tion. In contrast, Duck Pond, Massachusetts, a kettle lake located in an outwash plain composed of crystalline sands, has been very acidic, with a pH *ca.* 5.3 throughout its 12 000 year history (Winkler, 1988).

Post-glacial changes in catchment vegetation Whilst the dominant time-dependent acidifying process during the postglacial period is base-cation leaching of catchment soils, the rate and extent of acidification is also moderated by vegetation changes, and the impact of vegetation change on soils. In some cases the vegetation change may be driven by changing soil base status, but in other cases climate change, catchment disturbance, e.g., by fire or wind, or early human interference may be more important (see below). What is unclear at present is the extent to which studies of diatom assemblages in lake sediments can be used to discriminate between acidification as a result of base-cation leaching (as discussed above) from other water quality changes, involving organic acids and nutrients, that might occur as a result of shifts in the charac-ter of catchment vegetation and soils.

R. W. BATTARBEE ET AL.

For example, in the study of the Round Loch of Glenhead originally designed to assess the extent to which mid–postglacial catchment paludification and the spread of blanket mires could be the cause of lake acidification, diatom changes that were coincident with changes in the vegetation and soils, did not indicate any significant depression in lake water pH (Jones *et al.*, 1986). Changes in DOC may have been more important (Jones *et al.*, 1986; Birks *et al.*, 1990*b*).

In a similar study, Renberg *et al.* (1990) and Korsman *et al.* (1994) assessed the potentially acidifying influence of the colonization of lake catchments by spruce by examining the diatom response to the natural immigration of spruce about 3000 years ago into northern and central Sweden. Eight lakes were selected that are acidified or sensitive to acidification at the present time and that had minimal amounts of peatland in their catchments. Pollen analysis was used to identify the levels in the sediment cores that corresponded to the arrival of spruce in the region and diatom analysis to identify any consequent acidification. The results showed that, despite a major shift in vegetational composition towards spruce, no change in lake water pH was indicated by the diatoms.

Alternatively, there are examples where changes in catchment vegetation appear to have caused both decreased and increased acidity in lake water. Decreases in acidity are usually associated with transient catchment disturbances such as windthrow of forest trees or fire (Rhodes & Davis, 1995; Korhola *et al.*, 1996; Huvane & Whitehead, 1996; Korsman & Segerström, 1988). Huvane and Whitehead (1996) showed that a sudden drop in the pollen frequency for *Tsuga* in a core from North Pond, Massachusetts, coincided with an increase in charcoal particles and a marked peak of the planktonic diatoms *Asterionella formosa* Hassall, *Cyclotella comta* (Ehrenb.) Kütz. and *Synedra radians* Kütz. It seems probable that this was the result of a fire in the catchment causing the rapid release of nutrients and base cations to the lake as organic matter was burnt.

Increased lake acidity is usually the result of longer-term processes as soils acidify following changes in the composition of the catchment vegetation. For example, in the study of North Pond (Huvane & Whitehead, 1996), the catchment disturbance described above only temporarily reversed the long-term decrease in diatom-inferred pH as hemlock gradually expanded between 7500 and 3500 BP. In Nygaard's classic study of Lake Gribsø in Denmark early acidification was shown to have occurred at around AD 400 coincident with the spread of beech forest (Nygaard, 1956), and, in a Finnish study, the rapid expansion of the acidobiontic diatom taxa *Tabellaria binalis* (Ehrenb.) Grun. and *Semiorbis hemicyclus* (Ehrenb.) Patr., coincided with the appearance and increase in the pollen of spruce (*Picea*) from around 3700 BP (Salomaa & Alhonen, 1983). Whilst this is in contrast to the apparent lack of any acidifying impact of spruce described above from Swedish studies, such variable responses are to be expected, depending on specific lake-catchment details of geology, soils, vegetation and hydrology.

Diatom analysis has been used frequently to assess the extent to which human modification of catchment soils and vegetation has changed the acidity of lakes. There are some cases where there is clear evidence of an increase in pH following land-use change, cases where there has been no impact, and, in special circumstances involving modern forestry, cases where land-use change has exacerbated the acidifying impact of 'acid rain'.

With the exception of natural fires (e.g., Huvane & Whitehead, 1996; Virkanen *et al.*, 1997), there is little or no evidence that poorly buffered lakes have experienced an increase in pH during the Holocene that can be ascribed to natural causes. The only sustained increases in pH for such lakes appears to be associated with the impact of early agriculture, probably also involving fire (Renberg *et al.*, 1993*b*). Following the observation (Fig. 5.7) of the increased pH of Lilla Öresjön at about 2300 BP, Renberg *et al.* (1993*b*) carried out pH reconstruction for a further 12 lakes in Southern Sweden. In almost all cases they demonstrated that increases in diatom-inferred pH had occurred, in some cases by over 1 pH unit, and they showed, from pollen and charcoal analysis, that the increases were associated with evidence for agriculture dating from between 2300 BP and 1000 BP. Elevated pH was maintained, probably due to continued burning, preventing the accumulation of acid humus in catchment soils, until the abandonment of agriculture and subsequent reafforestation in the nineteenth century. The more recent acidification of many of these lakes, although mainly due to acid deposition, was also in part due to a reacidification process following the decline in agriculture. A similar conclusion was reached by Davis *et al.* (1994*b*), who studied the acidification history of 12 lakes in New England. In this case, it was catchment disturbance associated with logging that caused an increase in lake water pH. Acidification of these lakes in the last century was ascribed first to re-acidification following the regrowth of conifers and secondly, as pH values fell below prelogging levels, to acid deposition.

In the SWAP Paleolimnology Programme (Battarbee & Renberg, 1990), studies were designed specifically to test the hypothesis that land-use change was an alternative cause of surface water acidification to acid deposition. Sites, such as Lilla Öresjön, where both catchment land-use change and acid deposition could be acting together were avoided. Consequently, some studies focused on sites where land-use change could be eliminated as a potential cause (see below p. 109) and some were concerned with sites where the influence of acid deposition could be eliminated.

For this latter approach it was necessary to choose sites situated in sensitive areas with low levels of acid deposition at the present time, or to use historical analogues. Anderson and Korsman (1990) argued that a good historical analogue for modern land-use changes is provided by the effects of the depopulation of farms and villages in Hälsingland, N. Sweden during Iron Age times, an event well established by archeologists. Following abandonment of the area from *ca.* 500 AD, resettlement did not occur before the Middle Ages (*ca.* 1100 AD).

Although evidence for regeneration of forest vegetation during this period can be identified in pollen diagrams from lakes in this region, there is no evidence from diatom analysis for lake acidification during this period at the two sites studied, Sjösjön and Lill Målsjön (Anderson & Korsman, 1990). In a similar way, Jones *et al.* (1986) showed that there was no evidence for pH decline at the Round Loch of Glenhead in Scotland at the time when pollen evidence indicated deforestation and an increase in catchment peat cover during Neolithic and Bronze Age times.

An exception to the proposition that land-use change is unimportant in causing surface water acidification is the apparent increased acidity of lakes and streams in the UK associated with modern forestry plantations. Comparisons between adjacent afforested and moorland streams in the uplands of the UK have shown that afforested streams have lower pH, higher aluminium and sulfate concentrations, and poorer fish populations than moorland streams (Harriman & Morrison, 1982; Stoner *et al.*, 1984). These differences have been mainly attributed either to the direct effect of forest growth (Nilsson *et al.*, 1982) or to the indirect effect of the forest canopy enhancing ('scavenging') dry and occult sulfur deposition (e.g., Unsworth, 1984), or to a combination of effects. In an attempt to separate these factors in time and space, Kreiser *et al.* (1990) devised a paleolimnological study comparing the acidification histories of afforested (L. Chon) and non-afforested (L. Tinker) sites in the Trossachs region of Scotland, an area of high S deposition, with afforested (L. Doilet) and non-afforested (Lochan Dubh) sites in the Morvern region, an area of relatively low sulfur deposition. All four sites showed evidence of acidification over the last century or so, and all show contamination over this time period by trace metals and carbonaceous particles, although the concentrations of these contaminants in the northwest region of Scotland are very low (Kreiser *et al.*, 1990).

In the high sulfur deposition region the main acidification at the afforested site, L. Chon, occurred after afforestation (from the mid-1950s) at a time when pH at the adjacent moorland 'control' site remained largely unchanged. Since both these sites have very similar base cation chemistry, and both had very similar diatom-inferred pH trends prior to the afforestation of L. Chon, the data suggest that the large difference in present-day water chemistry is due to afforestation.

In the region of lower sulfur deposition, the main acidification of the afforested site, L. Doilet, also occurs after afforestation, but the degree of acidification is slight and not significantly different than for the 'control' site, Lochan Dubh. For these reasons and because L. Doilet is more sensitive to acidification (Ca^{2+} 40 μeq/l) than L. Chon (Ca^{2+} 80 μeq/l), Kreiser *et al.* (1990) concluded that any acidification caused by forest growth alone had been minimal, and that the data strongly support the 'scavenging' hypothesis. In other words, afforestation of sensitive catchments can seriously exacerbate surface water acidification but only in regions receiving high levels of sulfur deposition (see p. 106 below).

Unless significant quantities of organic acids are present, bicarbonate alkalinity generation in stream and lake catchments is usually sufficient to maintain the pH of naturally acid, or naturally acidified, waters above pH 5.0–5.3. Unusually acidic waters, with pH, 5.0, consequently only occur where there are natural sources of strong mineral acids (Yoshitake & Fukushima, 1995; Renberg, 1986), or where sites are polluted. In the latter case, pollution sources are either associated with 'acid rain' (sulfur and nitrogen deposition derived from the combustion of fossil fuels and/or emissions from metal ore smelters) or, more locally, associated with acid mine drainage (Fritz & Carlson, 1982; Brugam & Lusk, 1986) and direct discharge of industrial effluents (van Dam & Mertens 1990).

Lake-catchment sources of sulfur Unusually acidic waters can occur where lake waters are influenced directly by sulfur, as in the case of some volcanic lakes (Yoshitake & Fukushima 1994), or where catchment soils are sulfur rich (Renberg, 1986). In a study of six volcanic lakes in Japan with pH ranging from 2.2–6.2, Yoshitake & Fukushima (1994) observed high standing crops of diatoms even at pH 2.2 to 3.0. At these extremely low values, *Pinnularia braunii* (Grun.) Cleve was the dominant taxon. Other common diatoms included *Eunotia exigua* (Breb. ex Kutz.) Rabenh., *Aulacoseira distans* (Ehrenb.) Simonsen and *Anomoeoneis brachysira* (= *Brachysira brebissonii*). Renberg (1986) describes an example of a lake in N. Sweden recently isolated from the Bothnian Bay (Baltic Sea) with a modern pH of 3. Diatom analysis of a core from the lake shows that, shortly after isolation, the lake was dominated by a largely circumneutral diatom flora of *Fragilaria* spp. and *Melosira* (= *Aulacoseira*) spp. However, a rapid change in the flora took place with these taxa being completely replaced, and the sole dominant becoming *Eunotia exigua*. Renberg ascribes the acidification of the lake to the exposure and oxidation of sulfides that had previously been deposited in the Baltic Sea, exacerbated by drainage activities that took place in the catchment to enhance cultivation of the soils.

Acid deposition As already pointed out, it was the 'acid rain' issue that stimulated the use of diatoms as indicators of water acidity. In this context the earliest use of diatom analysis in Europe was by Miller (1973), and in Almer *et al.* (1974), Berge (1976, 1979), Davis & Berge (1980), van Dam and Kooyman-van Blokland (1978), van Dam *et al.* (1981), and Flower and Battarbee (1983). In North America similar studies were under way, most notably by Del Prete and Schofield (1981) and Charles (1985) in the Adirondacks, by Davis *et al.* (1983) in New England, and Dixit (1986) and Dixit *et al.* (1987) in Canada. Whilst most of these studies focused on sediment cores, Berge (1976) and van Dam *et al.* (1981) demonstrated the usefulness of comparing old diatom samples from herbarium collections with present-day samples collected from the same locations. In the study by van Dam *et al.* (1981), old diatom samples, collected in 1920 from 16 moorland ponds were used to infer the past range of pH in the pools. They estimated that pH amongst the 16 ponds ranged between 4 and 6 in the 1920s, but this had narrowed to a range from pH 3.7 to 4.6 at the present day owing to acidification. Floristically, there had been a major increase in the abundance of *Eunotia exigua* in the humic-poor pools, and decreases in a range of taxa including *E. veneris* (= *incisa*), *Fragilaria virescens* Ralfs and *Frustulia rhomboides* var. *saxonica* (Rabenh.) De Toni.

The early studies of sediment cores cited above did not have the benefit of fully developed transfer functions, but the various techniques of pH reconstruction that were used showed beyond doubt that rapid, recent acidification had taken place in many areas of N. America and Europe that were both sensitive to acidification and that were receiving significant amounts of acid, especially sulfur, deposition. In particular, it was clear that planktonic floras of *Cyclotella* taxa were disappearing, and that acid tolerant taxa such as *T. binalis*, *T. quadriseptata*, *Navicula subtilissima* Cleve and *Eunotia incisa* were becoming more abundant. Moreover, inspection of the dates of these changes (Battarbee, 1984) showed that the onset of acidification at these lakes approximately coincided with the history of fossil fuel combustion in the countries concerned, with the earliest acidification taking place in the UK (mid-nineteenth century). In North America and Eastern Scandinavia, the same changes did not occur until the twentieth century.

A typical acidification sequence can be illustrated at the Round Loch of Glenhead in Scotland that has been intensively studied (Flower & Battarbee, 1983; Flower *et al.*, 1987; Jones *et al.*, 1989). The present pH of the lake is about 4.8. A small number of indigenous brown trout (*Salmo trutta*) is still present in the lake but some adjacent lakes are fishless. The lake and its catchment are situated entirely on granitic bedrock and the catchment has mainly peaty soils and a moorland vegetation that is grazed by sheep.

The core has been dated by the ^{210}Pb method (Appleby *et al.*, 1986) and the proportion of the main diatom species at successive sediment levels has been calculated. The diagram (Fig. 5.8) shows that there has been little change in the diatom flora of this lake until the mid-nineteenth century when diatoms characteristic of circumneutral water such as *Brachysira vitrea* (Grun.) R. Ross began to decline and be replaced by more acidophilous species. Acidification continued through the twentieth century, and by the early 1980s the diatom flora of the lake was dominated only by acid tolerant taxa such as *Tabellaria quadriseptata* Knudson and *T. binalis*. The reconstructed pH curve since about 1700 (Fig. 5.8) shows a stable pH of about 5.4 until about 1850, followed by a reduction of almost 1 pH unit in the following 130 years to 1980.

Despite the excellent correspondence between the incidence of acid deposition and this kind of evidence for the timing and extent of surface water acidification, there was far from universal acceptance that the acute decline in fish populations in many upland lakes was caused by acid deposition.

Consequently, many diatom studies that were carried out to address the

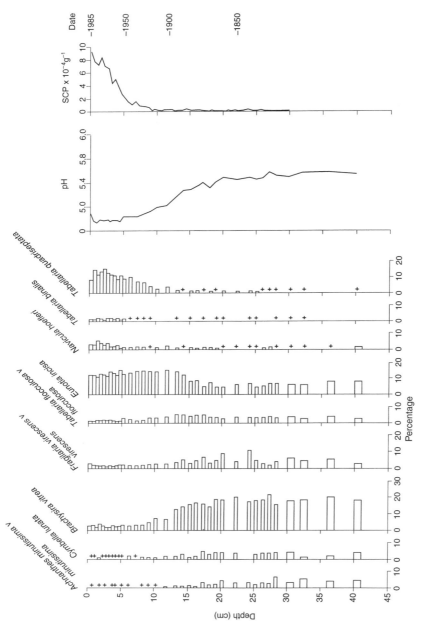

Fig. 5.8. Summary diatom diagram from the Round Loch of Glenhead with ^{210}Pb dates, diatom-inferred pH and spheroidal carbonaceous particles (modified from Jones *et al.*, 1989).

acid rain problem focused on research designs to test alternative hypotheses, especially those outlined above relating to catchment influences. In the case of the Round Loch of Glenhead, and similar lakes in the UK, it was possible to argue that there were no alternative explanations as catchment burning and grazing intensity had not decreased over this period, and there was no afforestation of the catchment. Moreover, whilst base-cation leaching was an important acidifying process in some catchments, especially in the early phases of lake development, this was a very slow process, and for most lakes the rate of base-cation loss is balanced by primary mineral weathering.

Perhaps the most conclusive evidence for the impact of acid rain comes from the diatom analysis of remote lakes perched on hilltops, where only changes in the acidity of precipitation can account for lake acidification.

Many such sites occur in the Cairngorm mountains of Scotland. The high corrie lakes in this area have steep-sided boulder-strewn catchments with little soil and sparse vegetation. Despite their isolation and lack of catchment change, these sites are all acidified, and their sediments contain high levels of atmospheric contaminants (Battarbee *et al.*, 1988*a*; Jones *et al.*, 1993). Similar results were obtained in Norway, where two hilltop lakes, Holetjörn and Ljosvatn, in Vest-Agder, SW Norway, were studied (Birks *et al.*, 1990*c*). Although the sites were naturally acidic, both showed a rapid drop in reconstructed pH values in the early part of the twentieth century, coincident with increasing contamination of the lake by trace metals and fossil-fuel derived carbonaceous particles.

Evidence that acid deposition could not only cause lake acidification, but was the primary reason for it, was established when the distribution of acidified sites was examined and compared to the regional pattern of sulfur deposition, both in Europe (e.g., Davis *et al.*, 1983; Renberg & Hellberg, 1982; Flower *et al.*, 1987; Battarbee *et al.*, 1988; Arzet *et al.*, 1986*a,b*; van Dam 1988), and in North America (Charles *et al.*, 1989; Davis *et al.*, 1990; 1994*b*; Sweets *et al.*, 1990, Dixit *et al.*, 1992*a,b*; Cumming *et al.*, 1992, 1994). Strongly acidified sites are only found in areas of high sulfur deposition, and, in areas of very low sulfur deposition, acidification on a regional scale only occurs in the most sensitive areas (Battarbee *et al.*, 1988*a*; Whiting *et al.*, 1989; Holmes *et al.*, 1989; Charles 1990b). Such a dose–response relationship has been demonstrated especially in the case of the UK where Battarbee (1990) was able to quantify the relationship and show how the incidence of lake acidification could be predicted using an empirical model relating sulfur deposition at a site to lake water calcium concentration (Battarbee *et al.*, 1996).

For some regions of the United States, studies of lake acidification have been carried out simply by comparing the diatom-inferred pH of samples taken from the bottoms (preacidification) and tops (modern) of sediment cores. This approach allows many lakes to be studied in a relatively short period of time, and by selecting sites on a statistical basis population estimates of the incidence of acidification can be made (Cumming *et al.*, 1992, Sullivan *et al.*, 1990, Dixit & Smol 1994).

Acid mine drainage and industrial wastewater discharge Drainage from mines, especially abandoned mines, can cause extreme surface water acidification through the bacterial oxidation of pyrite (FeS$_2$) to sulfuric acid. pH values as low as 2 can occur. Although most diatom species cannot survive such low pH values, *Eunotia exigua* in particular is often abundant.

In order to assess whether acid mine waters progressively neutralize over time, Brugam and Lusk (1986) studied 48 mine drainage lakes in the US Midwest and carried out diatom analysis on sediment cores from 20. Six of the lakes were initially acidic, characterized by species such as *Pinnularia biceps* Greg., but became alkaline during their postmining history, indicated by dramatic changes to strongly alkaliphilous taxa such as *Rhopalodia gibba* (Ehrenb.) O. Müll. and *Navicula halophila* (Grun *ex* Van Heurck). These floristic changes represent changes in water pH of 4 or 5 pH units, and are far more striking than the changes associated with the effects of acid deposition where changes of 1 pH unit or less are more usual.

The best known example of acidification from industrial discharges is the case of Lake Orta in Northern Italy, where nitrification processes following the release of ammonia from a cupro-ammonia rayon factory since 1926 caused pH levels to fall to *ca.* pH 4 (Mosello *et al.*, 1986). Liming began in 1989 and the lake pH is now about 5.2. Van Dam & Mertens (1990) collected epilithic samples from the lake, and showed that the present flora is dominated by acid and metal-tolerant taxa, especially *Eunotia exigua* and *Pinnularia subcapitata* var. *hilseana* (Janisch *ex* Rabenh.) O. Müll. Examination of earlier descriptions of diatom samples from the lake in 1884 and 1915 showed that these taxa were absent and that the lake was dominated by circumneutral and alkaliphilous taxa.

Diatoms and the management of acidified waters

AMELIORATION STRATEGIES

Mitigating the effects of surface water acidification caused by 'acid rain' or other pollution sources requires either increasing the base status of the water body (by liming) or reducing acid inputs, e.g., by controlling emissions of acidifying sulfur and nitrogen compounds from fossil-fueled power stations. Whilst liming has been carried out systematically in Sweden, and more occasionally in other countries (see below), long-term management of acidification requires reduction at source.

Diatoms and critical loads Much recent research, especially in Europe has been concerned with attempts to identify the theoretical level at which acid deposition causes biological 'harm'. Such a level, or critical load can then be used as a management target (Bull, 1991; Battarbee, 1995).

There are a number of critical load models (Nilsson & Grennfelt, 1988) but a model that uses the diatom-based calibration function described above has been extensively used to set critical loads values for acidified UK waters (Battarbee *et al.*, 1996). As diatoms are amongst the most sensitive organisms to acidification, it can be argued that the first point of change in a sediment core towards a more acid-tolerant assemblage represents the time at which the critical load for the site is first exceeded. After calibration, the model enables the probable acid deposition at any site at the time of critical load exceedance to be calculated given a knowledge of the contemporary acid deposition and the calcium concentration of the water. If the critical load is then subtracted from the actual load, an exceedance value is derived, that can be used as a measure of damage to an individual site, or can be applied to many sites to produce maps of critical load exceedance (CLAG, 1995).

Liming Whilst liming is not regarded as a solution to the acid rain problem, it has been carried out systematically in Sweden and experimentally in a number of countries especially as a method for alleviating the impact of acidification on fish and other vertebrate populations.

There have been a number of studies concerned with the response of diatoms to liming (e.g., Round, 1990; Flower *et al.*, 1990; Ohl *et al.*, 1990; Rhodes, 1991; Bellemakers & van Dam, 1992; Renberg & Hultberg, 1992; Cameron, 1995). In the study by Flower *et al.* (1990), it was shown that the liming of Loch Fleet created chemical conditions that allowed fish populations to survive, but in so doing generated a diatom assemblage with a significantly different species composition than had previously occurred in the lake. Most notably the planktonic diatoms *Asterionella formosa* and *Synedra acus* Kütz. appeared, not *Cyclotella kützingiana* Thwaites, the dominant planktonic taxon prior to acidification.

A similar response was demonstrated by Renberg and Hultberg (1992) for Lysevatten in southwestern Sweden. Liming resulted in the expansion of *Achnanthes minutissima* Kütz., *Cymbella microcephala* Grun. and *Synedra acus*, but, as in the case of Loch Fleet, the *Cyclotella* flora was not restored (Renberg & Hultberg, 1992). Again at Holmes Lake in the Adirondacks, Rhodes (1991) showed that liming caused the proliferation of *Achnanthes minutissima*, but not the re-establishment of *Cyclotella stelligera* (Cleve & Grun.), the dominant pre-acidification planktonic diatom.

RECOVERY AND REVERSIBILITY

Monitoring acid streams and lakes and the role of diatoms As a consequence of the dominant role played by diatom analysis in 'acid rain' studies, diatoms are now recognized as the premier biological indicators of surface water acidity and are being used throughout Europe and North America for water quality monitoring (van Dam, 1996; van Dam *et al.*, 1996; van Dam & Mertens, 1995; Patrick *et al.*, 1996; Coring, 1996; Dixit & Smol, 1994). In the UK a monitoring network of 22 sites was established in 1988 (Patrick *et al.*, 1995, 1996). This network includes sensitive streams and lakes in areas of high and low acid deposition and at each

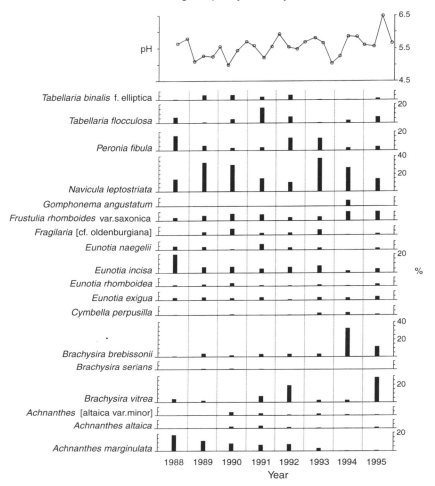

Fig. 5.9. Trends in the composition of epilithic diatom communities and measured pH from Loch Chon, Scotland from the UK Acid Waters Monitoring Network (from Monteith *et al.*, 1997).

site diatoms, in addition to other key chemical and biological indicators, are monitored. Diatoms are only collected once per year, but by sampling at the same time of year and ensuring high-quality control procedures and collecting from the same habitat (the epilithon) underlying trends in water quality can be sensitively detected (Monteith *et al.*, 1997). In the case of Loch Chon (Figure 5.9) the gradual decrease of *Achnanthes marginulata* Grun. and *Eunotia incisa* W. Sm. ex Greg. and their replacement by *Brachysira* spp. indicates slight recovery in the status of the lake.

Evidence of recovery from high resolution studies of recent sediments and repeat coring
Although sediment core analysis of diatoms from acid lakes has been used

R. W. BATTARBEE *ET AL.*

mainly for reconstructing acidification trends over relatively long time periods, sediments can also potentially be used as a substitute for monitoring. The advantages of using the uppermost sediment record is that contemporary trends can be observed at almost all affected lakes, not only those selected as part of a monitoring network, and sediment samples tend to give a more integrated representation of the whole lake than individual samples from live communities. The disadvantage is that precise dating of uppermost sediments is difficult, sediments do not often accumulate sufficiently rapidly, and bioturbation and resuspension can sometimes destroy or mask trends, at least over short time periods.

The above method has been assessed at a number of European (Battarbee *et al.*, 1988b; Allott *et al.*, 1992; Ek *et al.*, 1995; Juggins *et al.*, 1996) and N. American sites (Dixit *et al.*, 1992a, c), and Cumming *et al.* (1994) have used the approach for Adirondack lakes using chrysophyte scale records rather than diatoms. At the Round Loch of Glenhead, Allott *et al.* (1992) were able to show that the slight recovery in water column mean pH from 4.7 to 4.9 could be detected in the recent sediment diatom record, but only in cores with a sediment accumulation rate higher than 0.7 mm/yr. As the sediment accumulation rate at acidified lakes is normally not any more rapid, this approach only has value if lakes are recovering rapidly or if a relatively long time (perhaps 10 years) is allowed between repeat coring dates.

In Örvattnet the accumulation rate of the sediment was sufficiently rapid for Ek *et al.* (1995) to demonstrate the beginning of recovery using this approach. A detailed diatom analysis of the uppermost sediments of a core from the lake showed a decrease in the relative abundance of *Tabellaria binalis*, indicating a slight rise in pH, dating from the 1970s. Although there was little clear increase in measured water pH during this time, acid episodes in spring have become less severe and there has been a decrease in sulfate deposition by between 30 and 40%. Ek *et al.* (1995) consequently argue that diatom analysis of the core is a more sensitive measure of recovery than water chemistry monitoring.

A further example of this approach related to lakes in the Sudbury region of Ontario, Canada, acidified mainly by SO_2 emissions from local metal mining and smelting activities. Until the 1970s this region was recognized as the world's largest point source of SO_2 (Keller 1992). Paleolimnological studies of lakes in this region provide convincing evidence of the potential speed of lake recovery in response to reductions in sulfur deposition and in the power of the diatom record in tracking short-term water quality changes (Dixit *et al.*, 1992a, c). Baby Lake lies about 1 km southwest of the Coniston Smelter near Sudbury. Marked changes have occurred over the last century (Fig. 5.10), showing acidification and then recovery as emissions from the smelter declined.

The sensitivity of diatom assemblages to water quality change of this kind has also been illustrated by Vinebrooke (1996) using diatoms transplanted amongst Ontario lakes to assess the relative importance of changing water chemistry and biotic interactions.

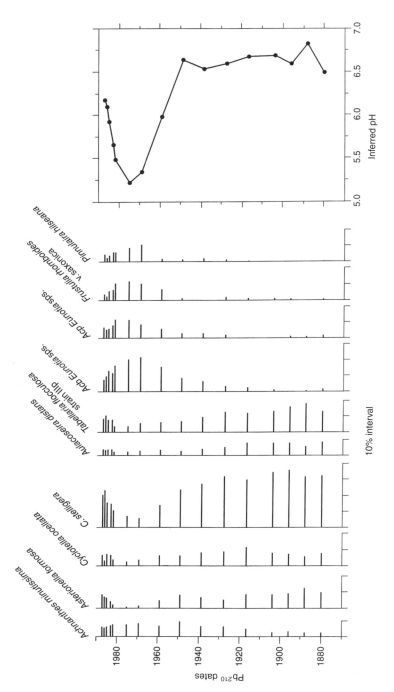

Fig. 5.10. Summary diatom diagram and diatom-inferred pH changes of a sediment core from Baby Lake, Ontario, showing acidification and recovery (from Dixit *et al.*, 1992*a*).

Future prospects: dynamic modelling and the role of diatom analysis in model validation
Whilst diatom records cannot be used to predict how specific sites will respond to a reduction in acid deposition in the future, they can be used to validate dynamic models that are designed for such predictions. The most frequently used model is MAGIC (Model of Acidification of Groundwaters in Catchments) which simulates the physical and chemical processes that occur in response to sulfur loading on catchments (Cosby *et al.*, 1985).

The model can be set up for any individual site given a knowledge of rainfall and runoff quantity and quality, sulfur deposition history, and soil characteristics (especially bulk density, depth, cation exchange capacity, sulfate adsorption rates and base saturation, temperature and carbon dioxide levels). Once set up and evaluated, the model is run to reconstruct past acidification trends and to predict future pH and water chemistry according to specified future sulfur deposition scenarios.

The output of MAGIC can be validated by comparing model-based pH hindcasts with diatom-based reconstructions (Jenkins *et al.*, 1990, 1997; Sullivan *et al.*, 1996). The relatively close agreement between these two independent methods gives some confidence in the use of the model to predict future changes following planned reductions in sulfur emissions in Europe and North America. However, model-diatom comparisons do show discrepancies either due to inaccurate or incomplete parameterization of the model (e.g., Sullivan *et al.*, 1996) or to deficiencies in the diatom–pH inferences described above.

Chemical and biological targets Because lake acidification was not identified as a serious environmental problem until after significant ecological change had taken place, and because acidified lakes are predominantly remote, there are few extant instrumental or documentary records that can be used to indicate the preacidification status of these systems. Diatom analysis can play an important role in assessing both the past chemical and biological status of acidified lakes and thereby help to define targets for recovery.

One of the most useful roles for diatom analysis in the management of acidified lakes is its capacity to define pre-acidification baseline pH levels that can be used as targets for restoration (Battarbee, 1997). This is especially important if liming is a management option (Dixit *et al.*, 1996; Korsman 1993). In a study of Daisy Lake, a lake acidified by the emissions from metal smelting in the Sudbury region of Ontario, Canada, Dixit *et al.* (1996) carried out replicate diatom analysis and pH reconstruction of four cores and concluded that the current liming programme on the lake needed to raise pH by about 1.5 pH units to restore preacidification pH levels.

In a contrasting exercise Korsman (1993) was able to show from diatom analysis that many lakes in northern Sweden with low pH values at present and that were thought to be acidified were naturally acidic. No pH reduction had taken place, and liming was not necessary. Likewise Winkler (1988) points out that the indiscriminate liming of acid ponds can change the biota of naturally acid lake ecosystems that have evolved over millenia, and management decisions about liming need to have a knowledge of the long-term pH history of a lake.

A more complete lake restoration for acidified lakes can be planned if targets for the future biological structure of a site can be derived. However, this requires not only a knowledge of past baseline pH but also of preacidification biology. Some elements of the preacidification flora and fauna can be inferred from the lake sediment record, but this is restricted principally to diatoms, chrysophytes, cladocerans and chironomids. An alternative approach is to identify lakes that can be used as analogues for the preacidification status of currently acidified lakes. Flower *et al.* (1997) used this approach for two acidified lakes in Scotland. By using a Chi-square dissimilarity measure, they compared diatom assemblages from different depths of the sediment cores from the two lakes with the surface sediment diatom assemblages of lakes in the diatom–pH modern training set. Although the results suggested that a larger range of non-acidified low calcium sites are needed in the training set to find very close analogues, the study illustrated the potential of the technique.

Summary

Although the use of diatoms as indicators of pH has a long history, it was the emergence of surface water acidification as a major environmental problem that brought diatoms to prominence and that rapidly accelerated the development of the robust statistical methodology that is currently used not only for pH reconstruction from sediment cores but also for the reconstruction of other hydrochemical variables such as total phosphorus (Bennion, 1994; Hall & Smol, this volume) salinity (Fritz *et al.*, 1991; and this volume), and other variables.

Traditionally diatom analysis has been used to show how lakes have developed over the Holocene period in relation to base cation leaching and catchment paludification, fire, and the impact of early agriculture. In the 'acid rain' controversy of the 1980s, diatom analysis played a pivotal role in demonstrating how the recent acidifying impacts of acid deposition could be differentiated in space and time from the influence of these largely natural catchment-based processes.

Diatom techniques are now playing important roles in the management of acidified waters. In particular, diatom analysis is being used to differentiate between acid and recently acidified surface waters, to determine critical loads of acidity, to evaluate models of acidification, and to monitor the response of acidified waters to reduced acid deposition following mitigation policies.

In less acidified regions, diatom-based pH reconstruction is important in identifying pristine and relatively pristine sites where variations in lakewater pH are the result of natural forces, including climatic variability. Psenner & Schmidt (1992) have demonstrated the correspondence between pH reconstruction and instrumental air temperature series for lakes in the

Austrian Alps, suggesting that, in the absence of acid rain, lake acidity variation as reflected by diatom assemblages may be a good surrogate for lake temperature and climate change. Future work on diatoms as indicators of pH needs to evaluate these observations and assess the way in which variations in lake water temperature can influence pH in low alkalinity lakes.

Acknowledgments

We are most grateful to John Birks, John Kingston, Herman van Dam, Roger Flower and Tim Allott as well as to the editors and an anonymous reviewer for helpful comments on an early draft of this manuscript. RWB would also like to thank Catherine Dalton and Cath Pyke for their contributions towards the technical production of the manuscript.

References

Alhonen, P. (1968). On the late-glacial and early post-glacial diatom succession in Loch of Park, Aberdeenshire, Scotland. *Memoranda Societatis Pro Fauna et Flora Fennica, 44,* 13–20.

Allott, T. E. H. (1992). The reversibility of lake acidification: A diatom study from the Round Loch of Glenhead, Galloway, Scotland. Unpublished PhD thesis, University College London.

Allott, T. E. H., Harriman, R., & Battarbee, R. W. (1992). Reversibility of acidification at the Round Loch of Glenhead, Galloway, Scotland. *Environmental Pollution, 77,* 219–25.

Almer, B., Dickson, W., Ekström, C., Hörnström, E., & Miller, U. (1974). Effects of acidification on Swedish lakes. *Ambio, 3,* 30–6.

Anderson, D. S., Davis, R. B., Miller, N. G., & Stuckenrath, R. (1986). History of late- and post-glacial vegetation and disturbance around Upper South Branch Pond, northern Maine. *Canadian Journal of Botany 64,* 1977–86.

Anderson, D. S., Davis, R. B., & Berge, F. (1986). Relationships between diatom assemblages in lake surface-sediments and limnological characteristics in southern Norway. In *Diatoms and Lake Acidity,* ed. J. P. Smol, R. W. Battarbee, R. B. Davis, & J. Meriläinen, pp. 97–113. Dordrecht, The Netherlands: Dr. W. Junk.

Anderson, N. J. (1990). Variability of sediment diatom assemblages in an upland, wind-stressed lake (Loch Fleet, Galloway, S.W. Scotland). *Journal of Paleolimnology, 4,* 43–59.

Anderson, N. J., & Korsman, T. (1990). Land-use change and lake acidification: Iron-age de-settlement in northern Sweden as a pre-industrial analogue. *Philosophical Transactions of the Royal Society, London,* **B 327,** 373–6.

Appleby, P. G., Nolan, P. J., Gifford, D. W., Godfrey, M. J., Oldfield, F., Anderson, N. J., & Battarbee, R. W. (1986). ^{210}Pb dating by low background gamma counting. *Hydrobiologia,* **143,** 21–27.

Arzet, K., Steinberg, C., Psenner, R., & Schulz, N. (1986a). Diatom distribution and diatom inferred pH in the sediment of four alpine lakes. *Hydrobiologia,* **143,** 247–54.

Arzet, K., Krause-Dellin, D., & Steinberg, C. (1986b). Acidification of four lakes in the Federal Republic of Germany as reflected by diatom assemblages, cladoceran remains, and sediment chemistry. In *Diatoms and Lake Acidity,* ed. J. P. Smol, R. W. Battarbee, R. B. Davis, & J. Meriläinen, pp. 227–50. Dordrecht, The Netherlands: Dr. W. Junk.

Atkinson, K. M., & Haworth, E. Y. (1990). Devoke Water and Loch Sionascaig: Recent environmental changes and the post-glacial overview. *Philosophical Transactions of the Royal Society, London,* **B 327,** 349–55.

Baker, J. P., & Schofield, C. L. (1982). Aluminium toxicity to fish in acidic waters. *Water, Air and Soil Pollution*, **18**, 289–309.

Barker, P. (1992). Differential diatom dissolution in Late Quaternary sediments from Lake Manyara, Tanzania: An experimental approach. *Journal of Paleolimnology*, **7**, 235–51.

Baron, J., Norton, S. A., Beeson, D. R., & Herrmann, R. (1986). Sediment diatom and metal stratigraphy from Rocky Mountain lakes with special reference to atmospheric deposition. *Canadian Journal of Fisheries and Aquatic Sciences*, **43**, 1350–62.

Battarbee, R. W. (1984). Diatom analysis and the acidification of lakes. *Philosophical Transactions of the Royal Society, London*, **B 305**, 451–77.

(1990). The causes of lake acidification, with special reference to the role of acid deposition. *Philosophical Transactions of the Royal Society, London*, **B 327**, 339–47.

(1991). Recent palaeolimnology and diatom-based reconstruction. In *Quaternary Landscapes,* ed. L. C. K. Shane & E. J. Cushing, pp. 129–74. Minneapolis: University of Minnesota Press.

(ed.) (1995). *Acid Rain and its Impact: the Critical Loads Debate* London: Ensis Publishing. 183pp.

(1997). Freshwater quality, naturalness and palaeolimnology. In *Freshwater Quality: Defining the Undefinable,* ed. P. J. Boon & D. L. Howell, pp. 155–71. Edinburgh: The Stationery Office.

Battarbee, R. W., & Charles, D. F. (1986). Diatom-based pH reconstruction studies of acid lakes in Europe and North America: A synthesis, *Water Air and Soil Pollution*, **31**, 347–54.

(1987). The use of diatom assemblages in lake sediments as a means of asessing the timing, trends, and causes of lake acdification. *Progress in Physical Geography*, **11**, 552–80.

(1994). Lake acidification and the role of paleolimnology. In *Acidification of Freshwater Ecosystems: Implications for the Future,* ed. C. Steinberg & R. Wright, pp. 51–65. Dahlem Workshop Environmental Sciences Research Report 14. Chichester: Wiley.

Battarbee, R. W., & Renberg, I. (1990). The Surface Water Acidification Project (SWAP) Palaeolimnology Programme. *Philosophical Transactions of the Royal Society, London*, **B 327**, 227–32.

Battarbee, R. W., Allott, T. E. H., Juggins, S., Kreiser, A. M., Curtis C., & Harriman, R. (1996). An empirical critical loads model for surface water acidification, using a diatom-based palaeolimnological approach. *Ambio*, **25**, 366–9.

Battarbee, R. W., Anderson, N. J., Appleby, P. G., Flower, R. J., Fritz, S. C., Haworth, E. Y., Higgitt, S., Jones, V. J., Kreiser, A., Munro, M. A. R., Natkanski, J., Oldfield, F., Patrick, S. T., Richardson, N. G., Rippey, B., & Stevenson, A. C. (1988*a*). *Lake Acidification in the United Kingdom 1800–1986: Evidence From Analysis of Lake Sediments.* London: Ensis. 68pp.

Battarbee, R. W., Flower, R. J., Stevenson, A. C., Jones, V. J., Harriman, R., & Appleby, P. G. (1988*b*). Diatom and chemical evidence for reversibility of acidification of Scottish lochs. *Nature*, **332**, 530–2.

Battarbee, R. W., Flower, R. J., Stevenson, A. C., & Rippey, B. (1985). Lake acidification in Galloway: A palaeoecological test of competing hypotheses. *Nature*, **314**, 350–2.

Battarbee, R. W., Mason, J., Renberg, I., & Talling, J. F. (eds.) (1990). *Palaeolimnology and lake acidification.* London: The Royal Society, 219 pp.

Battarbee, R. W., Stevenson, A. C., Rippey, B., Fletcher, C., Natkanski, J., Wik, M., & Flower, R. J. (1989). Causes of lake acidification in Galloway, south-west Scotland: A palaeoecological evaluation of the relative roles of atmospheric contamination and catchment change for two acidified sites with non-afforested catchments. *Journal of Ecology*, **77**, 651–72.

Beamish, J., & Harvey, H. H. (1972). Acidification of La Cloche Mountain lakes, Ontario, and resulting fish mortalities. *Journal of the Fisheries Research Board of Canada*, **29**, 1131–43.

Bellemakers, M. J. S., & van Dam, H. (1992). Improvement of breeding success of the moor frog (*Rana arvalis*) by liming of acid moorland pools and the consequences of liming for water chemistry and diatoms. *Environmental Pollution*, **78**, 165–71.

Bennion, H. (1994). A diatom-phosphorus transfer function for shallow, eutrophic ponds in southeast England. *Hydrobiologia,* **275/276,** 391–410.

Bennion, H., Anderson, N. J., & Juggins, S. (1996). Predicting epilimnetic phosphorus concentrations using an improved diatom-based transfer function and its application to lake eutrophication management. *Environmental Science and Technology,* **30,** 2004–7.

Berge, F. (1976). Kiselalger og pH i noen elver(og innsjöer i Agder og Telemark. En sammenlikning mellom årene 1949 og 1975. [Diatoms and pH in some rivers and lakes in Agder and Telemark (Norway): A comparison between the years 1949 and 1975.] *Sur Nedbörs Virkning på Skog og Fisk.* IR18/76. Norwegian Forest Research Institute, Aas, Norway.

(1979). Kiselalger og pH i noen innsjöer i Agder og Hordaland. [Diatoms and pH in some lakes in the Agder and Hordaland Counties, Norway.] IR42/79. *Sur Nedbörs Virkning på Skog og Fisk.* IR18/76. Norwegian Forest Research Institute, Aas, Norway.

Birks, H. J. B. (1987). Methods for pH calibration and reconstruction from palaeolimnological data: procedures, problems, potential techniques. Surface Water Acidification Programme, Mid-term Review Conference, Bergen, Norway, pp. 370–80.

(1994). The importance of pollen and diatom taxonomic precision in quantitative palaeoenvironmental reconstructions. *Review of Palaeobotany and Palynology,* **83,** 107–17.

(1995). Quantitative palaeoenvironmental reconstructions. In *Statistical Modelling of Quaternary Science Data,* ed. D. Maddy & J. S. Brew, pp. 161–254. Quaternary Research Association Technical Guide 5, Cambridge.

Birks, H. J. B., Line, J. M., Juggins, S., Stevenson, A. C., & ter Braak, C. J. F. (1990a). Diatoms and pH reconstruction. *Philosophical Transactions of the Royal Society, London,* **B 327,** 263–78.

Birks, H. J. B., Juggins, S., & Line, J. M. (1990b). Lake surface-water chemistry reconstructions from palaeolimnological data. In *The Surface Waters Acidification Programme,* ed. B. J. Mason, pp. 301–13. Cambridge: Cambridge University Press.

Birks, H. J. B., Berge, F., Boyle, J. F., & Cumming, B. F. (1990c). A palaeoecological test of the land-use hypothesis for recent lake acidification in South-West Norway using hilltop lakes. *Journal of Paleolimnology,* **4,** 69–85.

Brugam, R. B., & Lusk, M. (1986). Diatom evidence for neutralization in acid surface mine lakes. In *Diatoms and Lake Acidity,* ed. J. P. Smol, R. W. Battarbee, R. B. Davis, & J. Meriläinen, pp. 115–29. Dordrecht, The Netherlands: Dr. W. Junk.

Bull, K. (1991). The critical load/levels approach to gaseous pollutant emission control. *Environmental Pollution,* **69,** 105–23.

Camburn, K. E., & Kingston, J. C. (1986). The genus *Melosira* from soft-water lakes with special reference to northern Michigan, Wisconsin and Minnesota. In *Diatoms and Lake Acidity,* ed. J. P. Smol, R. W. Battarbee, R. B. Davis, & J. Meriläinen, pp. 17–34. Dordrecht, The Netherlands: Dr. W. Junk.

Camburn, K. E., Kingston, J. C., & Charles, D. F. (1986). PIRLA Diatom Iconograph. PIRLA Unpublished Report Number 3. Bloomington: Indiana University.

Cameron, N. G. (1995). The representation of diatom communities by fossil assemblages in a small acid lake. *Journal of Paleolimnology,* **14,** 185–223.

Charles, D. F. (1985). Relationships between surface sediment diatom assemblages and lake-water characteristics in Adirondack lakes. *Ecology,* **66,** 994–1011.

(1990a). A checklist for describing and documenting diatom and chrysophyte calibration data sets and equations for inferring water chemistry. *Journal of Paleolimnology,* **3,** 175–8.

(1990b). Effects of acidic deposition on North American lakes: Paleolimnological evidence from diatoms and chrysophytes. *Philosophical Transactions of the Royal Society of London,* **B327,** 403–12.

Charles, D. F., & Norton, S. A. (1986). Paleolimnological evidence for trends in atmospheric deposition of acids and metals. In *Acid Deposition: Long-Term Trends,* pp. 335–435. Washington: National Academy Press.

Charles, D. F., & Smol, J. P. (1988). New methods for using diatoms and chrysophytes to infer past pH of low-alkalinity lakes. *Limnology and Oceanography*, **33**, 1451–62.

(1994*a*). Long-term chemical changes in lakes: Quantitative inferences from biotic remains in the sediment record. In *Environmental Chemistry of Lakes and Reservoirs*, ed. L. Baker. pp. 3–31, Washington: American Chemical Society.

Charles, D. F., & Whitehead, D. R. (1986). The PIRLA project: Paleoecological investigation of recent lake acidification. *Hydrobiologia*, **143**, 13–20.

Charles, D. F., Battarbee, R. W., Renberg, I. van Dam, H., & Smol, J. P. (1989). Paleoecological analysis of lake acidification trends in North America and Europe using diatoms and chrysophytes. In *Acid Precipitation*, Vol. 4. *Soils, Aquatic Processes, and Lake Acidification*, ed. S. A. Norton, S. E. Lindberg, & A. L. Page, pp. 207–76. New York: Springer-Verlag.

Charles, D. F., Binford, M. W., Furlong, E. T., Hites, R. A., Mitchell, M. J., Norton, S. A., Oldfield, F., Paterson, M. J., Smol, J. P., Uutala, A. J., White, J. R., Whitehead, D. R., & Wise, R. J. (1990). Paleoecological investigation of recent lake acidification in the Adirondack Mountains, N.Y. *Journal of Paleolimnology*, **3**, 195–241.

Charles, D. F., Dixit, S. S., Cumming, B. F., & Smol, J. P. (1991). Variability in diatom and chrysophyte assemblages and inferred pH: Paleolimnological studies of Big Moose Lake, New York (USA). *Journal of Paleolimnology*, **5**, 267–84.

Charles, D. F., Smol, J. P., & Engstrom, D. R. (1994*b*). Paleolimnological approaches to biological monitoring. In *Biological Monitoring of Aquatic Systems,* ed. L. L. Loeb & A. Spacie, pp. 233–93. Boca Raton: CRC Press.

CLAG (1995). *Critical Loads of Acid Deposition for United Kingdom Freshwaters.* Critical Loads Advisory Group sub-group report on freshwaters, 79 pp. Penicuik: Institute of Terrestrial Ecology.

Cleve-Euler, A. (1951–1955). Die Diatomeen von Schweden und Finland. Kungliga Vetenskopsakademiens Handlingar Series 4, **2**(1), 3–163; **4**(1), 3–158; **4**(5), 3–355; **5**(4), 3–231; **3**(3), 3–153.

Coring, E. (1996). Use of diatoms for monitoring acidification in small mountain rivers in Germany with special emphasis on 'diatom assemblage type analysis' (DATA). In *Use of Algae for Monitoring Rivers II*, ed. B. A. Whitton & E. Rott, pp. 7–16. Proceedings of an international symposium, Innsbruck.

Cosby, B. J., Hornberger, J. N., Galloway, J. N., & Wright, R. F. (1985). Freshwater acidification from atmospheric deposition of sulfuric acid: A quantitative model. *Environmental Science and Technology*, **19**, 1144–9.

Crabtree, K. (1969). Post-glacial diatom zonation of limnic deposits in north Wales. *Mitteilungen Internationale Vereinigung für Theoretische und Angewandte Limnologie*, **17**, 165–71.

Cumming, B. F., Davey, K. A., Smol, J. P., & Birks, H. J. B. (1994). When did acid-sensitive Adirondack lakes (New York, USA) begin to acidify and are they still acidifying? *Canadian Journal of Fisheries and Aquatic Sciences*, **51**, 1550–68.

Cumming, B. F., Smol, J. P., Kingston, J. C., Charles, D. F., Birks, H. J. B., Camburn, K. E., Dixit, S. S., Uutala, A. J., & Selle, A. R. (1992). How much acidification has occurred in Adirondack region lakes (New York, USA) since preindustrial times? *Canadian Journal of Fisheries and Aquatic Sciences*, **49**, 128–41.

Davis, R. B. (1987). Paleolimnological diatom studies of acidification of lakes by acid rain: an application of quaternary science. *Quaternary Science Reviews*, **6**, 147–63.

Davis, R. B., & Anderson, D. S. (1985). Methods of pH calibration of sedimentary diatom remains for reconstructing history of pH in lakes. *Hydrobiologia*, **120**, 69–87.

Davis, R. B., & Berge, F. (1980). Atmospheric deposition in Norway during the last 300 years as recorded in SNSF lake sediments, U.S.A. and Norway; II, Diatom stratigraphy and inferred pH. In *Ecological Impact of Acid Precipitation; Proceedings of an International Conference*, ed. D. Drabloes & A. Tollan, pp. 270–271. SNSF-Project, Oslo.

Davis, R. B., Anderson, D. S., & Berge, F. (1985). Palaeolimnological evidence that lake acidification is accompanied by loss of organic matter. *Nature*, **316**, 436–38.

Davis, R. B., Anderson, D. S., Norton, S. A., Ford, J., Sweets, P. R., & Kahl, J. S. (1994*a*).

Sedimented diatoms in northern New England lakes and their use as pH and alkalinity indicators. *Canadian Journal of Fisheries and Aquatic Sciences,* **51,** 1855–76.

Davis, R. B., Anderson, D. S., Norton, S. A., & Whiting, M. C. (1994b). Acidity of twelve northern New England (U. S. A.) lakes in recent centuries. *Journal of Paleolimnology,* **12,** 103–54.

Davis, R. B., Anderson, D. S., Whiting, M. C., Smol, J. P., & Dixit, S. S. (1990). Alkalinity and pH of three lakes in northern New England, U. S. A. over the past 300 years. *Philosophical Transactions of the Royal Society, London,* **B327,** 413–21.

Davis, R. B., Norton, S. A., Hess, C. T., & Brakke, D. F. (1983). Paleolimnological reconstruction of the effects of atmospheric deposition of acids and heavy metals on the chemistry and biology of lakes in New England and Norway. *Hydrobiologia,* **103,** 113–23.

Del Prete, A., & Schofield, C. (1981). The utility of diatom analysis of lake sediments for evaluating acid precipitation effects on dilute lakes. *Archiv für Hydrobiologie,* **91,** 332–40.

DeNicola, D. M. (1986). The representation of living diatom communities in deep-water sedimentary diatom assemblages in two Maine (U. S. A.) lakes. In *Diatoms and Lake Acidity,* ed. J. P. Smol, R. W. Battarbee, R. B. Davis, & J. Meriläinen, pp. 73–85. Dordrecht, The Netherlands: Dr. W. Junk.

Dixit, A. S., Dixit, S. S., & Smol, J. P. (1992a). Long-term trends in lake water pH and metal concentrations inferred from diatoms and chrysophytes in three lakes near Sudbury, Ontario. *Canadian Journal of Fisheries and Aquatic Sciences,* **49,** 17–24.

(1992c). Algal microfossils provide high temporal resolution of environmental change. *Water, Air and Soil Pollution,* **62,** 75–87.

(1996). Setting restoration goals for an acid and metal-contaminated lake: A paleolimnological study of Daisy Lake (Sudbury, Canada). *Journal of Lake and Reservoir Management,* **12,** 323–30.

Dixit, S. S. (1986). Diatom-inferred pH calibration of lakes near Wawa, Ontario. *Canadian Journal of Botany,* **64,** 1129–33.

Dixit, S. S., & Smol, J. P. (1994). Diatoms as indicators in the Environmental Monitoring and Assessment Program – Surface Waters (EMAP – SW). *Environmental Monitoring and Assessment,* **31,** 275–306.

Dixit, S. S., & Smol, J. P. (1995). Diatom evidence of past water quality changes in Adirondack seepage lakes (New York, USA). *Diatom Research,* **10,** 113–129.

Dixit, S. S., Cumming, B. F., Birks, H. J. B., Smol, J. P., Kingston, J. C., Uutala, A. J., Charles, D. F., & Camburn, K. E. (1993). Diatom assemblages from Adirondack lakes (New York, USA) and the development of inference models for retrospective environmental assessment. *Journal of Paleolimnology,* **8,** 27–47.

Dixit, S. S., Dixit A. S., & Evans, R. D. (1987). Paleolimnological evidence of recent acidification in two Sudbury (Canada) lakes. *Science of the Total Environment,* **67,** 53–63.

(1988). Sedimentary diatom assemblages and their utility in computing diatom-inferred pH in Sudbury Ontario lakes. *Hydrobiologia,* **169,** 135–48.

Dixit, S. S., Dixit, A. S., & Smol, J. P. (1991). Multivariate environmental inferences based on diatom assemblages from Sudbury (Canada) lakes. *Freshwater Biology,* **26,** 251–66.

(1992b). Assessment of changes in lake water chemistry in Sudbury area lakes since pre-industrial times. *Canadian Journal of Fisheries and Aquatic Sciences,* **49,** 8–16.

Driscoll, C. T. (1984). A procedure for the fractionation of aqueous aluminium in dilute acidic waters. *International Journal of Environmental Analytical Chemistry,* **16,** 267–83.

Driscoll, C. T., Baker, J. P., Bisogni, J. J., & Schofield, C. L. (1980). Effect of aluminium speciation on fish in dilute acidified waters. *Nature,* **284,** 161–4.

Driscoll, C. T., Yatsko, C. P., & Unangst, F. J. (1987). Longitudinal and temporal trends in the water chemistry of the North Branch of the Moose River. *Biogeochemistry,* **3,** 37–61.

Ek, A., Grahn, O., Hultberg, H., & Renberg, I. (1995). Recovery from acidification in lake Örvattnet, Sweden. *Water, Air and Soil Pollution,* **85,** 1795–800.

Emmett, B., Charles, D. F., Feger, K. H., Harriman, R., Hemond, H. F., Hultberg, H., Lessmann, D., Ovalle, O., Van Miegroet, H., & Zoettl, H. W. (1994). Group Report: Can we differentiate between natural and anthropogenic acidification? In *Acidification of Freshwater Ecosystems: Implications for the Future,* ed. C. Steinberg & R.

Wright, pp. 118–140, Dahlem Workshop Environmental Sciences Research Report 14. Chichester: Wiley.

Evans, G. H. (1970). Pollen and diatom analysis of late-Quaternary deposits in the Blelham Basin, North Lancashire. *New Phytologist*, **69**, 821–74.

Evans, G. H., & Walker, R. (1977). The late-Quaternary history of the diatom flora of Llyn Clyd and Llyn Glas, two small oligotrophic high mountain tarns in Snowdonia (Wales). *New Phytologist*, **78**, 221–36.

Flower, R. J. (1986). The relationship between surface sediment diatom assemblages and pH in 33 Galloway lakes: Some regression models for reconstructing pH and their application to sediment cores. *Hydrobiologia*, **143**, 93–103.

Flower, R. J., & Battarbee, R. W. (1983). Diatom evidence for recent acidification of two Scottish lochs. *Nature*, **20**, 130–3.

(1985). The morphology and biostratigraphy of *Tabellaria quadriseptata* Knudson (Bacillariophyta) in acid waters and lake sediments in Galloway, south-west Scotland. *British Phycological Journal*, **20**, 69–79.

Flower, R. J., Battarbee, R. W., & Appleby, P. G. (1987). The recent palaeolimnology of acid lakes in Galloway, south-west Scotland: Diatom analysis, pH trends and the role of afforestation. *Journal of Ecology*, **75**, 797–824.

Flower, R. J., Cameron, N. G., Rose, N., Fritz, S. C., Harriman, R., & Stevenson, A. C. (1990). Post-1970 water chemistry changes and palaeolimnology of several acidified upland lakes in the U. K. *Philosophical Transactions of the Royal Society, London*, **B 327**, 427–33.

Flower, R. J., Juggins, S., & Battarbee, R. W. (1997). Matching diatom assemblages in lake sediment cores and modern surface samples: the implications for lake conservation and restoration with special reference to acidified systems. *Hydrobiologia*, **344**, 27–40.

Foged, N. (1977). *Freshwater Diatoms in Ireland*. Germany: A. R. Gantner Verlag KG, p. 221.

Ford, M. S. (1990). A 10,000 year history of natural ecosystem acidification. *Ecological Monographs*, **60**, 57–89.

Fritz, S. C., & Carlson, R. E. (1982). Stratigraphic diatom and chemical evidence for acid strip-mine lake recovery. *Water, Air and Soil Pollution*, **17**, 151–63.

Fritz, S. C., Juggins, S., Battarbee, R. W., & Engstrom, D. R. (1991). A diatom-based transfer function for salinity, water-level, and climate reconstruction. *Nature*, **352**, 706–8.

Fritz, S. C., Stevenson, A. C., Patrick, S. T., Appleby, P. G., Oldfield, F., Rippey, B., Natkanski, J., & Battarbee, R. W. (1989). Paleolimnological evidence for the recent acidification of Llyn Hir, Dyfed, Wales. *Journal of Paleolimnology*, **2**, 245–62.

Harriman, R., & Morrison, B. R. S. (1982). The ecology of streams draining forested and non-forested catchments in an area of central Scotland subject to acid precipitation. *Hydrobiologia*, **88**, 251–63.

Haworth, E. Y. (1969). The diatoms of a sediment core from Blea Tarn, Langdale. *Journal of Ecology*, **57**, 429–39.

Holmes, R. W., Whiting, M. C., & Stoddard, J. L. (1989). Changes in diatom-inferred pH and acid neutralizing capacity in a dilute, high elevation, Sierra Nevada lake since AD 1825. *Freshwater Biology*, **21**, 295–310.

Hustedt, F. (1937–1939). Systematische und ökologische Untersuchungen über den Diatomeen-Flora von Java, Bali, Sumatra. *Archiv für Hydrobiologie* (Suppl.), 15 & 16.

Huttunen, P., & Meriläinen, J. (1983). Interpretation of lake quality from contemporary diatom assemblages. In *Palaeolimnology* ed. J. Meriläinen, P. Huttunen, & R. W. Battarbee, pp. 91–7. The Hague: Junk.

Huttunen, P., & Turkia, J. (1990). Estimation of palaeoalkalinity from diatom assemblages by means of CCA. In *Proceedings of the 10th international diatom symposium*, pp. 443–50.

Huttunen, J., Meriläinen, J., Cotten, C., & Rönkkö, J. (1988). Attempts to reconstruct lake water pH and colour from sedimentary diatoms and Cladocera. *Verhandlungen Internationale Vereinigung für Theoretische und Angewandte Limnologie*, **23**, 870–3.

Huttunen, P., Meriläinen, J., & Tolonen, K. (1978). The history of a small dystrophied forest lake, Southern Finland. *Polskie Archiwum Hydrobiologii*, 25, 189–202.

Huvane, J. K., & Whitehead, D. R. (1996). The paleolimnology of North Pond: Watershed–lake interactions. *Journal of Paleolimnology*, **16**, 323–54.

Iversen, J. (1958). The bearing of glacial and interglacial epochs on the formation and extinction of plant taxa I. In *Systematics of Today* ed. O. Hedberg, pp. 210–15. Uppsala Universitets Årsskrift 6.

Jenkins, A., Renshaw, M., Helliwell, R., Sefton, C., Ferrier, R., & Swingewood, P. (1997). *Modelling Surface Water Acidification in the UK: Application of the MAGIC Model to the Acid Waters Monitoring Network*. IH Report No. 131, Institute of Hydrology, UK.

Jenkins, A., Whitehead, P. G., Cosby, B. J., & Birks, H. J. B. (1990). Modelling long-term acidification: A comparison with diatom reconstructions and the implications for reversibility. *Philosophical Transactions of the Royal Society, London*, B 327, 209–14.

Jensen, K. W., & Snekvik, E. (1972). Low pH level wipe out salmon and trout populations in southernmost Norway. *Ambio*, 1, 223–5.

Jones, V. J., & Flower, R. J. (1986). Spatial and temporal variability in periphytic diatom communities: Palaeoecological significance in an acidified lake. In *Diatoms and Lake Acidity*, ed. J. P. Smol, R. W. Battarbee, R. B. Davis, & J. Meriläinen, pp. 87–94. Dordrecht, The Netherlands: Dr. W. Junk.

Jones, V. J., Flower, R. J., Appleby, P. G., Natkanski, J., Richardson, N., Rippey, B., Stevenson, A. C., & Battarbee, R. W. (1993). Palaeolimnological evidence for the acidification and atmospheric contamination of lochs in the Cairngorms and Lochnagar areas of Scotland. *Journal of Ecology*, 81, 3–24.

Jones, V. J., Stevenson, A. C., & Battarbee, R. W. (1986). Lake acidification and the land-use hypothesis: A mid-post-glacial analogue. *Nature*, 322, 157–8.

(1989). Acidification of lakes in Galloway, south west Scotland: A diatom and pollen study of the post-glacial history of the Round Loch of Glenhead. *Journal of Ecology*, 77, 1–23.

Juggins, S., Flower, R. J., & Battarbee, R. W. (1996). Palaeolimnological evidence for recent chemical and biological changes in U. K. Acid Waters Monitoring Network sites. *Freshwater Ecology*, 36, 203–19.

Keller, W. (1992). Introduction and overview to aquatic acidification studies in the Sudbury, Ontario, Canada, area. *Canadian Journal of Fisheries and Aquatic Sciences*, 49 (Suppl. 1), 3–7.

Kingston, J. C., & Birks, H. J. B. (1990). Dissolved organic carbon reconstructions from diatom assemblages in PIRLA project lakes, North America. *Philosophical Transactions of the Royal Society, London*, B 327, 279–288.

Kingston, J. C., Cumming, B. F., Uutala, A. J., Smol, J. P., Camburn, K. E., Charles, D. F., Dixit, S. S., & Kreis, R. G. (1992a). Biological quality control and quality assurance: A case study in paleolimnological biomonitoring. In *Ecological Indicators*, ed. D. H. McKenzie, D. E. Hyatt, & V. J. McDonald, pp. 1542–3. London and New York: Elsevier Applied Science.

Kingston, J. C., Birks, H. J. B., Uutala, A. J., Cumming, B. F., & Smol, J. P. (1992b). Assessing trends in fishery resources and lake water aluminum from paleolimnological analyses of siliceous algae. *Canadian Journal of Fisheries and Aquatic Sciences*, 49, 116–27.

Kingston, J. C., Cook, R. B., Kreis, R. G. Jr., Camburn, K. E., Norton, S. A., Sweets, P. R., Binford, M. W., Mitchell, M. J., Schindler, S. C., Shane, L. C. K., & King, G. A. (1990). Paleoecological investigation of recent lake acidification in the northern Great Lakes states. *Journal of Paleolimnology*, 4, 153–201.

Korhola, A. A., & Tikkanen, M. J. (1991). Holocene development and early extreme acidification in a small hilltop lake in southern Finland. *Boreas*, 20, 333–56.

Korhola, A., Virkanen, J., Tikkanen, M., & Gilmour, B. S. (1996). Fire-induced pH rise in a naturally acid hilltop lake in southern Finland: A palaeoecological survey. *Journal of Ecology* 84, 257–66.

Korsman, T. (1993). Acidification trends in Swedish lakes: An assessment of past water conditions using lake sediments. PhD thesis, Umeå University.

Korsman, T., & Birks, H. J. B. (1996). Diatom-based water chemistry reconstructions from northern Sweden: A comparison of reconstruction techniques. *Journal of Paleolimnology*, 15, 65–77.

Korsman, T., & Segerström (1988). Forest fire and lake-water acidity in a northern Swedish boreal area: Holocene changes in lake-water qulaity at Makkassj(n. *Journal of Ecology,* **86**, 113–24.

Korsman, T., Renberg, I., & Anderson, N. J. (1994). A palaeolimnological test of the influence of Norway spruce (*Picea abies*) immigration on lake-water acidity. *The Holocene,* **4**, 132–40.

Kreiser, A. M., & Battarbee, R. W. (1988). Analytical Quality Control (AQC) in diatom analysis. *Proceedings of Nordic Diatomist Meeting,* University of Stockholm, Department of Quaternary Geology Research Report 12, pp. 41–44.

Kreiser, A. M., Appleby, P. G., Natkanski, J., Rippey, B., & Battarbee, R. W. (1990). Afforestation and lake acidification: A comparison of four sites in Scotland. *Philosophical Transactions of the Royal Society, London,* **B 327**, 377–83.

Krug, E. C., & Frink, C. R. (1983). Acid rain on acid soil: A new perspective. *Science,* **221**, 520–5.

Lancaster, J., Real, M, Juggins, S., Monteith, D. T., Flower, R. J., & Beaumont, W. R. C. (1996). Monitoring temporal changes in the biology of acid waters. *Freshwater Biology,* 179–202.

Line, J. M., & Birks, H. J. B. (1990). WACALIB version 2.1 – A computer program to reconstruct environmental variables from fossil assemblages by weighted averaging. *Journal of Paleolimnology,* **3**, 170–3.

Line, J. M., ter Braak, C. J. F., & Birks, H. J. B. (1994). WACALIB version 3.3 – A computer program to reconstruct environmental variables from fossil assemblages by weighted averaging and to derive sample-specific errors of prediction. *Journal of Paleolimnology,* **10**, 147–52.

Lundquist, G. (1924). Utvecklingshistoriska insjöstudier i Syd-sverige. *Sveriges Geologiska Undersökning* Ser. C **330**, 1–129.

Meriläinen, J. (1967). The diatom flora and the hydrogen ion concentration of the water. *Annales Botanici Fennici,* **4**, 51–8.

Miller, U. (1973). Diatoméundersökning av bottenproppar från Stora Skarsjön, Ljungskile. *Statens Naturvårdsverk Publikationer,* **7**, 43–60.

Monteith, D. T., Renshaw, M., Beaumont, W. R. C., & Patrick, S. T. (1997). *The United Kingdom Acid Waters Monitoring Network Data Report for 1995–1996 (Year 8).* Report to the Department of the Environment (Contract No. EPG 1/3/73) and the Department of the Environment Northern Ireland, London: ENSIS 134 pp.

Mosello, R., Bonacina, C., Carollo, A., Libera, V., & Tartari, G. A. (1986). Acidification due to in-lake ammonia oxidation: an attempt to quantify the proton production in a highly polluted subalpine Italian lake. *Memorie dell'Istituto Italiano di Idrobiologia,* **44**, 47–71.

Muniz, I. P., & Leivestad, H. (1980). Acidification – effects on freshwater fish. In *Proceedings of the International Conference on the Ecological Impacts of Acid Precipitation,* ed. D. Drabløs & A. Tollan, pp. 84–92. Oslo Norway: NSF Project.

Munro, M. A. R., Kreiser, A. M., Battarbee, R. W., Juggins, S., Stevenson, A. C., Anderson, D. S., Anderson, N. J., Berge, F., Birks, H. J. B., Davis, R. B., Flower, R. J., Fritz, S. C., Haworth, E. Y., Jones, V. J., Kingston, J. C., & Renberg, I. (1990). Diatom quality control and data handling. *Philosophical Transactions of the Royal Society, London,* **B 327**, 257–61.

Nilsson, J., & Grennfelt, P. (eds.) (1988). *Critical Loads for Sulphur and Nitrogen.* UNECE/Nordic Council workshop report, Skokloster, Sweden. March 1988. Copenhagen: Nordic Council of Ministers.

Nilsson, I. S., Miller, H. G., & Miller, J. D. (1982). Forest growth as a possible cause of soil and water acidification: An examination of the concepts. *Oikos,* **39**, 40–9.

Nygaard, G. (1956). Ancient and recent flora of diatoms and chrysophyceae in Lake Gribsö. Studies on the humic acid lake Gribsö. *Folia Limnologica Scandinavica,* **8**, 32–94.

Odén, S. (1968). *The Acidification of Air Precipitation and its Consequences in the Natural Environment Energy Committee Bulletin, 1.* Stockhom: Swedish Natural Sciences Research Council.

Oehlert, G. W. (1988). Interval estimates for diatom inferred lake pH histories. *Canadian Journal of Statistics,* **16**, 51–60.

Ohl, L. E., Gont, R. A., & Dibble, E. D. (1990). Diatom response to liming of a temperate, brown water lake. *Canadian Journal of Botany*, **68**, 347–53.

Patrick, S. T., Monteith, D. T., & Jenkins, A. (eds.) (1995). *UK Acid Waters Monitoring Network: The First Five Years. Analysis and Interpretation of Results April 1988–March 1993*. London: ENSIS.

Patrick, S. T., Battarbee, R. W., & Jenkins, A. (1996). Monitoring acid waters in the U. K.: An overview of the U. K. Acid Waters Monitoring Network and summary of the first interpretative exercise. *Freshwater Biology*, **36**, 131–50.

Pennington, W. (1984). Long-term natural acidification of upland sites in Cumbria: Evidence from post-glacial lake sediments. *Freshwater Biological Association Annual Report*. **52**, 28–46.

Pennington, W., Haworth, E. Y., Bonny, A. P., & Lishman, J. P. (1972). Lake sediments in northern Scotland. *Philosophical Transactions of the Royal Society, London*, **B 264**, 191–294.

Psenner, R., & Schmidt, R. (1992). Climate-driven pH control of remote alpine lakes and effects of acid deposition. *Nature*, **356**, 781–3.

Renberg, I. (1976). Palaeolimnological investigations in Lake Prästsjön. *Early Norrland*, **9**, 113–60.

(1978). Palaeolimnology and varve counts of the annually laminated sediment of Lake Rudetjärn, northern Sweden. *Early Norrland*, **11**, 63–92.

(1986). A sedimentary diatom record of severe acidification in Lake Blåmissusjön, N. Sweden, through natural processes. In *Diatoms and Lake Acidity*, ed. J. P. Smol, R. W. Battarbee, R. B. Davis, & J. Meriläinen, pp. 213–19. Dordrecht, The Netherlands: Dr. W. Junk.

(1990). A 12,600 year perspective of the acidification of Lilla Öresjön, southwest Sweden. *Philosophical Transactions of the Royal Society, London*, **B 327**, 357–61.

Renberg, I., & Battarbee, R. W. (1990). The SWAP Palaeolimnology Programme: A synthesis. In *The Surface Waters Acidification Programme*, ed. B. J. Mason, pp. 281–300. Cambridge: Cambridge University Press, Cambridge.

Renberg, I., & Hellberg, T. (1982). The pH history of lakes in southwestern Sweden, as calculated from the subfossil diatom flora of the sediments. *Ambio*, **11**, 30–3.

Renberg, I., & Hultberg, H. (1992). A paleolimnological assessment of acidification and liming effects on diatom assemblages in a Swedish lake. *Canadian Journal of Fisheries and Aquatic Sciences*, **49**, 65–72.

Renberg, I., Korsman, T., & Anderson, N. J. (1990). Spruce and surface water acidification: An extended summary. *Philosophical Transactions of the Royal Society, London*, **B 327**, 371–2.

(1993a). A temporal perspective of lake acidification in Sweden. *Ambio*, **22**, 264–71.

Renberg, I., Korsman, T., & Birks, H. J. B. (1993b). Prehistoric increases in the pH of acid-sensitive Swedish lakes caused by land-use changes. *Nature*, **362**, 824–6.

Rhodes, T. E. (1991). A paleolimnological record of anthropogenic disturbances at Holmes Lake, Adirondack Mountains, New York. *Journal of Paleolimnology*, **5**, 255–62.

Rhodes, T. E., & Davis, R. B. (1995). Effects of late Holocene forest disturbance and vegetation change on acidic Mud Pond, Maine, USA. *Ecology*, **76**, 734–46.

Rosenqvist, I. T. (1977). *Acid soil – acid water*. Oslo: Ingeniörforlaget.

(1978). Alternative sources for acidification of river water in Norway. *The Science of the Total Environment*, **10**, 39–49.

Round, F. E. (1957). The late-glacial and post-glacial diatom succession in the Kentmere Valley deposit: I. Introduction, Methods and Flora. *New Phytologist*, **56**, 98–126.

(1961). Diatoms from Esthwaite. *New Phytologist*, **60**, 98–126.

(1990). The effect of liming on the benthic diatom populations in three upland Welsh lakes. *Diatom Research*, **5**, 129–40.

(1991). Epilithic diatoms in acid water streams flowing into the reservoir Llyn Brianne. *Diatom Research*, **6**, 137–45.

Ryves, D. B. (1994). Diatom dissolution in saline lake sediments: An experimental study in the Great Plains of North America. Unpublished PhD thesis, University of London.

Salomaa, R., & Alhonen, P. (1983). Biostratigraphy of Lake Spitaalijärvi: an ultraoligotrophic small lake in Lauhanvuori, western Finland. *Hydrobiologia*, **103**, 295–301.

Steinberg, C., & Putz, R. (1991). Epilithic diatoms as bioindicators of stream acidification. *Verhandlungen der Internationalen Vereinigung für Theoretische und Angewandte Limnologie*, **24**, 1877–80.

Steinberg, C., Hartmann, H., Arzet, K., & Krause-Dellin, D. (1988). Paleoindication of acidification in Kleiner Arbersee (Federal Republic of Germany, Bavarian Forest) by chydorids, chrysophytes, and diatoms. *Journal of Paleolimnology*, **1**, 149–57.

Stevenson, A. C., Birks, H. J. B., Flower, R. J., & Battarbee, R. W. (1989). Diatom-based pH reconstruction of lake acidification using canonical correspondence analysis. *Ambio*, **18**, 228–33.

Stevenson, A. C., Juggins, S., Birks, H. J. B., Anderson, D. S., Anderson, N. J., Battarbee, R. W., Berge, F., Davis, R. B., Flower, R. J., Haworth, E. Y., Jones, V. J., Kingston, J. C., Kreiser, A. M., Line, J. M., Munro, M. A. R., & Renberg, I. (1991). *The Surface Waters Acidification Project Palaeolimnology Programme: Modern Diatom/Lake-Water Chemistry Data-Set.* London: Ensis Ltd, 86pp.

Stoner, J. H., Gee, A. S., & Wade, K. R. (1984). The effects of acidification on the ecology of streams in the upper Tywi catchment in west Wales. *Environmental Pollution*, **35**, 125–57.

Sullivan, T. J. (1990). Historical changes in surface water acid–base chemistry in response to acidic deposition. In *National Acid Precipitation Assessment Program State of Science and Technology Report 11.* Washington, DC.

Sullivan, T. J., Charles, D. F., Smol, J. P., Cumming, B. F., Selle, A. R., Thomas, D., Bernert, J. A., & Dixit, S. S. (1990). Quantification of changes in lakewater chemistry in response to acidic deposition. *Nature*, **345**, 54–8.

Sullivan, T. J., Cosby, B. J., Driscoll, C. T., Charles, D. F., & Hemond, H. (1996). Influence of organic acids on model projections of lake acidification. *Water, Air and Soil Pollution*, **91**, 271–82.

Sullivan, T. J., McMartin, B., & Charles, D. F. (1996). Re-examination of the role of landscape change in the acidification of lakes in the Adirondack Mountains, New York. *The Science of the Total Environment*, **183**, 231–48.

Sullivan, T. J., Turner, R. S., Charles, D. F., Cumming, B. F., Smol, J. P., Schofield, C. S., Driscoll, C. T., Cosby, B. J., Birks, H. J. B., Uutala, A. J., Kingston, J. C., Dixit, S. S., Bernert, J. A., Ryan, P. F., & Marmorek, D. R. (1992). Use of historical assessment for evaluation of process-based model projections of future environmental change – lake acidification in the Adirondack Mountains, New York, USA. *Environmental Pollution*, **77**, 253–262.

Sweets, P. R., Bienert, R. W., Crisman, T. L., & Binford, M. W. (1990). Paleoecological investigations of recent lake acidification in northern Florida. *Journal of Paleolimnology*, **4**, 103–37.

ter Braak, C. J. F. (1987a). CANOCO – a FORTRAN program for canonical community ordination by [partial] [detrended] [canonical] correspondence analysis, principal components analysis and redundancy analysis (version 2.1) ITI-TNO, Wageningen, 95 pp.

(1987b). Unimodal models to relate species to environment. Unpublished PhD thesis, University of Wageningen.

ter Braak, C. J. F., & Juggins, S. (1993). Weighted averaging partial least squares regression (WA-PLS): An improved method for reconstructing environmental variables from species assemblages. *Hydrobiologia*, **269/270**, 485–502.

ter Braak, C. J. F., & van Dam, H. (1989). Inferring pH from diatoms: a comparison of old and new calibration methods. *Hydrobiologia*, **178**, 209–23.

Turner, M. A., Howell, E. T., Summerby, M., Hesslein, R. H., Findlay, D. L., & Jackson, M. B. (1991). Changes in epilithon and ephiphyton associated with experimental acidification of a lake to pH 5. *Limnology and Oceanography*, **36**, 1390–405.

Unsworth, M. H. (1984). Evaporation from forests in cloud enhances the effect of acid deposition. *Nature*, **312**, 262–4.

Uutala, A. J. (1990). *Chaoborus* (Diptera: Chaoboridae) mandibles – paleolimnological indicators of the historical status of fish populations in acid sensitive lakes. *Journal of Paleolimnology*, **4**, 139–51.

van Dam, H. (1988). Acidification of three moorland pools in The Netherlands by acid precipitation and extreme drought periods over seven decades. *Freshwater Biology*, **20**, 157–76.

(1996). Partial recovery of moorland pools from acidification: indications by chemistry and diatoms. *Netherlands Journal of Aquatic Ecology*, **30**, 203–18.

van Dam, H., & Kooyman-van Blokland, H. (1978). Man-made changes in some Dutch moorland pools, as reflected by historical and recent data about diatoms and macrophytes. *Internationale Revue gesamte Hydrobiologie*, **63**, 587–607.

van Dam, H., & Mertens, A. (1990). A comparison of recent epilithic diatom assemblages from the industrially acidified and copper polluted Lake Orta (Northern Italy) with old literature data. *Diatom Research*, **5**, 1–13.

(1995). Long-term changes of diatoms and chemistry in headwater streams polluted by atmospheric deposition of sulfur and nitrogen compounds. *Freshwater Biology*, **34**, 579–600.

van Dam, H., Suurmond, G., & ter Braak, C. J. F. (1981). Impact of acidification on diatoms and chemistry of Dutch moorland pools. *Hydrobiologia*, **83**, 425–59.

van Dam, H., Houweling, H., Wortelboer, F. G., & Erisman, J. W. (1996). Long-term changes of chemistry and biota in moorland pools in relation to changes of atmospheric deposition. AquaSense TEC, Wageningen/DLO-Institute for Forestry and Nature Research (IBN-DLO), Wageningen/National Institute for Health and Environmental Protection, Bilthoven.

Vinebrooke, R. D. (1996). Abiotic and biotic regulation of periphyton in recovering acidified lakes. *Journal of the North American Benthological Society*, **15**, 318–31.

Virkanen, J., Korhola, A., Tikkanen, M., & Blom, T. (1997). Recent environmental changes in a naturally acidic rocky lake in southern Finland, as reflected in its sediment geochemistry and biostratigraphy. *Journal of Paleolimnology*, **17**, 191–213.

Vyverman, W., Vyverman, R., Rajendran, V. S., & Tyler, P. (1996). Distribution of benthic diatom assemblages in Tasmanian highland lakes and their possible use as indicators of environmental changes. *Canadian Journal of Fisheries and Aquatic Sciences*, **53**, 493–508.

Whitehead, D. R., Charles, D. F., Jackson, S. T., Smol, J. P., & Engstrom, D. R. (1989). The developmental history of Adirondack (N.Y.) lakes. *Journal of Paleolimnology*, **2**, 185–206.

Whitehead, D. R., Charles, D. F., Reed, S. E., Jackson, S. T., & Sheehan, M. C. (1986). Late-glacial and holocene acidity changes in Adirondack (NY) lakes. In *Diatoms and Lake Acidity*, ed. J. P. Smol, R. W. Battarbee, R. B. Davis, & J. Meriläinen, pp. 251–74. Dordrecht, The Netherlands: Dr. W. Junk.

Whiting, M. C., Whitehead, D. R., Holmes, R. W., & Norton, S. A. (1989). Paleolimnological reconstruction of recent acidity changes in four Sierra Nevada lakes. *Journal of Paleolimnology*, **2**, 285–304.

Williams, D. M., Hartley, B., Ross, R., Munro, M. A. R., Juggins, S., & Battarbee, R. W. (1988). *A Coded Checklist of British Diatoms.* London: ENSIS Publishing.

Winkler, M. G. (1988). Paleolimnology of a Cape Cod Kettle Pond: Diatoms and reconstructed pH. *Ecological Monographs*, **58**, 197–214.

Yoshitake, S., & Fukushima, H. (1994). Distribution of attached diatoms in inorganic acid lakes in Japan. In *Proceedings of the Thirteenth International Diatom Symposium*, ed. D. Marino & M. Montresor, pp. 321–33. Bristol: Biopress Limited.

6 Diatoms as indicators of lake eutrophication

ROLAND I. HALL AND JOHN P. SMOL

Introduction

Eutrophication refers to enrichment of aquatic systems by inorganic plant nutrients (Wetzel, 1983; Mason, 1991). Lake eutrophication occurs when nutrient supplies, usually phosphorus (P) and nitrogen (N), are elevated over rates that occur in the absence of any system perturbation, and results in increased lake productivity. Causes of eutrophication include human (anthropogenic eutrophication) and non-human (natural eutrophication) disturbances. Marked natural eutrophication events are rare and may result from dramatic episodes, such as forest fire (e.g., Hickman *et al.*, 1990) and tree die-off (Boucherle *et al.*, 1986; Hall & Smol, 1993). Climatic shifts, such as droughts, may also concentrate lakewater nutrients or give rise to an increased contribution of nutrient-rich groundwater (e.g., Webster *et al.*, 1996). In most cases, however, water-quality problems are caused by anthropogenic nutrient inputs from domestic and industrial sewage disposal, farming activities and soil erosion.

Eutrophication is the most widespread form of lake pollution on a global scale, and has many deleterious impacts on aquatic systems (Harper, 1992). In addition to increasing overall primary production, eutrophication causes considerable changes to biochemical cycles and biological communities. Marked changes occur at all levels in the food web and entire communities can change or die out. For example, changes in the ratio of N:P often results in primary production shifting from primarily diatoms and other smaller edible algae towards larger cyanobacteria that are better competitors for N (Tilman *et al.*, 1986), and more resistant to grazing (Reynolds, 1984). As the light climate changes with increased phytoplankton turbidity, many shallow lakes may lose their submerged aquatic macrophyte communities, which in turn impact higher trophic levels (e.g., fish, waterfowl). Decomposition of plant and algal biomass reduces oxygen availability in deep waters, causing declines in available fish habitat and, under extreme conditions, massive fish kills. A major concern is that lake ecosystems may become unstable and biodiversity may be lost during eutrophication (Margalef, 1968).

Eutrophication is a costly economic problem. For example, massive algal blooms increase water treatment costs and sometimes cause treatment facilities to malfunction (Vaughn, 1961; Hayes & Greene, 1984). Furthermore, algal blooms (including diatoms) can create taste and odor problems (Mason, 1991), and algal breakdown products may chelate with iron and aluminum to increase metal contamination of drinking water (Hargesheimer & Watson, 1996). Toxins produced by cyanobacteria (e.g., microcystin; Kotak *et al.*, 1995) pose risks to human health, livestock and wildlife.

In many regions of the world, humans have been impacting lakes for a long time and over a variety of time scales. For example, in agricultural regions of Europe, Asia and Africa, lakes may have been disturbed over long time scales (centuries to millennia) in response to forest clearance and the onset of agriculture (e.g., Fritz, 1989). In many regions, however, major human impacts became noticeable more recently (e.g., last 50–150 years).

The long history of human impacts complicates remediation efforts in at least two ways. First, internal nutrient loads build up in lake sediments over time, causing delays in recovery following removal of external nutrient sources (e.g., Marsden, 1989). Second, long-term data are unavailable for most lakes over the full timescale of human impacts (usually <20 years; Likens, 1989). We rarely undertake extensive studies until 'after the fact' (Smol, 1992, 1995). None the less, preventative as well as restorative management programmes require detailed guidelines as to how much ecological and water-quality change has occurred (i.e., to set realistic target conditions). Consequently, alternative approaches must be developed to assess when individual lakes began to change, what the causal mechanisms were, and what the natural productivity and variability was before impact.

Fortunately, lakes accumulate vast amounts of ecological and chemical information in their deep-water sediments which can be used to reconstruct past changes in lake ecosystems (Anderson & Battarbee, 1994). Diatoms are well preserved in most lake sediments and are useful bioindicators in applied eutrophication studies.

In this chapter, we focus on the use of diatom algae for applied eutrophication studies in freshwater lakes. We further focus on paleolimnological applications. Our review is primarily on lake studies, as other chapters in this book discuss different aquatic environments (e.g., streams and rivers (Stevenson & Pan, this volume), marine and estuarine ecosystems (Cooper, this volume; Denys & de Wolf, this volume; Snoeijs, this volume; Sullivan, this volume).

Why diatoms are useful indicators of lake eutrophication

Diatoms are an abundant, diverse and important component of algal assemblages in freshwater lakes. They comprise a large portion of total algal biomass over a broad spectrum of lake trophic status (Kreis *et al.*, 1985). Because diatoms are an abundant, high-quality food source for herbivores, they may play an important role in aquatic food-web structure and function (Round *et al.*, 1990).

As a consequence, responses of diatoms to lake eutrophication can have important implications for other components of aquatic ecosystems.

While diatoms collectively show a broad range of tolerance along a gradient of lake productivity, individual species have specific habitat and water-chemistry requirements (Patrick & Reimer, 1966; Werner, 1977; Round *et al.*, 1990). In addition, distinct diatom communities live in open waters of lakes (plankton), or primarily in association with plants (epiphyton), rocks (epilithon), sand (epipsammon) or mud (epipelon) in littoral, nearshore habitats. Consequently, diatoms can be used to track shifts from littoral to planktonic algal production and macrophyte declines which may occur during eutrophication (Osborne & Moss, 1977).

Diatoms are well suited to studies of lake eutrophication, because individual species are sensitive to changes in nutrient concentrations, supply rates and ratios (e.g., Si:P; Tilman, 1977; Tilman *et al.*, 1982). Each taxon has a specific optimum and tolerance for nutrients which can usually be quantified to a high degree of certainty (e.g., P: Hall & Smol, 1992; Reavie *et al.*, 1995; Fritz *et al.*, 1993; Bennion, 1994, 1995; Bennion *et al.*, 1996; N: Christie & Smol 1993). This ability to quantify responses of individual taxa to nutrient concentrations has provided diatomists with a powerful tool for quantifying environmental changes that accompany eutrophication and recovery.

Diatom assemblages are typically species-rich. It is common to find over a hundred taxa in a single sediment sample. This diversity of diatoms contains considerable ecological information concerning lake eutrophication. Moreover, the large numbers of taxa provide redundancies of information and important internal checks in datasets which increase confidence of environmental inferences (Dixit *et al.*, 1992).

Diatoms respond rapidly to eutrophication and recovery (e.g., Zeeb *et al.*, 1994). Because diatoms are primarily photoautotrophic organisms, they are directly affected by changes in nutrient and light availability (Tilman *et al.*, 1982). Their rapid growth and immigration rates and the lack of physical dispersal barriers ensure there is little lag-time between perturbation and response (Vinebrooke, 1996). Consequently, diatoms are early indicators of environmental change.

The taxonomy of diatoms is generally well documented (Hustedt, 1930; Patrick & Reimer, 1966, 1975; Krammer & Lange-Bertalot, 1986–1991). Species identifications are largely based on cell wall morphology. Because the cell wall is composed of resistant opaline silica (Si), diatom valves are usually well preserved in most samples, including lake sediments. Consequently, by taking sediment cores and analyzing diatom assemblages, it is possible to infer past environmental conditions using paleolimnological techniques.

Below, we present an overview of approaches that employ diatoms to study lake eutrophication, and then illustrate the contributions of diatoms to our knowledge base by presenting applications, or case studies, of the different approaches.

Approaches

Lake eutrophication operates at different spatial and temporal scales. Consequently, scientists and lake managers must rely on a number of different approaches to study the effects of nutrient enrichment and to identify the best possible management strategies. Applied diatom studies generally fall into two broad categories – experiments and field observations. Some studies effectively utilize both approaches (e.g., Tilman *et al.*, 1986).

EXPERIMENTS

Experimental approaches have been useful for documenting the importance of resource limitation as a mechanism determining algal species composition in lakes, and for identifying the specific resource most strongly limiting growth of diatom species (e.g., Tilman *et al.*, 1982). In applied studies, often the first piece of information lake managers need to know is exactly which resource (e.g., nutrient) most strongly limits nuisance algal growth, in order to devise effective control strategies.

Controlled experiments can be performed in a laboratory or in lakes, and can vary in scale from small bottles and chemostats (e.g., Schelske *et al.*, 1974; Tilman, 1977), to mesocosms (e.g., Schelske & Stoermer, 1971, 1972; Lund & Reynolds, 1982), and even to entire lakes (e.g., Schindler, 1977; Yang *et al.*, 1996). In general, larger-scale experiments provide information that is more realistic of aquatic ecosystems, because they may include important ecological processes, such as biogeochemical exchanges with the sediments and the atmosphere, and food-web interactions. Smaller-scale experiments, however, provide more precise experimental control, and are cheaper and easier to perform, but they are oversimplified systems compared to real lakes (Schindler, 1977). Whole-lake experiments provide the most realistic responses of aquatic communities to eutrophication, but the rate of eutrophication is often greatly accelerated above rates at which anthropogenic eutrophication usually operate.

One important contribution of experimental bioassay research has been the production of detailed physiological data for individual diatom species (e.g., Tilman *et al.*, 1982). By combining physiological data with resource-based competition models, researchers have begun to provide an important bridge between modern limnological studies and paleolimnology (sediment-based studies), because they can provide a mechanistic link to interpret observed species shifts in lakes (e.g., Kilham *et al.*, 1996). However, resource competition theory and models will be useful to applied eutrophication studies only after detailed physiological information becomes available for most of the common diatom species, and when experiments have been performed across the full range of nutrient concentrations (and nutrient ratios) that accompany lake eutrophication.

Field observation is a widely used approach to gather ecological information concerning distributions of individual diatom species and entire assemblages in lakes. Field observation studies usually involve lake surveys along ecological gradients to describe correlative relationships between species composition and water chemistry. While these studies do not measure the causal factors to which diatoms respond, they provide information on the net outcome of diatom assemblages to the complex interactions of all important control factors. Consequently, field surveys provide a useful approach for describing nutrient and habitat preferences of diatom species along a gradient of lake productivity.

One example of the above approach is the survey of planktonic diatoms along a nutrient gradient in Lake Michigan by Stoermer and Yang (1970). They related the distributions of diatom species in algae samples collected over ~100 years from polluted and unpolluted sites with measured limnological variables. Using this approach, Stoermer and Yang were able to identify dominant diatom taxa characteristic of highly oligotrophic open-lake environments, as well as eutrophic small-basin environments. Ecological generalizations provided by this survey have been extremely useful for interpreting past and present lake trophic status in many other regions (Bradbury, 1975).

With the development of multivariate statistics (i.e., canonical ordination) and weighted-averaging numerical approaches that quantify and assess the statistical significance of species–environment relationships, the use of lake surveys has increased (Charles & Smol, 1994; Charles *et al.*, 1994). In particular, large surveys of surface sediment diatom assemblages along productivity gradients have provided considerable ecological information concerning the relationships between diatoms and nutrients (e.g., Hall & Smol, 1992; Bennion, 1994; Fig. 6.1).

Long-Term Monitoring Programs (LTMPs) of lakes are increasingly used to monitor regional and national water quality, with the aim to assess and provide an early warning for eutrophication and other water-quality problems (Smol, 1990). Water chemistry forms the foundation of most LTMPs, but biological indicators such as diatoms are increasingly incorporated into these programs (e.g., US Environmental Protection Agency's Environmental Monitoring and Assessment Program; Whittier & Paulsen, 1992). This is because most LTMPs work on a tiered system, with a few sites monitored intensively (e.g., weekly), others less frequently (e.g., seasonally), and the majority of lakes sampled infrequently (e.g., ≤ once a year). Because much of the information collected by LTMPs are based on a few 'snapshots' of the system, it is important to use biological assemblages such as diatoms. Diatoms integrate environmental conditions over a longer time frame than a single water-chemistry sample, and can integrate information from several different habitats (e.g., surface sediment samples integrate diatom assemblages living

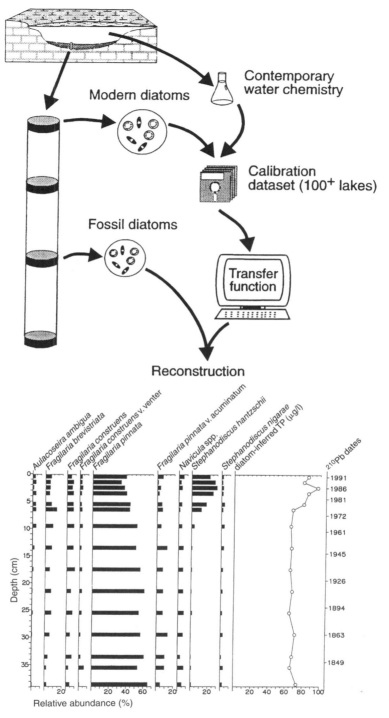

Fig. 6.1. Outline of the modern training set approach and development of quantitative diatom transfer functions (modified from Juggins *et al.*, 1995).

Diatoms as indicators of lake eutrophication

in littoral and pelagic habitats, over the previous few years). Consequently, the cost-to-benefit ratio is low for LTMPs using diatoms.

Responses of periphytic diatoms to eutrophication

Diatoms are often an important component of algal communities growing attached to substrata, known as the periphyton. Most experimental and field studies have focused on planktonic algae, and responses of periphytic diatom communities to eutrophication are less well understood (Lowe, 1996). However, periphyton can dominate primary production in many lakes (Hargrave, 1969; Wetzel, 1964; Wetzel *et al.*, 1972; Anderson 1990*a*), especially the numerous small, shallow lakes (i.e., where a large portion of the bottom receives sufficient light for photosynthesis; Wetzel, 1996) that dominate many landscapes. Consequently, periphytic diatom communities may provide information concerning littoral zone responses to eutrophication.

The use of periphytic diatoms in applied diatom studies has been hampered in part because mechanisms regulating periphyton are still poorly understood (Burkhardt, 1996; Lowe, 1996). Nutrient availability plays a strong role in determining the quantity and distribution of periphyton, but periphyton are subject to complex, multiple controls (Fairchild & Sherman, 1992, 1993; Niederhauser & Schanz, 1993). For example, steep resource gradients of space and light availability, water turbulence, and grazing (Lowe & Hunter, 1988) are important factors, but may not respond in a predictable or direct manner to eutrophication. Substratum and algal mat chemistry appears to play a stronger role on periphyton communities than water column chemistry (Cattaneo, 1987; Bothwell, 1988; Hansson, 1988, 1992), and chemical conditions may be strikingly different between the water column and algal mats (Revsbeck *et al.*, 1983; Revsbeck & Jørgensen, 1986). Consequently, periphyton may respond weakly or indirectly to water column nutrient additions, especially in meso- to eutrophic systems where nutrient limitation of periphyton may not be severe (Hansson, 1992; Fairchild & Sherman, 1992; Turner *et al.*, 1994). Lowe (1996) and Burkhardt (1996) review factors regulating benthic algal communities, and Bennion (1995) discusses associated problems for diatom-P transfer functions.

In some instances, periphytic communities may actually respond more rapidly to eutrophication than phytoplankton (Kann & Falter, 1989). Goldman (1981) was able to relate increased nutrient loading into oligotrophic Lake Tahoe, USA to increases in benthic algal biomass. Significant increases in periphytic diatoms were also observed in the littoral zone of Lake Taupo, New Zealand following sewage loading (Hawes & Smith, 1992). However, the effects of P enrichment on periphyton communities is not always predictable – sometimes it results in dominance by cyanobacteria (e.g., Fairchild *et al.*, 1985), but not always (Stockner & Shortreed, 1978). Nitrogen limitation appears to favor some diatom species (e.g., *Epithemia* spp., *Rhopalodia gibba* (Ehrenberg) O.

Müller; Fairchild *et al.*, 1985), presumably due in part to the ability of their cyanobacterial endosymbionts to fix atmospheric N (DeYoe *et al.*, 1992).

Periphytic diatoms have supplied information concerning changes in littoral zone conditions during eutrophication and recovery (e.g., Anderson, 1990a). Abundances of some epiphytic diatoms, for example *Achnanthes minutissima* Kützing, may be used to assess changes in littoral vegetation. For example, in a paleolimnological study of a lake in Norfolk, UK, Moss (1978) used changing abundances of epiphytic diatom taxa in a sediment core to track past changes in macrophyte abundance as a consequence of cultural activities. Under extreme eutrophication, periphytic algae can be reduced due to shading by phytoplankton (e.g., Björk-Ramberg & Änell, 1985; Hansson, 1992).

Using diatoms to investigate past eutrophication

Water-quality management and lake restoration have become important aspects of environment and natural resource agencies. The susceptibility of lakes to eutrophication, however, varies with a number of site- and region-specific factors (e.g., suitability of the catchment for agriculture, depth and quality of soil, industrialization and urbanization, geology, climate, lake depth, natural trophic status), and the magnitude and duration of past pollution. The timescales of eutrophication also vary enormously between lakes, from years to millennia (Fritz, 1989; Anderson 1995a,b). Consequently, the lack of long-term data for most lakes (>20 years; Likens, 1989) creates a clear need for approaches which can assess past environmental conditions. For example, scientists and aquatic managers need to know if, when, and how much a lake has changed through time (Smol, 1992). They must also establish the pre-impact conditions and natural variability in order to set realistic goals for restoration and management decisions (Ford, 1988; Smol, 1992, 1995). Research that quantifies the responses of a variety of lake types to past nutrient pollution (which varied in duration, nutrient load, and nutrient combinations) will undoubtedly be invaluable for predicting lake responses to likely future scenarios (Smol, 1992). It is in this area that sedimentary diatom records and paleolimnology can be particularly valuable (Smol, 1992; Anderson, 1993).

Our ability to infer eutrophication trends quantitatively using diatoms has increased greatly during the past decade or so (Hall & Smol, 1992; Anderson, 1993). Prior to the 1980s, most diatom-based environmental assessments of lake eutrophication were qualitative, based on the ecological interpretation of shifts in the abundance of individual species in sediment cores (e.g., Bradbury, 1975; Battarbee, 1978). Information on ecological preferences of individual indicator taxa were largely gleaned from contemporary phycological surveys that described patterns in diatom distributions among lakes of differing productivity (e.g., Stoermer & Yang, 1970), or anecdotal descriptions in taxonomic books (e.g., Hustedt, 1930; Patrick & Reimer, 1966, 1975; Cholnoky, 1968). Using this approach, investigators were able to infer, albeit qualitatively, anthropogenic eutrophication trajectories in lakes and correlate them with known changes in sewage inputs and land-use activities, such as agriculture, road construction and urbanization (e.g., Bradbury, 1975; Smol & Dickman, 1981; Engstrom *et al.*, 1985). An historical overview of some of the developments has been summarized by Battarbee (1986).

The development of diatom indices was among the first attempts to develop more rigorous evaluations of lake eutrophication from sedimentary diatom assemblages. For example, Nygaard (1949) proposed that the number of planktonic Centrales species to the number of planktonic Pennales species, known as the C:P ratio, could be used as an index of lake productivity. On a broad scale, Nygaard considered Centrales to indicate eutrophic conditions, whereas Pennales were more eurytopic, although he was aware that some taxa do not strictly follow this classification. Later, Stockner and Benson (1967) observed decreases in the relative abundances of the centric diatoms 'Stephanodiscus astrea var. minutula' (Kützing) Grunow and *Aulacoseira italica* (Ehrenberg) Simonsen (then *Melosira italica*), and increases in the araphidinate diatom *Fragilaria crotonensis* Kitton in Lake Washington, in response to increasing sewage discharge from Seattle to the lake. Contrary to Nygaard's original C:P ratio, and on the basis of observations in Lake Washington and other temperate lakes, Stockner (1971, 1972) proposed that the ratio of the two planktonic diatom groups, Araphidinieae and Centrales (the A:C ratio), could be used as an index of eutrophication.

Centrales:Penales and A:C ratios were overly simplistic, and it was not long before investigators commonly found situations that contradicted these diatom indices (e.g., Brugam, 1979; Brugam & Patterson, 1983). The shortcoming of diatom ratios is that ecological requirements of diatoms do not strictly follow frustule morphology. For example, some centric diatoms (e.g., *Stephanodiscus hantzschii* Grunow, *Cyclostephanos tholiformis* Stoermer, Håkansson & Theriot, *Aulacoseira granulata* (Grunow) Simonsen) are reliable indicators of eutrophic conditions (e.g., Brugam, 1979; Hall & Smol, 1992; Bennion, 1994, 1995; Reavie *et al.*, 1995), whereas other centrics are more common in oligotrophic lakes (e.g., *Cyclotella ocellata* Pantocsek, *C. kuetzingiana* Thwaites, *C. comensis* Grunow, *A. distans* (Grunow) Simonsen; Bradbury, 1975; Hall & Smol, 1992, 1996). Similarly, araphate and pennate diatoms are common across the full spectrum of lake trophic status, from oligotrophic (e.g., *Synedra filiformis* Cleve-Euler, *S. ulna* var. *chaseana* Thomas; Bradbury, 1975) to eutrophic conditions (e.g., *Fragilaria capucina* Desmazières and *F. capucina* var. *mesolepta* (Rabenhorst) Rabenhorst; Bradbury, 1975; Reavie *et al.*, 1995). Quite simply, too much ecological information is lost by clumping diatoms into just two categories. Consequently, it was not long before the widespread use of these indices came to a halt.

The next stage in the development of quantitative methods for reconstructing lake eutrophication employed linear regression models (Agbeti & Dickman, 1989; Whitmore, 1989). These models involved grouping diatom taxa into a number of ecological categories, based either on information from literature sources (Agbeti & Dickman, 1989) or on their present-day distributions in surface sediment calibration lakes (Whitmore, 1989), followed by the development of indices based on these categories. While linear regression models were useful, they had several shortcomings (ter Braak & van Dam, 1989; Birks *et al.*, 1990*a*). These models assumed a linear relationship between groups of diatom taxa and trophic status, which was inconsistent with observations that species respond to environmental gradients in a unimodal fashion (Gause, 1930; Whittaker, 1956; ter Braak & van Dam, 1989). The coefficients of multiple linear regression equations were often unstable, because the abundances of ecologically similar taxa tended to be highly correlated (Montgomery & Peck, 1982; ter Braak & van Dam, 1989). Furthermore, ecological information was lost when a large number of taxa were clumped into categories. Finally, the relationships between the indices generated by this approach (e.g., DITI: Agbeti & Dickman, 1989; TROPH 1: Whitmore, 1989) and water quality (e.g., chlorophyll *a*, total P (TP) concentration) could not be easily communicated to lake managers.

MODERN QUANTITATIVE METHODS

Recent advances in methodology and statistical techniques have vastly improved our ability to infer quantitatively lakewater chemistry, and hence eutrophication trends, from diatoms (Birks *et al.*, 1990*a,b*; Stevenson *et al.*, 1991; Charles & Smol, 1994). Instead of relying on individual indicator species, or ratios and ecological categories of just a few diatom taxa, we can now generate and utilize more precise ecological data for all the common diatom species within an assemblage to quantify past and contemporaneous changes in lake trophic status.

An important first step is to obtain quantitative data on the ecological optima and tolerances of diatoms along a gradient of lake trophic status. These data are usually generated from datasets of surface sediment diatom assemblages and corresponding present-day limnological conditions for a set of lakes, typically 50 or more in a given region – in general the more the better (Charles & Smol, 1994) (Fig. 6.1). These so-called training sets, or calibration sets, should span the environmental gradients of interest with multiple water-chemistry measurements sufficient to characterize average annual or seasonal conditions (e.g., spring-overturn or ice-free period). The surface sediment samples (usually top 0.5 to 1.0 cm) represent the last few years' accumulation of diatoms deposited from a variety of habitats within the lake. Typically, multivariate direct–gradient (or canonical) ordination (e.g., canonical correspondence analysis (CCA) or redundancy analysis (RDA)) is used to explore diatom-environment relationships in lake training sets. In particular, these statistical methods allow investigators to determine which environmental factors are most strongly correlated with species distributions (see also Battarbee *et al.*, this volume).

Using a diatom training set from 41 lakes in Michigan, Fritz *et al.* (1993)

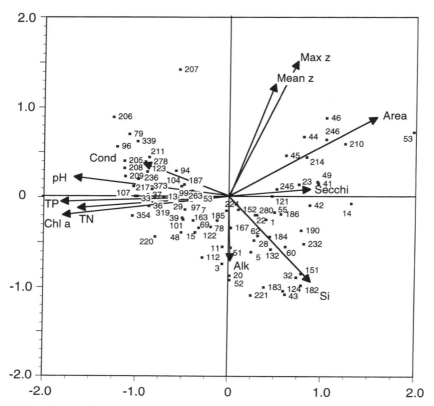

Fig. 6.2. Canonical correspondence analysis plot of diatom species in 41 Michigan lakes. Arrows represent the direction of maximum variance of the measured environmental variables. Species codes are presented in Fritz *et al.* (1993) (redrawn from Fritz *et al.*, 1993, with permission from Blackwell Science Ltd).

demonstrated that distributions of diatom taxa in surface sediments were correlated with measured lake trophic status variables (TP, TN, chlorophyll *a*; Fig. 6.2). A CCA plot of diatom taxa and environmental variables in the 41 lakes clearly shows that the major gradient of variation in the taxonomic composition of diatom assemblages is related to lake trophic status, as indicated by the long arrows separated by small angles along axis 1 for TP, TN and chlorophyll *a* in Fig. 6.2. In the ordination diagram of species and environmental variables, axis 1 separates planktonic indicators of productive lakes, such as the small *Stephanodiscus* species (# 205, 206, 208; plotted on the left side of the diagram) from oligotrophic *Cyclotella* taxa characteristic of large, deep, oligotrophic lakes (# 41–46, 53; plotted on the right side; Fig. 6.2). Monte Carlo permutation tests demonstrated that correlations between diatoms and TP were statistically significant.

Once the dominant environmental variables that determine species distributions have been identified in the above way, transfer functions can be generated to infer environmental characteristics from diatom assemblages. Transfer functions are generated using a two-step process (Birks *et al.*, 1990a; Birks, 1995). First, the relationship between surface sediment diatom distributions and present-day water chemistry are modelled using the modern training dataset. In this step (in mathematical terms, regression), the optima and tolerances of all common diatom taxa are estimated for the variable(s) under consideration. These estimated optima and tolerances form a so-called 'transfer function' or 'inference model' which can then be used in a second (calibration) step to infer past P concentration, for example, from fossil diatom assemblages (Fig. 6.1).

Weighted averaging (WA) has become the popular method for solving the regression and calibration steps, described above. WA assumes non-linear, unimodal response surfaces of diatom species along environmental gradients, which is observed for many diatom taxa in nature (ter Braak & van Dam, 1989; Birks *et al.*, 1990a,b). WA performs very well compared to other methods with noisy, species-rich, compositional data, with species that are absent in many samples, and with long ecological gradients – properties that characterize many diatom training sets (ter Braak & van Dam, 1989; Birks *et al.*, 1990b; ter Braak & Juggins, 1993; ter Braak *et al.*, 1993). Advantages of WA are that it is based on simple, ecologically realistic models which have good empirical predictive power, and produce lower prediction errors than other techniques. Newly developed computer programs, for example, WACALIB (Line & Birks, 1990; Line *et al.*, 1994) and CALIBRATE (Juggins & ter Braak, 1992), have greatly facilitated the use of diatom transfer functions. Consequently, WA has been used to construct TP transfer functions for a number of regions, including western Canada (Hall & Smol, 1992; Reavie *et al.*, 1995), eastern Canada (Agbeti, 1992; Hall & Smol, 1996), midwestern USA (Fritz *et al.*, 1993), northeastern USA (Dixit & Smol, 1994), Northern Ireland (Anderson *et al.*, 1993), southern England (Bennion, 1994, 1995), Denmark (Anderson & Odgaard, 1994), European Alps (Wunsam & Schmidt, 1995), and northwestern Europe (Bennion *et al.*, 1996). Christie and Smol (1993) developed a transfer function for estimating TN concentrations from diatoms in southeastern Ontario lakes, Canada. Pan and Stevenson (1996) recently produced a diatom-TP inference model for wetlands.

Because of problems inherent in trying to estimate past lakewater TP concentrations from geochemical data (Engstrom & Wright, 1984), diatom-based inference models provide one of the few methods to estimate historical TP concentrations, quantify rates of lake eutrophication, and establish goals for water-quality remediation (Anderson *et al.*, 1993). As a consequence, diatom inference models are increasingly used by scientists and lake managers to study lake eutrophication. As their use becomes more widespread, there is an increasing need for critical ecological and statistical evaluation of all reconstructions.

ASSESSMENT OF DIATOM-TP MODEL PERFORMANCE, ERRORS AND EVALUATION

Weighted averaging regression and calibration is a powerful technique for quantitative reconstructions of lake water chemistry and eutrophication, but it is not without problems and limitations. As with any model, diatom inference models based on WA will always produce a result. However, it is important to know how reliable the resulting value is. In principle, the performance of WA models is set by the quality of the training dataset upon which it is based, as well as the ability of the empirical formulae to adequately model ecological distributions of the biota. As a basic rule, quantitative reconstructions are most useful to limnologists and lake managers when diatom-inferred TP values are quantitatively similar to measured TP (accurate), have narrow error estimates of prediction (precise), and perform well in most lakes (robust). A comprehensive review by Birks (1995) provides information concerning the basic biological and statistical requirements of quantitative reconstruction procedures, as well as methods for assessing their ecological and statistical performance.

Weighted averaging is sensitive to the distribution of the environmental variables (e.g., TP) in training sets (Birks, 1995). The accuracy of the WA method depends largely on reliable estimates of WA coefficients, namely the optima and tolerances, derived from modern diatom distributions in training sets. However, WA estimates of species optima will be biased when the distribution of lakes along the gradient is highly uneven, or when species are not sampled over their entire range. The bias at the ends of the gradient (e.g., at very high or low TP) is caused by truncated species responses, or so-called 'edge effects', and results in inaccurate estimates of species optima.

Perhaps the best solution to avoid problems associated with edge-effects is to sample a sufficient number of lakes along the full environmental gradient. Unfortunately, this may not always be possible in some geographic regions. For example, eutrophic lakes (i.e., with TP >30 μg/l) are relatively rare in many regions (e.g., Precambrian Shield region of central Canada; Hall & Smol, 1996). Consequently, some diatom-TP training sets have high representation of oligotrophic lakes (Hall & Smol, 1992, 1996; Reavie et al., 1995). In contrast, datasets from southern England, Wales and Denmark are dominated by eutrophic and hypereutrophic lakes, and few lakes have TP <100 μg/l (Anderson & Odgaard, 1994; Bennion, 1994). One method to extend the range of TP is to combine several regional training sets into one large database. Researchers in Europe have done precisely this. By combining regional datasets from England, Northern Ireland, Denmark and Sweden, they were able to extend the range of TP (5–1000 μg/l) and reduce the bias towards high TP concentrations (Bennion et al., 1996). An added benefit of amalgamating regional datasets is that it increases the chance of having good modern analogues for fossil diatom assemblages. For example, by adding the Swedish lakes into the combined European dataset, transfer functions include analogues for diatom assemblages in lakes with very low TP concentration that can be used to more accurately estimate preagricultural trophic status.

The performance of some regional training sets may be improved by using WA-partial least squares (WA-PLS; ter Braak & Juggins, 1993; ter Braak et al., 1993; Bennion et al., 1996). WA-PLS is an extension of WA that uses residual structure contained in correlations between taxa and other environmental variables, in addition to TP, to improve the predictive power of the WA coefficients. Bennion *et al.* (1996) provide an application of WA-PLS to infer lake eutrophication, and discuss the merits of this technique.

To evaluate model performance, it is important to consider both random and systematic errors produced by the modern training set as a whole (Oksanen *et al.*, 1988, ter Braak & Juggins, 1993, Birks, 1995). Birks (1995) provides an excellent review of methods for assessing the performance of modern training sets, and space limitations permit only a brief description here. Random errors, or model *precision*, are usually defined by the strength of the relationship between inferred and measured TP in the training set, and are best evaluated by comparing correlation coefficients (r^2, measures strength of the relationship) and root mean squared errors of prediction (RMSE, measures predictive ability). Due to the inherent circularity of developing diatom transfer functions and then testing them with the same training set of lakes, RMSE is always underestimated and r^2 is always overestimated. As a consequence, we generally rely on some form of cross-validation to generate more reliable and realistic error estimates. Most WA approaches use computer-intensive, randomized resampling procedures that are based on either bootstrapping (Efron & Gong, 1983) or jack-knifing (ter Braak & Juggins, 1993) to estimate random errors. Systematic differences, or *errors* (*sensu* Altman & Bland, 1983), may be assessed by the mean and maximum bias in a training set (ter Braak & Juggins, 1993; Birks, 1995). *Mean bias* may be defined as the mean difference between inferred and measured values in the training set, and estimates the tendency of WA models to over- or under-estimate measured values.

While it is important to assess the performance of the training set as a whole, it is even more critical to assess how well diatom inference models perform in the specific lake under investigation. Ultimately, the best way to assess the accuracy of TP inferences from fossil diatom assemblages is to directly compare the inferred values against available long-term records of measured lakewater TP concentration (so-called model evaluation or ground-truthing). Critical assessments of diatom transfer functions should include this type of evaluation in lakes across the entire spectrum of trophic status and human eutrophication intensity. This research is under way, but because diatom-TP transfer functions are a relatively recent development (since 1992), and because there are few lakes for which good long-term chemical data exist, this evaluation approach has been performed in only a few cases (Bennion *et al.*, 1995; Marchetto & Bettinetti, 1995; Hall *et al.*, 1997). These few examples indicate that diatom-inferred TP compares reasonably well to measured values, but that estimates may be sensitive to the absence of modern analogues and may tend to underestimate high TP values (Bennion *et al.*, 1995).

The combination of long-term water-chemistry records and excellent sediment chronology allowed Bennion *et al.* (1995) to undertake the most detailed evaluation of a diatom-TP transfer function to date. They inferred TP from diatoms in an annually laminated sediment sequence from Mondsee, Austria, and compared inferred epilimnetic TP against 20 years of annual mean TP

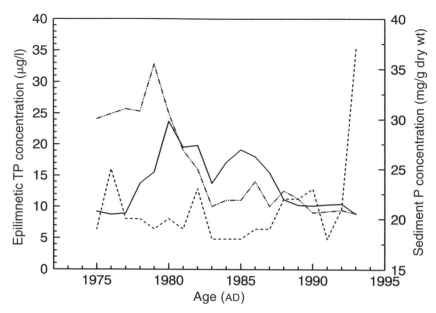

Fig. 6.3. Diatom inference model evaluation from Mondsee, showing close correspondence between measured annual mean epilimnetic TP concentration (dashed and dotted line) and diatom-inferred epilimnetic TP concentration (solid line). Sediment P content (dashed line) does not reflect changes in water column P concentrations. Sediments from Mondsee are annually laminated, permitting comparisons at an annual time-step since 1975 when water chemistry monitoring began (modified from Bennion *et al.*, 1995 with permission of Blackwell Science Ltd).

measurements. Using this approach, they showed that diatom-TP inferences closely tracked measured changes in epilimnetic TP concentration (Fig. 6.3). In comparison, sediment P concentrations did not reflect changes in water column P, illustrating that redox changes and post-depositional P mobility hamper the ability to determine water-chemistry changes from sedimentary P records (Anderson *et al.*, 1993; Bennion *et al.*, 1995). A slight mis-match between the timing of the TP peak (1979 for water chemistry vs. 1980 for diatom model) was attributed to sediment dating errors due to difficulties in distinguishing some varves. The model underestimated measured TP prior to 1980, possibly as a result of poor diatom assemblage analogues. Nevertheless, the diatom-inference model closely mirrored the major trends in measured TP, namely increasing TP during a eutrophication phase in the 1970s and declining TP during the 1980s since the installation of sewage treatment (Bennion *et al.*, 1995; Fig. 6.3).

In the absence of direct comparisons against measured water chemistry, several indirect methods are available to assess whether diatom transfer functions are likely to produce realistic estimates of past conditions. Bootstrapping can be used to estimate the magnitude of random prediction errors (RMSE) for individual fossil samples (Birks *et al.*, 1990a). Bootstrapped-derived errors can vary depending on the taxonomic composition of fossil diatom assemblages, due to differences in the abundances of taxa with stronger or weaker relation-

R. I. HALL AND J. P. SMOL

ships to the environmental variable being reconstructed. Confidence is also increased when fossil diatom assemblages show a high degree of similarity to the modern training set (i.e., possess good modern analogues), and this can be quantified using a dissimilarity index (e.g., χ^2) calculated for comparisons between the taxonomic composition of all modern and fossil diatom assemblages (Birks et al., 1990a). Inferences are likely to be more reliable if fossil assemblages comprise taxa that show a strong relationship to the variable of interest (i.e., show good 'fit' to the environmental variable being reconstructed, sensu Birks et al., 1990a), and this can be assessed by examining the residual fit to the first axis in canonical ordinations that are constrained to the variable being considered (see Birks et al., 1990a; Hall & Smol, 1993).

A large research effort since 1988 has developed and refined the numerical basis of diatom inference models, with the result that they now perform well and provide realistic error estimates. However, some additional factors may affect the performance of diatom-TP inference models. For example, TP concentrations can vary substantially during the course of a year or even a growing season. Consequently, training sets must employ a water-chemistry sampling frequency sufficient to provide an accurate estimate of the annual or seasonal mean. However, the annual range of TP concentrations is greater in more eutrophic lakes (Gibson et al., 1996). Because different diatom species attain peak populations at different times of an annual cycle, they may indicate quite different TP concentrations. This feature likely introduces unavoidable noise into diatom inference models, with the result that prediction errors will increase under more eutrophic conditions.

Within-lake variability in diatom assemblages could introduce error into diatom inferences (Anderson, 1990a,b,c). However, the good agreement between TP inferences from six cores taken across a range of water depths (3–14 m) in a wind-stressed, shallow lake suggests that a single core located in the central deep-water basin is likely representative of the whole lake (Anderson, 1998). Variability may be higher in surface sediment diatom assemblages than in down-core samples, indicating that surface sediment variability may be a source of error in modern training sets and, hence, in transfer functions (Anderson, 1998). Further research is required to assess errors introduced by within-lake variability in surface sediment diatom distributions.

Taxonomic inconsistencies may also impair the predictive ability of diatom inference models (Birks, 1994). It is therefore important that researchers maintain the highest level of taxonomic knowledge, consistency, and between-lab harmonization possible.

An obstacle to accurate inferences of historical TP is the observation that diatom community composition is under multifactorial control. TP is only one of a group of covariates (e.g., N, Si, chlorophyll a) that measure trophic status (Fritz et al., 1993; Jones & Juggins, 1995). Potentially, diatom-inferred reconstructions can be affected by independent historical changes in covariates that are unrelated to alterations in TP (e.g., other nutrients, light, turbulence, herbivory). One way to minimize misinterpreting TP trends from diatom assemblages is to analyze other bioindicators that have different susceptibilities to covariates. For example, fossil pigments can be used to track changes in other algal groups (Leavitt *et al.*, 1989; Leavitt, 1993), which may also differ in their sensitivity to shifts in N, Si, and light availability. A shift from diatoms to N-fixing cyanobacteria may supply important information concerning changes in N and Si availability that were unrelated to P. Analysis of fossil zooplankton may indicate that taxonomic shifts were also caused by changes in herbivory (Jeppesen *et al.*, 1996). Furthermore, chironomid assemblages can be used to quantify deep-water oxygen availability (Quinlan *et al.*, 1998), which may provide important information on fish habitat availability and food-web changes, as well as the potential for increased internal P supply.

Case Studies

Diatoms have contributed to our knowledge of lake eutrophication in a number of different ways, and new applications are being developed at a rapid rate. In the following section, we present some of the different types of studies that have used diatoms to investigate lake eutrophication and recovery, pinpoint the underlying cause(s), assess regional-scale water-quality changes, and evaluate empirical water-quality models.

INDIVIDUAL LAKE STUDIES

The most common paleolimnological application of diatoms has been to investigate eutrophication trends in individual lakes. Such studies have successfully identified pre-impact conditions, and the timing, rate, extent and probable causes of eutrophication and recovery. Also, where eutrophication has been well documented, individual lake scenarios have been used to study the responses of diatom communities. In some lakes that contain annually laminated sediments (varves), a very high degree of temporal resolution can be attained (Simola, 1979).

Short-term (decade to century scale) eutrophication trends Intensive paleolimnological studies of Lough Augher, in Northern Ireland, provide an excellent example of the detailed information that can be obtained by studying sedimentary diatom assemblages (Anderson, 1989, 1995a; Anderson *et al.*, 1990; Anderson & Rippey, 1994). By analyzing diatoms in 12 cores from this relatively small and simple basin, Anderson and co-workers have shown in great detail how Lough Augher became eutrophic as a result of untreated sewage effluent from a local creamery during the period 1900–1972. Diatom-inferred TP (DI-TP) and diatom accumulation rates increased synchronously from 1920–1950, in response to initial increases in nutrient loads from the creamery (Fig. 6.4(*a*)). However, during the most eutrophic period, DI-TP and diatom accumulation rates became uncoupled. Diatom accumulation rates indicated

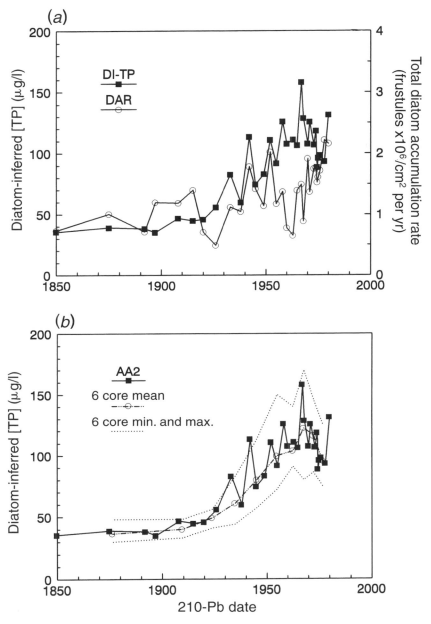

Fig. 6.4. Indices of lake responses to point-source eutrophication at Lough Augher, Northern Ireland. (a) Diatom accumulation rate (DAR), and diatom-inferred epilimnetic TP concentration (DI-TP) from a central deep-water core (AA2) using weighted-averaging and a surface sediment diatom training set from Northern Ireland. (b) Comparison of diatom-inferred epilimnetic TP concentration from a central deep-water core (AA2) and mean, minimum and maximum values from six cores taken from 3–14 m water depth, to illustrate that TP inferences based on diatoms in a single deep-water core are representative of the lake basin as a whole. Chronology is derived by correlation to a dated master core. (Original: redrawn from data in Anderson, 1989; Anderson & Rippey, 1994; Anderson 1998.)

that planktonic diatom production peaked during 1940–1957, before the highest inferred P levels, and hence highest algal productivity, were reached (*ca.* 1950–1970; Anderson, 1989, 1995*a*). The apparent contradiction of lower diatom production at higher rates of nutrient supply could be interpreted in terms of changes in Si:P ratios as the supply of P increased. As Si:P ratios declined, increasing silica limitation would have placed diatoms at a competitive disadvantage relative to non-siliceous algae (Tilman *et al.*, 1986). Consistent with this hypothesis, planktonic diatom communities became dominated by lightly silicified diatoms (small *Stephanodiscus* species) which are better competitors at low Si availability (Lund, 1950; Tilman *et al.*, 1982). In contrast, littoral diatom communities were not affected to the same extent, likely because they have alternative sources of nutrients (P, Si) than the water column (e.g., sediments, macrophytes, detrital P cycling; Anderson, 1989). Consequently, littoral diatoms showed maximum accumulation rates during the productivity peak (1950–1970).

The response of diatoms in Lough Augher to reduced nutrient loads following redirection of sewage effluent in the late 1970s showed an improvement in lake water quality from eutrophic to mesotrophic conditions (Fig. 6.4; Anderson & Rippey, 1994). Diatom-inferred TP and diatom accumulation rates declined. Despite indications that water quality improved and evidence suggesting that P attained new equilibrium levels in Lough Augher, Anderson *et al.* (1990) showed that diatom species composition had not yet reached a new equilibrium even 5 years after sewage redirection. Thus, biological recovery may lag behind chemical recovery of eutrophied lakes (see also Stoermer *et al.*, 1996). The process of biological recovery in eutrophied lakes has not yet been studied in any great detail. Consequently, the influence of factors such as the duration and extent of eutrophication, lake depth, flushing, and food-web structure in determining the rate and end-point of biological recovery remain poorly understood.

Anderson's research on Lough Augher also provides insights on the merits and limitations of using absolute and relative diatom abundances as measures of lake eutrophication. Accumulation rates (as numbers of valves or biovolume per unit sediment area per year) provide the most direct assessment of changes in diatom production, but they are difficult to estimate accurately. Accumulation rates are highly variable, both spatially and temporally, even in small lakes with simple morphometry (Anderson, 1989). This high variability is due in part to changes in dry mass accumulation rates (sediment focusing: Davis & Ford, 1982; Anderson, 1990*a,b,c*) and Si dissolution rates (Rippey, 1983), meaning that diatom accumulation rates in a single core may not reflect changes in historical diatom production in the lake. Also, because changes in water column Si limitation and grazing can reduce the contribution of diatoms to total production, diatom accumulation rates may not be an accurate index of primary production in lakes (Battarbee, 1986; Anderson, 1998). A consequence of high variability in diatom accumulation rates is that large numbers of cores and diatom counts are required. The high costs associated with multiple-core studies may prevent the widespread use of accumulation rate data in most applied diatom studies.

Table 6.1. *Mean and range of diatom-inferred lake water total phosphorus concentrations at Diss Mere, Norfolk for different historical periods during the past 7000 years*

Historical Period	Age (BP)	Mean (μg TP/l)	Range (μg TP/l)
Modern times	0–150	495	385–675
Post-Medieval	150–500	343	297–400
Roman/Anglo-Saxon/Medieval	500–2200	117	71–179
Iron Age/Roman	2200–2500	93	58–119
Late Bronze Age	2500–3000	90	36–114
Early Bronze Age	3000–3500	221	221
Neolithic	3500–5000	33	6–97
Mesolithic	5000–7000	20	4–35

Note:
Modified from Birks *et al.* (1995), with permission.

In contrast, Anderson *et al.* (1990) demonstrated that diatom percent abundances were considerably less variable than accumulation rate data in 12 sediment cores from Lough Augher. Diatom TP inferences which utilize percentage data, therefore, are less affected by coring location than accumulation rates (Anderson, 1995a). Anderson has shown good agreement between TP inferences for six cores across a range of water depth (3–14 m) in Lough Augher, and concluded that TP inferences based on diatoms in a single deep-water core are representative of the lake basin as a whole (Fig. 6.4(*b*); Anderson, 1998). Thus, diatom-TP inference models appear to be a robust, time-efficient, and cost-effective tool for monitoring lake eutrophication and recovery.

Long-term human impacts (millennium scale) In many regions of the world with a long history of human settlement and land use (e.g., Europe, Africa, Asia, Middle East), lakes have been altered by anthropogenic eutrophication over long-term, millennial timescales, and not simply since the Industrial Revolution. For example, combined diatom and pollen studies at Diss Mere, England (Fritz, 1989; Peglar *et al.*, 1989) have shown that anthropogenic activities began in the Neolithic period (5000–3500 BP) with the creation of small forest clearances. The first incidence of marked eutrophication occurred during the Bronze Age (3500–2500 BP) with marked increases in the abundance of *Synedra acus* Kützing, in response to forest clearance and the onset of cereal agriculture (Fritz, 1989; Peglar *et al.*, 1989). Using a modern diatom-TP training set, Birks *et al.* (1995) inferred that early Bronze Age activities increased mean TP from 33 to >200 μg/l, altering the lake from a mesotrophic to hypereutrophic condition (Table 6.1). After this eutrophication phase, the lake never returned to preimpact TP concentrations.

Mixed pasture and crop agriculture during the Roman, Anglo-Saxon and

Medieval periods maintained the above lake in a eutrophic state (mean TP *ca.* 100 µg/l, Table 6.1). Major eutrophication occurred during post-Medieval periods with the expansion of the town of Diss in the fifteenth and sixteenth centuries, and diatom-inferred TP rose to 343 µg/l. Decline and the eventual loss of macrophytes coincided with this large increase in DI-TP to over 200 µg/l (Peglar, 1993; Birks *et al.*, 1995). Analysis of diatom assemblages and reconstructions of TP show the close relationship between human activities and eutrophication in this lake.

The long history of human perturbations on lakes such as Diss Mere has implications for restoration and lake management. Any attempt to restore lakes with a long history of human impacts may need to consider aquatic conditions that existed thousands of years ago in order to define preimpact or natural conditions. Such knowledge of the preimpact state can only be obtained using the sediment record, and diatoms are likely the best indicators to use. Furthermore, restoration to the pristine condition may be extremely difficult, if not impossible, because P stored in the sediments for centuries to millennia can maintain a eutrophic state.

In the above section, we have illustrated how diatoms have been used to document eutrophication and recovery of lakes to both relatively short-term (last 100 years) and long-term (past 3500 years) human impacts, including agriculture and industry. Many other studies have used diatoms to determine the effects of, for example, land clearance (e.g., Bradbury, 1975; Manny *et al.*, 1978; Fritz *et al.*, 1993; Reavie *et al.*, 1995), canal construction (Christie & Smol, 1996), peat drainage (e.g., Simola, 1983; Sandman *et al.*, 1990), agriculture (e.g., Håkansson & Regnéll, 1993; Digerfeldt & Håkansson, 1993; Walker *et al.*, 1993; Anderson, 1997), industry (e.g., Ennis *et al.*, 1983; Engstrom *et al.*, 1985; Yang *et al.*, 1993), urbanization and sewage treatment (e.g., Haworth, 1984; Brugam & Vallarino, 1989; Anderson *et al.*, 1990, Brenner *et al.*, 1993; Anderson & Rippey, 1994; Bennion *et al.*, 1995), and road construction (Smol & Dickman, 1981) on lake trophic status. Donar *et al.* (1996) used diatoms to demonstrate eutrophication following reservoir construction, and the subsequent impacts of industry and urbanization.

Although a major focus has been on paleolimnological studies of temperate lake systems, diatoms can also be used to study eutrophication in other systems. For example, diatom-based methods can be used to study both recent (Hecky, 1993) and long-term (Whitmore *et al.*, 1996) cultural eutrophication in tropical lakes. These approaches can also be readily transferred to studies of marine and coastal environments (Anderson & Vos, 1992), wetlands (Pan & Stevenson, 1996), and rivers (Reavie, 1997). Van Dam and Mertens (1993) provide an example of how diatoms attached to herbarium macrophytes can be used to evaluate water-quality changes. Unfortunately, space limitations do not allow us to discuss these applications further.

Natural eutrophication events Not all eutrophication events are caused by humans. In some cases natural disturbances may accelerate nutrient supplies

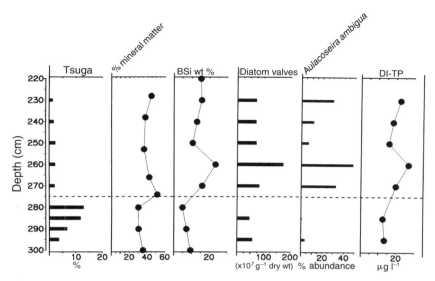

Fig. 6.5. Sedimentary evidence of a natural eutrophication event in Flower Round Lake, Ontario in response to the hemlock decline 4800 BP. The hemlock decline (*Tsuga*) increased catchment erosion (indicated by increased mineral content), leading to increased diatom production (panels with biogenic silica (BSi) and diatom valve concentration), and an increase in the percentage abundance of meso-eutrophic diatom species *Aulacoseira ambigua* (Grunow in Van Heurck) Simonsen. Diatom-inferred TP (DI-TP) increased from 14 μg/l to 30 μg/l (modified from Hall & Smol, 1993 with permission of Kluwer Academic Publishers).

to lakes, though these are usually rare events, such as fire or other natural forest disturbances. For example, several studies have linked increased eutrophication to the sudden mid-Holocene decline in hemlock trees (*Tsuga canadensis*) in eastern North America (Whitehead *et al.*, 1973; Jones *et al.*, 1984; Boucherle *et al.*, 1986). It is widely believed that a forest pathogen caused the widespread, rapid and synchronous death of hemlock throughout its range in North America *ca.* 4800 years ago (Davis, 1981; Allison *et al.*, 1986). Increased erosional supply of nutrients, increased runoff due to lower evapotranspiration, and increased input of deciduous leaf litter are all cited as possible processes by which the hemlock decline increased aquatic productivity (Whitehead *et al.*, 1973; Boucherle *et al.*, 1986). Hall and Smol (1993) demonstrated that diatom communities in five Ontario lakes responded to the decline in hemlock, but that only one lake showed dramatic eutrophication (Fig. 6.5). The magnitude of lake response appeared to be due to differences in catchment area and slope, which are consistent with the idea that increased nutrient supplies resulted from elevated erosion and runoff rates. For example, diatom-inferred TP increased from 14 μg/l (oligotrophic) to 30 μg/l in the lake with the largest and steepest catchment as a result of this natural forest disturbance (Fig. 6.5).

Naturally productive lakes As noted earlier, estimates of preimpact or 'natural' lake trophic status can provide realistic targets for mitigation programs.

Diatom-based paleolimnological techniques have also been used to identify naturally-productive lakes (Reavie *et al.*, 1995; Karst, 1996). Such data have important management implications, as mitigation efforts are unlikely to 'restore' these lakes to oligotrophic conditions.

REGIONAL ASSESSMENTS USING DIATOMS

It is often instructive to assess water-quality changes on a broader, regional scale, rather than simply on a lake by lake basis. Regional assessments using diatoms can provide an effective tool to quantify the magnitude of water-quality changes due to human activities, and to map or identify problem areas where lakes have been most severely affected.

An example of a large-scale regional monitoring project is the United States Environmental Protection Agency's Environmental Monitoring and Assessment Program (EMAP; Whittier & Paulsen, 1992). The diatom component of this project used a 'top and bottom' sediment core approach to quantify lake water-quality changes since preindustrial times in the northeastern USA (Dixit & Smol, 1994; S. S. Dixit *et al.*, *unpublished data*). In Year 1 (1991), diatoms were investigated in surface samples of deep-water sediment cores in order to generate transfer functions for quantifying important water quality variables (TP, Secchi, ph, Cl^-), and to estimate present-day conditions. The lakes were randomly selected from a population of lakes so that results could be extrapolated to the target population of 8610 lakes in the northeastern USA. Diatom assemblages from sediments deposited during approximately the early 1800s (i.e., the bottom samples of each sediment core) were used to estimate preindustrial conditions. Difference between inferences for bottom and top sediment samples estimated the magnitudes of water-quality changes that occurred since preindustrial times.

Using the above approach, Dixit & Smol (1994) were able to show that inferred TP increased in lakes which are presently eutrophic (i.e., with present-day TP > 30 μg/l), and that inferred TP changes were lower in lakes that are presently oligotrophic (Fig. 6.6). When viewed as cumulative frequency distributions, DI-TP clearly indicates that the percentages of oligo- and mesotrophic lakes declined, while the proportion of eutrophic lakes increased in northeastern USA since preindustrial times (Fig. 6.6). Dixit and Smol (1994) also developed an index of disturbance based on DI-TP, DI-Cl^-, DI-Secchi, species richness and diversity and detrended correspondence analysis (DCA) sample scores. By mapping the distribution of their disturbance index, they demonstrated that the highest inferred disturbances occurred in lakes in regions with high population densities (Fig. 6.6). Thus, human activities are likely an important factor controlling water quality and eutrophication in this region. The inference models developed in this study can be used to continue monitoring water-quality trends on an ongoing basis. Several other publications are now being prepared dealing with this large project.

This sediment core 'top and bottom' approach provides an efficient method

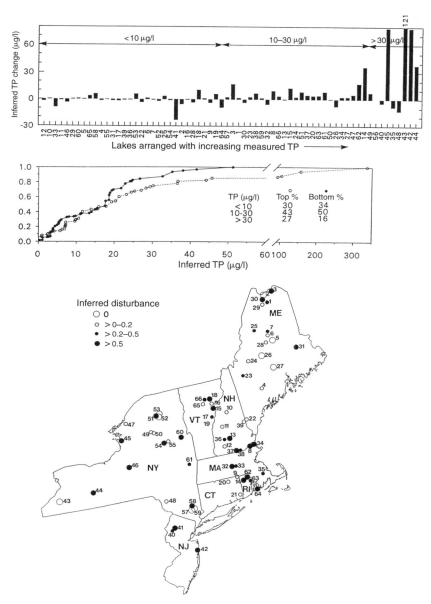

Fig. 6.6. Diatom-based assessments of lake eutrophication and disturbance in EMAP-Surface Waters Program lakes (NE USA) since preindustrial times. Upper panel shows the inferred changes in diatom-inferred lake water TP concentration since preindustrial times, from differences in inferred values between top (0–1 cm) and bottom (usually >25 cm) sediment core samples. Middle panel shows cumulative proportions of present-day (top core samples) and preindustrial (bottom samples) distributions of diatom-inferred TP concentration for the target population of lakes in the NE USA. Lower panel shows the spatial distribution of diatom-inferred disturbance index (based on DI-TP, DI-Cl⁻, DI-Secchi, species richness and diversity, and detrended correspondence analysis sample scores). Diatoms inferred high disturbances in lakes that are in high population density areas (modified from Dixit & Smol, 1994 with permission of Kluwer Academic Publishers).

for assessing regional water-quality changes in large numbers of lakes because it requires analysis of far fewer samples than studies employing continuous temporal samples. Hall and Smol (1996) used a similar approach to assess how Precambrian Shield lakes in a relatively remote, rural region of central Ontario have responded to a combination of shoreline development (mainly cottages and past logging) and acidic deposition since preindustrial times. Lakes in this region are generally unproductive with present-day TP ranging from 2–30 µg/l. Diatoms inferred relatively small changes in lake water TP in most lakes, despite considerable cottage development and past logging (Fig. 6.8). Surprisingly, diatoms inferred TP declines in about half of the lakes with relatively high present-day TP (10–24 µg/l), indicating that presently mesotrophic lakes tended to have higher natural TP prior to cottage development. Based on these findings, Hall and Smol (1996) concluded that TP loads to lakes have been reduced by regional acidification and past logging, and that the magnitude of these reductions exceeded any increases in TP loads from cottages and other lakeshore developments.

EUTROPHICATION MANAGEMENT USING DIATOMS IN COMBINATION WITH OTHER APPROACHES

Studies combining diatoms with other methods (e.g., historical records, empirical P models) provide an effective approach to assess causes of eutrophication and to guide restoration. For example, analysis of diatoms, sediment geochemistry, and historical records from two coastal freshwater lakes in southwest England demonstrated that the lakes have become increasingly eutrophic since 1945 due to increases in agricultural activity and municipal sewage loading (O'Sullivan, 1992). Nutrient export coefficient models (Vollenweider, 1975) were then used to estimate historical changes in P loading from long-term records of land use, fertilizer application, and livestock (1905–1985). In this way, O'Sullivan (1992) demonstrated that P loads were near permissible levels for maintaining oligotrophic conditions in 1905, but by 1945 P loads had increased to dangerous levels. Since 1945, P loads have gradually increased and now are ten times the acceptable limit for oligotrophic lakes. Finally, the same nutrient models were used to evaluate the effects of various restoration strategies on annual N and P loads to the lakes. Total removal of P from sewage and detergents would still result in P loads two-fold above OECD permissible limits. Consequently, agricultural contributions of P also needed to be reduced. Based on these findings, O'Sullivan (1992) suggested that P loads could be reduced to permissible levels by treating or diverting sewage effluent, and by establishing buffer strips of woodland at least 15 m wide along the rivers draining into the lakes.

Rippey *et al.* (1998) combined a clever use of diatoms, DI-TP, sediment chemistry, and mass-balance nutrient models to demonstrate that flushing rate and internal nutrient loading can play a strong role in regulating eutrophication in some lakes. The use of DI-TP and sediment calcium concentrations showed a rapid period of eutrophication during 1973–1979 in White Lough, a small (7.4 ha) eutrophic lake in an agricultural catchment in Northern Ireland. Historical land-use records suggested the eutrophication event was not triggered by increased external nutrient loads. Using a simple empirical nutrient model, Rippey *et al.* (1997) demonstrated that the eutrophication event was, in fact, initiated by a reduction in surface runoff and stream flow during warm, dry summers in the early 1970s. They identified two factors responsible for the eutrophication event. First, P concentrations increased when longer hydraulic residence time caused increased retention within the lake basin of P released from the sediments. Second, rapid eutrophication resulted when sediment P release rates increased as a consequence of elevated water column P concentration and increased anoxia. An important implication arising from this study is that dry climatic periods, by altering the hydrological regime, may cause significant eutrophication in meso- to eutrophic lakes. Lakes with water residence times of about one year may be most susceptible (Rippey *et al.*, 1997).

USING DIATOMS TO EVALUATE EMPIRICAL EUTROPHICATION MODELS

Empirical or mass-balance nutrient models are increasingly used by managers and scientists to quantify lake responses to changes in land use and to develop policy tools for shoreline development (e.g., Dillon & Rigler, 1975; Reckow & Simpson, 1980; Canfield & Bachmann 1981; Hutchinson *et al.*, 1991; Soranno *et al.*, 1996). Predictive abilities of most models, however, have never been rigorously assessed. Diatom transfer functions provide one of the only quantitative methods available to independently evaluate the ability of empirical models to estimate predisturbance conditions in lakes (Smol, 1992, 1995).

Diatoms were used to evaluate the ability of the Ontario Ministry of Environment & Energy's Trophic Status Model (TSM: Dillon *et al.*, 1986; Hutchinson *et al.*, 1991) to infer preindustrial TP concentrations in lakes by comparing TSM estimates with values inferred from sedimentary diatom assemblages deposited before 1850, based on Hall and Smol's (1996) diatom-TP transfer function. Model evaluation was performed in 47 lakes in the area where the TSM was developed. The ability of diatoms ($r^2 = 0.41$) and the TSM ($r^2 = 0.55$) to estimate present-day measured TP was comparable (Fig. 6.7(a),(b)), and there was a similarly strong relationship between present-day TP inferred from the diatom model and the TSM ($r^2 = 0.36$; Fig. 6.7(c)). However, the correspondence between diatom and TSM estimates of preindustrial TP concentrations was extremely poor ($r^2 = 0.08$; Fig. 6.7(d)). Diatoms consistently inferred higher background TP than the TSM (Fig. 6.7(d), Fig. 6.8).

The diatom transfer function undoubtedly produced some errors, but the poor agreement between preindustrial TP estimates provided reason to reconsider some key assumptions and parameters in the TSM. For example, paleolimnological studies demonstrated that acid deposition and past logging activities reduced P concentrations in many of the lakes (Hall & Smol, 1996; Fig.

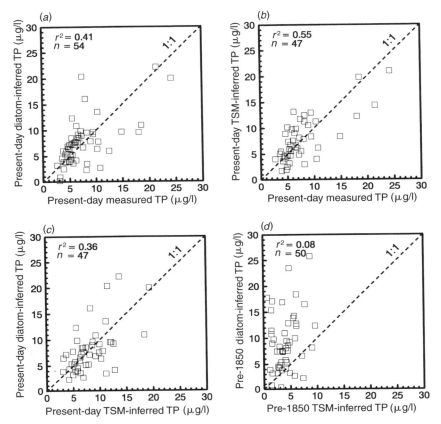

Fig. 6.7. Using diatom inference models to assess predictive abilities of empirical models to estimate preindustrial TP. (a) Relationship between present-day diatom-inferred and measured TP in 54 lakes in south-central Ontario, (b) relationship between Trophic Status Model (TSM) and measured present-day TP in 47 of the lakes, (c) relationship between TP estimates from surface sediment diatoms and the TSM, and (d) the relationship between preindustrial estimates of TP concentrations using diatoms and the TSM (original figure).

6.8). However, the TSM only considered human activities that increase lake water TP, and did not consider changes in regional acidification or forestry that could have caused TP to decline (Fig. 6.8). As a consequence, the TSM appeared to overestimate preindustrial TP in Ontario lakes. Based on comparisons with diatoms, the TSM adequately models present-day P concentrations in lakes, but assumptions that P dynamics have remained constant during the past 150 years appear to be too simplistic to permit reliable estimates of preindustrial conditions.

USING DIATOMS TO QUANTIFY FACTORS REGULATING LAKE EUTROPHICATION

Lake eutrophication is under complex, multifactorial control – regulated to a large extent by human activities, climate, morpho-edaphic conditions of the

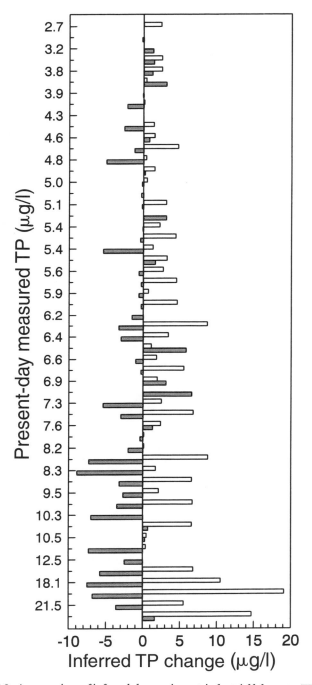

Fig. 6.8. A comparison of inferred changes in post-industrial lake water TP concentrations along a productivity gradient using two independent approaches; a diatom–TP inference model (shaded bars) and an empirical TP model (TSM; open bars). The empirical model does not infer a decline in TP in any of the lakes, whereas the diatom model infers TP declines in many of the more productive lakes in the region (original figure).

watershed, and food-web interactions. Consequently, eutrophication still poses major challenges to applied ecologists and managers, and mechanisms that regulate long-term water quality in lakes have proven difficult to quantify using experimental and standard monitoring practices. Lake sediment diatom records can supply long-term information on water-quality trends, but without long-term records of possible control factors, cause and effect relationships cannot be rigorously and quantitatively assessed. For example, indirect gradient ordination, rate-of-change numerical techniques (e.g., Lotter *et al.*, 1992), and diatom transfer functions provide effective numerical approaches to identify and describe temporal patterns of community change and inferred water quality, but they do not lend much insight into the factors that cause the observed changes.

Factors regulating water quality, however, can be quantified using an approach developed by Borcard *et al.* (1992), known as variance partitioning analysis. Variance partitioning employs canonical ordination (e.g., CCA, RDA), and can be used to examine correlations between fossil diatom assemblages from sediment cores and historical records of possible control factors (e.g., geographic position, climate, resource use and urbanization). As with all correlational approaches, variance partitioning cannot prove causal mechanisms, but it can provide an efficient screening technique to identify potential management strategies or to generate testable hypotheses. Variance partitioning has been performed with two (e.g., Borcard *et al.*, 1992; Zeeb *et al.*, 1994) or three (e.g., Qinghong & Bråkenhielm, 1995; Jones & Juggins, 1995) categories of explanatory variables.

In one example, Zeeb *et al.* (1994) used variance partitioning to quantify the influence of measured water-chemistry changes on diatom assemblages in annually-laminated sediments in Lake 227 (Experimental Lakes Area, Canada) during 20 years of whole-lake eutrophication experiments (1969–1989). In their analyses, 79.1% of the total variance in sediment diatom assemblages was explained by changes in water chemistry and time. Water chemistry, independent of the effects of time, explained 60.5% of the total variance in diatom distributions, indicating a strong relationship between diatoms and water chemistry. Time data, independent of water chemistry, explained only 6.3% of the variation, and indicated that no fundamental, temporally structured environmental processes were missed by their analyses. The joint effects of water chemistry and time explained a further 12.3% of the total variance, indicating diatoms tracked temporally-structured water-chemistry changes that occurred during the course of experimental nutrient additions. Overall, these analyses indicate there is a strong relationship between changes in water chemistry and diatom assemblages.

Future directions

Tremendous progress has been made in the ways we have learned to glean information concerning lake eutrophication from diatoms. Undoubtedly, the momentum of the past decade will continue to spur many interesting and novel developments of eutrophication research, and the use of diatoms in lake management will continue to increase. Below, we briefly describe some of the exciting new research areas that we believe will likely develop in the future.

MULTIPLE INDICATOR STUDIES

Ecologists and ecosystem managers are becoming increasingly aware of the importance of complex food-web interactions in determining ecosystem structure and function (e.g., Polis & Winemiller, 1996). Food-web interactions undoubtedly play a role in lake responses to eutrophication (Carpenter *et al.*, 1995). One important question for applied diatom studies is: To what extent are diatoms regulated by resources versus herbivores along a lake eutrophication gradient? Studies of diatoms alone will not be able to determine the role of complex interactions among different communities and trophic levels, because diatoms are only one of many algal groups in lakes, and comprise only one of several trophic levels. Fortunately, many other aquatic and terrestrial biota leave fossil remains in lake sediments. Future studies of lake eutrophication will undoubtedly benefit by analyzing diatoms in combination with other bioindicators that encompass multiple trophic levels.

The combined use of fossil pigments and diatoms provides one of the only ways to identify when algal communities become N-limited rather than P-limited during eutrophication, because pigments from N-fixing cyanobacteria are preserved in direct proportion with the standing crop of past populations which produced them (Leavitt, 1993). Furthermore, pigments from most groups of algae and photosynthetic bacteria preserve well in lake sediments, and can be used to reconstruct the biomass of all major algal groups, as well as total algal biomass (Leavitt *et al.*, 1989, 1994; Leavitt & Findlay, 1994). In conjunction with diatom studies, analysis of biogenic Si (Schelske *et al.*, 1986) and fossil pigments may help elucidate the relative roles of Si, P and N limitation during eutrophication.

Zooplankton fossils may be used to quantify past changes in the density of planktivorous fish (Jeppesen *et al.*, 1996). By combining diatom and zooplankton analyses in sediment cores, it is now possible to investigate food-web changes during eutrophication and recovery. This area of research will be of considerable interest to ecologists, as well as lake and fisheries managers.

Inference models are now being developed to infer deep-water oxygen availability from lake sediment chironomid (Diptera) assemblages (Quinlan *et al.*, 1998). By combining diatom and chironomid analyses, researchers will undoubtedly gain a better understanding of the relationships between changes in upper and lower regions of the water column during eutrophication and recovery (Walker *et al.*, 1993).

Empirical P models (e.g., Vollenweider, 1969) have been widely used by lake managers to calculate the P loading required to achieve a desired drop in lake P content, using the following basic formula:

$$TP = L/z \times (\sigma + \rho)$$

where TP is the mean annual lake TP concentration (mg/m³), L is the areal mean P load to the lake (mg/m²/y), z is the mean lake depth (m), σ is the P sedimentation coefficient (1/y), and ρ is the hydraulic flushing coefficient (1/y).

Rippey and Anderson (1996) recently presented an interesting and novel approach that can be used to estimate long-term P loading by employing diatom-TP inference models. When used in conjunction with Vollenweider-type models and whole basin estimates of P sedimentation, DI–TP can be used to estimate past changes in P loading. In this way, DI–TP from a central sediment core provides estimates of historical changes in mean annual lake TP concentration. Stream flow can be estimated from hydrological models (Thornthwaite & Mather, 1957) or from regional runoff information (e.g., Foy *et al.*, 1982). Assuming that the outflow equals the inflow, the loss of P through the outflow is calculated from the outflow and DI–TP. Whole-lake sediment P fluxes provide a direct measure of changes in P retention in a lake. Historical P loading is then calculated as the sum of the areal P flux to the sediments and the areal P flux through the outflow. Using ^{210}Pb sediment dating methods, this approach can be used to estimate P loads to most lakes during the past 150 years. This approach provides one of the only ways to estimate historical P loadings, because few lakes in the world have P loading data for more than a decade or two (Rippey, 1995). As a consequence, this approach provides an important new tool for lake managers to identify past critical levels of P loading that caused lakes to eutrophy beyond acceptable standards.

The above method can also be used to investigate how the P sedimentation coefficient changes in lakes during eutrophication, which is an important step for assessing key assumptions of Vollenweider-type P models (Rippey, 1995). Empirical P models rely on our ability to estimate the P sedimentation coefficient accurately. However, these models assume that P sedimentation coefficients remain constant through time, despite strong evidence that they vary with changes in hypolimnetic oxygen availability which occur during eutrophication (Nürnberg, 1984). Fortunately, the ability to infer past lake water TP concentrations using diatoms now makes it possible to calculate temporal changes in the P sedimentation coefficient in any lake, and thus describe how it varies during lake eutrophication and recovery. According to Rippey (1995), the P sedimentation coefficient can be calculated in three basic steps. First, historical lakewater P concentrations are estimated using the diatom-TP transfer functions and fossil diatoms from a central deep-water sediment core. Second, changes in P retention flux in the sediment (F, mg/m²/y) are quantified using whole-basin mean sedimentary P fluxes. Finally, historical changes in lake P sedimentation coefficients are calculated using the formula $\sigma = F/z \times TP$ (1/y). This approach can be used to assess the relative contributions of shallow vs. deep sediments to P recycling and retention during lake eutrophication and recovery (Rippey, 1995). It may also provide a useful diagnostic tool for identifying the important factors that regulate P retention in lake sediments. Such information is essential in order to quantify the rate of lake recovery following a reduction in external P load, but so far changes in the sediment P coefficient have remained difficult to predict (e.g., Jeppesen *et al.*, 1991).

Summary

Diatoms respond rapidly to eutrophication and provide detailed information on ecological changes that occur during eutrophication and recovery, in both deep-water and near-shore habitats. Because diatoms preserve well in most lake sediments, they can be used to generate long time-series of data concerning lake responses to eutrophication and recovery. Moreover, recent advances in quantitative methods (e.g., canonical ordination, weighted-averaging, bootstrapping, jack-knifing) permit reliable estimates of past and recent water-chemistry changes during eutrophication from diatom assemblages. Diatoms in sediment cores can be used to establish baseline or pre-disturbance water chemistry as targets for lake rehabilitation. Natural variability can also be assessed. In this chapter, we provided some of the many examples of how diatoms are being used as powerful tools for eutrophication research and management. The past decade was largely devoted to developing and refining diatoms as quantitative indicators of lake eutrophication. The numeric and experimental tools are now well developed and sophisticated. The stage is set for these methods to be applied in novel ways that undoubtedly will further our ecological understanding of lake eutrophication and improve water-quality management.

Acknowledgments

Most of our research has been funded by grants from the Natural Sciences and Engineering Research Council of Canada. This chapter was written while Roland Hall was employed in Dr P. Leavitt's laboratory, with partial support by a Postdoctoral Fellowship from the University of Regina. We thank Dr H. Bennion and members of our laboratories for helpful comments on the manuscript.

References

Agbeti, M. D. (1992). Relationship between diatom assemblages and trophic variables: A comparison of old and new approaches. *Canadian Journal of Fisheries and Aquatic Sciences*, **49**, 1171–5.

Agbeti, M. D., & Dickman, M. (1989). Use of fossil diatom assemblages to determine historical changes in trophic status. *Canadian Journal of Fisheries and Aquatic Sciences*, **46**,
1013–21.

Allison, T. D., Moeller, R. E., & Davis, M. B. (1986). Pollen in laminated sediments provides
evidence for a mid-Holocene forest pathogen outbreak. *Ecology*, **64**, 1101–5.

Altman, D. G., & Bland, J. M. (1983). Measurement in medicine: The analysis of method
comparison studies. *The Statistician*, **32**, 307–17.

Anderson, N. J. (1989). A whole-basin diatom accumulation rate for a small eutrophic lake
in Northern Ireland and its palaeoecological implications. *Journal of Ecology*, **77**,
926–46.

(1990a). Variability of diatom concentrations and accumulation rates in sediments of
a small lake basin. *Limnology and Oceanography*, **35**, 497–508.

(1990b). Spatial pattern of recent sediment and diatom accumulation in a small, monomictic, eutrophic lake. *Journal of Paleolimnology*, **3**, 143–60.

(1990c). Variability of sediment diatom assemblages in an upland, wind-stressed lake
(Loch Fleet, Galloway, Scotland). *Journal of Paleolimnology*, **4**, 43–59.

(1993). Natural versus anthropogenic change in lakes: The role of the sediment record.
Trends in Ecology and Evolution, **8**, 356–61.

(1995a). Using the past to predict the future: Lake sediments and the modelling of
limnological disturbance. *Ecological Modelling*, **78**, 149–72.

(1995b). Naturally eutrophic lakes: Reality, myth or myopia? *Trends in Ecology and
Evolution*, **10**, 137–8.

(1997). Reconstructing historical phosphorus concentrations in rural lakes using
diatom models. In *Phosphorus Loss to Water From Agriculture*, ed. H. Tunney, P. C.
Brookes, & A. E. Johnson, pp. 95–118. Wallingford, UK: CAB International.

(1998). Variability of diatom-inferred phosphorus profiles in a small lake basin and
its implications for histories of lake eutrophication. *Journal of Paleolimnology*, **20**,
47–55.

Anderson, N. J., & Battarbee, R. W. (1994). Aquatic community persistence and variability:
A palaeoecological perspective. In *Aquatic Ecology: Scale, Pattern and Process*, ed. P. A.
Giller, A. Hildrew, & D. Rafelli, pp. 233–59. Oxford, UK: Blackwell Scientific Press.

Anderson, N. J., & Odgaard, B. V. (1994). Recent palaeoecology of three shallow Danish lakes.
Hydrobiologia, **275/276**, 411–22.

Anderson, N. J., & Rippey, B. (1994). Monitoring lake recovery from point-source eutrophication: The use of diatom-inferred epilimnetic total phosphorus and sediment chemistry. *Freshwater Biology*, **32**, 625–39.

Anderson, N. J., & Vos, P. (1992). Learning from the past: Diatoms as palaeoecological indicators of changes in marine environments. *Netherlands Journal of Aquatic Ecology*, **26**,
19–30.

Anderson, N. J., Rippey, B., & Stevenson, A. C. (1990). Change to a diatom assemblage in a
eutrophic lake following point source nutrient re-direction: A paleolimnological
approach. *Freshwater Biology*, **23**, 205–17.

Anderson, N. J., Rippey, B., & Gibson, C. E. (1993). A comparison of sedimentary and diatom-
inferred phosphorus profiles: Implications for defining pre-disturbance nutrient
conditions. *Hydrobiologia*, **253**, 357–66.

Battarbee, R. W. (1978). Biostratigraphical evidence for variations in the recent patterns of
sediment accumulation in Lough Neagh, Northern Ireland. *Internationale
Vereinigung für theoretische und angewandte Limnologie, Verhandlungen*, **20**, 625–9.

(1986). Diatom analysis. In *Handbook of Holocene Palaeoecology and Palaeohydrology*, ed. B. E.
Berglund, pp. 527–70. Chichester, UK: John Wiley.

Bennion, H. (1994). A diatom–phosphorus transfer-function for shallow, eutrophic ponds
in southeast England. *Hydrobiologia*, **275/276**, 391–410.

(1995). Surface-sediment diatom assemblages in shallow, artificial, enriched ponds,
and implications for reconstructing trophic status. *Diatom Research*, **10**, 1–19.

Bennion, H., Wunsam, S., & Schmidt, R. (1995). The validation of diatom-phosphorus transfer functions: An example from Mondsee, Austria. *Freshwater Biology*, **34**, 271–83.

R. I. HALL AND J. P. SMOL

Bennion, H., Juggins, S., & Anderson, N. J. (1996). Predicting epilimnetic phosphorus concentrations using an improved diatom-based transfer function and its application to lake management. *Environmental Science and Technology*, **30**, 2004–7.

Birks, H. J. B. (1994). The importance of pollen and diatom taxonomic precision in quantitative palaeoecological reconstructions. *Review of Palaeobotany and Palynology*, **83**, 107–17.

(1995). Quantitative palaeoenvironmental reconstructions. In *Statistical Modelling of Quaternary Science Data*, Technical Guide **5**, ed. D. Maddy & J. S. Brew, pp. 161–254. Cambridge, UK: Quaternary Research Association

Birks, H. J. B., Juggins, S., & Line, J. M. (1990a) Lake surface-water chemistry reconstructions from palaeoecological data. In *The Surface Waters Acidification Programme*, ed. B. J. Mason, pp. 301–11. Cambridge, UK: Cambridge University Press.

Birks, H. J. B., Line, J. M., Juggins, S., Stevenson, A. C., & ter Braak, C. J. F. (1990b). Diatoms and pH reconstructions. *Philosophical Transactions of the Royal Society of London, series B*, **327**, 263–78.

Birks, H. J. B., Anderson, N. J., & Fritz, S. C. (1995). Post-glacial changes in total phosphorus at Diss Mere, Norfolk inferred from fossil diatom assemblages. In *Ecology and Palaeoecology of Lake Eutrophication*, ed. S. T. Patrick & N. J. Anderson, pp. 48–9. Copenhagen, DK: Geological Survey of Denmark DGU Service Report no. 7

Bjöjk-Ramberg, S., & Änell, C. (1985). Production and chlorophyll concentration of epipelic and epilithic algae in fertilized and nonfertilized subarctic lakes. *Hydrobiologia*, **126**, 213–9.

Borcard, D., Legendre, P., & Drapeau, P. (1992). Partialling out the spatial component of ecological variation. *Ecology*, **73**, 1045–55.

Bothwell, M. L. (1988). Growth rate response of lotic periphytic diatoms to experimental phosphorus additions. *Canadian Journal of Fisheries and Aquatic Sciences*, **45**, 261–70.

Boucherle, M. M., Smol, J. P., Oliver, T. C., Brown, S. R., & McNeely, R. (1986). Limnological consequences of the decline in hemlock 4800 years ago in three Southern Ontario lakes. *Hydrobiologia*, **143**, 217–25.

Bradbury, J. P. (1975). *Diatom Stratigraphy and Human Settlement*. The Geological Society of America Special Paper no. 171. Boulder, CO.

Brenner, M., Whitmore, T. S., Flannery, M. S., & Binford M. W. (1993). Paleolimnological methods for defining target conditions in lake restoration: Florida case studies. *Lake and Reservoir Management*, **7**, 209–17.

Brugam, R. B. (1979). A re-evaluation of the Araphidineae/Centrales index as an indicator of lake trophic status. *Freshwater Biology*, **9**, 451–60.

Brugam, R. B., & Patterson, C. (1983). The A/C (Araphidineae/Centrales) ratio in high and low alkalinity lakes in eastern Minnesota. *Freshwater Biology*, **13**, 47–55.

Brugam, R. B., & Vallarino, J. (1989). Paleolimnological investigations of human disturbance in Western Washington lakes. *Archiv für Hydrobiologie Beiheft Ergebnisse der Limnologie*, **116**, 129–59.

Burkhardt, M. A. (1996). Nutrients. In *Algal Ecology: Freshwater Benthic Ecosystems*, ed. R. J. Stevenson, M. L. Bothwell, & R. L. Lowe, pp. 184–228. San Diego, CA: Academic Press.

Canfield, D. E., Jr., & Bachmann, R. W. (1981). Prediction of total phosphorus concentration, chlorophyll *a* and Secchi depths in natural and artificial lakes. *Canadian Journal of Fisheries and Aquatic Sciences*, **38**, 414–23.

Carpenter, S. R., Christensen, D. L., Cole, J. J., Cottingham, K. L., He, X., Hodgson, J. R., Kitchell, J. F., Knight, S. E., Pace, M. L., Post, D. M., Schindler, D. E., & Voichick, N. (1995). Biological control of eutrophication in lakes. *Environmental Science and Technology*, **29**, 784–6.

Cattaneo, A. (1987). Periphyton in lakes of different trophy. *Canadian Journal of Fisheries and Aquatic Sciences*, **44**, 296–303.

Charles, D. F., & Smol, J. P. (1994). Long-term chemical changes in lakes: Quantitative inferences using biotic remains in the sediment record. In *Environmental Chemistry of Lakes and Reservoirs: Advances in Chemistry, Series 237*, ed. L. Baker, pp. 3–31. Washington, DC: American Chemical Society.

Charles, D. F., Smol, J. P., & Engstrom, D. R. (1994). Paleolimnological approaches to biomonitoring. In *Biological Monitoring of Aquatic Systems*, S. Loeb & D. Spacie, pp. 233–93.
Ann Arbor, MI: Lewis Press.

Cholnoky, B. J. (1968). *Die Okologie der Diatomeen in Binnengewässern*. Weinheur: J. Cramer.

Christie, C. E., & Smol, J. P. (1993). Diatom assemblages as indicators of lake trophic status
in southeastern Ontario lakes. *Journal of Phycology*, **29**, 575–86.

(1996). Limnological effects of 19th century canal construction and other disturbances
on the trophic state history of Upper Rideau Lake, Ontario. *Lake and Reservoir
Management*, **12**, 78–90.

Davis, M. B. (1981). Outbreaks of forest pathogens in Quaternary history. *Proceedings of the
IV International Palynology Conference, Lucknow, India (1976–77)*, **3**, 216–27.

Davis, M. B., & Ford, M. S. (1982). Sediment focussing in Mirror Lake, New Hampshire.
Limnology and Oceanography, **27**, 137–50.

DeYoe, H. R., Lowe, R. L., & Marks, J. C. (1992). Effects of nitrogen and phosphorus on the
endosymbiont load of *Rhopalodia gibba* and *Epithemia turgida* (Bacillariophyceae).
Journal of Phycology, **28**, 773–7.

Digerfeldt, G., & Håkansson, H. (1993). The Holocene paleolimnology of Lake Såmbosjön,
Southwestern Sweden. *Journal of Paleolimnology*, **8**, 189–210.

Dillon, P. J., & Rigler, F. H. (1975). A simple method for predicting the capacity of a lake for
development based on lake trophic status. *Journal of the Fisheries Research Board of
Canada*, **32**, 1519–31.

Dillon, P. J., Nicholls, K. H., Scheider, W. A., Yan, N. D., & Jeffries, D. S. (1986). *Lakeshore
Capacity Study, Trophic Status*. Research and Special Projects Branch, Ontario Ministry
of Municipal Affairs and Housing. Toronto, ON: Queen's Printer for Ontario.

Dixit, S. S., & Smol, J. P. (1994). Diatoms as indicators in the Environmental Monitoring and
Assessment Program-Surface Waters (EMAP-SW). *Environmental Monitoring and
Assessment*, **31**, 275–306.

Dixit, S. S., Smol, J. P., Kingston, J. C., & Charles, D. F. (1992). Diatoms: Powerful indicators
of environmental change. *Environmental Science and Technology*, **26**, 23–33.

Donar, C. M., Neely, R. K., & Stoermer, E. G. (1996). Diatom succession in an urban reservoir
system. *Journal of Paleolimnology*, **15**, 237–43.

Efron, G., & Gong, G. (1983). A leisurely look at the bootstrap, the jackknife, and cross validation. *American Statistician*, **37**, 36–48.

Engstrom, D. R., & Wright, H. E., Jr. (1984). Chemical stratigraphy of lake sediments as a
record of environmental change. In *Lake Sediments and Environmental History*, ed. E. Y.
Haworth & J. W. G. Lund, pp. 1–67. Minneapolis, MN: University of Minnesota Press.

Engstrom, D. R., Swain, E. B., & Kingston, J. C. (1985). A palaeolimnological record of human
disturbance from Harvey's Lake, Vermont: Geochemistry, pigments and diatoms.
Freshwater Biology, **15**, 261–88.

Ennis, G. L., Northcote, T. G., & Stockner, J. G. (1983). Recent trophic changes in Kootenay
Lake, British Columbia, as recorded by fossil diatoms. *Canadian Journal of Botany*, **61**,
1983–92.

Fairchild, G. W., & Sherman, J. W. (1992). Linkage between epilithic algal growth and water
column nutrients in softwater lakes. *Canadian Journal of Fisheries and Aquatic Sciences*,
49, 1641–9.

(1993). Algal periphyton response to acidity and nutrients in softwater lakes: Lake
comparison vs. nutrient enrichment approaches. *Journal of the North American
Benthological Society*, **12**, 157–67.

Fairchild, G. W., Lowe, R. L., & Richardson, W. T. (1985). Algal periphyton growth on nutri-
ent-diffusing substrates: An *in situ* bioassay. *Ecology*, **66**, 465–72.

Ford, J. (1988). The effects of chemical stress on aquatic species composition and community
structure. In *Ecotoxicology: Problems and Approaches*, ed. S. A. Lewin, M. A. Harwell, J. R.
Kelly, & K. D. Kimball, pp. 99–114. New York, NY: Springer-Verlag.

Foy, R. H., Smith, R. V., & Stevens, R. V. (1982). Identification of the factors affecting nitrogen
and phosphorus loading to Lough Neagh. *Journal of Environmental Management*, **15**,
109–29.

Fritz, S. C. (1989). Lake development and limnological response to prehistoric and historic land-use in Diss, Norfolk, U. K. *Journal of Ecology*, **77**, 182–202.

Fritz, S. C., Kingston, J. C., & Engstrom, D. R. (1993). Quantitative trophic reconstructions from sedimentary diatom assemblages: A cautionary tale. *Freshwater Biology*, **30**, 1–23.

Gause, G. F. (1930). Studies on the ecology of the Orthoptera. *Ecology*, **11**, 307–25.

Gibson, C. E., Foy, R. H., & Bailey-Watts, A. E. (1996). An analysis of the total phosphorus cycle in temperate lakes: The response to enrichment. *Freshwater Biology*, **35**, 525–32.

Goldman, C. R. (1981). Lake Tahoe: Two decades of change in a nitrogen deficient oligotrophic lake. *Internationale Vereinigung für theoretische und angewandte Limnologie, Verhandlungen*, **24**, 411–5.

Håkansson, H., & Regnéll, R. (1993). Diatom succession related to land use during the last 6000 years: A study of a small eutrophic lake in southern Sweden. *Journal of Paleolimnology*, **8**, 49–69.

Hall, R. I., & Smol, J. P. (1992). A weighted-averaging regression and calibration model for inferring total phosphorus concentration from diatoms in British Columbia (Canada) lakes. *Freshwater Biology*, **27**, 417–34.

(1993). The influence of catchment size on lake trophic status during the hemlock decline and recovery (4800 to 3500 BP) in southern Ontario lakes. *Hydrobiologia*, **269/270**, 371–90.

(1996). Paleolimnological assessment of long-term water-quality changes in south-central Ontario lakes affected by cottage development and acidification. *Canadian Journal of Fisheries and Aquatic Sciences*, **53**, 1–17.

Hall, R. I., Leavitt, P. R., Smol, J. P., & Zirnhelt, N. (1997). Comparison of diatoms, fossil pigments and historical records as measures of lake eutrophication. *Freshwater Biology*, **38**, 401–17.

Hansson, L. A. (1988). Effects of competitive interactions on the biomass development of planktonic and periphytic algae in lakes. *Limnology and Oceanography*, **33**, 121–8.

(1992). Factors regulating periphytic algal biomass. *Limnology and Oceanography*, **37**, 322–8.

Hargesheimer, E. E., & Watson, S. B. (1996). Drinking water treatment options for taste and odour control. *Water Research*, **30**, 1423–30.

Hargrave, B. T. (1969). Epibenthic algal production and community respiration in the sediments of Marion Lake. *Journal of the Fisheries Research Board of Canada*, **26**, 2003–26.

Harper, D. (1992). *Eutrophication of Freshwaters*. London, UK: Chapman Hall.

Hawes, I. H., & Smith, R. (1992). Effect of localised nutrient enrichment on the shallow epilithic periphyton of oligotrophic Lake Taupo. *New Zealand Journal of Marine and Freshwater Research*, **27**, 365–72.

Haworth, E. Y. (1984). Stratigraphic changes in algal remains (diatoms and chrysophytes) in the recent sediments of Blelham Tarn, English Lake District. In *Lake Sediments and Environmental History*, ed. E. Y. Haworth & J. W. G. Lund, pp. 165–90. Leicester, UK: Leicester University Press.

Hayes, C. R., & Greene, L. A. (1984). The evaluation of eutrophication impact in public water supply reservoirs in East Anglia. *Water Pollution Control*, **83**, 45–51.

Hecky, R. E. (1993). The eutrophication of Lake Victoria. *Internationale Vereinigung für theoretische und angewandte Limnologie, Verhandlungen*, **25**, 39–48.

Hickman, M., Schweger, C. E., & Klarer, D. M. (1990). Baptiste Lake, Alberta – a late Holocene history of changes in a lake and its catchment in the southern boreal forest. *Journal of Paleolimnology*, **4**, 253–67.

Hustedt, F. (1930). *Bacillariophyta (Diatomeae). Die Süsswasserflora Mitteleuropas*, 2nd edn, vol. 10. Stuttgart: Gustav Fischer Verlag.

Hutchinson, N. J., Neary, B. P., & Dillon, P. J. (1991). Validation and use of Ontario's Trophic Status Model for establishing lake development guidelines. *Lake and Reservoir Management*, **7**, 13–23.

Jeppesen, E., Kristensen, P., Jensen, J. P., Sondergaard, M., Mortensen, E., & Lauidsen, T. (1991). Recovery resilience following a reduction in external phosphorus loading of shallow, eutrophic Danish lakes: Duration, regulating factors and methods for overcoming resilience. *Memoire dell'Istituto Italiano di Idrobiologia*, **48**, 127–48.

Jeppesen, E., Madsen, E. A., & Jensen, J. P. (1996). Reconstructing the past density of planktivorous fish and trophic structure from sedimentary zooplankton fossils: A surface sediment calibration data set from shallow lakes. *Freshwater Biology*, **36**, 115–27.

Jones, R., Dickman, M. D., Mott, R. J., & Ouellet, M. (1984). Late Quaternary diatom and chemical profiles from a meromictic lake in Quebec, Canada. *Chemical Geology*, **44**, 267–86.

Jones, V. J., & Juggins, S. (1995). The construction of a diatom-based chlorophyll *a* transfer function and its application at three lakes on Signy Island (maritime Antarctic) subject to differing degrees of nutrient enrichment. *Freshwater Biology*, **34**, 433–45.

Juggins, S., & ter Braak, C. J. F. (1992). *CALIBRATE – a program for species-environment calibration by [weighted averaging] partial least squares regression*. Environmental Change Research Centre, Unpublished computer program. London, UK: University College London.

Juggins, S., Anderson, N. J., & Bennion, H. (1995). Quantitative phosphorus reconstructions from sedimentary diatom remains. In *Ecology and Palaeoecology of Lake Eutrophication*, ed. S. T. Patrick & N. J. Anderson, pp. 48–9. Copenhagen, DK: Geological Survey of Denmark DGU Service Report no. 7.

Kann, J., & Falter, C. M. (1989). Periphyton indicators of enrichment in Lake Pend Oreille, Idaho. *Lake and Reservoir Management*, **5**, 39–48.

Karst, T. L. (1996). Paleolimnological analyses of lake eutrophication patterns in Collins Lake and Lake Opinicon, Southeastern Ontario, Canada. MSc Thesis, Queen's University, Department of Biology, Kingston, Ontario.

Kilham, S. S., Theriot, E. C., & Fritz, S. C. (1996). Linking planktonic diatoms and climate change in the large lakes of the Yellowstone ecosystem using resource theory. *Limnology and Oceanography*, **41**, 1052–62.

Kotak, B. G., Lam, A. K-Y., Prepas, E. E., Kenefick, S. L., & Hrudey, S. E. (1995). Variability of the hepatotoxin microcystin-LR in hypereutrophic drinking water lakes. *Journal of Phycology*, **31**, 248–63.

Krammer, K., & Lange-Bertalot, H. (1986–1991). *Süsswasserflora von Mitteleuropa*, Band 2(vols. 1–4). Stuttgart: Gustav Fischer Verlag.

Kreis, R. G., Jr., Stoermer, E. F., & Ladewski, T. B. (1985). *Phytoplankton Species Composition, Abundance, and Distribution in Southern Lake Huron, 1980; Including a Comparative Analysis with Conditions in 1974 Prior to Nutrient Loading Reductions*. Great Lakes Research Division Special Report no. 107. Ann Arbor, MI: The University of Michigan.

Leavitt, P. R. (1993). A review of factors that regulate carotenoid and chlorophyll deposition and fossil pigment abundance. *Journal of Paleolimnology*, **9**, 109–27.

Leavitt, P. R., & Findlay, D. L. (1994). Comparison of fossil pigments with 20 years of phytoplankton data from eutrophic Lake 227, Experimental Lakes Area, Ontario. *Canadian Journal of Fisheries and Aquatic Sciences*, **51**, 2286–99.

Leavitt, P. R., Carpenter, S. R., & Kitchell, J. F. (1989). Whole-lake experiments: The annual record of fossil pigments and zooplankton. *Limnology and Oceanography*, **34**, 700–17.

Leavitt, P. R., Hann, B. J., Smol, J. P., Zeeb, B. A., Christie, C. E., Wolfe, B., & Kling, H. (1994). Paleolimnological analysis of whole-lake experiments: An overview of results from Experimental Lakes Area Lake 227. *Canadian Journal of Fisheries and Aquatic Sciences*, **51**, 2322–32.

Likens, G. E. (ed.) (1989). *Long-Term Studies in Ecology*. New York, NY: Springer-Verlag.

Line, J. M., & Birks, H. J. B. (1990). WACALIB version 2.1 – A computer program to reconstruct environmental variables from fossil assemblages by weighted averaging. *Journal of Paleolimnology*, **3**, 170–3.

Line, J. M., ter Braak, C. J. F., & Birks, H. J. B. (1994). WACALIB version 3.3 – a computer program to reconstruct environmental variables from fossil assemblages by weighted averaging and to derive sample-specific errors of prediction. *Journal of Paleolimnology*, **10**, 147–52.

Lotter, A. F., Ammann, B., & Sturm, M. (1992). Rates of change and chronological problems during the late-glacial period. *Climate Dynamics*, **6**, 233–9.

Lowe, R. L. (1996). Periphyton patterns in lakes. In *Algal Ecology: Freshwater Benthic Ecosystems*, ed. R. J. Stevenson, M. L. Bothwell, & R. L. Lowe, pp. 57–77. San Diego, CA: Academic Press.

Lowe, R. L., & Hunter, R. D. (1988). Effect of grazing by *Physa integra* on periphyton community structure. *Journal of the North American Benthological Society*, **7**, 29–36.

Lund, J. W. G. (1950). Studies on *Asterionella* Hass. I. The origin and nature of the cells producing seasonal maxima. *Journal of Ecology*, **38**, 1–35.

Lund, J. W. G., & Reynolds, C. S. (1982). The development and operation of large limnetic enclosures in Blelham Tarn, English lake District, and their contribution to phytoplankton ecology. *Progress in Phycological Research*, **1**, 1–65.

Manny, A., Wetzel, R. G., & Bailey, R. E. (1978). Paleolimnological sedimentation of organic carbon, nitrogen, phosphorus, fossil pigments, pollen and diatoms in a hypereutrophic hardwater lake: A case history of eutrophication. *Polskie Archiwum Hydrobiologii*, **25**, 243–67.

Marchetto, A & Bettinetti, R. (1995). Reconstruction of the phosphorus history of two deep, subalpine Italian lakes from sedimentary diatoms, compared with long-term chemical measurements. *Memoire dell'Istituto Italiano di Idrobiologia*, **53**, 27–38.

Margalef, R. (1968). *Perspectives in Ecological Theory*. Chicago, IL: University of Chicago Press.

Marsden, M. W. (1989). Lake restoration by reducing external phosphorus loading: the influence of sediment phosphorus release. *Freshwater Biology*, **21**, 139–62.

Mason, C. F. (1991). *Biology of Freshwater Pollution*, 2nd edn. Essex: Longman Group (FE) Ltd.

Montgomery, D. C., & Peck, E. A. (1982). *Introduction to Linear Regression Analysis*. New York, NY: John Wiley.

Moss, B. (1978). The ecological history of a medieval man-made lake, Hickling Broad, Norfolk, United Kingdom. *Hydrobiologia*, **60**, 23–32.

Niederhauser, P., & Schanz, F. (1993). Effects of nutrient (N, P, C) enrichment upon the littoral diatom community of an oligotrophic high-mountain lake. *Hydrobiologia*, **269/270**, 453–62.

Nürnberg, G. K. (1984). The prediction of internal phosphorus load in lakes with anoxic hypolimnia. *Limnology and Oceanography*, **29**, 111–24.

Nygaard, G. (1949). Hydrobiological studies on some Danish ponds and lakes. II: The Quotient hypothesis and some new or little known phytoplankton organisms. *Det Kongelinge Dansk Videnskabernes Selskab Biologiske Skrifter*, **7**, 1–193.

Oksanen, J., Läärä, E., Huttunen, P., & Meriläinen, J. (1988). Estimation of pH optima and tolerances of diatoms in lake sediments by the methods of weighted averaging, least squares and maximum likelihood, and their use for the prediction of lake acidity. *Journal of Paleolimnology*, **1**, 39–49.

Osborne, P. L., & Moss, B. (1977). Palaeolimnology and trends in the phosphorus and iron budgets of an old man-made lake, Barton Broad, Norfolk. *Freshwater Biology*, **7**, 213–33.

O'Sullivan, P. E. (1992). The eutrophication of shallow coastal lakes in Southwest England – understanding and recommendations for restoration, based on palaeolimnology, historical records, and the modelling of changing phosphorus loads. *Hydrobiologia*, **243/244**, 421–34.

Pan, Y., & Stevenson, R. J. (1996). Gradient analysis of diatom assemblages in western Kentucky wetlands. *Journal of Phycology*, **32**, 222–32.

Patrick, R., & Reimer, C. (1966). *The Diatoms of the United States*, vol. 1, Monograph 3, pp. 1–668. Philadelphia, PA: Academy of Natural Sciences.

(1975). *The Diatoms of the United States*, vol. 2 part 1, Monograph 13, pp. 1–213. Philadelphia, PA: Academy of Natural Sciences.

Peglar, S. M. (1993). The development of the cultural landscape around Diss Mere, Norfolk, UK, during the past 7000 years. *Review of Palaeobotany and Palynology*, **76**, 1–47.

Peglar, S. M., Fritz, S. C., & Birks, H. J. B. (1989). Vegetation and land-use history in Diss, Norfolk, England. *Journal of Ecology*, **77**, 203–22.

Polis, G. A., & Winemiller, K. O. (ed.) (1996). *Food Webs: Integration of Patterns and Dynamics*. New York, NY: Chapman and Hall.

Qinghong, L., & Bråkenhielm, S. (1995). A statistical approach to decompose ecological variation. *Water, Air and Soil Pollution*, **85**, 1587–92.

Quinlan, R., Smol, J. P., & Hall, R. I. (1998). Quantitative inferences of past hypolimnetic anoxia in south-central Ontario lakes using fossil chironomids (Diptera: Chironomidae). *Canadian Journal of Fisheries and Aquatic Sciences*, **55**, 587–96.

Reavie, E. D. (1997). *Diatom Ecology and Paleolimnology of the St. Lawrence River*. PhD thesis, Biology Department, Queen's University, Kingston, Ontario, 224 pp.

Reavie, E. D., Hall, R. I., & Smol, J. P. (1995). An expanded weighted-averaging model for inferring past total phosphorus concentrations from diatom assemblages in eutrophic British Columbia (Canada) lakes. *Journal of Paleolimnology*, **14**, 49–67.

Reckow, K. H., & Simpson, J. T. (1980). A procedure using modelling and error analysis for the prediction of lake phosphorus concentration from land use information. *Canadian Journal of Fisheries and Aquatic Sciences*, **37**, 1439–48.

Revsbeck, N. P., & Jørgensen, B. B. (1986). Microelectrodes: Their use in microbial ecology. *Advances in Microbial Ecology*, **9**, 749–56.

Revsbeck, N. P., Jørgensen, B. B., Blackburn, T. H., & Cohen, Y. (1983). Microelectrode studies of photosynthesis and O_2, H_2S and pH profiles of a microbial mat. *Limnology and Oceanography*, **28**, 1062–74.

Reynolds, C. S. (1984). *The Ecology of Freshwater Phytoplankton*. Cambridge, UK: Cambridge University Press.

Rippey, B. (1983). A laboratory study of the silicon release process from a lake sediment (Lough Neagh, Northern Ireland). *Archiv für Hydrobiologie Beiheft Ergebnisse der Limnologie*, **96**, 417–33.

(1995). Lake phosphorus models. In *Ecology and Palaeoecology of Lake Eutrophication*, ed. S. T. Patrick & N. J. Anderson, pp. 58–60. Copenhagen, DK: Geological Survey of Denmark DGU Service Report no. 7

Rippey, B., & Anderson, N. J. (1996). Reconstruction of lake phosphorus loading and dynamics using the sedimentary record. *Environmental Science and Technology*, **30**, 1786–8.

Rippey, B., Anderson, N. J., & Foy, R. H. (1997). Accuracy of diatom-inferred total phosphorus concentrations, and the accelerated eutrophication of a lake due to reduced flushing and increased internal loading. *Canadian Journal of Fisheries and Aquatic Sciences*, **54**, 2637–46.

Round, F. E., Crawford, R. M., & Mann, D. G. (1990). *The Diatoms: Biology and Morphology of the Genera*. Cambridge, UK: Cambridge University Press.

Sandman, O., Lichu, A., & Simola, H. (1990). Drainage ditch erosion history as recorded in the varved sediment of a lake in East Finland. *Journal of Paleolimnology*, **3**, 161–9.

Schelske, C. L., & Stoermer, E. F. (1971). Eutrophication, silica depletion and predicted changes in algal quality in Lake Michigan. *Science*, **173**, 423–4.

(1972). Phosphorus, silica and eutrophication of Lake Michigan. In *Nutrients and Eutrophication*, ed. G. E. Likens, pp. 157–71. American Society of Limnology and Oceanography Special Symposium, Volume 1.

Schelske, C. L., Rothman, E. D., Stoermer, E. F., & Santiago, M. A. (1974). Responses of phosphorus limited Lake Michigan phytoplankton to factorial enrichments with nitrogen and phosphorus. *Limnology and Oceanography*, **19**, 409–19.

Schelske, C. L., Stoermer, E. F., Fahnenstiel, G. L., & Haibach, G. L. (1986). Phosphorus enrichment, silica utilization, and biogeochemical silica depletion in the Great Lakes. *Canadian Journal of Fisheries and Aquatic Sciences*, **43**, 407–15.

Schindler, D. W. (1977). Evolution of phosphorus limitation in lakes. *Science*, **195**, 260–2.

Simola, H. (1979). Micro-stratigraphy of sediment laminations deposited in a chemically stratifying eutrophic lake during the years 1913–1976. *Holarctic Ecology*, **2**, 160–8.

(1983). Limnological effects of peatland drainage and fertilization as reflected in the varved sediment of a deep lake. *Hydrobiologia*, **106**, 43–57.

Smol, J. P. (1990). Paleolimnology: Recent advances and future challenges. *Memoire dell'Istituto Italiano di Idrobiologia* **47**, 253–76.

(1992). Paleolimnology: An important tool for effective ecosystem management. *Journal of Aquatic Ecosystem Health*, **1**, 49–58.

(1995). Paleolimnological approaches to the evaluation and monitoring of ecosystem health: Providing a history for environmental damage and recovery. In *Evaluating and Monitoring the Health of Large-Scale Ecosystems: NATO ASI Series, Vol. 128*, ed. D. J. Rapport, C. L. Gaudet, & P. Calow, pp. 301–18. Berlin: Springer-Verlag.

Smol, J. P., & Dickman, M. D. (1981). The recent histories of three Canadian Shield lakes: A paleolimnological experiment. *Archiv für Hydrobiologie Beiheft Ergebnisse der Limnologie*, **93**, 83–108.

Soranno, P. A., Hubler, S. L., Carpenter, S. R., & Lathrop, R. C. (1996). Phosphorus loads to surface waters: A simple model to account for spatial pattern of land use. *Ecological Applications*, **6**, 865–78.

Stevenson, A. C., Juggins, S., Birks, H. J. B., Anderson, D. S., Anderson, N. J., Battarbee, R. W., Berge, F., Davis, R. B., Flower, R. J., Haworth, E. Y., Jones, V. J., Kingston, J. C., Kreiser, A. M., Line, J. M., Munro, M. A. R., & Renberg, I. (1991). *The Surface Waters Acidification Project Palaeolimnology Programme: Modern Diatom/Lake-water Chemistry Data-set.* London, UK: ENSIS Publishing.

Stockner, J. G. (1971). Preliminary characterization of lakes in the Experimental Lakes Area, north-western Ontario using diatom occurrence in lake sediments. *Journal of the Fisheries Research Board of Canada*, **28**, 265–75.

(1972). Paleolimnology as a means of assessing eutrophication. *Internationale Vereinigung für theoretische und angewandte Limnologie, Verhandlungen*, **18**, 1018–30.

Stockner, J. G., & Benson, W. W. (1967). The succession of diatom assemblages in the recent sediments of Lake Washington. *Limnology and Oceanography*, **12**, 513–32.

Stockner, J. G., & Shortreed, K. R. S. (1978). Enhancement of autotrophic production by nutrient addition in a coastal rainforest stream on Vancouver Island. *Journal of the Fisheries Research Board of Canada*, **35**, 28–34.

Stoermer, E. F., & Yang, J. J. (1970). *Distribution and Relative Abundance of Dominant Planktonic Diatoms in Lake Michigan.* Great Lakes Research Division Publication no. 16. Ann Arbor, MI: University of Michigan.

Stoermer, E. F., Emmert, G., Julius, M. L., & Schelske, C. L. (1996). Paleolimnologic evidence of rapid change in Lake Erie's trophic status. *Canadian Journal of Fisheries and Aquatic Sciences*, **53**, 1451–8.

ter Braak, C. J. F., & Juggins, S. (1993). Weighted averaging partial least squares regression (WA-PLS): an improved method for reconstructing environmental variables from species assemblages. *Hydrobiologia*, **269/270**, 485–502.

ter Braak, C. J. F., & van Dam, H. (1989). Inferring pH from diatoms: A comparison of old and new calibration methods. *Hydrobiologia*, **178**, 209–23.

ter Braak, C. J. F., Juggins, S., Birks, H. J. B., & van der Voet, H. (1993). Weighted averaging partial least squares regression (WA-PLS): Definition and comparison with other methods for species–environment calibration. In *Multivariate Environmental Statistics*, ed. G. P. Patil & C. R. Rao, pp. 525–60. Amsterdam, NL: Elsevier Science Publishers.

Thornthwaite, C. W., & Mather, J. R. (1957). Instructions and tables for computing the potential evapotranspiration and the water balance. *Publications in Climatology*, **10**, 185–311.

Tilman, D. (1977). Resource competition between planktonic algae: An experimental and theoretical approach. *Ecology*, **58**, 338–48.

Tilman, D., Kilham, S. S., & Kilham, P. (1982). Phytoplankton community ecology: the role of limiting nutrients. *Annual Review of Ecology and Systematics*, **13**, 349–72.

Tilman, D., Kiesling, R., Sterner, R., Kilham, S. S., & Johnson, F. A. (1986). Green, bluegreen and diatom algae: Taxonomic differences in competitive ability for phosphorus, silicon and nitrogen. *Archiv für Hydrobiolgie Beiheft Ergebnisse der Limnologie*, **106**, 473–85.

Turner, M. A., Howell, E. T., Robinson, G. G. C., Campbell, P., Hecky, R. E., & Schindler, E. U. (1994). Role of nutrients in controlling growth of epilithon in oligotrophic lakes of low alkalinity. *Canadian Journal of Fisheries and Aquatic Sciences*, **51**, 2784–93.

Vaughn, J. C. (1961). Coagulation difficulties of the south district filtration plant. *Pure Water*, **13**, 45–9.

van Dam, H., & Mertens, A. (1993). Diatoms on herbarium macrophytes as indicators for water quality. *Hydrobiologia*, **269/270**, 437–46.

Vinebrooke, R. D. (1996). Abiotic and biotic regulation of periphyton in recovering acidified lakes. *Journal of the North American Benthological Society*, **15**, 318–31.

Vollenweider, R. A. (1969). Möglichkeiten und Grenzen elementarer Modelle der Stoffbilanz von Seen. *Archiv für Hydrobiologie Beiheft Ergebnisse der Limnologie*, **66**, 1–36.

(1975). Input–output models with special reference to the phosphorus loading concept. *Schweizer Zeitschrift Hydrologie*, **37**, 58–83.

Walker, I. R., Reavie, E. D., Palmer, S., & Nordin. R. D. (1993). A palaeoenvironmental assessment of human impact on Wood Lake, Okanagan Valley, British Columbia, Canada. *Quaternary International*, **20**, 51–70.

Webster, K. E., Kratz, T. K., Bowser, C. J., & Magnusson, J. J. (1996). The influence of landscape position on lake chemical responses to drought in northern Wisconsin. *Limnology and Oceanography*, **41**, 977–84.

Werner, D. (ed.) (1977). *The Biology of Diatoms*. Berkely, CA: University of California Press.

Wetzel, R. G. (1964). A comparative study of the primary productivity of higher aquatic plants, periphyton and phytoplankton in a large shallow lake. *Internationale Revue der Gesamten Hydrobiologie* **48**, 1–61.

(1983). *Limnology*, 2nd edn. New York, NY: CBS College Publishing.

(1996). Benthic algae and nutrient cycling in lentic freshwater ecosystems. In *Algal Ecology: Freshwater Benthic Ecosystems*, ed. R. J. Stevenson, M. L. Bothwell & R. L. Lowe, pp. 641–69. San Diego, CA: Academic Press.

Wetzel, R. G., Rich, P. H., Miller, M. C., & Allen, H. L. (1972). Metabolism of dissolved and particulate detrital carbon in a temperate hard-water lake. *Memoire dell'Istituto Italiano di Idrobiologia Supplement* **29**, 253–76.

Whitehead, D. R., Rochester, H., Jr., Rissing, S. W., Douglas, C. B., & Sheehan, M. C. (1973). Late glacial and postglacial productivity changes in a New England pond. *Science*, **181**, 744–7.

Whitmore, T. J. (1989). Florida diatom assemblages as indicators of trophic state and pH. *Limnology and Oceanography*, **34**, 882–95.

Whitmore, T. J., Brenner, M., Curtis, J., Dahlin, B. H., & Leyden, B. W. (1996). Holocene climatic and human influences on lakes of the Yukatan Peninsula, Mexico: An interdisciplinary, palaeolimnological approach. *The Holocene*, **6**, 273–87.

Whittaker, H. (1956). Vegetation of the Great Smoky Mountains. *Ecological Monographs*, **26**, 1–80.

Whittier, T. R., & Paulsen, S. G. (1992). The surface waters component of the Environmental Monitoring and Assessment Program (EMAP): An overview. *Journal of Aquatic Ecosystem Health*, **2**, 119–26.

Wunsam, S., & Schmidt, R. (1995). A diatom-phosphorus transfer function for alpine and pre-alpine lakes. *Memoire dell'Istituto Italiano di Idrobiologia*, **53**, 85–99.

Yang, J-R., Duthie, H. C., & Delorme, L. D. (1993). Reconstruction of the recent environmental history of Hamilton Harbour (Lake Ontario, Canada) from analysis of siliceous microfossils. *Journal of Great Lakes Research*, **19**, 55–71.

Yang, J-R., Pick, F. R., & Hamilton, P. B. (1996). Changes in the planktonic diatom flora of a large mountain lake in response to fertilization. *Journal of Phycology*, **32**, 232–43.

Zeeb, B. A., Christie, C. E., Smol, J. P., Findlay, D. L., Kling, H., & Birks, H. J. B. (1994). Responses of diatom and chrysophyte assemblages in Lake 227 sediments to experimental eutrophication. *Canadian Journal of Fisheries and Aquatic Sciences*, **51**, 2300–11.

7 Continental diatoms as indicators of long-term environmental change

J. PLATT BRADBURY

Introduction

It is curious that diatoms, whose short lifespans and capacity for rapid regeneration make them especially suitable for short-term paleoenvironmental studies, would also have a significant role as indicators of long-term environmental change. This chapter explores the nature of long diatom records, their relation to global environmental changes, guidelines for their interpretion, and problems common to such records.

Definitions and concepts

To examine the use of continental diatoms as proxies of long-term environmental change, it is first necessary to define what is meant by 'long-term'. 'Long' for this paper refers to lake records that encompass several glacial / interglacial cycles (e.g., IGBP, 1992). At a minimum, long records reach the preceding interglacial, the Eemian or Sangamon, and correlate to the marine oxygen isotope stage 5e, about 125 000 years ago.

It is also relevant to discriminate between 'long-term environmental change' as opposed to short-term environmental changes that have occurred over long time periods. Long-term environmental change results from processes which operate along uninterrupted trends of thousands or millions of years. Only three basic processes actually operate at this scale: (i) Deep-seated lithosphere convection caused by radioactive decay that drives continental drift, tectonism and volcanism; (ii) Variations in orbital relations between the earth and sun that govern insolation and long-term climate change; (iii) Slowing of the earth's rotation and the increasing distance between earth and moon; very long-term (10^9 years) changes irrelevant to the scope of this chapter.

Interactions between the gradual processes of lithosphere convection and global insolation ultimately force both climate and biologic change (evolution and biogeography). However, the results are neither gradual nor unidirectional. Stratigraphic studies of biologic, climatic and tectonic processes reveal histories of short-term, abrupt and often drastic responses. In geology, the once preferred concept of gradual change over long time periods has been replaced by 'incremental catastrophism' or short-term changes that often occur rapidly and represent significant deviations from earlier environmental modes. The short-term nature of environmental change could reflect thresholds imposed by longer-term changes. Examples include earthquakes and climate-changing volcanism as responses to progressive strain imposed by continental drift. Nevertheless, not all short-term changes are permutations of gradual long-term processes that interact to produce the observed shifts. The quasibiennial oscillation, sunspot cycles, and secular variations in solar activity are examples of such short-term changes that affect terrestrial environments but are apparently not related to the long-term root processes mentioned above.

To summarize, diatoms do not exhibit long-term, gradual changes. Their habitats vary at short intervals, often at subseasonal scales and diatoms rapidly change and adapt to new conditions. The fossil diatom assemblages in long records of environmental change are actually environmental snapshots that reflect a summation of local conditions such as weather, nutrient fluxes, biotic and limnologic interactions. The time–stratigraphic scale represented by a fossil assemblage may often be much larger than the environmental time scale which determined the presence, absence or abundance of a given diatom species in that assemblage (e.g., Anderson, 1995). Nevertheless, fossil diatom assemblages from long records do record long-term changes because such records come from persistent, stable, long-lived, well-buffered lake systems suitable for recording long-term changes.

What sites are available for long-term studies?

Records of continental diatoms that extend far into the past may be derived from outcrops of diatomaceous sediments or from long cores from ancient, long-lived basins (Fig. 7.1). Outcrops have often been exploited for descriptive studies (e.g., VanLandingham, 1964), but isolated deposits have limited potential for either evolutionary or paleoenvironmental questions. The advent of radiometric dating methods for intercalated volcanic rocks in outcrops of Tertiary diatoms in combination with ages derived from fossil mammals or vascular plants in these sediments allowed Krebs et al. (1987) to examine a number of sites and establish the general chronological appearance of some widespread centric diatom genera (Fig. 7.2).

Paleoenvironmental studies of isolated outcrops, especially of Tertiary age, are compromised by disruptive tectonism and erosion that have confounded the relations of the sediments to the basin where they were originally deposited. It is impossible, from a single outcrop or even scattered outcrops, to learn much about basin depth, areal extent, groundwater relations, and the location and nature of the watershed that fed and drained the basin. In addition, comparative studies of outcrops will suffer from the likelihood that the outcrops are seldom exactly coeval or necessarily from the same basin. Consequently, differences in diatom floras from separated outcrops may relate

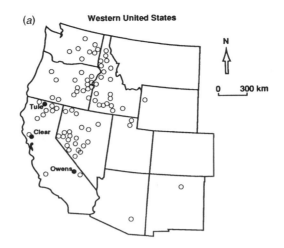

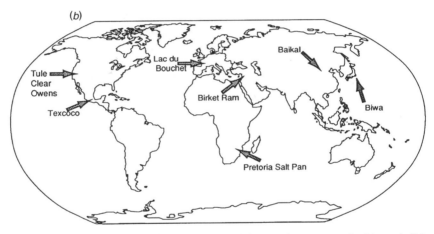

Fig. 7.1. (*a*). Locations of long records of continental diatoms in western United States (solid circles) and of outcrops of diatomites and diatomaceous sediments (open circles) of Tertiary and Quaternary age examined for construction of a continental diatom biochronology (generalized from Krebs & Bradbury, 1995). (*b*). World map showing approximate locations of selected long, continental diatom records.

to inherent limnological differences rather than broad climatic, biologic, or other environmental changes.

Modern basins and lakes with long diatom records avoid these problems to some extent. Long sediment records of diatoms will be preferentially found in large lake basins within down-warping or graben-forming tectonic regimes. Such regimes are necessary to maintain basin capacity and accommodate the sediment load over long periods. Otherwise, the basins would fill and cease to exist as lakes. Unfortunately, tectonism and graben formation is more often episodic than continual. As a consequence, long paleolimnologic sequences may record episodes of tectonic quiescence as shallow, marsh-like

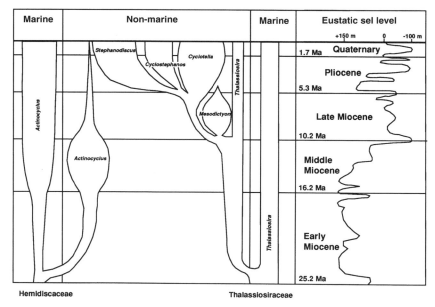

Marine	Non-marine	Marine	Eustatic sel level

Fig. 7.2. The approximate chronological appearance of widespread non-marine centric diatom genera in the western United States (modified from Krebs *et al.*, 1987) and general Neogene eustatic sea level curve (Haq *et al.*, 1987). Balloon width reflects species diversity.

or even fluvial conditions that result from basin filling rather than from climatic changes affecting the hydrologic balance. A good example of this apparently occurred between 650 ka and 510 ka at Owens Lake, California (36.5° N × 118.0° W) (Fig. 7.3) where sand-rich sediments indicate basin filling as a result of reworked volcanic ash and massive deposition of glacial detritus during that part of the mid-Pleistocene (Smith *et al.*, 1997).

Lake size is an important factor affecting the value of long-term diatom records. Large and comparatively deep lakes are likely to be buffered against short-term climatic and geomorphic changes that eliminate small lake, pond and marsh systems and thereby compromise the continuity of their records. Large lakes will have slower accumulation rates that average or dampen small changes, thus providing records at a scale compatible with their long records. Otherwise, assuming continual deposition, environmentally noisy records would prevail that could obscure correlation to climatic changes at regional scales.

How are diatoms interpreted in long records?

Using diatoms to interpret paleoenvironmental conditions must rely upon specific, often short-term ecological and physiological characteristics of individual species. Fossil diatom assemblages that accumulate over significant periods of time and that are sampled as such will contain many species that

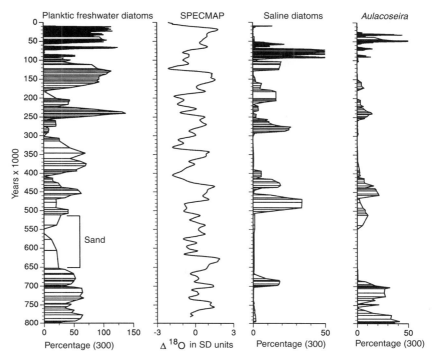

Fig. 7.3. Selected taxa and ecological groups of diatoms from Owens Lake, California and the marine oxygen isotope record (SPECMAP) for the past 800 ky.

never lived together in either time or space. The larger the stratigraphic sample thickness and the slower the sedimentation rate, the more time will be averaged in the prepared material and the more difficult it will be to relate the fossil assemblage to modern diatom communities and populations. Unless very coarse environmental discrimination (e.g., fresh vs. saline water) is satisfactory, there is no justification for taking long, 'channel' samples from a core. Even the stratigraphically small samples of long records will probably homogenize more than a decade of sediment accumulation that may already be mixed by bioturbation to some extent.

A wide range of specific environmental interpretations can be made with diatoms on the basis of their life histories, habitats, and ecological preferences (Bradbury & Krebs, 1995). Specific interpretations, many of which are very short-term in nature (e.g., the seasonal supply ratios of Si and P) could be applied to long records, although they probably have little long-term significance. However, such interpretations do provide insights into the nature of environmental variability that constrain more general interpretations of past environments. For example, *Aulacoseira ambigua* (Grunow) Simonsen and *A. granulata* (Ehrenberg) Simonsen are important summer-season planktonic diatoms today. Their appearance in the Owens Lake core (Fig. 7.3) (Bradbury, 1997) indicates fresh, nutrient-rich water and high-light environments in the past that probably reflect increased summer precipitation in an area where today winter precipitation dominates.

Despite the relevance of specific ecological interpretations of diatoms to long-term records, most long-term interpretations from diatoms are quite general. In the western United States, and in other areas where arid and wet climates have alternated, the most basic interpretation relies on the proportion of saline vs. freshwater diatoms which reflect arid and moist climates respectively. In regions of greater moisture, the most basic interpretation often relies on the proportion of planktonic to benthic diatoms that respectively indicate higher lake levels (or more open water conditions) vs. lower lake levels (or marsh environments). Occasionally it is useful to separate the planktonic diatoms into seasonal groups, although in modern lakes the seasonal distribution of many planktonic species is quite variable and lessens the paleoecologic value of this discrimination. Major sub-groups of benthic diatoms (motile, loosely attached, stalked, adnate) can provide insights into the contribution of rivers or near-shore habitats to large lakes (e.g., Bradbury et al., 1989). These 'large-scale' interpretations approximately match the long-term variations in large lake systems and are therefore appropriate for unraveling the paleolimnologic history of long lacustrine records of large lakes.

Interpretations of diatoms from long records tend to match the research questions asked of such records. Should very specific questions be asked, it is logical that specific interpretations could be made, just as they are in shorter, high-resolution lacustrine records.

Research directions

Investigations of continental diatoms over long periods of time fall into three basic approaches:

(i) Geology and tectonism;
(ii) Biogeography, ecology and evolution;
(iii) Paleoenvironments and paleoclimates.

GEOLOGY AND TECTONISM

Although continental drift has certainly been relevant to the present and past behavior of climate and oceans, relative movement of the continents has been comparatively minor over the geologic history of freshwater diatoms (Eocene to present). Consequently, it is unlikely that the study of diatoms will reveal much about progressive, large-scale changes in continental spatial relations. On the other hand, because ancient lake deposits appear in geological contexts of diastrophism, volcanism, and profound geomorphic change, diatom records can focus on geological issues. As examples, changes in regional drainage systems and the proximity of marine embayments was deduced as early as 1855 by Okeden in Wales and much later in the trough of the Colorado River by Bradbury & Blair (1979).

J. PLATT BRADBURY

Large-scale Tertiary paleogeographic, tectonic and eustatic sea level changes account for distinctive diatom assemblages in the western United States. Here, the abundance of freshwater, middle Miocene *Actinocyclus* species (e.g., Bradbury & Krebs, 1982; Bradbury *et al.*, 1985; Bradbury & Krebs, 1995) reflects the past presence of low-gradient rivers and extensive lakes in Idaho, Nevada, and Oregon produced by Basin and Range tectonic extension. The absence or low elevation of the Sierra Nevada and Cascade mountain ranges provided access for this mostly marine and brackish-water genus to enter and adapt to freshwater systems (Krebs, 1994). Modern analogues for this scenario may occur in coastal England (Belcher & Swale, 1979), and northern Germany (Hustedt, 1957), where *Actinocyclus* species have been found in lakes connected to the sea by estuaries or slow-moving rivers.

BIOGEOGRAPHY, ECOLOGY, AND EVOLUTION

For the most part, biological questions have been limited to description of diatoms from long extinct lake systems. A few such studies have proposed phylogenetic relations and hinted at hypothetical evolutionary trends manifested by extinct diatoms (Kociolek & Stoermer, 1989, 1990). Actual documentation of progressive morphological change throughout a sequence of lake deposits has not often been pursued. One exception (Theriot, 1992) tracks the development of *Stephanodiscus yellowstonensis* Theriot & Stoermer from its progenitor, *S. niagarae* Ehrenberg, in a core from Yellowstone Lake, Wyoming (44.4° N × 110.4° W) during the early Holocene. The morphologic change had been completed in about 3000 years. This study and observations of the morphological variability and rapid appearance of unique *Stephanodiscus* species in Lake Baikal (52.0–56.0° N × 104.0–110.0° E) (Bradbury *et al.*, 1994) indicate that diatoms can perhaps evolve as rapidly as climatic and limnologic changes allow. Long lacustrine records from sites like Lake Baikal have an excellent potential for documenting morphological diatom evolution, but understanding causal mechanisms requires analysis of additional paleolimnological and paleoecological proxies from these records.

Some long-term biogeographic and ecological attributes of diatoms and diatom floras have geological ramifications. For example, the appearance of extinct taxa in the geological record, or the first appearance of extant taxa may provide biochronologic data useful to determine the age of lacustrine deposits (e.g., VanLandingham, 1985; Krebs, 1994; Krebs & Bradbury, 1995). In the western United States and in several other parts of the world, centric diatom genera have appeared roughly synchronously (Fig. 7.2), whether by evolution or by biogeographic linkages under climatic and tectonic (sea level) control.

PALEOENVIRONMENTS AND PALEOCLIMATES

At the present time, long paleoenvironmental diatom records are typically investigated as a means of documenting past climate changes because of considerable global interest in this subject. This approach involves interpreting past limnological conditions from diatoms and subsequently extrapolating to the climatic environments responsible for determining past lacustrine conditions. Ultimately, the past climates result from orbital relations between the earth and sun that govern insolation (Hayes *et al.*, 1976).

Long diatom records have been assembled from outcrops of diatomaceous sediments in the African Rift Valley and interpreted in terms of lake evolution and climate (Gasse, 1975, 1980). However, the continuity and correlation of such records may be compromised as a result of erosion and incomplete exposures. Chronological control under such circumstances is often ambiguous. Consequently, paleolimnologic and paleoclimatic histories from outcrop sequences are difficult to compare with more continuous marine and ice core records.

A cored, 90-m record from the Pretoria Salt Pan, an impact crater in South Africa (25.6° S × 28.1° E), is estimated to be 200 ka in age at the base, according to extrapolation of sedimentation rates in the radiocarbon-dated upper 20 m (Partridge *et al.*, 1993). Diatoms exist throughout much of the record, although their stratigraphic distribution is discontinuous. Diatom analyses of the upper 50 m of the core indicate fluctuations between a saline, high pH assemblage (*Nitzschia pusilla* Grunow and *Chaetoceros muelleri* Lemmermann) and shallower and fresher water assemblages that include *Achnanthes exigua* Grunow, *Nitzschia palea* (Kützing) Wm. Smith, *N. liebethruthii* Rabenhorst and others. *Pseudostaurosira brevistriata* (Grunow) Williams & Round characterized the record between 40 m and 50 m. Lack of a firm chronology makes interpretation difficult, but it is clear that glacial periods are at least partly represented by saline conditions whereas interglacial limnological environments were fresher, at least seasonally.

Other crater lake sites, such as Lac du Bouchet, a volcanic crater or maar in France (44.9° N × 3.8° E) (Pailles, 1989) and Birket Ram in the Golan Heights, Syria (33.23° N × 35.77° E) (Ehrlich & Singer, 1976) have produced long diatom records. Dating control on many such records, however, is usually inadequate even though diatoms are abundant and well preserved. Such sites merit additional study.

Interest in long paleoclimate histories has generated funding for raising long cores from large extant basins and lakes. One renowned, multi-proxy, 200 m record is from Lake Biwa, Japan (35.3° N × 136.1° E) which encompasses about 500 000 years (Nishimura & Yokoyama, 1975). Unfortunately, the diatom stratigraphy (Mori, 1974) has never been fully published. Nevertheless, *Aulacoseira solida* (Eulenstein) Krammer lives in the lake today and characterized the Holocene of Lake Biwa whereas *Stephanodiscus* assemblages dominate the late glacial coeval with cool climate pollen indicators. However, preceding glacial – interglacial cycles are not clearly documented by the diatom stratigraphy or other climate proxies examined, and correlation to global climate cycles represented by marine oxygen isotope records has not been convincing

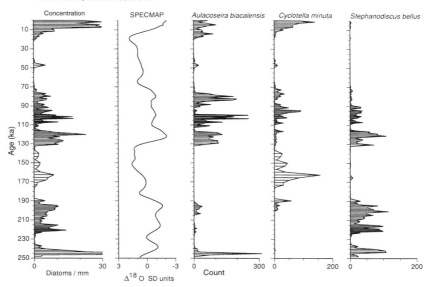

Fig. 7.4. Profiles of selected taxa from Lake Baikal core 340. Diatom concentration (number of valves per mm along a microscope traverse) suggests a correlation to the marine oxygen isotope record (SPECMAP) that provides age control for this record.

(Wright, 1985). Perhaps this reflects an inconsistent age model or simply that earlier responses of this low elevation lake to climate change are muted or somehow obscured by changes in basin morphology in this tectonically active region. Nevertheless, at a more general level, broad peaks in biogenic silica profiles appear to correlate with warm climate intervals of the last interglacial stage (Xiao *et al.*, 1997).

Lake Baikal, Russia, is known to have a long, if discontinuous record of diatoms (Bradbury *et al.*, 1994). A defensible chronology based on correlation of biogenic silica profiles to marine oxygen isotope stages (Colman *et al.*, 1995) has been proposed that spans the past 250 ky. Endemic, extinct centric diatoms from Lake Baikal such as *Stephanodiscus bellus* Chursevich & Loginova that characterize earlier interglacials (Fig. 7.4) offer little help in interpretation of the paleolimnology and paleoclimate of the core because of lack of modern analogs. Nevertheless, at a most general level, it appears that high concentrations of diatoms and of biogenic silica in the core indicate elevated diatom productivity that characterize interglacial climates. Diatom productivity was low during glacial periods because of increased ice cover and turbidity from silt and clay introduced into the lake from glaciated drainages surrounding the basin.

In western North America, four sites contain long diatom records of climate change: Lake Texcoco, central Mexico (19.5° N × 99.0° W), and Clear Lake (39.0° N × 122.8° W), Tule Lake (42.0° N × 121.5° W) and Owens Lake (36.5° N. × 118.0° W), all in California (Fig. 7.1). The Lake Texcoco record (Bradbury,

1971, 1989) suffers from a virtual lack of chronology below the lowest finite radiocarbon date (27.2 ka). Most of the analyzed record is characterized by diatoms of moderately saline, alkaline lakes. Holocene, low lake level conditions are represented by alkaline, spring-fed marsh taxa. During the late Pleistocene, saline diatoms dominate. For example, *Nitzschia frustulum* (Kützing) Grunow indicates shallow but extensive lakes during the full glacial and *Cyclotella quillensis* Bailey and allies suggest deeper water but still moderately saline conditions during the mid Wisconsin. The deepest levels in the 35–m core contain Holocene-like assemblages of alkaline marsh diatoms that may represent the last interglacial.

The Clear Lake diatom record (Bradbury, 1988), like the Lake Biwa record, shows a Holocene dominance of *Aulacoseira* (*A. granulata* (Ehrenberg) Simonsen in this case) preceded by late-to-full glacial levels characterized by *Stephanodiscus*. The base of the core apparently extends beyond 400 ka according to tephrochronologic correlations. Assemblages of *Aulacoseira*, *Stephanodiscus* or benthic freshwater diatoms (*Pseudostaurosira brevistriata*) characterize the deeper intervals of the core, but the age model is disrupted by hiatuses and it is probable that the basin morphology has been much modified by tectonism through past millennia. Consequently, although chronologically controlled diatom assemblages apparently correlate to past glacial and interglacial environments, their continuity and significance is unclear.

Tule Lake, in northern California is the longest, continuously analyzed record of continental diatoms. The core extends to 3 Ma and the chronology is controlled by radiocarbon, paleomagnetic stratigraphy and tephrochronology. The diatom record (Bradbury, 1991, 1992) exhibits 400 ky-cycles of *Aulacoseira solida* (Eulenstein) Krammer that alternate with marsh diatoms in the late Pliocene and early Pleistocene. These cycles apparently reflect winter-warm, moist climates alternating with cooler and drier climates controlled by orbital insolation changes at that scale. Marsh species (*Staurosira* and *Pseudostaurosira* assemblages) between 2.6 and 2.3 Ma probably correlate to Pliocene glacial climates of the Praetiglian (Zagwin, 1986). By 800 ka, 100 ky cycles of marsh diatom assemblages suggest a progression of cooler and drier environments as moisture sources shifted south in response to high-latitude glaciations. Interglacial climates are usually characterized by increased abundance of *Aulacoseira* and *Stephanodiscus* species that indicate more open water conditions resulting from increased annual precipitation as ice sheets waned and storm tracks returned northward.

The 323-m Owens Lake core contains an 800 ky record (Bradbury, 1997). The chronology is provided by radiocarbon in the upper part and by the Brunhes / Matuyama paleomagnetic reversal (780 ka) near the base. Alternations of freshwater planktonic and saline benthic diatom assemblages (Fig. 7.3) track glacial and interglacial limnic environments respectively and approximately correlate to the marine oxygen isotope stratigraphy (Bradbury, 1997). Glacial climates at Owens Lake are wet, in contrast to Tule Lake, because Owens Lake received more winter precipitation as westerly storm tracks (the Aleutian Low) was forced south by growing ice sheets.

Problems with long diatom records

The circumstances that determine the accumulation and preservation of long-term diatom records (large, long-lived lakes) carry problems that affect the interpretation of such diatom records.

   (i) Long-lived lakes require tectonism to maintain basin configuration and space for sediment accumulation. The cycle of lake aging, as it applies to basin filling at least, is physically inescapable and ensures that lakes become marshes or river channels unless tectonism-caused down warping or graben formation exists to maintain the basin. Such tectonic events and trends will leave their own limnologic signal that may be very difficult to distinguish from a climatic or biologic change in the record (Fig. 7.3).

  (ii) Lake basins in areas of extreme climates (glacial or arid) will preserve lacustrine records only during suitable climates. In arid regions, lacustrine records are often deflated when lakes desiccate and remnants of high-stand lacustrine episodes may be eroded and redeposited in subsequent filling cycles. Lacustrine deposits cannot survive alpine or continental glaciation intact. Most lakes in glaciated areas are less than 20 ka in age.

 (iii) Because long-lived lakes are most common and continuous in comparatively stable lacustrine / climatic environments, diatoms in such systems evolve to well-adapted, distinctive forms. This has certainly happened in the past at Lake Baikal (Bradbury *et al.*, 1994) and has produced large and distinctive centric diatoms during the penultimate and earlier interglacials. During the last glaciation, these diatoms became extinct. As a consequence, their ecological preferences and tolerances are unknown by modern analog, and their specific paleolimnologic interpretation is speculative.

 (iv) Interpretation of long diatom records almost always suffers from inadequate chronological control, especially for levels older than the range of radiocarbon ($\sim$ 30 ka). Correlation (curve matching) of diatom paleoenvironmental profiles to other proxies of past climate change (e.g., oxygen isotope records) to establish a chronology involves circular reasoning that may detract from the value of diatom or any other proxy record of environmental change.

The several problems that plague interpretation of long diatom records can only be resolved by study of additional paleoenvironmental proxies in the same core. Pollen, ostracodes, sedimentology, mineralogy, geochemistry and stable isotopes, as well as many other biological and geological data, are required to properly interpret long records of geologic, biologic and climatic change. The need to refine chronological tools to date preradiocarbon records is paramount.

Summary

Diatoms from long records in large, stable, long-lived and well-buffered lake systems can record broad, low-frequency environmental changes that encompass several glacial/interglacial cycles. The most appropriate sites for such records are tectonic basins, often grabens, whose continued depression preserves accumulated sediments over long periods of time. Sites with very small drainage areas (volcanic and impact craters) with slow deposition rates may also contain comparatively long lacustrine records.

Because diatoms have short life cycles and respond rapidly to environmental changes, core or outcrop samples for diatom study should represent a time-restricted stratigraphic interval to allow opportunity for analog interpretation of fossil assemblages. Although paleolimnological interpretations of long-term diatom records may be quite specific, usually interpretations are broad, reflecting major limnological changes and trends. Such interpretations have been applied to geologic, evolutionary, ecological and climate research themes.

Long lacustrine records with abundant diatoms are known from many parts of the world. Nevertheless, few have been analyzed and dated with critical detail. The potential for long-term diatom studies from such records is great and will depend in part on improved chronological techniques as well as methods for extracting long, continuous cores from existing large lakes.

References

Anderson, N. J. (1995). Temporal scale, phytoplankton ecology and paleolimnology. *Freshwater Biology*, **34**, 367–78.

Belcher, J. H., & Swale, E. M. F. (1979). English freshwater records of *Actinocyclus normanii* (Greg.) Hustedt (Bacillariophyceae). *British Phycological Journal*, **14**, 225–9.

Bradbury, J. P. (1971). Paleolimnology of Lake Texcoco, Mexico. Evidence from diatoms. *Limnology and Oceanography*, **16**, 180–200.

(1988). Diatom biostratigraphy and the paleolimnology of Clear Lake, Lake County, California. In *Late Quaternary Climate, Tectonism, and Sedimentation in Clear Lake, Northern California Coast Ranges*, ed. J. D. Sims, pp. 97–129. Geological Society of America, Special Paper 214, Boulder, Colorado.

(1989). Late Quaternary lacustrine paleoenvironments in the Cuenca de México. *Quaternary Science Reviews*, **8**, 75–100.

(1991). The late Cenozoic diatom stratigraphy and paleolimnology of Tule Lake, Siskiyou County, California. *Journal of Paleolimnology*, **6**, 205–55.

(1992). Late Cenozoic lacustrine and climatic environments at Tule Lake, northern Great Basin, USA. *Climate Dynamics*, **6**, 275–85.

(1997). A diatom-based paleohydrologic record of Owens Lake sediments from core OL-92. In *An 800,000-year paleoclimatic record from core OL-92, Owens Lake, southeast California*, ed. G. I. Smith & J. L. Bischoff, pp. 99–112. Geological Society of America Special Paper 317, Boulder, Colorado.

Bradbury, J. P., & Blair, W. N. (1979). Paleoecology of the upper Miocene Hualapai Limestone Member of the Muddy Creek Formation, northwestern Arizona. In *Rocky Mountain Association of Geologists and Utah Geological Association, Basin and Range Symposium, Ely, Nevada.*, pp. 293–303.

Bradbury, J. P., & Krebs, W. N. (1982). Neogene and Quaternary lacustrine diatoms of the western Snake River Basin, Idaho – Oregon, U. S. A. *Acta Geologica Academiae Scientiarum Hungaricae*, **25**, 97–122.

(1995). The diatom genus *Actinocyclus* in the western United States. US Geological Survey Professional Paper 1543 A-B, 73 p.

Bradbury, J. P., Dieterich K. V., & Williams, J. L. (1985). Diatom flora of the Miocene lake beds near Clarkia in northern Idaho. In *Late Cenozoic History of the Pacific Northwest*, ed. C. J. Smiley, pp. 33–59. San Francisco: American Association for the Advancement of Science.

Bradbury, J. P., Forester, R. M., & Thompson, R. S. (1989). Late Quaternary paleolimnology of Walker Lake, Nevada. *Journal of Paleolimnology*, **1**, 249–67.

Bradbury, J. P., Bezrukova, Y. V., Chernyaeva, G. P., Colman, S. M., Khursevich, G., King, J. W., & Likoshway, Y. V. (1994). A synthesis of post-glacial diatom records from Lake Baikal. *Journal of Paleolimnology*, **10**, 213–52.

Colman, S. M., Peck, J. A., Karabanov E. B., Carter, S. J., Bradbury, J. P., King, J. W., & Williams, D. F. (1995). Continental climate response to orbital forcing from biogenic silica records in Lake Baikal. *Nature*, **378**, 769–71.

Ehrlich, A., & Singer, A. (1976). Late Pleistocene diatom succession in a sediment core from Birket Ram, Golan Heights. *Israel Journal of Earth Sciences*, **25**, 138–51.

Gasse, F. (1975). L'évolution des lacs de l'Afar Central (Éthiopie et T. F. A. I.) du Plio-Pléistocène à l'Acutel. Thèse présentée à L'Université de Paris VI, vol. 1, 406 p., vol. 2, 103 p., vol. 3, pl. 1–59.

(1980). Les diatomées lacustres Plio-Pléistocènes de Gadeb (Éthiopie) Systématique, Paléoécologie, Biostratigraphie. *Revue Algologique*, Mémoire hors-série no. 3, 249 p.

Haq, B. U., Hardenbol, J., & Vail, P. R. (1987). Chronology of fluctuating sea levels since the Triassic. *Science*, **235**, 1156–67.

Hayes, J. D., Imbrie, J., & Shackleton, N. J. (1976) Variations in the earth's orbit: Pacemaker of the ice ages. *Science*, **194**, 1121–32.

Hustedt, F. (1957). Die Diatomeenflora des Flußsystems der Weser im Gebiet der Hansestadt Bremen. *Abhandlungen des Naturwissenschaftichen Vereins Bremen*, **34**, 181–440.

International Geosphere-Biosphere Programme (IGBP). (1992). Past Global Changes Project: Proposed implementation plans for research, Report No. 19. ed. J. A. Eddy, 110 p. Bern, Switzerland, IGBP.

Kociolek J. P., & Stoermer, E. F. (1989) Phylogenetic relationships and evolutionary history of the diatom genus *Gomphoneis*. *Phycologia*, **28**, 438–54.

(1990). Diatoms from the upper Miocene Hot Springs Limestone, Snake River Plain, Idaho (U. S. A.). *Micropaleontology*, **36**, 331–52.

Krebs, W. N. (1994). The biochronology of freshwater planktonic diatom communities in western North America. In *Proceedings of the 11th International Diatom Symposium, San Francisco*, ed. J. P. Kociolek, pp. 485–99. Memoirs of the California Academy of Sciences 17.

Krebs, W. N., Bradbury, J. P., & Theriot, E. (1987). Neogene and Quaternary lacustrine diatom biochronology, western USA. *Palaios*, **2**, 505–13.

Krebs, W. N., & Bradbury, J. P. (1995). *Geologic Ranges of Lacustrine Actinocyclus Species, Western United States*. US Geological Survey Professional Paper, 1543–B, 53–73.

Mori, S. (1974) Diatom succession in a core from Lake Biwa. In *Paleolimnology of Lake Biwa and the Japanese Pleistocene*, vol. 2, ed. S. Horie, pp. 247–54. Kyoto.

Nishimura, S., & Yokoyama, T. (1975). Fission-track ages of volcanic ashes of core samples of Lake Biwa and the Kobiwako group. In *Paleolimnology of Lake Biwa and the Japanese Pleistocene*, vol. 3, ed. S. Horie, pp. 138–42. Kyoto.

Okeden, F. (1855). On the deep diatomaceous deposits of the mud of Milford Haven and other localities. *Quekett Journal of Microscopical Science*, **3**, 25–30.

Pailles, C. (1989). Les diatomées du lac de maar du Bouchet (Massif Central, France) Reconstruction des paleoenvironnements au cours des 120 derniers millenaires. These Universite d'Aix-Marseille II., pp. 274.

Partridge, T. C., Kerr, S. J., Metcalfe, S. E., Scott, L., Talma, A. S., & Vogel, J. C. (1993). The Pretoria Salt Pan, a 200,000 year Southern African lacustrine sequence. *Palaeogeography, Palaeoclimatology, Paleoecology*, **101**, 317–37.

Smith, G. I., Bischoff, J. L., & Bradbury, J. P. (1997). Synthesis of the paleoclimatic record of the Owens Lake core OL-92. In *An 800,000-year paleoclimate record from core OL-92, Owens Lake, southeast California*, eds. G. I. Smith & J. L. Bischoff, pp.143–60. Geological Society of America Special Paper 317, Boulder, Colorado.

Theriot, E. (1992). Clusters, species concepts and morphological evolution of diatoms. *Systematic Biology*, **41**, 141–57.

Wright, H. E., Jr. (1985). An ancient lake. *Science*, **228**, 1082–3.

VanLandingham, S. L. (1964). *Miocene non-marine diatoms from the Yakima region in south central Washington*. Beihefte zur Nova Hedwigia, 14, 78 p.

(1985). Potential Neogene diagnostic diatoms from the western Snake River Basin, Idaho and Oregon. *Micropaleontology*, **31**, 167–74.

Xiao, J., Inouchi, Y., Kumai, H., Yoshikawa, S., Kondo, Y., Liu, T., & An, Z. (1997). Biogenic silica record in Lake Biwa of central Japan over the past 145,000 years. *Quaternary Research*, **47**, 277–83.

Zagwin, W. H. (1986). Plio-Pleistocene climate change: Evidence from pollen assemblages. *Memorias de la Sociedad Geologica Italiana*, **31**, 145–52.

J. PLATT BRADBURY

8 Diatoms as indicators of water level change in freshwater lakes

JULIE A. WOLIN AND HAMISH C. DUTHIE

Introduction

Water level changes result from a variety of geological, biological or climatic processes. Many of these changes occur over long periods of time; however, some may be rapid or result from catastrophic events. In glaciated regions, water level changes are influenced by the geological process of isostatic rebound. As the weight of glacial ice is removed from the Earth's surface, depressed surfaces adjust upward, while adjacent areas may subside (i.e., Larsen, 1987). Lakes are often formed as embayments become isolated from larger water bodies, and existing lakes can reinvade subsiding regions. Differential rates of rebound within large lake basins may affect drainage patterns and water levels. Transport of sediments across river mouths or embayments can also create lakes and increase water levels.

In most cases, diatom microfossils from lake sediments can be used to identify the above changes. The clearest diatom signals are found in areas where lakes are isolated from marine or brackish waters (Denys & de Wolf, this volume). In freshwater systems, however, salinity gradients are absent, and signals are generally recorded as increases of deep-water or planktonic forms. Water level changes in lakes affected by isostatic rebound and sediment-transport isolation are common in the Laurentian Great Lakes region of North America (e.g., Yang & Duthie, 1995a; Wolin, 1996) and in the North Sea and Baltic region of Europe (e.g., Digerfeldt, 1988).

One of the most important factors controlling water level is a change in hydrological conditions. Lake-levels are determined by changes in the balance between moisture gains and losses. Inputs include: stream inflow, basin runoff, groundwater inflow, and lake surface precipitation. Losses occur through stream outflow, evaporation from the lake surface, groundwater outflow, and in some cases, deep seepage. Most of these hydrological responses are associated with climatic or ecological change (e.g., Dearing & Foster, 1986; Mason et al., 1994; Street-Perrot & Harrison, 1985).

Biological processes such as successional changes in vegetation can also alter drainage patterns and groundwater flow, which in turn affect water levels (Dearing & Foster 1986). In the early development of lakes from glacial regions, establishment of vegetation usually alters surface runoff into the lake. This acts to moderate water level fluctuations on a seasonal or short-term basis. As lakes age, deposition of plant and animal remains and sediment inputs from the drainage basin cause infilling. As a result, the lake becomes more shallow over time (Wetzel, 1983). These natural processes can be accelerated by human activities such as forest clearance, farming and nutrient inputs (e.g., Dearing, 1983; Fritz, 1989; Gaillard *et al.*, 1991). Damming and channelizing activities by humans or other animals can also result in rapid water level changes (e.g., Bradbury, 1971).

The influence of climate on water level fluctuations is particularly important for reconstructing and projecting climate changes. During wet periods lake levels generally increase while, under dry conditions, water levels decline. The strength of the corresponding sedimentary signal is dependent on the type of lake. Richardson (1969) and Street-Perrott and Harrison (1985) discuss various types of lakes and their sensitivity to climatic changes. Shallow, closed-basin lakes should have the strongest signals, whereas deep, small surface area lakes with drainage would have the weakest. As described by Fritz *et al.* (this volume), saline (closed basin) lakes provide very good proxies for climate reconstructions with diatoms. Unfortunately, freshwater (open drainage) lakes predominate in many regions of the world and diatom reconstructions of water level changes for these lakes are more problematic. Freshwater diatom assemblages respond indirectly to changes in the aquatic environment which result from fluctuations in water level, and these same environmental conditions can be caused by other perturbations.

As water levels fluctuate, corresponding changes occur in available habitat, light, chemical conditions, stratification and mixing regimes. As lake levels fall, benthic and epiphytic habitats generally increase. The area of lake bottom within the region of light penetration (photic zone) also increases. Exceptions are found in steep-sided lakes or in lakes where shallow conditions lead to greater turbidity or productivity-driven light limitation. Thermal stratification is affected as water levels decline in lakes subject to high wind activity. In shallow systems, this can result in a total loss of stratification and frequent resuspension of bottom sediments (e.g., Brenner *et al.*, 1990; Odgaard, 1993). Nutrient and chemical salt concentrations increase and corresponding changes in pH can occur. High-water levels generally result in the opposite effects.

These above changes have an effect on the biota within the lake. As discussed in other chapters of this volume, diatoms are extremely sensitive to different aquatic environments and can be used as indicator organisms to reflect these changes. In freshwater lakes, changes in the planktonic to non-planktonic ratio is often used to indicate water level fluctuations (e.g., Gasse *et al.*, 1989; Marciniak, 1990; Owen *et al.*, 1982). Planktonic or free-floating diatoms live in open water. Non-planktonic or littoral forms are composed of benthic taxa which live on the lake bottom, epiphytes which live on other plants, or tycoplanktonic forms which, although generally associated with the benthic and nearshore community, can be transported into the planktonic

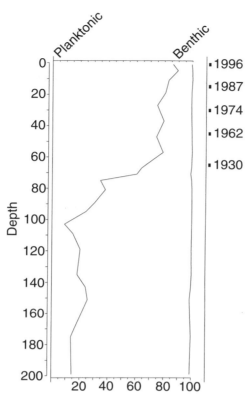

Fig. 8.1. Relative abundance (%) of diatom life forms in a sediment core from Søgaard Sø.
Dates based on ^{210}Pb.

community. At low-lake levels, the increase in shallow water habitat and
macrophyte growth should result in a greater amount of benthic and epiphytic
forms. Higher percentages should then be seen in lake sediments. During high
water levels, the percentage of planktonic forms is expected to increase. The
difficulty in using this ratio is that the same signals may be caused by other
factors. For example, in the Danish shallow lake, Søgaard Sø (Fig. 8.1), a shift
from benthic to planktonic forms results from hypereutrophic nutrient conditions (mean total phosphorus >200 μg/l) and corresponds to sedimentary evidence of submerged macrophyte loss. Increased nutrient inputs stimulate
phytoplankton and epiphytic growth of algae. This results in light limitation
of submerged aquatic plants inhibiting their ability to photosynthesize and,
in a system dominated by phytoplankton as macrophyte loss occurs (Irvine *et
al.*, 1989).

Other types of diatom signals have also been used; however, these require
special insight into specific lake processes. In general, the freshwater diatom
signal in and of itself, is not sufficient for water level reconstruction. In certain
cases, planktonic freshwater forms may indicate lower water stages than saline
species. In an early lake phase of Walker Lake, Nevada, the freshwater taxon

Stephanodiscus niagarae Ehrenb. is replaced by *Cyclotella quillensis* Bailey, a moderately saline species (Bradbury, 1987). Although this transition occurs during a period of high water level, the sedimentation rate curve suggests that the rate of filling was slowed and balanced by evaporation. The result was an increase in salinity not related to lower water levels. In Lake Valencia, Venezuela following a period of low lake-levels, a dominance of freshwater littoral forms indicated inundation of distant littoral environments by high water levels (Bradbury *et al.*, 1981).

It is important to use multiple lines of evidence to support any proposed water level changes and to exclude other factors which may result in the same signals (Digerfeldt, 1986). Additional supporting evidence useful in reconstructing lake-level fluctuations include: sediment stratigraphy, fine-grained particle analysis, stable isotope analysis, chemical content, charcoal, macrofossils, pollen, and ostracodes. In many cases, human activities can alter or obscure climatic signals present in the record (e.g., Gaillard *et al.*, 1991). This is particularly true in areas where human activity has existed for centuries. An advantage of North and South American regions is that human influence is generally low prior European settlement.

Given corresponding information about a particular lake, various combinations of changes in the diatom flora can then be used to trace changes in lake-levels of freshwater systems. The recent development and application of quantitative models using diatoms for water level reconstructions are an important step, but these methods need to be applied to more studies. Indeed, it is not an easy task, but as the following examples should indicate, with proper investigation it is possible to show lake-level changes in freshwater systems. Given that a large number of lakes exist in regions where the only hydrologic and water level signals found will be those in the sedimentary record, it is important to utilize this information for the reconstruction of geological events, human manipulations, or past climatic signals.

Types of diatom indicators

LIFE FORM

Numerous paleolimnological studies have used changes in life form (benthic, epiphytic and planktonic diatoms) to determine water level fluctuations. In this chapter, we concentrate on a few examples, although many others exist. Two extensive compilations of lake-level climate records are available from the National Oceanic and Atmospheric Administration (NOAA) Paleoclimatology Program (World Wide Web: http://www.ngdc.noaa.gov/paleo/paleo.html). The majority of studies use changes in the percentages of life forms to indicate high or low lake-levels. In most cases, quantitative methods for water level reconstructions have not been attempted. Approximately half of the 63 investigations from lakes in northern Eurasia (Tarasov *et al.*, 1994) use fresh-

J.A. WOLIN AND H.C. DUTHIE

water diatom stratigraphic evidence. English summaries of these investigations, data and references are listed for each lake in the database. The majority of the diatom studies in the Baltic and other countries of the former Soviet Union have been investigated by N. Davydova (e.g., Davydova, 1986, Davydova & Raukas, 1986) and by Dorofeyuk (1988) in Mongolia. In the European data base (Yu & Harrison, 1995), only about 30% of the 115 lake studies use freshwater diatom assemblages. The majority of these are in Finland, Germany, Great Britain, and Norway. Two of these studies, Bertzen (1987) in Tegel See and Lotter (1988) in Rotsee, attempt to use diatom life-form data to reconstruct actual lake depths. These compilations are by no means complete in terms of freshwater diatom investigations. The studies were chosen based on standards of dating control and consistency of climatic indicators. Lake-levels records influenced by non-climatic factors were excluded.

In addition to changes in the percentage of benthic or planktonic diatoms, temporary increases in littoral material can also record a rise or fall in lake-level. This signal is more likely found in deep-water core sites and is created by erosion of the sediment limit (Digerfeldt, 1986). As lake levels fall, material in the littoral region is eroded and transported into deeper waters through the process of sediment focusing. As lake-levels rise, a similar signal can occur, although transport may not be as extensive. Wolin (1996) reported such a signal in Lower Herring Lake, Michigan, where lake-level changes were investigated from a deep-water core taken at 15.25 m. Fine-grained sediment distribution, sediment chemistry and diatoms were analyzed to determine long-term cycles in water level change. As lake-levels fell, an increase in benthic/epiphytic forms occurred. This increase did not persist during low water periods, but was repeated as water levels rose again. The signal, in itself, was not enough to indicate high or low water level; however, corresponding changes in the fine-grained sediments showed an increase in coarse particles during low-water periods as well as increases in sand and carbonate content. Increases in fine-particle sediments occurred during high water stages.

PHYSICAL AND CHEMICAL ENVIRONMENT

Other signals may be seen in diatom assemblages as a result of shifts in environmental conditions caused by changes in water levels. These signals may reflect physical changes such as stability of the metalimnion, susceptibility to turbulence, nutrient or thermal conditions, or changes in pH. Most studies that have utilized these signals have also used some form of littoral/planktonic changes as a basis for interpreting lake-level change.

Rather than changes in the planktonic/non-planktonic ratio, water level changes may be signalled by changes in dominant planktonic forms which respond to physical or chemical variables. This is particularly true in deep water cores from large lake systems where the influence of littoral regions is minimal.

The presence of diatoms associated with turbulence regimes has been useful for inferences of water level change. The genus *Aulacoseira* is a particularly good example. A heavily silicified diatom with high sinking rates, its ecology requires turbulence to maintain its presence in the water column (e.g., Bradbury 1975). Increased turbulence and corresponding nutrient increases during low water stages in a lake can favor this genus over other planktonic species.

Owen and Crossley (1992) reported that frequent upwelling conditions in southern Lake Malawi favoured *Aulacoseira* filaments. Due to its shallower depth, the southern portion of the lake is more susceptible to wind mixing. Additional studies by Pilskaln and Johnson (1991) in Lake Malawi found that *Aulacoseira* dominated assemblages during dry-windy periods where it takes advantage of the high-nutrient high-turbulence conditions. Almost equal amounts of *Aulacosiera* and *Stephanodiscus* were favored during wet, low-wind velocity periods. The presence of this particular diatom signal may be useful in determining long-term climatic trends particularly in high-resolution studies, such as the recent study by Stager *et al.* (1997) in Lake Victoria.

The 10000 year record of varved lake sediments from Elk Lake, Minnesota is dominated by planktonic diatoms (Dean *et al.*, 1984). Within this community, increases in the percentage of *Aulacoseira* (= *Melosira*) are used to indicate low-water conditions during a dry prairie period between 8500 and 4000 years ago. The combination of low-water conditions and high wind exposure provide the turbulent, high-nutrient conditions favorable to this taxon. A similar *Aulacoseira* signal was seen during a low-lake phase in Pickerel Lake (Haworth, 1972) and Bradbury (1975) postulated that the same conditions present in Elk Lake were responsible for the dominance of this taxon.

Though some diatoms benefit from turbulence, for others, it is a disadvantage. Increased wind-driven turbulence in lakes can lead to destabilization of the metalimnion. This provides an advantage for large, heavy diatoms with high nutrient requirements. However, in clear lakes, certain diatom species benefit from stable stratification conditions. These forms are adapted to lower light levels and obtain nutrients which diffuse from the hypolimnion (Fahnenstiel & Scavia 1987). During high water levels, increased stability of the metalimnion favors the diatoms adapted for this environment. The presence of taxa associated with this 'deep chlorophyll layer' were used by Wolin (1996) to indicate stable metalimnion conditions during high water periods in Lower Herring Lake.

NUTRIENT ENVIRONMENT

This signal is probably the most problematic for use in reconstruction of water level changes. Nutrient signals in diatom assemblages are primarily used to indicate human impacts or erosional changes within a lake basin. However, increases in nutrient concentrations also occur during low water periods. If one can separate anthropogenic signals, these same indicators may be useful for climatic signals as well. Rippey *et al.* (1997) found that, during the 1970s, a period of dry conditions resulted in nutrient increases in White Lough, a small eutrophic lake in Northern Ireland. Evidence indicates that reduced flushing and an increase in sediment phosphorus release were the source of this change. An increase in diatom-inferred total phosphorus reflected this nutrient signal which resulted from climatic rather than anthropogenic forcings. A nutrient signal related to lake-level change was also found in Lower Herring Lake (Wolin, 1992). Increases in percentages of *Cyclotella stelligera* (Cl., & Grun.) V. H., a taxon known to respond to nutrient inputs (Schelske *et al.*, 1974), occurred during a known low-water period of the 1930s and was accompanied by a similar increase in benthic forms.

pH

Climate changes, or more precisely wetter/drier shifts, are usually inferred indirectly in paleolimnological studies from evidence for water level changes. Climate change in Batchawana Lake, Ontario (Delorme *et al.*, 1986), inferred from pollen analysis, was connected with water level changes and pre-anthropogenic pH changes. Tree pollen data was used to determine changes in wet and dry habitat and compared with prehistoric, diatom-inferred pH changes (see Battarbee *et al*, this volume). The occurrence of pH shifts prior to any possible atmospheric-induced changes eliminated the possibility of anthropogenic causes and was interpreted as resulting from lake-level changes. Decreases in moisture (i.e., lake-level) correspond with increases in humification and lower diatom-inferred pH. Increases in water level, probably the result of a beaver dam, caused increases in circumneutral and alkaliphilous (higher pH) diatom taxa. A recent study by Krabbenhoft and Webster (1995) gives insight into possible mechanisms for climate influence on pH changes. They showed that drought conditions reduced groundwater inflow, water high in major base cation concentrations (Ca^{2+}, Mg^{2+}), and most likely resulted in the acidification of Nevins Lake, Michigan. Sweets and Stevensen (1997) also linked changes in inferred diatom pH with corresponding changes in diatom-inferred dissolved organic carbon (DOC), pollen profiles and stratigraphic evidence as a proxy indicator of lake-level decline in Lake Barco, Florida.

An opposite effect was found by Bradbury (1986) in two lakes of western Tasmania where increases in acidophilous (low pH) taxa were interpreted as resulting from increased flushing of the lake basin with low pH, low conductivity water during snowmelt or increased precipitation. Alkaliphilic forms were interpreted as representing seasonal or short-term dry periods, which caused an increase in alkaline components.

Quantitative reconstructions

Knowledge of the depth distribution of diatoms in lakes has received less attention from researchers than many other aspects of algal ecology, and consequently the quantitative use of freshwater diatoms as proxy indicators of water level changes in lakes has, until recently, been neglected. Diatom / depth distributions in lakes, such as that contained in Round's (1961) pioneering study in the English Lake District, are uncommon. However, Earle *et al.* (1988), in an intensive study to determine settling patterns of sedimentary diatoms in three chemically similar but morphometrically dissimilar Canadian Shield lakes, concluded that spatial heterogeneity in sample composition was a function of both water depth and basin morphometry. In a recent review of the literature, Yang and Duthie (1995*b*) proposed that depth, by integrating physical factors, is a strong controlling variable on sedimentary diatom distribution. As the following examples show, several quantitative methods using freshwater diatom assemblages have recently been applied in the reconstruction of lake-level changes, however, further work is needed.

REGRESSION

Barker *et al.* (1994) used diatom assemblages in their reconstruction of water level changes in Lake Sidi Ali, Morocco. Although present in an arid region, the lake does not accumulate salts and thus use of changes in salinity for lake-level changes are not possible. Modern diatom assemblages collected from six surface samples along a depth transect were used to develop a model for lake-level reconstruction. Diatoms were classified into three habitat types: planktonic, tycoplanktonic and littoral, based on modern ecological data from Morocco. Analysis of the modern assemblages showed a logarithmic relationship between depth (D) and the Planktonic/Littoral ratio (P/L). However, classification of individual taxa into planktonic or littoral categories proved problematic. Surface samples were useful in determining classification of the major problem taxa. One centric taxon, *Cyclotella* sp. 1, was found in greatest percentages in the littoral zone. This genus is generally classified as planktonic, although several species are know to exist as tycoplanktonic forms. Another problem group were members of the pennate genus *Fragilaria* (this study uses the older classification by H. C. Lyngbye) which may be classified as littoral, tycoplanktonic or sometimes planktonic. In order to explore the effect of different classifications on the final regression equation, four versions of the P/L ratio were constructed using different classifications and including or excluding *Fragilaria* species. The final model chosen for reconstruction was a simple linear regression: $\mathrm{Log}_{10}(P/L) = a\,D - b$ with an adjusted r^2 of 0.88 in which two major species of *Cyclotella* were classified as planktonic, all other species as littoral, and *Fragilaria* were excluded from both groups.

An important assumption made for the reconstructions is that diatom zonal sedimentation patterns have not altered with water depth. Barker (per-

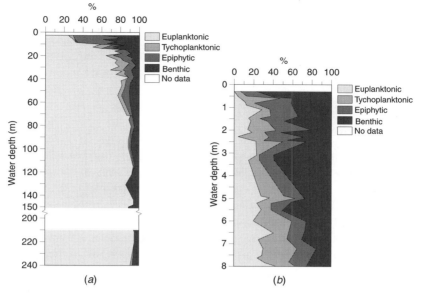

Fig. 8.2. Relative abundance (%) of diatom life forms from surficial sediment samples collected along depth gradients in (a) Lake Ontario, and (b) East Lake (Yang & Duthie, 1995b).

sonal communication) emphasized that this reconstruction should be viewed as a working model and that further testing of the assumptions made in its formation should be done. The present model was applied to Lake Sidi Ali, Morocco and works for reconstruction of specific basins; however, further development needs to occur for more universal application.

Relatively few publications exist on benthic diatom/depth distributions for the North American Great Lakes, possibly due to technical difficulties involved in sampling such large waterbodies. In Lake Michigan, diatom assemblages along depth gradients have been described by Stoermer (1968, 1975), Stevenson and Stoermer (1981), and Kingston *et al.* (1983), and surficial inshore and offshore diatom assemblages have been described in Lake Ontario (Duthie & Sreenivasa, 1972).

Recently, Yang and Duthie (1995b) investigated the depth distribution of four life forms: euplanktonic, tychoplanktonic, epiphytic and benthic, in 41 samples over a 240 m depth gradient in western Lake Ontario, and in an 8 m depth gradient in East Lake, an embayment isolated from the northeastern shoreline of Lake Ontario by a baymouth bar (Fig. 8.2).

Multiple regression was used to explore the relationship between measured water depth of surface samples and sample life-form composition. Both sites exhibited strong relationships resulting in an r^2 of 0.73 for East Lake, and 0.91 for Lake Ontario. The stronger relationship in Lake Ontario is probably due to its greater depth, distance from the shoreline of the deeper samples, and a larger number of samples. The main problem with this approach is the subjectivity involved in assignment of diatom taxa to life-form categories. Category assignments for each taxon were based on literature references and not all references agree on life-form designations. When such disagreements arose, emphasis was given to diatom studies based on the Laurentian Great Lakes (e.g., Stoermer, 1968, 1975; Stevenson & Stoermer, 1981; Kingston *et al.*, 1983; Stoermer *et al.*, 1985*a,b*).

ORDINATION

Rather than assigning life-form categories, Lotter (1987) utilized a different approach in his study of water level changes in Lake Rotsee, Switzerland. Diatom analysis was conducted on 19 modern surface samples taken along a longitudinal transect through Lake Rotsee. These samples were then combined with 46 subfossil samples from a mid-depth level core and analyzed using principal components analysis (PCA). Ordination of the modern diatom assemblages resulted in three distinct groups as characterized by current their water depth (<5 m, 5–9 m, and >9 m). The 46 subfossil samples grouped well with the modern samples and corresponding categories of the subfossil samples were then used to infer water depth categories within the core. Changes in percentages of life-form groups matched well with the inferred depth categories. These water-depth classes were then compared with life form categories and pollen zones to reconstruct lake-level changes during the Holocene.

WEIGHTED AVERAGE ANALYSIS

The above three examples have used linear models to define the distribution of diatom assemblages. However, the optimum abundance of a diatom or other organisms along an environmental gradient such as temperature, pH, or trophic status may be determined by the Gaussian response model (e.g., Stoermer & Ladewski, 1976) or by weighted average (WA) analysis (Charles, 1985; Stevenson *et al.*, 1989; ter Braak & van Dam, 1989; Line & Birks, 1990; Hall & Smol, 1992; Kingston *et al.*, 1992; Yang & Dickman, 1993).

Yang and Duthie (1995*b*) proposed a protocol for using WA analysis to determine optimum water depths for diatom species in sample sets collected along a depth gradient. Optima are determined from the frequency of each species in surface sediment samples along the depth gradient using the following equation:

$$W_k = \sum_{k}^{m} D_i P_{ik} / \sum_{k}^{m} P_{ik}$$

where W_k = weighted average (optimum) value of water depth for diatom taxon k ($k = 1 \ldots m$), D_i = measured water depth in meters for sample i ($i = 1 \ldots 49$), P_{ik} = percentage value of k taxon at sample i.

The resulting species water depth optima may then be used as a data set to infer water depth by WA unimodal regression (Stevenson *et al.*, 1989) from diatom analysis of sediment samples from the same lake or region. For example, water depth at time of deposition may be inferred from sediment core diatom assemblages. The following equation is used:

$$ID_i = \sum_k^m W_k P_{ik} / \sum_k^m P_{ik}$$

where ID_i = inferred water depth at core sample i.

Yang and Duthie (1995b) compared the differences between inferred and measured water depth in a surficial sediment sample set from Lake Ontario using both multiple regression and WA models. Differences between inferred and measured water depths were calculated from:

$$RMSE = \{[\sum_{i=1}^n (MD_i - ID_i)^2]/n - 1\}^{1/2}$$

where $RMSE$ = root mean squared error, MD_i = measured water depth for sample i, ID_i = inferred water depth for sample i ($i = 1 \ldots n$).

The results (Fig. 8.3) suggest that both methods produce good relationships between measured and inferred depths. However, there is a trend evident in the WA analysis for higher values to be underestimated and lower values to be overestimated. This trend, reported in other studies using WA, has been attributed to functional interactions among different environmental variables, which are not accounted for in simple inferences (Birks $et\,al.$, 1990; Kingston $et\,al.$, 1992). Nevertheless, WA analysis produces a high correlation between measured and inferred depth ($r^2 > 0.9$).

A major disadvantage of the method used in Barker $et\,al.$ (1994) arises from limiting classification to two habitat groups, something that Yang and Duthie solve by use of multiple classifications, but this does not solve the problem of possible ambiguities and subjectivity in classification. A different approach taken by Lotter (1987) and Yang and Duthie (1995b) is to directly explore the relationship of individual assemblages or taxa with depth without pre-assigned categories. Lotter used whole assemblages in his reconstructions, whereas Yang and Duthie take this a step further and define depth optima for individual taxa utilizing the weighted averaging regression approach. Further study needs to be conducted to include multiple environmental variables such as light, temperature, nutrient, bottom type, etc. and through canonical correspondence analysis (CCA) (ter Braak, 1986) to explore the strength of each variable with respect to benthic diatom distributions.

Applications of diatom-inferred depth changes in lakes

ISOSTATIC REBOUND

The postglacial water level history of Lake Ontario (Canada / USA), the lowest lake in the Laurentian Great Lakes, has been reconstructed by Anderson and

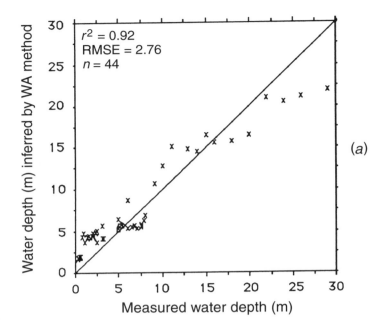

(a)

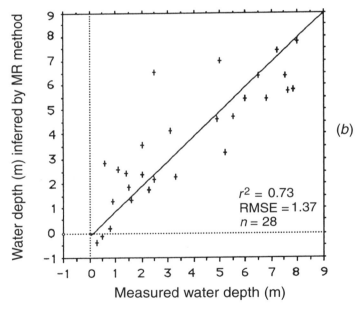

(b)

Fig. 8.3. Measured vs. diatom-inferred depth for a surficial sediment sample set from Lake Ontario, based on (a) weighted averaging, and (b) multiple regression models (Yang & Duthie, 1995b).

Lewis (1985), who based their arguments mainly on geological and sedimentological evidence. They postulated changes in the level of Lake Ontario in the order of 100 m over the past 11 000 years as a consequence of deglaciation and changing drainage patterns in the upper Great Lakes, and isostatic rebound in eastern Lake Ontario and the upper St Lawrence River.

Three recent paleolimnological studies on shoreline embayments have focused on limnological and hydrological events associated with the changing water levels of Lake Ontario over the Holocene. Diatom-inferred water depth changes have been investigated by Duthie *et al.* (1996) for Hamilton Harbour in western Lake Ontario, by Yang and Duthie (1995*a*) for East Lake in eastern Lake Ontario and by I. Kaczmarska-Ehrman (personal communication) for Grenadier Pond, Toronto.

Hamilton Harbour (Burlington Bay) In deriving diatom transfer functions from modern data a problem arises in that, since many sites are anthropogenically impacted, the application of these transfer functions to prehistoric sediment core data may not be valid. The highly polluted Hamilton Harbour (Yang *et al.*, 1993) is a case in point. To solve this problem, Duthie *et al.* (1996) used WA values of water depth for 91 diatom species derived from 49 surface sediment samples from western Lake Ontario and from East Lake (Fig. 8.4).

It is evident that as a result of isostatic rebound, Hamilton Harbour has changed over the past 8300 yr from a shallow separate waterbody to a deep embayment of Lake Ontario. The earliest evidence from about 8350 BP to about 7000 BP is of shallow pond dominated by benthic and epiphytic diatoms. An influx of euplanktonic diatoms at about 7000 BP signaled an initial, temporary connection to the rising waters of western Lake Ontario. Permanent confluence was established at about 6000 BP. The Nipissing Flood (Anderson & Lewis, 1985) at about 4400–4000 BP was evidenced in the core by a decline in total microfossils and by supporting isotopic evidence, but was not clearly delineated by trends in diatom-inferred depth. The final 3200 yr record shows a gradual deepening of the embayment with continued isostatic uplift in the upper St Lawrence outlet of Lake Ontario.

In general, the diatom-inferred depths agreed with independent physical estimates, especially during periods of environmental stability. However, the diatom-inferred water depths were unreliable during periods of rapid change or hydrological instability, such as the Nipissing Flood.

East Lake This site near the eastern end of Lake Ontario is separated from Athol Bay, Lake Ontario, by a permanent baymouth bar. Diatom analysis of a sediment core provided a 5400 yr record of limnological change, including evidence for two shallow wetland phases separating three lacustrine phases (Fig. 8.5). Water depths were estimated by weighted average calibration (pre-isolation) and by multiple regression (post-isolation).

The earliest samples were dominated by a planktonic diatom assemblage suggesting an open connection with Lake Ontario. Isolation by the formation

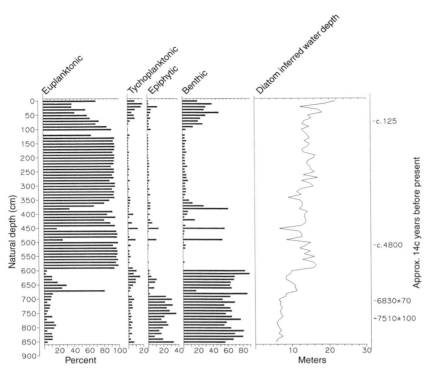

Fig. 8.4. Stratigraphy (% abundance) of four diatom life forms, and the diatom-inferred water depth in a sediment core from Hamilton Harbour, Lake Ontario. Redrawn from Duthie *et al.* (1996).

of a baymouth bar was probably initiated around 5100 BP, reflecting shoreline erosion associated with rising water levels in Lake Ontario caused by isostatic uplift in the upper St. Lawrence outlet. Shallow wetland and benthic diatoms dominated until the Nipissing Flood between 4500 BP and 3700 BP increased inferred water levels to 6–8 m. A second wetland phase accompanied the post-Nipissing decline in water level, but continuing isostatic adjustment of the Lake Ontario outlet has increased the elevation of East Lake to modern levels.

Grenadier Pond (Personal communication; I. Kaczmarska-Ehrman, Mount Allison University, New Brunswick) The sedimentary record reveals three distinct diatom assemblages reflecting three stages of pond development. A lacustrine stage dated from around 4200 BP coincides with the Nipissing Flood, and correlates with the sedimentology (McCarthy & McAndrews, 1988). The second assemblage (3200–1000 BP), characterized by oscillations in periphytic, benthic and metaphytic diatoms, imply short-term changes in water levels possibly related to formation of the baymouth bar. The final assemblage (1000 BP– present), is distinguished by planktonic forms reflecting formation of the modern, 6 m deep pond. Thus Grenadier Pond

J.A. WOLIN AND H.C. DUTHIE

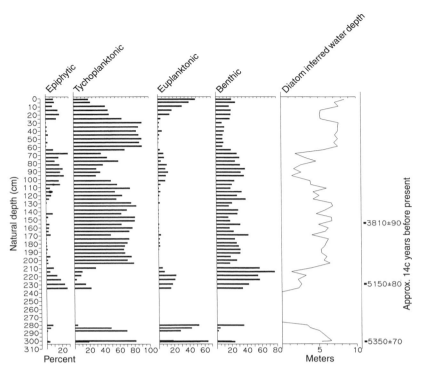

Fig. 8.5. Stratigraphy (% abundance) of four diatom life forms, and the diatom-inferred water depth in a sediment core from East Lake, Lake Ontario. Data between 280 cm and 235 cm are missing. Redrawn from Yang & Duthie (1995a).

corroborates the diatom-inferred water-depth changes in the other two Lake Ontario sites.

DIATOM EVIDENCE FOR WATER LEVEL CHANGES IN GREAT LAKES WETLANDS

Although coastal wetlands are important to the health of the North American Great Lakes ecosystem, in the past 50 years two-thirds of Great Lakes wetlands have been lost and many of those that remain have been degraded. Diatoms have been used in paleoecological studies of wetlands in Lake Ontario and Lake Erie to identify natural and anthropogenic impacts, including responses to long and short term water level changes.

A multivariate analysis of the 3000 yr diatom record and physical and chemical factors in a sediment core from Second Marsh, a Lake Ontario coastal wetland (Earle & Duthie, 1986), revealed a history of recurring cycles of two basic diatom communities with intermediate phases, consistent with repeated cycles of marsh build-up followed by degradation and erosion. Earle and Duthie attributed these cycles to fluctuating water levels in Lake Ontario. Additionally, a recent study of diatom assemblages in a sediment core from Big

Creek Marsh, Long Point, Lake Erie (Bunting *et al.*, 1997), also suggests alternating cycles of marshland and open water over the past millennium in response to fluctuating water levels.

Summary

The ability to determine water level changes in freshwater systems is an important tool for use in climatic, geologic and environmental studies. Signals of water level change can be successfully investigated with the use of freshwater diatom assemblages. Although these signals are not as clear cut as those from other systems (i.e., saline lakes), multiple types of freshwater responses can be utilized. Life-form or habitat group changes based on depth distributions (i.e. benthic vs. planktonic) produce the most reliable evidence of water level change. Under certain conditions, however, other diatom signals based on changes in the physical–chemical environment can also be used. The development of quantitative diatom-inferred depth models for reconstruction of water level changes in freshwater systems has greatly improved our understanding of past climatic and environmental changes. It must, however, be emphasized that quantitative or qualitative use of freshwater diatom data for reconstruction purposes must be used with caution. Many of the authors cited have stressed the importance of understanding the system being investigated and use of independent corroborating evidence before valid reconstructions can be made. The diatom signals observed are open to alternative interpretations. This difficulty, in part, is responsible for the low numbers of investigations attempting quantitative reconstructions in freshwater systems. More ecology investigations of diatom nutrient, light, and temperature requirements in relation to depth distribution are needed, as are more quantitative studies in general. As more data bases become available, water level information from lakes can be combined into regional climate reconstructions such as those developed for southern Sweden (Digerfeldt, 1988) and regional data assembled into more wide-ranging continental reconstructions (e.g., Harrison & Digerfeldt, 1993; Harrison *et al.*, 1996).

Acknowledgments

We thank Dr Jingrong Yang, Department of Biology, University of Ottawa, for the use of published and unpublished information. Additional thanks to John Anderson, Geological Survey of Denmark and Greenland, Philip Barker, Department of Geography, Lancaster University, UK, J. Platt Bradbury, US Geological Survey, Sandy Harrison, Department of Physical Geography, Uppsala University, Regine Jahn, Berlin Botanical Garden and Museum, André Lotter, Swiss Federal Institute for Environmental Science and

J.A. WOLIN AND H.C. DUTHIE

Technology – EAWAG, and Curt Stager, Paul Smith's College, for supplying information, reprints and discussions on their work.

References

Anderson, T. W., & Lewis, C. F. M. (1985). Postglacial water level history of the Lake Ontario basin. In *Quaternary evolution of the Great Lakes*, ed. P. F. Karrow & P. E. Calkin, pp. 231–51. Geological Association of Canada Special Paper 30.

Barker, P. A., Roberts, N., Lamb, H. F., van der Kaars, S., & Benfaddour, A. (1994). Interpretation of Holocene lake-level change from diatom assemblages in Lake Sidi Ali, Middle Atlas, Morocco. *Journal of Paleolimnology*, **12**, 223–34.

Bertzen, G. (1987). *Diatomeenanalytische Untersuchungen an spätpleistozänen und holozänen Sedimenten des Tegeler Sees*. Berliner Geographische Abhandlungen 45. Berlin, Germany.

Birks, H. J. B., Line, J. M., Juggins, S., Stevenson, A. G., & ter Braak, C. J. F. (1990). Diatoms and pH reconstruction. *Philosophical Transactions of the Royal Society of London*, **327**, 263–78.

Bradbury, J. P. (1971). Paleolimnology of Lake Texcoco, Mexico. Evidence from diatoms. *Limnology and Oceanography*, **16**, 180–200.

(1975). *Diatom Stratigraphy and Human Settlement in Minnesota*. Geological Society of America, Special Paper 171.

(1986). Late Pleistocene and Holocene paleolimnology of two mountain lakes in western Tasmania. *Palaios*, **1**, 381–8.

(1987). Late Holocene diatom paleolimnology of Walker Lake, Nevada. *Archive für Hydrobiologie. Supplement*, **79**, 1–27.

Bradbury, J. P., Layden, B., Saldago-Labouriau, M.., Lewis, W. M., Schubert, C., Binford, M. W., Frey, D. G., Whitehead, D. R., & Weibezahn, F. H. (1981). Late Quaternary environmental history of Lake Valencia, Venezuela. *Science*, **214**, 1299–305.

Brenner, M., Binford, M. W., & Deevey, E. S. (1990). Lakes. In *Ecosystems of Florida*, ed. R. L. Myers & J. J. Ewel, pp. 364–391. Orlando, Florida: University of Central Florida Press.

Bunting, M. J., Duthie, H. C., Campbell, D. R., Warner, B. G., & Turner, L. J. (1997). A paleo-ecological record of recent environmental change at Big Creek Marsh, Long Point, Lake Erie. *Journal of Great Lakes Research*, **23**, 349–68.

Charles, D. F. (1985). Relationships between surface sediment diatom assemblages and lake water characteristics in Adirondack lakes. *Ecology*, **66**, 994–1011.

Coakley, J. P., & Karrow, P. F. (1994). Reconstruction of post-Iroquois shoreline evolution in western Lake Ontario. *Canadian Journal of Earth Sciences*, **31**, 1618–29.

Davydova, N. N. (1986). Diatom analysis. In *Istoriya ozer SSSR* (lake History of the USSR), Leningrad, Nauka.

Davydova, N., & Raukas, A. (1986). Geological development of large lakes of the humid zone in the European part of the Soviet Union, and Holocene climatic changes on the basis of lake sediment data. *Journal of Biogeography*, **13**, 173–80.

Dean, W. E., Bradbury, J. P., Andersen, R. Y., & Barnosky, C. W. (1984). The variability of Holocene climate change: evidence from varved lake sediments. *Science*, **226**, 1191–4.

Dearing, J. A. (1983). Changing patterns of sediment accumulation in a small lake in Scania, southern Sweden. *Hydrobiologia*, **103**, 59–64.

Dearing, J. A., & Foster, I. D. L. (1986). Lake sediments and palaeohydrological studies. In *Handbook of Holocene Palaeoecology and Palaeohydrology*, ed. B. E. Berglund, pp. 67–90. New York: John Wiley.

Delorme, L. D., Duthie, H. C., Esterby, S. R., Smith, S. M., & Harper, N. S. (1986). Prehistoric inferred pH changes in Batchawana Lake, Ontario from sedimentary diatom assemblages. *Archiv für Hydrobiologie*, **108**, 1–22.

Digerfeldt, G. (1986). Studies on past lake-level fluctuations. In *Handbook of Holocene Palaeoecology and Palaeohydrology*, ed. B. E. Berglund, pp. 127–43. New York: John Wiley.

(1988). Reconstruction and regional correlation of Holocene lake-level fluctuations in Lake Bysjön, South Sweden. *Boreas*, **7**, 165–82.

Dorofeyuk, N. I. (1988). Holocene palaeogeography of MPR by diatom record from lake bottom sediments. In *Prirodnye usloviya, rastitelnyi pokrov i zhivotnyi mir Mongolii* (Nature conditions, vegetation cover and animals of Mongolia). pp. 61–82. Pushchino.

Duthie, H. C., & Sreenivasa, M. R. (1972). The distribution of diatoms on the superficial sediments of Lake Ontario. In *Proceedings of the 15th Conference of Great Lakes Research*, pp. 45–52. International Association for Great Lakes Research.

Duthie, H. C., Yang, J-R., Edwards, T. W. D., Wolfe, B. B., & Warner, B. G. (1996). Hamilton Harbour, Ontario: 8300 years of environmental change inferred from microfossil and isotopic analyses. *Journal of Paleolimnology*, **15**, 79–97.

Earle, J. C., & Duthie, H. C. (1986). A multivariate statistical approach for interpreting marshland diatom succession. In *Proceedings of the Eighth International Diatom Symposium*, ed. M. Ricard, pp. 441–58. Koenigstein: Koeltz Scientific Books.

Earle, J. C., Duthie, H. C., Glooschenko, W. A., & Hamilton, P. B. (1988). Factors affecting the spatial distribution of diatoms on the surface sediments of three Precambrian Shield lakes. *Canadian Journal of Fisheries and Aquatic Sciences*, **45**, 469–78.

Fahnenstiel, G. L., & Scavia, D. (1987). Dynamics of Lake Michigan phytoplankton: The deep chlorophyll layer. *Journal of Great Lakes Research*, **13**, 285–95.

Fritz, S. C. (1989). Lake development and limnological response to prehistoric and historic land-use in Diss, Norfolk, U. K. *Journal of Ecology*, **77**, 182–202.

Gaillard, M-J., Dearing, J. A., El-Daoushy, F., Enell, M., & Håkansson, H. (1991). A late Holocene record of land-use history, soil erosion, lake trophy and lake-level fluctuations at Bjäresjösjön (South Sweden). *Journal of Paleolimnology*, **6**, 51–81.

Gasse, F., Lédée, V., Massault, M., & Fontes, J-C. (1989). water level fluctuations of Lake Tanganyika in phase with oceanic changes during the last galciation and deglaciation. *Nature*, **342**, 57–9.

Hall, R. I., & Smol, J. P. (1992). A weighted averaging regression and calibration model for inferring total phosphorus concentration from diatoms in British Columbia (Canada) lakes. *Freshwater Biology*, **27**, 417–34.

Harrison, S. P., & Digerfeldt, G. (1993). European lakes as palaeohydrological and palaeoclimatic indicators. *Quaternary Science Reviews*, **12**, 233–48.

Harrison, S. P., Yu, G., & Tarasov, P. E. (1996). Late Quaternary lake-level record from Northern Eurasia. *Quaternary Research*, **45**, 138–59.

Haworth, E. Y. (1972). Diatom Succession in a core from Pickerel Lake, North-eastern South Dakota. *Geological Society of America, Bulletin*, **83**, 157–72.

(1984). Stratigraphic changes in algal remains (diatoms and chrysophytes) in recent sediments of Blelham Tarn, English Lake District. In *Lake Sediments and Environmental History*, ed. E. Y. Haworth & J. W. G. Lund, pp. 165–90. Minneapolis: University of Minnesota Press.

Irvine, K., Moss, B., & Balls, H. (1989). The loss of submerged plants with eutrophication II. Relationships between fish and zooplankton in a set of experimental ponds, and conclusions. *Freshwater Biology*, **22**, 89–107.

Kingston, J. C., Lowe, R. E., Stoermer, E. F., & Ladewski, T. B. (1983). Spatial and temporal distribution of benthic diatoms in northern Lake Michigan. *Ecology*, **64**, 1566–80.

Kingston, J. C., Birks, H. J. B., Utuala, A. C., Cumming, B. F., & Smol, J. P. (1992). Assessing trends in fishery and lake water aluminium from paleolimnological analyses of siliceous algae. *Canadian Journal of Fisheries and Aquatic Sciences*, **49**, 116–27.

Krabbenhoft, D. P., & Webster, K. E. (1995). Transient hydrological controls on the chemistry of a seepage lake. *Water Resources Research*, **31**, 2295–305.

Larsen, C. E. (1987). Geological history of glacial Lake Alqonquin and the Upper Great Lakes. US Geological Survey Bulletin 1801.

Line, J. M., & Birks, H. J. B. (1990). WACALIB version 2.1 – a computer program to reconstruct environmental variables from fossil assemblages by weighted averaging. *Journal of Paleolimnology*, **3**, 170–3.

Lotter, A. (1988). Past water level fluctuations at Lake Rotsee (Switzerland), evidenced by diatom analysis. In *Proceedings of Nordic Diatomist Meeting, Stockholm, 1987*, Department of Quaternary Research (USDQR) Report 12, 1988, pp. 47–55. Stockholm, Sweden: University of Stockholm.

McCarthy, F. M. G., & McAndrews, J. H. (1988). Water levels in Lake Ontario 4230–2000 years BP: Evidence from Grenadier Pond, Toronto, Canada. *Journal of Paleolimnology*, 1, 99–113.

Marciniak, B. (1990). Late glacial and Holocene diatoms in sediments of the Bledowo Lake (Central Poland). In *Proceedings of the 10th Diatom Symposium*, ed. H. Simola, pp. 379–90. Koenigstein: Koeltz Scientific Books.

Mason, I. M., Guzkowska, M. A. J., & Rapley, C. G. (1994). The response of lake-levels and areas to climatic change. *Climatic Change*, 27, 161–97.

Odgaard, B. (1993). Wind-determined sediment distribution and Holocene sediment yield in a small Danish, kettle lake. *Journal of Paleolimnology*, 8, 3–13.

Owen, R. B., & Crossley, R. (1992). Spatial and temporal distribution of diatoms in sediments of Lake Malawi, Central Africa, and ecological implications. *Journal of Paleolimnology*, 7, 55–71.

Owen, R. B., Barthelme, J. W., Renaut, R. W., & Vincens A. (1982). Paleolimnology and archaeology of Holocene deposits north-east of Lake Turkana, Kenya. *Nature*, 298, 523–9.

Pilskaln, C. H., & Johnson, T. C. (1991). Seasonal signals in Lake Malawi sediments. *Limnology and Oceanography*, 36, 544–57.

Richardson, J. L. (1969). Former lake-level fluctuations – their recognition and interpretation. *Mitteilungen International Vereiningung für Theoretische und Angewandte Limnologie*, 17, 78–93.

Rippey, B., Anderson, N. J., & Foy, R. H. (1997). Accuracy of diatom-inferred total phosphorus concentrations and the accelerated eutrophication of a lake due to reduced flushing. *Canadian Journal of Fisheries and Aquatic Sciences*, 54(11), 2637–46.

Round, F. E. (1961). Studies on bottom-living algae in some lakes of the English Lake District. Part VI. The effect of depth on the epipelic algal community. *Journal of Ecology*, 49, 245–54.

Schelske, C. L., Rothman, E. D., Stoermer, E. F., & Santiago, M. A. (1974). Responses of phosphorus limited Lake Michigan phytoplankton to factorial enrichments with nitrogen and phosphorus. *Limnology and Oceanography*, 19, 409–19.

Stager, J. C., Cumming, B., & Meeker, L. (1997). A high-resolution 11,400-year diatom record from lake Victoria, East Africa. *Quaternary Research*, 47, 81–9.

Stevenson, A. C., Birks, H. J., Flower, R. J., & Battarbee, R. W. (1989). Diatom-based pH reconstruction of lake acidification using canonical correspondence analysis. *Ambio*, 18, 44–52.

Stevenson, R. J., & Stoermer, E. F. (1981). Quantitative differences benthic algal communities along a depth gradient in Lake Michigan. *Journal of Phycology*, 17, 29–36.

Stoermer, E. F. (1968). Nearshore phytoplankton populations in the Grand Haven, Michigan, vicinity during thermal bar conditions. In *Proceedings of the 11th Conference of Great Lakes Research*, pp. 37–150. International Association for Great Lakes Research.

(1975). Comparison of benthic diatom communities in Lake Michigan and Lake Superior. *Internationale Vereinigung für Theoretische und Angewandte Limnologie*. Verhandlung 19, 932–8.

Stoermer, E. F., & Ladewski, T. B. (1976). *Apparent optimal temperatures for the occurrence of some common phytoplankton species in southern Lake Michigan*. Great Lakes Research Division Special Report No. 62. Ann Arbor, MI: University of Michigan.

Stoermer, E. F., Wolin, J. A., Schelske, C. L., & Conley, D. J. (1985a). An assessment of ecological changes during the recent history of Lake Ontario based on siliceous algal microfossils preserved in the sediments. *Journal of Phycology*, 21, 257–76.

(1985b). Post-settlement diatom succession in the Bay of Quinte, Lake Ontario. *Canadian Journal of Fisheries and Aquatic Sciences*, 42, 754–67.

Street-Perrott, F. A., & Harrison, S. P. (1985). Lake-levels and climate reconstruction. In

Paleoclimate Analysis and Modelling, ed. A. D. Hecht, pp. 291–340. New York: John Wiley.

Sweets, P. R., & Stevenson, R. J. (1997). Diatom inference of ionic concentration as a proxy for paleoclimate in seepage lakes and wetlands. In *Published Abstracts, American Society of Limnology and Oceanography Meeting, Santa Fe, New Mexico, 1997*, pp. 317.

Tarasov, P. E., Harrison, S. P., Saarse, L., Pushenko, M. Y., Andreev, A. A., Aleshinskaya, Z. V., Davydova, N. N., Dorofeyuk, N. I., Efremov, Y. V., Khomutova, V. I., Sevastyanov, D. V., Tamosaitis, J., Uspenskaya, O. N., Yakushko, O. F., & Tarasova, I. V. (1994). *Lake Status Records from the Former Soviet Union Mongolia: Data Base Documentation.* Paleoclimatology Publications Series Report no. 2. World Data Center-A for Paleoclimatology, NOAA Paleoclimatology Program. Boulder, Colorado.

ter Braak, C. J. F. (1986). Canonical correspondence analysis: A new eigenvector method for multivariate direct gradient analysis. *Ecology*, **76**, 1167–79.

ter Braak, C. J. F., & van Dam, H. (1989). Inferring pH from diatoms: A comparison of old and new calibration methods. *Hydrobiologia*, **178**, 209–23.

Wetzel, R. G. (1983). *Limnology* 2nd edn. New York: Saunders College Publishing.

Wolin, J. A. (1992). *Paleoclimatic implications of late Holocene lake-level fluctuations in Lower Herring Lake, Michigan.* PhD thesis. Ann Arbor, Michigan: University of Michigan.

(1996). Late Holocene lake-level fluctuations in Lower Herring Lake, Michigan, U.S.A. *Journal of Paleolimnology*, **15**, 19–45.

Yang, J-R., & Dickman, M. (1993). Diatoms as indicators of lake trophic status from central Ontario. *Diatom Research*, **8**, 179–93.

Yang, J-R., & Duthie, H. C. (1995a). Diatom paleoecology of East Lake, Ontario: A 5400 yr record of limnological change. *Proceedings 13th International Diatom Symposium*, ed., D. Marino & M. Montresor pp. 555–71. Bristol: Biopress Limited.

(1995b). Regression and weighted averaging models relating surficial diatom assemblages to water depth in Lake Ontario. *Journal of Great Lakes Research*, **21**, 84–94.

Yang, J-R., Duthie, H. C., & Delorme, L. D. (1993). Reconstruction of the recent environmental history of Hamilton Harbour from quantitative analysis of siliceous microfossils. *Journal of Great Lakes Research*, **19**, 55–71.

Yu, G., & Harrison, S. P. (1995). *Lake Status Records from Europe: Data Base Documentation.* Paleoclimatology Publications Series Report no. 3. World Data Center-A for Paleoclimatology, NOAA Paleoclimatology Program. Boulder, Colorado.

Part III
Diatoms as indicators in extreme environments

9 Diatoms as indicators of environmental change near arctic and alpine treeline

ANDRÉ F. LOTTER, REINHARD PIENITZ, AND ROLAND SCHMIDT

Introduction

Timber line represents the most prominent ecotone in mountainous and arctic regions. It is characterized by the transition from closed forest to the most advanced solitary trees (i.e., timber line), to single tree islands (i.e., treeline), and eventually to unforested vegetation. This biological boundary can vary in width between tens of meters and many kilometers. In northern Europe it is formed by deciduous trees (*Betula*, *Alnus*, *Populus*), whereas coniferous trees (*Pinus*, *Picea*, *Larix*) form treeline in the Alps, northern North America and Eurasia.

Treeline is primarily related to cold temperatures but a complex set of different climatic factors, as well as the specific adaptation of trees, actually defines the forest limit (e.g., Tranquillini, 1979). This is evident from the decrease in altitude of treeline from subtropical to arctic regions and, on a smaller scale, by the higher forest limit on southern slopes compared to northern slopes (Ellenberg, 1986). In the Alps, timber line represents the transition between the subalpine and the alpine belts (Ozenda, 1985; Ellenberg, 1986). The lower boundary of the alpine belt, however, is difficult to locate as human impact, grazing and climatic oscillations have lowered natural tree limit by several hundred meters in the last millennia (e.g., Lang, 1994; Tinner *et al.*, 1996).

In the north, physical and biotic features are sufficiently distinct to unequivocally separate 'arctic' from 'boreal' regions. However, great disparity exists among definitions as to where the Boreal region ends and the Arctic region begins (Larsen, 1989). The zone of transition between the two, the so-called 'forest–tundra' transition, or ecotone, can be identified with a certain degree of accuracy, but the drawing of lines on maps to delineate Arctic, transitional, and Boreal zones remains controversial. The Arctic as a geographical concept has been most usefully defined either as the area north of polar treeline or as the region north of the July mean daily isotherm of 10 °C (Hustich, 1979). The most pragmatic way to delimit the southern margin of the Arctic is to take the boundary of wooded vegetation that can be observed on aerial photographs or satellite images.

Aquatic habitats suitable for diatoms at these altitudes and latitudes are many-fold. The lakes are characterized by special limnological features, and are usually only ice free during a short period in summer, with cold water temperatures prevailing. Furthermore, light availability in the water column is strongly restricted by ice and snow cover as well as by the polar winter. During summer, however, the light conditions change significantly, especially at high latitude lakes: the photoperiod is long and the angle of the sun is high. The resulting UV (ultraviolet) radiation may have an inhibitory effect on periphytic littoral algae, such as *Achnanthes minutissima* Kützing, if dissolved organic matter is at low concentrations (e.g., Vinebrooke & Leavitt, 1996). Lakes that are influenced by glacial meltwater are turbid with high concentration of silt and clay. In the absence of glaciers, such lakes have either a high water transparency or, if mires are present in their catchment, the input of humic acids will result in dystrophic conditions.

Lakes on opposite sides of treeline exhibit striking differences in water chemistry and physical conditions (Pienitz *et al.*, 1997*a,b*; Vincent & Pienitz, 1996; Sommaruga & Psenner, 1997), which are reflected by abrupt changes in diatom community structure and composition. For example, differences in lake mixing regimes, nutrients, and lakewater transparency are most apparent at this ecotonal boundary marked by sharp changes in abiotic (e.g., changes in albedo and permafrost) and biotic variables (e.g., atmospheric deposition of macronutrients by pollen; Doskey & Ugoagwu, 1989; Lee *et al.*, 1996). Furthermore, sediment accumulation rates may change drastically across this ecotone: accumulation rates above treeline are often up to an order of magnitude lower than in comparable boreal or temperate lake basins.

The dearth of alpine and northern treeline diatom studies is primarily related to logistic problems. Nevertheless, early diatom studies from high altitude and high latitude regions usually provide extended taxon lists (e.g., Cleve-Euler, 1934; Krasske, 1932; Hustedt, 1942, 1943; Foged, 1955) and often classify taxa according to their plant–geographical distribution (e.g., nordic–alpine).

Freshwater diatoms, like most algae, are usually not considered sensitive indicators of temperature (e.g., Battarbee, 1991), even though clear latitudinal patterns in their distributions are apparent (Foged, 1964), and classifications according to their thermal requirements (Backman & Cleve-Euler, 1922; Hustedt, 1939, 1956) have long existed. Several experimental studies (e.g., Eppley, 1977; Patrick, 1971, 1977; Hartig & Wallen, 1986; Dauta *et al.*, 1990) indicate a temperature dependency of diatom growth and community composition. Under natural conditions, Stoermer & Ladewski (1976) and Kingston *et al.* (1983) found a relationship between temperature and the occurrence of certain diatoms. Kilham *et al.* (1996), however, associate the climatic distribution of diatoms with resource-related competitive interactions.

Diatoms as paleoenvironmental indicators at treeline

Treeline vegetation will be strongly impacted by future climatic warming (Houghton *et al.*, 1990; Smith *et al.*, 1992; Monserud *et al.*, 1993; Grabherr *et al.*,

1994; Kupfer & Cairns, 1996). However, aquatic environments and ecosystems, water supplies and fisheries will also be affected (Houghton *et al.*, 1996; Watson *et al.*, 1996). Global warming will result in increased heat uptake of lakes which, in turn, will lead to earlier break-up of ice and, consequently, to changes in stratification and oxygen regimes (Robertson & Ragotzkie, 1990; Hondzo & Stefan, 1993; Schindler *et al.*, 1990, 1996*a,b*; Sommaruga-Wögrath *et al.*, 1997; Livingstone, 1998).

Because the location of northern treeline results from and influences the mean position of major atmospheric boundaries (i.e., the Polar Front; Bryson, 1966; Pielke & Vidale, 1995), it is important to understand the linkages between climate and vegetation along the northern edge of the boreal forests. The sensitivity of arctic treeline position to climatic factors makes this ecotone an ideal location for investigating the effects and timing of climatic change. Paleoecological records provide a means of reconstructing the impact of past climatic variations on treeline vegetation, thereby allowing a better understanding of the causes and dynamics of past and future climatic change. Moreover, latitudinal changes in the position of the Polar Front through time provide important boundary conditions that may be used in testing and evaluating Global Circulation Models.

Paleolimnological data on past climatic changes in treeline regions are particularly important, as many of the more traditional, terrestrial-based paleoecological techniques, such as palynology, may reach their methodological limits at sites above treeline (e.g., Lang, 1994; Gajewski *et al.*, 1995; Smol *et al.*, 1995). These restrictions do not apply to diatoms. They are extremely abundant and ecologically diverse. Their short lifespan and fast migration rates enable them to respond quickly to environmental changes. Their potential as paleo-indicators mostly relies on their good preservation in lake sediments.

Many variables affect the size and species composition of diatom communities. Studies using diatom records to infer past climates have generally yielded qualitative results rather than quantitative paleoclimate estimates. Climate either directly (e.g., via changes in lake water temperature, mixing regime) or indirectly influences diatoms by controlling, for example, habitat availability, catchment and aquatic vegetation, water colour and transparency, or nutrient supply (e.g., Moser *et al.*, 1996).

Diatom studies in alpine, boreal, and arctic lakes (e.g., Hustedt, 1942; Cleve-Euler, 1951–1955; Sabelina *et al.*, 1951; Florin, 1957; Foged 1964, 1981; Mölder & Tynni, 1967–1975; Tynni, 1975, 1976; Koivo & Ritchie, 1978; Arzet, 1987; Niederhauser & Schanz, 1993) have shown relationships between diatom assemblages and aspects of water chemistry. These surveys also illustrated the need for more precise, quantitative diatom autecological data to refine or replace the classifications introduced by Kolbe (1927) and Hustedt (1939, 1956).

Considerable progress has recently been made in the development of inferential statistics useful for inferring past conditions from paleoecological data. Surface sediment calibration sets (see Charles & Smol, 1994) are currently widely used to estimate environmental optima and tolerances of diatoms and for the development of diatom-based transfer functions (see Hall & Smol, this volume). These transfer functions allow quantitative reconstructions of past limnological conditions, using a range of numerical regression and calibration techniques (Birks, 1995). Although only a few calibration sets are available for treeline regions (e.g., Pienitz & Smol, 1993; Pienitz *et al.*, 1995a; Wunsam *et al.*, 1995; Allaire, 1996; Rühland, 1996; Lotter *et al.*, 1997; Weckström *et al.*, 1997a), many more are currently being completed in different regions (Fig. 9.1).

The Arctic

NORTH AMERICA

Having studied modern diatom assemblages along a transect crossing three vegetation zones in northwestern Canada (Fig. 9.1–A), Pienitz *et al.* (1995a) found that summer surface water temperature and lake depth were the two best predictors of diatom distribution. Their weighted-averaging (WA) model for diatom–temperature relationships predicted temperatures that closely approximated the actual observed values (Fig. 9.2), with the final inference model yielding a bootstrapped root mean squared error of prediction (RMSEP) of 1.8 to 2.0 °C. The WA optima ranged from 15.4 °C (*Stauroneis smithii* var. *minima* Haworth) to 21.5 °C (*Synedra radians* Kützing). Planktonic diatoms generally displayed optima at the higher end of the temperature range, whereas small benthic taxa with 'nordic–alpine' affinities were positioned at the lower end of the temperature gradient. Variance partitioning tests showed that there was a statistically significant component of variation in the diatom data that was explained by temperature independent of chemistry, thus suggesting that fossil diatom data could be used to reconstruct temperature changes. Such reconstructions, however, are related to water rather than air temperature. Although these two variables are closely related, the timing and extent of stratification (which is dependent on the timing of ice break-up), atmospheric circulation and lake depth also affect water temperature and must be considered in any extrapolation to air temperature (Pienitz *et al.*, 1995a; Livingstone & Lotter, 1998).

The inclusion of an altitudinal gradient in the Pienitz *et al.* (1995a) calibration set revealed some similarities between alpine lakes and lakes located in arctic tundra regions, thereby indicating the potential of diatoms as paleoclimate proxies in alpine regions (Pienitz, 1993).

By studying oligotrophic lakes that spanned boreal forest, forest-tundra and arctic tundra in the central Northwest Territories (Fig. 9.1–C), Pienitz and Smol (1993) demonstrated that dissolved inorganic carbon (DIC) and dissolved organic carbon (DOC) concentrations explained significant variation in the diatom distributions. WA regression and calibration techniques were used to develop transfer functions relating diatom distributions to measured DIC and DOC. These transfer functions were later applied to fossil diatoms from two

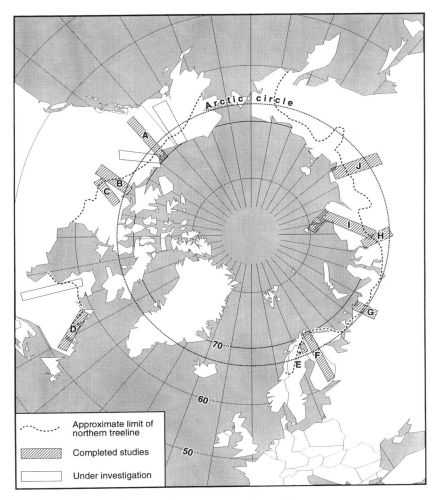

Fig. 9.1. Map showing diatom study regions at northern treeline.

treeline lakes. Diatom-inferred DIC and DOC and diatom concentrations revealed rapid increases in lake productivity associated with climate warming at *ca.* 5000 yrs BP and a contemporaneous shift from tundra to forest–tundra vegetation (MacDonald *et al.*, 1993; Pienitz, 1993; Pienitz *et al.*, 1999). Because lakewater DOC closely tracks the abundance of coniferous trees in a drainage basin (Pienitz *et al.*, 1997a,b; Vincent & Pienitz, 1996), diatom-based reconstructions of DOC can be used as a proxy for past vegetation shifts (Pienitz, 1993; Pienitz & Smol, 1993; Pienitz *et al.*, 1999).

Rühland's (1996) calibration set from the Canadian Northwest Territories (Fig. 9.1–B) similarly revealed distinct differences in diatom assemblages from boreal forest and arctic tundra sites. The most apparent trend was a shift from communities dominated by centric diatoms (e.g., *Cyclotella* spp.) in boreal forest lakes to those dominated by small, benthic, pennate taxa (e.g., *Fragilaria*,

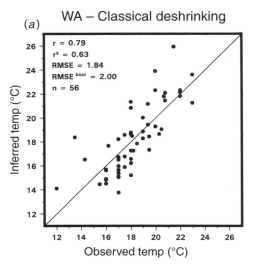
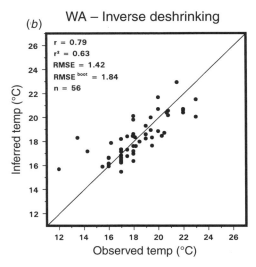

Fig. 9.2. Plots of observed vs. diatom-inferred summer surface water temperature (Temp) for the Yukon-Tuktoyaktuk Peninsula calibration set, based on weighted-averaging regression and calibration models using (a) classical deshrinking and (b) inverse deshrinking (after Pienitz *et al.*, 1995a).

Navicula and *Achnanthes*) in arctic tundra regions (Fig. 9.3). The comparatively low species diversity at tundra sites was mostly due to the high relative abundance of small *Fragilaria* spp. The results of the ordination analyses, together with the observed shifts in diatom assemblages across treeline and geological boundaries, suggested that alkalinity and conductivity exerted strong and significant controls on diatom taxa (Rühland, 1996).

A similar calibration set from Labrador (Fig. 9.1–D) showed that most diatom taxa were distributed along alkalinity and water colour gradients (Allaire, 1996), which together explained 30% of the total variance. Lake depth explained an additional 13% of the variation in the species data. The alkalinity optima ranged from 8.9 μeq/l (*Eunotia paludosa* var. *trinacria* (Krasske) Nörpel) to 306.5 μeq/l (*Cyclotella michiganiana* Skvortzow). An application of this inference model to fossil diatom data from southern Labrador revealed long-term natural acidification trends (Allaire, 1996), related to vegetational changes in the catchment.

SCANDINAVIA

Weckström *et al.* (1997a) studied the relationship between surface diatom assemblages and environmental variables in subarctic Fennoscandia (Fig. 9.1–E). Canonical Correspondence Analysis (CCA) suggested that pH, surface water temperature, conductivity, altitude, and sodium concentrations made significant contributions to explaining diatom distributions, with pH and temperature being the strongest predictor variables. Recently, this calibration set was expanded and diatom-based inference models for lakewater pH and

A. F. LOTTER *ET AL.*

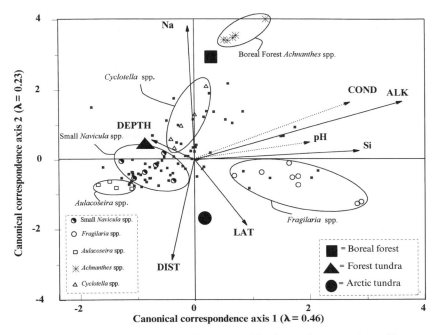

Fig. 9.3. Canonical correspondence analysis (CCA) of the Slave and Bear Province calibration set, showing general patterns in diatom distributions in relation to environmental variables (arrows) and major vegetational zones (centroids) (see Rühland, 1996).

temperature were developed (Weckström *et al.*, 1997*b*). The WA pH and temperature optima varied between pH 5.1 (*Tabellaria quadriseptata* Knudson) and 7.6 (*Neidium iridis* (Ehrenberg) Cleve), and between 10.3 °C (*Cyclotella rossii* Håkansson) and 14.7 °C (*Pinnularia legumen* Ehrenberg), respectively.

The diatom assemblages along a latitudinal transect extending from southern Finland to northern Norway (Fig. 9.1–F) were mainly composed of small acidophilic, periphytic taxa of boreo-alpine affinity, belonging to the genera *Achnanthes*, *Fragilaria* and *Navicula* (Pienitz *et al.*, 1995*b*). A regional comparison based on the biological composition of the Fennoscandian and Canadian subarctic sites showed similar trends in diatom assemblage composition with changing ecoclimatic zones.

SIBERIA

Information on freshwater diatoms from the Russian Arctic is scarce compared to its share of northern circumpolar treeline. Sabelina *et al.* (1951), Proschkina-Lavrenkov (1974), Aleksyuk & Bekman (1981), Chernyaeva (1981*a,b*), Antonov (1985) and Makarova (1992) published extensive reviews on freshwater diatoms from the former USSR. Besides providing information relating to diatom distributions, these references also include some ecological information. They represent the few comprehensive overviews of Russian diatoms, and are useful for identifying taxa commonly found in Siberia, the Kola Peninsula, Novaya Zemlya, and other arctic regions.

Recent research has focused on three areas spanning treeline in northern Siberia (Fig. 9.1). A first transect is located close to Norilsk near the Taymyr Peninsula (Fig. 9.1–H). Significant variables explaining the diatom distributions were conductivity, lake depth, surface water temperature, silica and DIC. *Pinnularia balfouriana* Grunow and *Thalassiosira pseudonana* Hasle & Heimdal were more abundant in tundra lakes, while *Stauroneis anceps* Ehrenberg was more commonly found in boreal forest lakes (Laing *et al.*, unpublished data).

A second transect in the Lena River region of north-central Siberia contains 31 oligotrophic lakes (Fig. 9.1–J). Multivariate techniques indicated that lake depth, DOC, particulate organic carbon (POC) and chloride were ecologically significant for diatoms. In particular, assemblages in forest lakes separated from tundra lakes along a combined depth and DOC gradient. Forest lakes were generally deeper, with higher DOC (Laing *et al.*, unpublished data).

Another 27 lakes spanning treeline in the Pechora River region have been sampled as a third Russian treeline transect (Fig. 9.1–G), but these analyses are still in progress (T. Laing, pers. comm.). Overall, these three transects indicate that forest lakes as compared to tundra lakes are associated with higher nutrient concentrations. Compiling all three Russian transects into one calibration set indicated that diatom assemblages were regionally distinct, reflecting local differences in geology, pollution and other factors. In general, however, the most common diatom taxa were similar to those found in northern Canada and Fennoscandia (Laing *et al.*, unpublished data).

Two sediment cores, both from lakes currently located in the tundra zone, were examined for changes in the fossil diatom assemblages (T. Laing, unpublished data). The abundance of *Picea* and *Larix* macrofossils and higher arboreal pollen values in the lower part of a sediment core taken northwest of Norilsk (between 4400 and 4000 yrs BP), as well as the presence of fossil rooted stumps within the catchment, revealed that trees were once locally present. Diatom assemblages showed a minor shift to taxa characteristic of cooler conditions (e.g., *Amphora pediculus* (Kützing) Grunow) concurrent with the disappearance of trees (T. Laing, unpublished data). The second sediment core taken in the Lena River region showed that trees existed in the area between 7500 and 3500 yrs BP. The major change in diatom assemblages occurred concurrently with the period of treeline advance, with their dominant taxa (e.g., *Achnanthes minutissima*) probably reflecting increases in lake productivity (T. Laing, unpublished data).

Several long sediment cores were sampled from lakes located along a 1400 km transect spanning forest–tundra (near Norilsk) through polar desert environments on the islands of Severnaya Zemlya (Fig. 9.1–I). Preliminary results from Lama Lake (east of Norilsk) show marked changes in community structure during the Holocene, especially with respect to ratios of planktonic/benthic taxa. The strong fluctuations observed within the planktonic taxa may be due to climate-induced variations in the thermal

A. F. LOTTER *ET AL.*

regime (e.g., a transition from dimictic to cold-monomictic conditions) of Lama Lake (Kienel, unpublished data).

The Alps

A STEEP ALTITUDINAL GRADIENT

Recent studies in mountainous regions have demonstrated a zonation of diatom assemblages along altitudinal gradients, which also incorporate gradients of water temperature (Servant-Vildary, 1982; Vyverman, 1992; Vyverman & Sabbe, 1995). Direct influences of temperature on the physiology of algal growth have been presented by, e.g., Raven & Geider (1988), whereas indirect effects due to prolonged ice-cover and changes in turbulent mixing have been discussed by Smol (1988) and Smol *et al.* (1991).

Sixty-eight Swiss lakes (Fig. 9.4) spanning an altitudinal gradient from 330 to 2350 m a.s.l. and a summer air temperature gradient of 7 to 21 °C demonstrate the importance of climate to diatoms and other aquatic organisms (chironomids, cladocera, chrysophytes), and have been used to develop multi-proxy temperature inference models (Lotter *et al.*, 1997). The power of reconstructing past climate change may be amplified by the use of several independent lines of evidence.

As air temperatures most closely correspond to surface water temperatures during summer (Livingstone & Lotter, 1998), instrumental air temperature time series have been used as a basis for the July temperature inference models. Only hardwater lakes with pH values well above 7 were chosen to rule out the strong influence of acid waters on the composition of aquatic organisms.

CCA of this calibration set showed that catchment (14.6%; geology, land use, vegetation type), climate (13.4%; temperature, precipitation), and limnological variables (11.5%; water depth, surface area, catchment area) had the largest, statistically significant independent explanatory powers, whereas water chemistry (conductivity, pH, alkalinity, DOC, nutrients, metals) explained also a large (14.6%) but not significant part of the total variance. A weighted averaging partial least squares model for diatoms and July temperature provided an apparent $r^2 = 0.96$, a jack-knifed $r^2 = 0.80$, and a jack-knifed RMSEP = 1.6 °C. A vast majority of the diatoms (72.5%) that occurred in 20% or more of the samples showed statistically significant relationships to July temperature, either as an unimodal or a sigmoidal response.

Below elevations of 1000 m a.s.l. planktonic diatoms were dominant, whereas above 1000–1500 m a.s.l. small periphytic (in some cases probably also tychoplanktonic) taxa such as *Fragilaria construens* (Ehrenberg) Grunow, *F. pinnata* Ehrenberg, *F. brevistriata* Grunow, and *Achnanthes minutissima* became more important (Fig. 9.4). There are, however, exceptions, mainly involving small centric taxa (e.g., *Cyclotella comensis*, *Thalassiosira pseudonana*). Low numbers of planktonic diatoms and increasing abundance of *Fragilaria* spp. is a

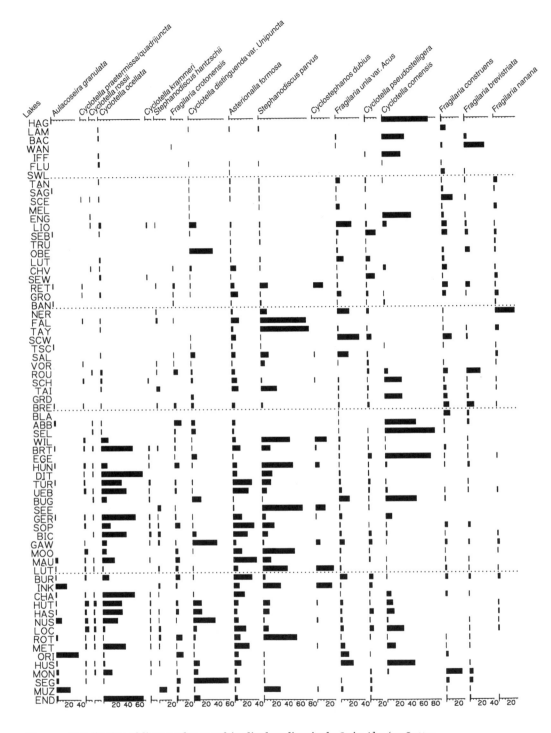

Fig. 9.4. Distribution of diatoms along an altitudinal gradient in the Swiss Alps (see Lotter
et al., 1997).

A. F. LOTTER *ET AL.*

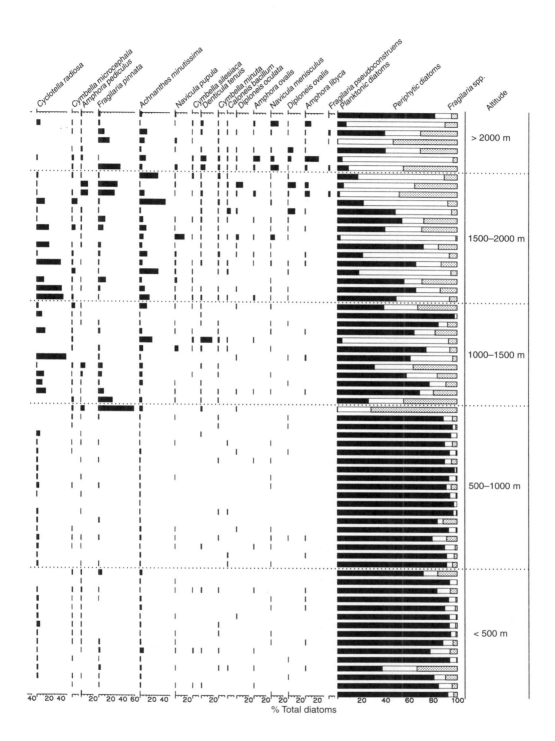

% Total diatoms

phenomenon often observed with increasing altitude or latitude. It may be related, on the one hand, to the fact that the growing season at these altitudes or latitudes is considerably shorter due to prolonged snow and ice-cover. As ice-melt starts at the shores, these marginal areas provide suitable habitats for the development of periphyton (Smol, 1988). Also, water transparency is generally high (Franz, 1979), thus favoring periphytic diatoms even in deeper lakes. On the other hand, alpine lakes are characterized by short-term fluctuations that may possibly favor *Fragilaria* spp. that are more adaptable and competitive.

Another diatom-temperature training set from the Alps, based exclusively on *Cyclotella* species, yielded an apparent $r^2 = 0.62$, with a bootstrap RMSEP = 1.3 °C (Wunsam *et al.*, 1995). Species temperature optima and tolerances were estimated by a WA model with *C. comensis* at the higher end (19.7 °C) and *C. styriaca* Hustedt at the lower end (11.9 °C) of the gradient. The enormous variability, especially within the small *Cyclotella* taxa, calls for more autecological studies of the various morphotypes.

Two calibration sets for nutrients (see also Hall & Smol, this volume), in particular, total phosphorus (TP), are also available from the Alps (Wunsam & Schmidt, 1995; Wunsam *et al.*, 1995; Lotter *et al.*, 1998). Yet, the majority of the calibration sites are located well below treeline. Application of these TP inference models to treeline lakes may, however, be limited by the generally low nutrient concentrations (Marchetto *et al.*, 1995; Müller *et al.*, in press) and the different diatom floras of the treeline lakes.

A CASE STUDY OF CHANGING ENVIRONMENTS:
THE LATE-GLACIAL

Although many inference models have recently been developed for use with arctic/alpine diatom assemblages, few of these have been applied as yet. Paleoenvironmental records from the Alps illustrate how the application of these models may help to better understand environmental changes at the end of the last Ice-Age.

Multidisciplinary paleolimnological techniques (Schmidt *et al.*, 1998) were applied to a long sediment core from Längsee (548 m a.s.l.), a small, meromictic kettle-hole lake situated in the southeastern Alps (Carinthia, Austria). Due to its location at the southern slope of the Alps, close to the Würm pleniglacial ice margin, the lake became ice-free early, probably already more than 18 000 years ago. Transfer functions (Wunsam & Schmidt, 1995; Wunsam *et al.*, 1995) have been applied to infer total phosphorus concentrations and summer lake surface temperature. Before 15 500 yr BP (Fig. 9.5), climatic warming after deglaciation allowed the immigration and expansion of shrubs (e.g., *Juniperus*) into the lake's catchment. Increase in lake temperature enhanced the development of the diatom plankton (*Cyclotella ocellata* Pantocsek, *C. comensis*, *C. cyclopuncta* Håkansson & Carter, *C. distinguenda* var. *unipunctata* (Hustedt) Håkansson & Carter, *Stephanodiscus alpinus*). For this phase a mean surface

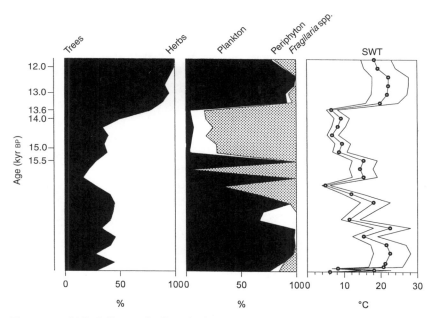

Fig. 9.5. Multidisciplinary paleolimnological approach to a late-glacial core section of Längsee (548 m a.s.l., Austria). Pollen percentages of trees and herbs, percentages of diatom life forms, and diatom-inferred summer surface water temperature (SWT; with sample-specific error band). Time-scale according to radiocarbon dates (see Schmidt *et al.*, 1998).

water temperature of 17.3 °C was inferred, suggesting that Längsee may have begun to stratify during this time. During the following climatic cooling, lake surface water temperature declined. Lower temperatures leading to prolonged ice-cover may be responsible for the low percentage of planktonic diatoms (Fig. 9.5). The change from a steppe to a tundra-like vegetation, rich in dwarf birch (*Betula nana*), and the still high amount of allochthonous material from catchment sources, indicate increasing precipitation. The contemporaneous increase in *Fragilaria* spp. (*F. construens*, including fo. *venter* (Ehrenberg) Hustedt, *F. pinnata*, all of which were more frequent in lakes with low summer surface temperature in the modern training set) explains the decrease in inferred summer water temperatures (mean inferred temperature = 10.7 °C). About 13 600 yr BP, reforestation by *Pinus cembra* began, which is also indicated by an increase in the amount of tree pollen and organic carbon content of the sediment. Diatom-inferred summer water temperatures increased at the same time towards modern levels (Fig. 9.5).

INDIRECT EFFECTS: A TEMPERATURE–PH RELATIONSHIP IN ALPINE LAKES

High altitude areas in the Alps are affected by precipitation with pH values between 4.8 and 5.2 and consequently acidification of crystalline bedrock sites has been reported. In these areas soil and vegetation have less influence on

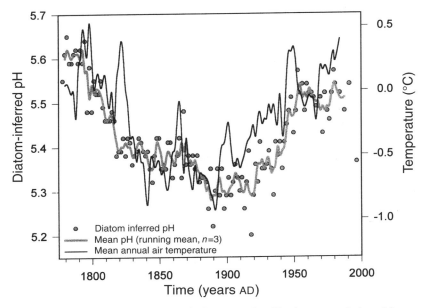

Fig. 9.6. Diatom-inferred pH history of Schwarzsee ob Sölden (2796 m a.s.l., Austria), during the last centuries, compared with the Austrian temperature curve (after Koinig *et al.*, 1997). The chronology is based on ^{210}Pb (constant rate of supply model) and ^{137}Cs dating (see Koinig, 1994; Koinig *et al.*, 1997; Sommaruga-Wögrath *et al.*, 1997).

biogenic acidification, because only a small fraction of the drainage area is covered by soil. High mountain lakes therefore react very sensitively to climatic and hydrological changes as well as to increasing acidity of precipitation (Arzet, 1987; Niederhauser, 1993; see also Battarbee *et al.*, this volume).

Paleolimnological investigations in Schwarzsee ob Sölden (2796 m a.s.l., Tyrol, Austria) showed a diatom-inferred pH decline from 6 to 5 (Arzet, 1987). Using a WA model based on a regional Alpine calibration set (Marchetto & Schmidt, 1993), the diatom-inferred pH changes in a core from Schwarzsee dating back to the eighteenth century were predominantly due to changes in abundances of the dominant *Aulacoseira* species (*A. alpigena* (Grunow) Krammer, *A. nygaardii* Camburn & Kingston, *A. distans*, including var. *nivalis* (W. Smith) Haworth, *A. perglabra* (Oestrup) Haworth; see Koinig, 1994 and Koinig *et al.*, 1997). The lowest inferred pH values occurred between 1880 and 1920. Towards 1970, the inferred pH values increased again to 5.55. A comparison of these pH reconstructions with the mean Austrian air temperature showed a strong correlation throughout the past 200 years (Fig. 9.6). These findings support the idea of a climate-driven pH control in high-alpine lakes (Psenner & Schmidt, 1992), i.e., that climatic cooling may cause decreases in pH, and rising temperatures increases in pH. The pH decline started at the onset of the Little Ice-Age, with glacier readvances in the Alps culminating around 1850. Koinig *et al.* (1997, 1998) and Sommaruga-Wögrath *et al.* (1997) suggested that pH increases during warm episodes may be a result of enhanced weather-

A. F. LOTTER ET AL.

ing rates, increased in-lake alkalinity production, longer water retention times, and larger amounts of dust deposition from the surroundings (see also Marchetto *et al.*, 1995). However, the hypothesis that the influence of modern acidification is counteracted by recent climatic warming is not valid for all lakes. At Lake Rassas (2682 m a.s.l., Southern Tyrol, Italy) the onset of anthropogenically derived acid precipitation at the beginning of the twentieth century (Schmidt & Psenner, 1992; Psenner & Schmidt, 1992) led to a decoupling of the temperature – pH relationship, and thus to lake acidification that could be traced by diatoms.

Common features to arctic and alpine diatom assemblages

Many of the above-mentioned calibration sets have revealed interesting similarities in diatom assemblage composition across the treeline, namely the presence of assemblages composed of large benthic and planktonic taxa in the more nutrient-enriched forest lakes, and assemblages dominated by small benthic taxa in the more dilute lakes above treeline. This distinct trend in diatom community structure and cell size with changing latitude and altitude, in patterns unrelated to water depth, may be related to differences in the physico-thermal properties of the lakes and the length of the growing season (Pienitz, 1993). For example, there is evidence that climatic warming, with its likely consequences of a longer ice-free season and enhanced thermal stratification, would give a competitive advantage to planktonic forms, keeping these diatoms in suspension in the photic zone for longer periods of time (Smol, 1988; Fee *et al.*, 1992).

The differences observed between diatom cell size in lakes below and above the treeline could be related to nutrient availability, length of growing season, and diatom growth rates. The maximum specific growth rates in algae decrease with increasing cell size (Raven & Geider 1988), and the high surface-to-volume ratio of small cells provides them with a competitive advantage under low nutrient conditions. According to Denys (1988), small benthic *Fragilaria* spp. fit into the classic model of organisms favored by *r*-selection (i.e., small size, high reproductive rate, large ecological amplitude). Because of their higher turnover rates, these opportunistic, small-celled diatoms may quickly form blooms and (temporarily) outcompete larger diatom species with slower growth rates during the brief growing season in tundra lakes (Pienitz, 1993).

With respect to patterns observed in planktonic/benthic forms and diatom cell size, availability of silica may also play an important role. As some of the studies revealed, silica concentrations explained a significant proportion of the variance in diatom distribution. Diatoms assimilate large quantities of silica, thereby modifying the flux rates of silica in freshwater ecosystems (e.g., Hutchinson, 1967; Wetzel, 1983; Schelske, this volume). Silica concentrations, on the other hand, are known to affect diatom growth rates, as well as the succession and composition of diatom communities (e.g., Lund, 1964; Servant-Vildary *et al.*, 1992). The small number of planktonic taxa observed in lakes above treeline may also be related to silica deficiency, which is a crucial factor especially for the growth of freshwater planktonic diatoms (e.g., Jørgensen, 1957; Kilham, 1971).

Summary

Diatoms in treeline regions are important and sensitive indicators of environmental change. The arctic and alpine treeline are important ecotones. Their present-day, past, and future positions depend largely on climatic factors. Climatic fluctuations at this ecotone either directly or indirectly influence diatom communities through alterations in physical and chemical limnetic properties. Therefore, diatoms have great potential as environmental indicators and contribute to and complement paleoecological studies in treeline regions. Reconstructing changes in past treeline position provides essential boundary conditions for Global Circulation Models to hindcast past or predict future climate change.

In all studies involving diatom calibration sets, pH or pH-related variables (e.g., alkalinity, ANC, DIC, Ca) have been shown to influence significantly diatom distributions. Apart from the predominant control exerted by pH and related variables, two major gradients appear to emerge from these studies: the concentration of lakewater DOC (which is related to catchment vegetation and treeline), and surface-water temperature (which is related to latitude and/or altitude). Temperature is a complex variable that is highly correlated with other characteristics of the environment, such as mixing regime and duration of ice-cover. The distinct influence exerted by the temperature gradient on diatom distribution as shown in the studies from northwestern Canada, Fennoscandia and the Alps, is of ecological and paleoecological significance as it strengthens the assumption that the relationship between organisms and climate is clearest at ecotonal boundaries, and, consequently, ecotones are optimal areas for studying climatic change (e.g., Smol *et al.*, 1991).

The potential for inferring past DOC levels from paleolimnological records combined with recent advances in bio-optical modelling in northern lakes (e.g., Laurion *et al.*, 1997), leads to the exciting prospect of reconstructing past underwater light regimes. Such analyses could include estimates of euphotic depth and spectral attenuation across the ultraviolet waveband. This new theme in paleolimnology ('lake paleo-optics') offers opportunities for integrating studies of the present-day with historical properties of lakes (Vincent & Pienitz, 1996). This approach is of special interest for lakes in treeline regions, since lakes above treeline may be more sensitive to small changes in DOC and rising UV-B (ultraviolet B) radiation associated with stratospheric ozone depletion than lakes of the boreal forest (Vincent & Pienitz, 1996).

Lake depth generally emerged as an important predictor for diatom communities, especially in arctic treeline regions. Its importance is known from many investigations and seems to support Smol's (1988) hypothesis that microhabitat availability significantly influences diatom assemblage composition (i.e., the relative percentage of planktonic and benthic taxa). Moreover, in paleoclimatological studies, lake depth becomes an important indicator with respect to lake-level changes (see Wolin & Duthie, this volume), e.g., through changes in atmospheric circulation patterns.

The results obtained from diatom studies in treeline regions seem to confirm a general relationship between the distribution of diatoms and ecoclimatic or vegetational zones. Despite the extremely high degree of floristic diversity that characterizes oligotrophic boreal lakes compared with lakes in temperate regions (e.g., Lange-Bertalot & Metzeltin, 1996), a surprisingly high degree of floristic similarity could be observed among calibration sets from different treeline regions. These regions may be floristically similar enough to allow the development of supra-regional data sets. A recently launched joint research initiative, the Circumpolar Treeline Diatom Database (CTDD), aims at expanding and improving existing diatom calibration sets by combining the existing data sets from North America, Scandinavia and Siberia into one large data set for the whole of circumpolar, northern hemispheric treeline regions. Nevertheless, all transfer function approaches have to be assessed critically (see, e.g., Birks, 1995) before using their results for hindcasting past environmental change at the arctic or alpine ecotone.

Acknowledgments

We would like to thank K. A. Koinig, A. Korhola, J. P. Smol, and I. R. Walker for valuable comments on earlier versions of the manuscript. Work presented here has been supported by Swiss NSF (grants 5001–34876 and 5001–44600) and by funding from the Natural Sciences and Engineering Research Council of Canada.

References

Aleksyuk, G. V., & Bekman, M. Y. (1981). Fitoplankton Putoranskikh ozer iusloviya ego razvitiya. In *Ozera Severo-Zapada Sibirskoj platformy*, ed. G. J. Galaziy & Y. P. Paramuzin, pp. 110–18. Novosibirsk.

Allaire, N. (1996). *Relation entre les assemblages de diatomées et les variables environnementales de 70 lacs du Labrador, et résultats préliminaires d'une étude paléolimnologique du Lac Hope Simpson*. MA thesis, Université Laval, Québec.

Antonov, S. E. (1985). Osobennosti fitoplanktona ozer Tajmyr, *Geografiya ozer Tajmyra*, ed. V. N. Adamenko & A. N. Egorov, pp. 128–30. Nauka, Leningrad.

Arzet K., (1987). *Diatomeen als pH-Indikatoren in subrezenten Sedimenten von Weichwasserseen*. Dissertation Universität Innsbruck.

Backman, A. L., & Cleve-Euler, A. (1922). Die fossile Diatomeenflora in Österbotten. *Acta Forestalia Fennica*, **22**, 5–73.

Battarbee, R. W. (1991). Paleolimnology and climate change. In *Evaluation of Climate Proxy*

Data in Relation to the European Holocene, ed. B. Frenzel, A. Pons & B. Gläser, Paläoklimaforschung, 6, Akademie der Wissenschaften und der Literatur Mainz, pp. 149–57.

Birks, H. J. B. (1995). Quantitative paleoenvironmental reconstructions. In *Statistical modelling of Quaternary science data*, ed. D. Maddy & J. S. Brew, Quaternary Research Association, Cambridge, pp. 161–254.

Bryson, R. A. (1966). Air masses, streamlines, and the boreal forest. *Geographical Bulletin*, 8, 228–69.

Charles, D. F., & Smol, J. P. (1994). Long-term chemical changes in lakes: Quantitative inferences using biotic remains in the sediment record. In *Environmental Chemistry of Lakes and Reservoirs*, ed. L. Baker, Advances in Chemistry Series, 237, pp. 3–31. American Chemical Society, Washington DC.

Chernyaeva, G. P. (1981a). Ekologo-sistematicheskaya kharakteristika diatomej ozera Nyakshinga. In *Ozera Severo-Zapada Sibirskoj platformy*, ed. G. J. Galaziy & Y. P. Paramuzin, pp. 118–23. Novosibirsk.

(1981b). Razpredelenie diatomej v ozadkakh I nekotorye voprosy paleogeografii ozer Agata Nizhnee I Agata Verkhnee. In *Istoriya bol'shikh ozer Tsentral'noj Subarktiki*, ed. G. J. Galaziy & Y. P. Paramuzin, pp. 122–30. Novosibirsk.

Cleve-Euler, A. (1934). The diatoms of Finnish Lapland. *Societas Scientiarum Fennica Commentationes Biologicae*, IV, 14, 1–154.

(1951–1955). Die Diatomeen von Schweden und Finnland. 1–5. *Kongliga Svenska Vetenskap Akademien Handlingar Serie 4*, 2/1, 1–163; 3/3, 1–153; 4/1, 1–158; 4/5, 1–225; 5/4, 1–232.

Dauta, A., Devaux, J., Piquemal, F., & Boumnich, L. (1990). Growth rate of four freshwater algae in relation to light and temperature. *Hydrobiologia*, 207, 221–6.

Denys, L. (1988). *Fragilaria* blooms in the Holocene of the western Belgian coastal plain. *Abstracts 10th International Diatom Symposium*, Joensuu, Finland.

Doskey, P. V., & Ugoagwu, B. J. (1989). Atmospheric deposition of macronutrients by pollen at a semi-remote site in northern Wisconsin. *Atmospheric Environment*, 23, 2761–6.

Ellenberg, H. (1986). *Vegetation Mitteleuropas mit den Alpen in ökologischer Sicht*. Stuttgart, E. Ulmer.

Eppley, R. W. (1977). The growth and culture of diatoms. In *The biology of diatoms*, ed. D. Werner, Botanical Monographs, 13, pp. 24–64. Oxford: Blackwell.

Fee, E. J., Shearer, J. A., DeBruyn, E. R., & Schindler, E. U. (1992). Effects of lake size on phytoplankton photosynthesis. *Canadian Journal of Fisheries and Aquatic Sciences*, 49, 2445–59.

Florin, M-B. (1957). Plankton of fresh and brackish waters in the Södertälje area (Sweden). *Acta Phytogeographica Suecica*, 37, 1–93.

Foged, N. (1955). Diatoms from Peary Island, north Greenland. *Meddelelser om Grønland*, 128, 1–90.

(1964). Freshwater diatoms from Spitsbergen. *Tromsö Museums Skrifter*, 11, 1–205.

(1981). *Diatoms in Alaska*. Vaduz: J. Cramer.

Franz, H. (1979). *Ökologie der Hochgebirge*. Stuttgart: E. Ulmer.

Gajewski, K., Garneau, M., & Bourgeois, J. C. (1995). Paleoenvironments of the Canadian High Arctic derived from pollen and plant macrofossils: Problems and potentials. *Quaternary Science Reviews*, 14, 609–29.

Grabherr, G., Gottfried, M., & Pauli, H. (1994). Climate effects on mountain plants. *Nature*, 369, 448.

Hartig, J. H., & Wallen, D. G. (1986). The influence of light and temperature on growth and photosynthesis of *Fragilaria crotonensis* Kitton. *Journal of Freshwater Ecology*, 3, 371–82.

Hondzo, M., & Stefan, H. G. (1993). Regional water temperature characteristics of lakes subject to climate change. *Climate Change*, 24, 187–211.

Houghton, J. T., Jenkins, G. J., & Ephraums, J. J. (ed.) (1990). *Climate Change: The IPCC Scientific Assessment*. Cambridge, UK: Cambridge University Press.

Houghton, J. J., Meiro Filho, L. G., Callender, B. A., Harris, N., Kattenberg, A., & Maskell, K. (ed.) (1996). *Climate change 1995 – the science of climate change. Contribution of Working Group I to the Second Assessment Report of the Intergovernmental Panel on Climate Change*.

Cambridge: Cambridge University Press.

Hustedt, F. (1939). Systematische und ökologische Untersuchungen über die Diatomeen-Flora von Java, Bali und Sumatra nach dem Material der Deutschen limnologischen Sunda-Expedition. *Archiv für Hydrobiologie Beiheft II*, **16**, 1–155, 274–394.

(1942). Diatomeen aus der Umgebung von Abisko in Schwedisch-Lappland. *Archiv für Hydrobiologie*, **39**, 82–174.

(1943). Die Diatomeenflora einiger Hochgebirgsseen der Landschaft Davos in den Schweizer Alpen. *Internationale Revue der gesamten Hydrobiologie*, **43**, 124–97.

(1956). *Kieselalgen (Diatomeen)*. Stuttgart: Kosmos.

Hustich, I. (1979). Ecological concepts and biogeographical zonation in the North: The need for a generally accepted terminology. *Holarctic Ecology*, **2**, 208–17.

Hutchinson, G. E. (1967). *A Treatise on Limnology*, vol. II, New York: John Wiley.

Jørgensen, E. G. (1957). Diatom periodicity and silicon assimilation. *Dansk Botanisk Arkiv*, **18**, 1–54.

Kilham, P. (1971). A hypothesis concerning silica and the freshwater planktonic diatoms. *Limnology and Oceanography*, **16**, 10–18.

Kilham, S. S., Theriot, E. C., & Fritz, S. C. (1996). Linking planktonic diatoms and climate change in the large lakes of the Yellowstone ecosystem using resource theory. *Limnology and Oceanography*, **41**, 1052–62.

Kingston, J. C., Lowe, R. L., Stoermer, E. F., & Ladewski, T. B. (1983). Spatial and temporal distribution of benthic diatoms in northern Lake Michigan. *Ecology*, **64**, 1566–80.

Koinig K. A., (1994). *Die pH-Geschichte des Schwarzsees ob Sölden. Eine Rekonstruktion der letzten 200 Jahre mittels benthischer Diatomeen*. Diplomarbeit Univ. Innsbruck/Institut für Limnologie Mondsee ÖAW. 1–62.

Koinig, K. A., Schmidt, R., & Psenner, R. (1998). Effects of air temperature changes and acid deposition on the pH history of three high alpine lakes. *Proceedings of the 14th International Diatom Symposium*. Koeltz Scientific Books, Koenigstein, 497–508.

Koinig, K. A., Schmidt, R., Wögrath, S., Tessadri, R., & Psenner, R. (1997). Climate change as the primary cause for pH shifts in a high Alpine lake. *Water, Air and Soil Pollution*, **104**, 167–80.

Koivo, L. K., & Ritchie, J. C. (1978). Modern diatom assemblages from lake sediments in the boreal–arctic transition region near the Mackenzie Delta, N.W.T., Canada. *Canadian Journal of Botany*, **56**, 1010–20.

Kolbe, R. W. (1927). Zur Ökologie, Morphologie und Systematik der Brackwasser-Diatomeen. *Pflanzenforschung*, **7**, 1–146.

Krasske, G. (1932). Beiträge zur Kenntnis der Diatomeenflora der Alpen. *Hedwigia*, **72**, 92–134.

Kupfer, J. A., & Cairns, D. M. (1996). The suitability of montane ecotones as indicators of global climatic change. *Progress in Physical Geography*, **20**, 253–72.

Lang, G. (1994). *Quartäre Vegetationsgeschichte Europas. Methoden und Ergebnisse*. Jena: G. Fischer.

Lange-Bertalot, H., & Metzeltin, D. (1996). Indicators of oligotrophy. 800 taxa representative of three ecologically distinct lake types. Carbonate buffered- oligo-trophic- weakly buffered soft water. *Iconographia Diatomologica*, **2**, 1–390.

Larsen, J. A. (1989). The Northern Forest Border in Canada and Alaska. *Ecological Studies*, **70**, 1–255.

Laurion, I., Vincent, W. F., & Lean, D. R. (1997). Underwater ultraviolet radiation: Development of spectral models for northern high latitude lakes. *Photochemistry and Photobiology*, **65**, in press.

Lee, E. J., Kenkel, N., & Booth, T. (1996). Atmospheric deposition of macronutrients by pollen in the boreal forest. *Écoscience*, **3**, 304–9.

Livingstone; D. M. (1998). Break-up dates of Alpine lakes as proxy data for local and regional mean surface air temperatures. *Climatic Change*, **37**, 407–39.

Livingstone, D. M., & Lotter, A. F. (1998). The relationship between air and water temperatures in lakes of the Swiss Plateau: A case study with paleolimnological implications. *Journal of Paleolimnology*, **19**, 181–98.

Lotter, A. F., Birks, H. J. B., Hofmann, W., & Marchetto, A. (1997). Modern diatom, Cladocera, chironomid, and chrysophyte cyst assemblages as quantitative indicators for the reconstruction of past environmental conditions in the Alps. I. Climate. *Journal of Paleolimnology*, **18**, 395–420.

(1998). Modern diatom, Cladocera, chironomid, and chrysophyte cyst assemblages as quantitative indicators for the reconstruction of past environmental conditions in the Alps. II. Nutrients. *Journal of Paleolimnology*, **19**, 443–63.

Lund, J. W. G. (1964). Primary production and periodicity of phytoplankton. *Verhandlungen der Internationalen Vereinigung für Theoretische und Angewandte Limnologie*, **15**, 37–56.

MacDonald, G. M., Edwards, T. W. D., Moser, K. A., Pienitz, R., & Smol, J. P. (1993). Rapid response of treeline vegetation and lakes to past climate warming. *Nature*, **361**, 243–6.

Makarova, I. W. (1992). *The Diatoms of the USSR – Fossil and Recent*, vol. II. (In Russian) Nauka, St. Petersburg.

Marchetto A., & R. Schmidt, (1993). A regional calibration data set to infer lakewater pH from sediment diatom assemblages in alpine lakes. *Memorie Istituto italiano di Idrobiologia*, **51**, 115–25.

Marchetto, A., Mosello, R., Psenner, R., Bendetta, G., Boggero, A., Tait, D., & Tartari, G. A. (1995). Factors affecting water chemistry of alpine lakes. *Aquatic Sciences*, **57**, 81–9.

Mölder, K., & Tynni, R. (1967–1975). Über Finnlands rezente und sub-fossile Diatomeen. *Bulletin of the Geological Society of Finland*, **40**, 151–70; **41**, 235–51; **42**, 129–44; **43**, 203–20; **44**, 141–59; **45**, 159–79.

Monserud, R. A., Tchebakova, N. M., & Leemans, R. (1993). Global vegetation change predicted by the modified Budyko model. *Climatic Change*, **25**, 59–83.

Moser, K. A., MacDonald, G. M., & Smol, J. P. (1996). Application of freshwater diatoms to geographical research. *Progress in Physical Geography*, **20**, 21–52.

Müller, B., Lotter, A. F., Sturm, M., & Ammann, A. (1998). Influence of catchment and altitude on the water and sediment composition of 68 small lakes in Central Europe. *Aquatic Sciences*, in press.

Niederhauser, P. (1993). *Diatomeen als Bioindikatoren zur Beurteilung der Belastung elektrolytarmer Hochgebirgsseen durch Säuren und Nährstoffe*. Dissertation, Universität Zürich.

Niederhauser, P., & Schanz, F. (1993). Effects of nutrient (N, P, C) enrichment upon the littoral diatom community of an oligotrophic high-mountain lake. *Hydrobiologia*, **269/270**, 453–62.

Ozenda, P. (1985). *La végétation de la chaîne alpine dans l'espace montagnard européen*. Paris: Masson.

Patrick, R. (1971). The effects of increasing light and temperature on the structure of diatom communities. *Limnology and Oceanography*, **16**, 405–21.

(1977). Ecology of freshwater diatoms and diatom communities. In *The Biology of Diatoms*, ed. D. Werner, Botanical Monographs, vol. **13**, pp. 284–332. Oxford: Blackwell.

Pielke, R. A., & Vidale, P. L. (1995). The boreal forest and the polar front. *Journal of Geophysical Research*, **100**, 25755–8.

Pienitz, R. (1993). *Paleoclimate proxy data inferred from freshwater diatoms from the Yukon and the Northwest Territories, Canada*. PhD thesis, Queen's University, Kingston (Ontario).

Pienitz, R., & Smol, J. P. (1993). Diatom assemblages and their relationship to environmental variables in lakes from the boreal forest–tundra ecotone near Yellowknife, Northwest Territories, Canada. *Hydrobiologia*, **269/270**, 391–404.

Pienitz, R., Smol, J. P., & Birks, H. J. B. (1995a). Assessment of freshwater diatoms as quantitative indicators of past climatic change in the Yukon and Northwest Territories, Canada. *Journal of Paleolimnology*, **13**, 21–49.

Pienitz, R., Douglas, M. S. V., Smol, J. P., Huttunen, P., & Meriläinen, J. (1995b). Diatom, chrysophyte and protozoan distributions along a latitudinal transect in Fennoscandia. *Ecography*, **18**, 429–39.

Pienitz, R., Smol, J. P., & Lean, D. R. S. (1997a). Physical and chemical limnology of 59 lakes located between the southern Yukon and the Tuktoyaktuk Peninsula, Northwest Territories (Canada). *Canadian Journal of Fisheries and Aquatic Sciences*, **54**, 330–46.

A. F. LOTTER ET AL.

(1997b). Physical and chemical limnology of 24 lakes located between Yellowknife and Contwoyto Lake, Northwest Territories (Canada). *Canadian Journal of Fisheries and Aquatic Sciences*, **54**, 347–58.

Pienitz, R., Smol, J. P., & MacDonald, G. M. (1999). Paleolimnological reconstruction of Holocene climatic trends from two boreal treeline lakes, Northwest Territories, Canada. *Arctic and Alpine Research*, in press.

Proschkina-Lavrenkov, A. I. (1974). *The Diatoms of the USSR–Fossil and Recent*, vol. I. (In Russian) Nauka, Leningrad.

Psenner R., & R. Schmidt, (1992): Climate-driven pH control of remote alpine lakes and effects of acid deposition. *Nature*, **356**, 781–3.

Raven, J. A., & Geider, R. J. (1988). Temperature and algal growth. *New Phytologist*, **110**, 441–61.

Robertson, D. M., & Ragotzkie, R. A. (1990). Changes in the thermal structure of moderate to large sized lakes in response to changes in air temperature. *Aquatic Sciences*, **52**, 360–80.

Rühland, K. (1996). *Assessing the use of diatom assemblages as paleoenvironmental proxies in the Slave and Bear geological provinces, NWT, Canada*. MSc thesis, Queen's University, Kingston (Ontario).

Sabelina, M. M., Kisselyev, I. A., Proschkina-Lavrenko, A. I., & Sheshukova, B. C. (1951). Diatomovye vodorosli. Opredelitely presnovodnykh vodoroslei sssr, Gosudarstvennoe Izdatelystvo. *Sovjetskaja Nauka*, **4**, Moscow.

Schindler, D. W., Bayley, S. E., Parker, B. R., Beaty, K. G., Cruikshank, D. R., Fee, E. J., Schindler, E. U., & Stainton, M. P. (1996a). The effects of climatic warming on the properties of boreal lakes and streams at the Experimental Lakes Area, north-western Ontario. *Limnology and Oceanography*, **41**, 1004–17.

Schindler, D. W., Beaty, K. G., Fee, E. J., Cruiksjank, D. R., DeBruyn, E. R., Findlay, G. A., Linsey, G. A., Shearer, J. A., Stainton, M. P., & Turner, M. A. (1990). Effects of climatic warming on lakes of the central boreal forest. *Science*, **250**, 967–70.

Schindler, D. W., Curtis, P. J., Parker, B. R., & Stainton, M. P. (1996b). Consequences of climate warming and lake acidification for UV-B penetration in North American boreal lakes. *Nature*, **379**, 705–8.

Schmidt, R., & Psenner, R. (1992). Climate changes and anthropogenic impacts as causes for pH fluctuations in remote high alpine lakes. *Documenta Istituto Italiano Idrobiologia*, **32**, 31–57.

Schmidt, R., Wunsam S., Brosch, U., Fott, J., Lami, A., Löffler, H., Marchetto, A., Müller, H. W., Prazakova, M., & Schwaighofer, B. (1998). Late and post-glacial history of meromictic Längsee (Austria), in respect to climate change and anthropogenic impact. *Aquatic Sciences*, **60**, 56–88.

Servant-Vildary, S. (1982). Altitudinal zonation of mountainous diatom flora in Bolivia: Application to the study of the Quaternary. *Acta Geol. Acad. Sci. Hungaricae*, **25**, 179–210.

Servant-Vildary, S., Melice, J. L., Sondag, F., Einarsson, A., Dickman, M., & Stewart, K. (1992). A transfer function diatom/silica displayed in Lake Myvatn (Iceland). *Abstracts 12th International Diatom Symposium*, Renesse, The Netherlands.

Smith, T. M., Shugart, H. H., Bonan, G. B., & Smith, J. B. (1992). Modeling the potential response of vegetation to global climate change. *Advances in Ecological Research*, **22**, 93–116.

Smol, J. P. (1988). Paleoclimate proxy from freshwater arctic diatoms. *Verhandlungen Internationale Vereinigung Limnologie*, **23**, 837–44.

Smol, J. P., Walker, I. R., & Leavitt, P. R. (1991). Paleolimnology and hindcasting climatic trends. *Verhandlungen Internationale Vereinigung Limnologie*, **24**, 1240–6.

Smol, J. P., Cumming, B. F., Douglas, M. S. V., & Pienitz, R. (1995). Inferring past climatic changes in Canada using paleolimnological techniques. *Geoscience Canada*, **21**, 113–8.

Sommaruga, R., & Psenner, R. (1997). Ultraviolet radiation in a high mountain lake of the Austrian Alps: Air and underwater measurements. *Photochemistry and Photobiology*, **65**, 957–63.

Sommaruga-Wögrath, S., Koinig, K., Schmidt, R., Sommaruga, R., Tessadri, R., & Psenner, R.

(1997). Temperature effects on the acidity of remote alpine lakes. *Nature*, **387**, 64–7.

Stoermer, E. F., & Ladewski, T. B. (1976). Apparent optimal temperatures for the occurrence of some common phytoplankton species in southern Lake Michigan. *University of Michigan, Great Lakes Research Division Publication*, **18**.

Tinner, W., Ammann, B., & Germann, P. (1996). Treeline fluctuations recorded for 12,500 years by soil profiles, pollen, and plant macrofossils in the Central Swiss Alps. *Arctic and Alpine Research*, **28**, 131–47.

Tranquillini, W. (1979). *Physiological ecology of the alpine timberline*. Berlin: Springer.

Tynni, R. (1975). Über Finnlands rezente und subfossile Diatomeen. VIII. *Bulletin Geological Survey of Finland*, **274**, 1–55.

(1976). Über Finnlands rezente und subfossile Diatomeen. IX. *Bulletin Geological Survey of Finland*, **284**, 1–37.

Vincent, W. F., & Pienitz, R. (1996). Sensitivity of high latitude freshwater ecosystems to global change: Temperature and solar ultraviolet radiation. *Geoscience Canada*, **23**, 231–6.

Vinebrooke, R. D., & Leavitt, P. R. (1996). Effects of ultraviolet radiation in an alpine lake. *Limnology and Oceanography*, **41**, 1035–40.

Vyverman, W. (1992). Altitudinal distribution of non-cosmopolitan desmids and diatoms in Papua New Guinea. *British Phycological Journal*, **27**, 49–63.

Vyverman, W., & Sabbe, K. (1995). Diatom–temperature transfer functions based on the altitudinal zonation of diatom assemblages in Papua New Guinea: A possible tool in the reconstruction of regional paleoclimatic changes. *Journal of Paleolimnology*, **13**, 65–77.

Watson, R. T., Zinyowera, M. C., Moss, R. H., & Dokken, D. J. (ed.) (1996). *Climate change 1995 – impacts, adaptations and mitigation of climate change: scientific-technical analyses. Contribution of Working Group II to the Second Assessment Report of the Intergovernmental Panel on Climate Change*. Cambridge: Cambridge University Press.

Weckström, J., Korhola, A., & Blom, T. (1997a). The relationship between surface-water temperature and diatom assemblages in 30 subarctic lakes from northern Fennoscandia: a potential tool for paleotemperature reconstructions. *Arctic and Alpine Research*, **29**, 75–92.

(1997b). Diatoms as quantitative indicators of pH and water temperature in subarctic Fennoscandian lakes. *Hydrobiologia*, **347**, 171–84.

Wetzel, R. G. (1983). *Limnology*, 2nd edn. Saunders College Publishing.

Wunsam, S., & Schmidt, R. (1995). A diatom–phosphorus transfer function for Alpine and prealpine lakes. *Memorie Istituto Italiano di Idrobiologia*, **53**, 85–99.

Wunsam, S., Schmidt, R., & Klee, R. (1995). *Cyclotella*-taxa (Bacillariophyceae) in lakes of the Alpine region and their relationship to environmental variables. *Aquatic Sciences*, **57**, 360–86.

10 Freshwater diatoms as indicators of environmental change in the High Arctic

MARIANNE S.V. DOUGLAS AND JOHN P. SMOL

Introduction

High arctic environments have received increased attention over recent years, as polar regions are considered to be especially sensitive to the effects of global climatic and other environmental changes (Walsh, 1991; Rouse *et al.*, 1997). For example, potential warming from 'greenhouse gases' is expected to be accentuated in the High Arctic (Roots, 1989). Other environmental changes, such as increased ultraviolet (UV-B) light penetration and deposition of airborne contaminants, have also been noted recently in high latitude regions (e.g., Landers, 1995).

There is considerable potential for using living and fossil diatom assemblages for tracking environmental trends in high arctic regions (Smol & Douglas, 1996). However, to date, relatively few studies have been completed on the taxonomy, ecology, and paleoecology of high arctic, freshwater diatoms, even though lakes and ponds are dominant features of most arctic landscapes. For example, about 18% (by area) of Canada's surface waters are situated north of 60° N (Statistics Canada, 1987), and Sheath (1986) estimates that tundra ponds cover approximately 2% of the Earth's surface. The heightened interest in high arctic environments, coupled with increased accessibility (e.g., with helicopter support) of these remote regions, has resulted in a recent surge of interest in arctic diatom research. Moreover, proxy techniques such as palynology and dendroecology have some serious limitations in high arctic regions due to the paucity of higher plants (Gajewski *et al.*, 1995). Consequently, paleolimnological approaches using diatoms may become especially important for studies of global environmental change.

In this chapter, we summarize some of the ways that diatoms have been used to track environmental changes in the High Arctic. We first present a brief description of arctic limnology and historical uses of diatoms in the High Arctic. Thereafter, we mainly discuss applications to studies of past climatic change, as recently most of the research has centered on this topic. However, other applications are also considered. Our primary geographic focus is on the Canadian Arctic Islands, Greenland, and Svalbard (Fig. 10.1). Lotter *et al.* (this volume) reviews diatom studies in lower arctic regions, including data currently available on the Russian Arctic. Very little work has been completed

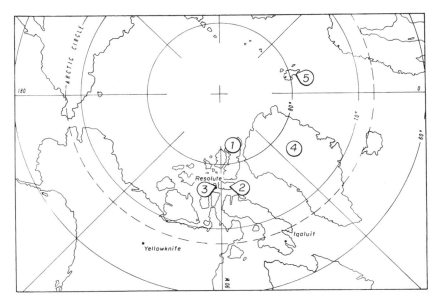

Fig. 10.1. Map showing a circumpolar projection of the high arctic islands referred to in this chapter. (1) Ellesmere, (2) Devon, (3) Cornwallis, (4) Greenland, (5) Spitsbergen (Svalbard).

on diatoms from high arctic streams and rivers (e.g., Petersen, 1924), and most accounts of lotic diatoms are associated with the few macroalgal surveys conducted on streams (e.g., Hamilton & Edlund, 1994; Sheath *et al.*, 1996).

Limnological setting

The High Arctic is characterized by cold temperatures, extended snow and ice cover, and extremes in irradiance, with the sun below the arctic horizon for several months during the polar night, followed by 24-hour periods of continuous sunlight during the summer months (Rouse *et al.*, 1997). Given the combined conditions of snow accumulation, permafrost and subsequent poor drainage, most summer snowmelt collects in lakes and ponds. Common lacustrine environments include shallow ponds inside ice-wedge polygons, thermokarst lakes, glacial basins, and water bodies associated with moraines (Hobbie, 1984).

Sheath (1986) distinguishes between arctic ponds and lakes on the basis of depth and ice thickness. Ponds are defined as water bodies that freeze completely to the bottom each winter, whereas lakes are sufficiently deep to maintain a layer of liquid water under the ice cover. As noted below, some deeper high arctic lakes maintain a permanent float of ice, even during summer. Ponds, on the other hand, with their lower heat capacities, thaw earlier and completely. Moreover, because of their shallow depths (e.g., 2 m and often <1 m deep), a tundra pond's entire water column and substrates can be

M. S. V. DOUGLAS AND J. P. SMOL

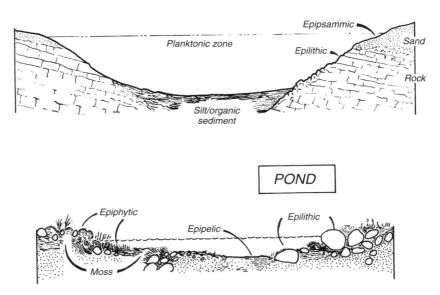

Fig. 10.2. Comparison of lake and pond habitats. All periphytic habitats can occur in both pond and lake systems. Planktonic diatoms are generally only found in lake systems.

exploited for algal growth (Fig. 10.2). Shallow ponds are especially common in high arctic regions, and are potentially sensitive bellwethers of environmental change (Douglas & Smol, 1994; Douglas *et al.*, 1994).

Several reviews have summarized the historical development of arctic limnology (e.g., Hobbie, 1984), and a few high arctic lakes (e.g., Rigler, 1978) and ponds (e.g., Douglas and Smol 1994) have been studied in considerable detail. None the less, we suspect that less limnological and phycological data exist for the High Arctic than for any other limnological region.

Historical review

The earliest records of high arctic diatoms come primarily from the late nineteenth century, when ships exploring the Arctic frequently returned with collections from a variety of habitats. Many of these early collections were examined by C. G. Ehrenberg (1853) and P. T. Cleve (e.g., 1864, 1873, 1883, 1896), which included samples from the Russian Arctic (e.g., Cleve, 1898).

Perhaps the earliest use of high arctic diatoms to help resolve a scientific question was by the Norwegian arctic explorer, Fridtjof Nansen (1897). At that time, knowledge of polar regions was still very limited, and Nansen set out to determine if ice drifted over the North Pole. One piece of evidence he used to bolster the drift theory was that both marine and freshwater diatoms (identified by P. T. Cleve) collected from the Greenland pack ice were the same species that had been collected earlier from ice near the Bering Strait.

The 1913–18 Canadian Arctic Expedition produced some of the earliest surveys of diatoms from the Western Canadian Arctic. Lowe (1923) examined material from small brackish ponds in both Alaska and the Northwest Territories. Because collections were not specifically oriented towards diatoms, the list is brief and he does not provide illustrations of taxa or discuss ecology in any detail. A contemporary, more detailed study was completed by Petersen (1924), who recorded 52 diatom taxa (several with illustrations and taxonomic notes) from the north coast of Greenland. Other early diatom work from Greenland includes the studies by Östrup (1897a,b, 1910).

A second Canadian Arctic Expedition was undertaken in 1938–1939, with a focus now on the Eastern Arctic, including several high arctic sites. Work on the material (collected by Nicholas Polunin) was delayed by World War II, but eventually Ross (1947) published descriptions of the 192 freshwater diatoms Polunin collected, along with their geographic distributions. His report included 11 plates of illustrations.

One of the most important pioneer researchers working on arctic diatoms was the Danish phycologist, Niels Foged. He published extensively on diatoms from many regions (see full list in Håkansson, 1988), but much of his research focused on polar regions, such as Greenland (1953, 1955, 1958, 1972, 1973, 1977), Iceland (1974), Spitsbergen (1964), and Alaska (1981).

Foged noted that earlier investigators (reviewed in his publications) provided short floral lists, but little information about ecology. In his studies from Greenland, Foged usually considered the pH and halobion spectra of diatom communities (e.g., Foged, 1953). Where possible, he provided a more detailed analysis of water chemistry, as it related to differences among petrographical regions (Foged, 1958). He also began using paleoecological approaches, and described changes in diatom assemblages from several postglacial deposits in Greenland (Foged, 1972, 1977, 1989). The main stratigraphic horizon of interest was the marine–lacustrine transition (see Denys & de Wolf, this volume), from which he deduced emergence patterns of lakes from the sea. Foged (1982) also attempted to use diatoms as part of a forensic investigation (see Peabody, this volume). He tried to determine if the eight so-called 'Greenland mummies' (who died around the year 1460, and whose mummified bodies were discovered in 1972) died as a result of drowning, by examining the diatoms contained in their tissues.

Following some early work by, for example, Cleve (1864) and Lagerstedt (1873), Foged (1964) produced a detailed volume on Spitsbergen (Svalbard) diatoms. He noted that many species were very variable and that their forms differed from those in other geographic areas. However, he considered the differences so slight that it would have been 'rash' to describe each one as a separate taxon. Foged listed temperature and pH as the major factors controlling populations.

Table 10.1. *Overall trends that are often recorded in high arctic ponds and lakes with colder or warmer temperatures*

Variable	Colder	Warmer
Ice cover	↑	↓
Growing season	↓	↑
Plankton	↓	↑
Mosses	↓	↑
Diversity	↓	↑
Nutrients	↓	↑
Production	↓	↑
pH	↓	↑
Conductivity	↓	↑

Note:
Most of these changes can be related to the extent of ice and snow cover. These are general trends, and exceptions do occur.

Following Foged's pioneering work, a few phycological surveys were conducted on the more accessible parts of the arctic islands, such as southern Baffin Island (e.g., Moore, 1974*a,b,c*; Hickman, 1974), Svalbard (e.g., Skulberg, 1996), Bear Island (Metzeltin & Witkowski, 1996), and southeastern Greenland (e.g., Denys & Beyens, 1987). Most high arctic diatom floras, however, because of their geographic isolation, are still poorly documented.

Applications and case studies

ICE COVER AND RELATED ENVIRONMENTAL VARIABLES

The high arctic environment imposes some important and overriding constraints on diatom populations. In deep lakes, a dominant feature is extended ice and snow cover, which dramatically influence lake systems by determining available habitats (Fig. 10.3) and other limnological variables (Table 10.1).

Ice cover on most arctic lakes begins to form in early to mid-September; the exact timing is mainly dependent on wind and cloud cover (Hobbie, 1984). The rate and duration of freezing is also related to air temperature and snow cover. Ice thicknesses exceeding 5 m have been observed (Blake, 1989), although thinner ice covers are more common (Hobbie, 1984). Many high arctic sites maintain their snow and ice cover throughout the brief summer, and only a shallow 'moat' of open water becomes ice-free. The depth and area of snow cover, as well as the ice type (i.e., black vs. white), affects the transmission of light, and hence the amount of light available for photosynthesis by planktonic and periphytic diatom communities.

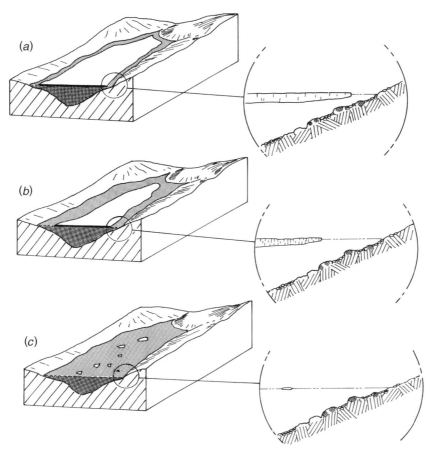

Fig. 10.3. Diagrammatic representation of ice and snow conditions on a high arctic lake during (a) cold, (b) moderate, and (c) warm summers. As temperatures increase, deeper areas of the lake are available for diatom growth. (From Smol, 1988, used with permission.)

The overriding influence of ice and snow cover on high arctic lakes prompted Smol (1983, 1988) to suggest that, because the extent of ice cover can influence which habitats are available for algal growth, past changes in diatom assemblages can be used to track past lake ice cover, and therefore infer past climatic changes (Fig. 10.3). A simple model was proposed: During colder years, ice cover is more extensive, and only a narrow moat develops in the littoral zone (Fig. 10.3(a)). Overall diatom production tends to be less, and taxa characteristic of very shallow littoral and semiterrestrial environments tend to be relatively more common. During warmer years (Fig. 10.3(b),(c)), ice cover is less extensive, overall production is higher, and taxa characteristic of deeper water substrates and planktonic habitats may be relatively more abundant.

The above model has been used to infer past climatic changes using diatom species composition from, for example, a core from Baird Inlet on eastern Ellesmere Island (Smol, 1983) and north-western Greenland (Blake *et al.*, 1992).

M. S. V. DOUGLAS AND J. P. SMOL

Lemmen *et al.* (1988) used diatoms to infer habitat and environmental conditions from a core in glacial Tasikutaaq Lake, on Baffin Island. As noted, changes in ice cover also affect total diatom biomass. During warmer years, ice cover is less extensive, and total algal production is often higher. Nutrients may also be higher from the surrounding drainage (see discussion below). Smol (1983) recorded higher diatom concentrations during proposed warmer periods in his study on Ellesmere Island, as did Blake *et al.* (1992) on Greenland. Similarly, Williams (1990*a*) used diatom concentration data to interpret past climatic changes (ice cover changes) on Baffin Island, and extended these ideas to interpreting past sea-ice conditions from diatoms in marine sediment cores (Andrews *et al.*, 1990; Williams, 1990*b*; Short *et al.*, 1994). Interestingly, Doubleday *et al.* (1995), working near the Alert military base on northern Ellesmere Island, noted the near total disappearance of diatoms in their core, despite generally good preservation. They speculated that, since this lake is deep (30 m) and currently supports an extensive ice cover even in summer, a slight cooling may have completely frozen over the lake, sealing it off, and precluding any sizeable diatom populations from growing.

The extent and duration of ice and snow cover has other implications for lake and pond environments (Table 10.1). Because many of these changes are accentuated in shallow ponds, we will use ponds as primary examples, although these changes are also evident in deeper lakes. Many of our comments are based on our long-term monitoring work from a series of 36 ponds on Cape Herschel (78°37′ N, 74°42′ W), on the east–central coast of Ellesmere Island (for review see Douglas *et al.*, 1998), as well as our ongoing studies on other high arctic islands (Fig. 10.1).

Although diatom diversity is often relatively low within a single Cape Herschel pond, a surprisingly diverse flora exists on the 2 × 5 km cape (Douglas & Smol, 1993, 1995*a*). High arctic ponds, such as those on Cape Herschel, typically only begin to thaw in late June or early July, and can begin to re-freeze by late August (Douglas & Smol, 1994). A slight cooling or warming can dramatically affect the length of time liquid water is present in these sites. With warmer summer temperatures, and hence a lengthened growing season, there is more opportunity for overall algal production to be higher, and for more complex and diverse aquatic communities to develop (Table 10.1). For example, with longer ice-free seasons, new substrates may become available, such as mosses, whilst during colder periods, only rock and sediment substrates may be available (Fig. 10.4). High arctic diatoms show some specificity to these different substrates (Douglas & Smol, 1995*a*; Hamilton *et al.*, 1994*a*; Wolfe, 1996*a*), and this information can be used to track past environmental shifts in paleolimnological studies (Douglas *et al.*, 1994).

A longer growing season also allows development of more complex diatom communities (e.g., Douglas *et al.*, 1994; Wolfe, 1994). For example, with extended growing seasons, secondary and tertiary growths of diatoms, including many stalked and tube-dwelling taxa, can grow attached to the original growth of adnate diatoms (Fig. 10.4). Overall diatom diversity is higher (Table 10.1).

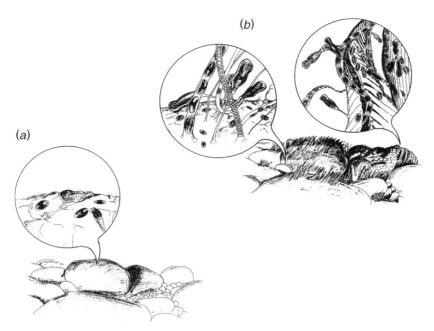

Fig. 10.4. Diagrammatic representation of habitat availability and diatom growth during (*a*) cold and (*b*) warmer growing seasons. In (*a*) very cold environments, extended ice cover and a very short growing season result in lower production of predominantly adnate diatoms living on rocks and sediment substrates. With (*b*) warmer conditions, additional substrates become available (e.g., mosses) and more complex diatom communities can develop.

With warmer temperatures, several changes also typically occur in water chemistry (Table 10.1). For example, nutrient inflows tend to be higher with increased snowmelt, a deeper active layer develops with permafrost melting, and runoff from the catchment increases. This often results in a corresponding increase in aquatic production. Because water in shallow arctic ponds closely tracks ambient air temperatures (Douglas & Smol, 1994), the higher water temperatures also tend to enhance overall production. Pond water pH tends to rise, partially in response to the increased photosynthesis, but also possibly through the effects on water renewal and relative yields of base cations and acid anions, as well as possible sulfate removal (Psenner & Schmidt, 1992; Schindler *et al.*, 1996; also see Wolfe & Härtling, 1996, and Wolfe, 1996*b*, for a more detailed discussion of arctic climate and pH relations). Conductivity also tends to increase (Table 10.1). As in many other lake regions, pH and conductivity exert strong influences on diatom assemblages in high arctic ponds (Douglas & Smol, 1993).

Because many tundra ponds are so shallow (e.g., <0.5 m deep), increased periods of warmer and drier climates (as are predicted by some General Circulation Models) would result in further lowerings in water levels, and even total desiccation of some shallower sites. For example, during the summer of 1995 on Cape Herschel, the warmest summer we had ever recorded (by *circa*

M. S. V. DOUGLAS AND J. P. SMOL

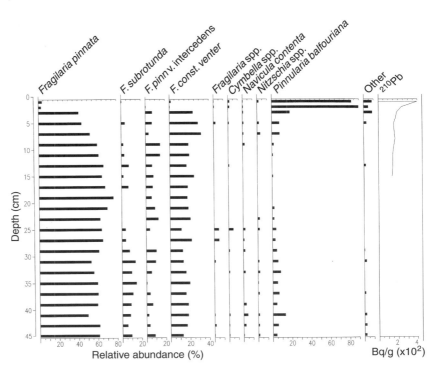

Fig. 10.5. Relative frequency diagrams of the dominant diatom taxa recorded in Elison Lake, Cape Herschel, Ellesmere Island (modified from Douglas *et al.*, 1994).

2 °C), some of the shallowest ponds dried up completely, whilst water levels were significantly lower in others. Such environmental changes could potentially be tracked in paleolimnological records by using diatoms (e.g., increases in aerophilic and terrestrial taxa).

High arctic pond sediments appear to be sensitive archives of past environmental shifts, tracking changes over several millennia. For example, some ponds on Cape Herschel have been accumulating sediments for over 8000 years, following their isolation from the sea due to isostatic uplift (Blake, 1992). Despite their shallowness (e.g., <2 m deep), and the fact that they freeze solid for about 10 months of the year, these ponds' sedimentary profiles are not greatly disturbed by cryoturbation or other mixing processes (discussed in Douglas *et al.*, 1998).

As one case study, we present the diatom species changes from a core from Elison Lake, Cape Herschel, which is discussed in more detail in Douglas *et al.* (1994). The basal sediments of this pond have been radiocarbon dated at 3850 ± 100 ^{14}C years (GSC-3170), and dating of the recent sediments indicate that the entire unsupported ^{210}Pb inventory occurs in the upper 3 cm of the core, reflecting the slow sedimentation rates characteristic of this region (Douglas *et al.*, 1994). The diatom flora exhibited marked species changes over the last 200 years (Fig. 10.5), characterized mainly by the striking relative increase in the moss epiphyte (Douglas & Smol, 1995a) *Pinnularia balfouriana* Grunow and

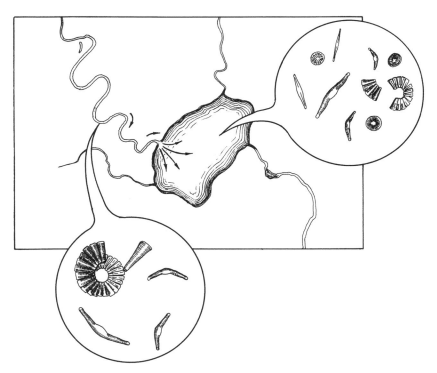

Fig. 10.6. Diagrammatic representation of river diatoms being deposited in lake basins. Changes in abundances of lotic diatoms in lake sediment profiles can be used to infer past river discharge, which may be related to climate.

other taxa, replacing the assemblage of small benthic *Fragilaria* species that thrived in that pond since its inception. Similarly, dramatic and synchronous stratigraphic changes occurred in all the other Cape Herschel cores that we studied (Douglas *et al.*, 1994; Douglas, 1993). These successional changes, which all occurred over the last two centuries, were unprecedented in the ponds' histories. The species changes are consistent with what would be expected with climatic warming in this region (see discussion above). Other researchers have now recorded similar changes in recent diatom assemblages in cores from other arctic regions (e.g., Wolfe, 1998; Hamilton *et al.*, 1998; and several studies in preparation).

RIVER INFLOWS

Another environmental variable that might be tracked in lake sediments using diatoms is past river inflow, which may be climatically driven (Fig. 10.6). For example, Ludlam *et al.* (1996) noted that certain taxa (e.g., *Meridion* and *Hannaea* species) characterized river environments (i.e., lotic taxa), whilst other diatoms characterized the littoral zone (i.e., lentic taxa) of a lake on Northern Ellesmere Island. They proposed a Lotic Index, which was a ratio of the percentage of

M. S. V. DOUGLAS AND J. P. SMOL

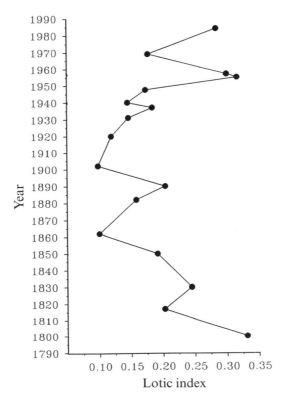

Fig. 10.7. Profile of the Lotic Index in a sediment core from Lake C2, Taconite Inlet, Northern Ellesmere Island (modified from Ludlam *et al.*, 1996).

Hannaea plus *Meridion* diatoms, divided by the total number of pennates. They used the Lotic Index to infer past river discharge from the diatoms preserved in a varved lake sediment core, covering the last two centuries (Fig. 10.7). They noted a period of declining runoff beginning about two centuries ago that ended in the late 1800s. The Lotic Index showed a clear positive relationship to sedimentation rate, as recorded in the varves (Ludlam *et al.*, 1996).

HIGH RESOLUTION STUDIES

A potential problem with paleolimnological studies of high arctic lakes is relatively coarse temporal resolution: the sedimentation rates in many lakes and ponds are comparatively slow (e.g., only about 0.5 m of sediment accumulation occurred over 9000 years in a lake on Ellesmere Island; Smol, 1983). There are, however, exceptions, and high-resolution studies are possible. For example, Bradley (1996) compiled a series of ten papers documenting the environmental history (including two papers dealing with diatoms: Douglas *et al.*, 1996; Ludlam *et al.*, 1996) from annually laminated lake sediment cores from Northern Ellesmere Island. On Devon Island, Gajewski *et al.* (1997) established the annual nature of laminae in their sediment core over the last *ca.* 150

years from a non-meromictic lake. Diatom concentrations increased by two orders of magnitude during the twentieth century, with major increases in the 1920s and 1950s. Varve thickness also increased at these times, coincident with increases in snow melt percentages in the Devon Ice Core (Koerner, 1977), both proxies for climate warming. Hughen *et al.* (1996) discuss the potential of using diatoms in laminated sediment cores from tidewater lakes on Baffin Island.

Although annually laminated sediments are ideal for high-resolution studies, important data on past environmental changes can still be deciphered from non-laminated sediments, especially if careful geochronological control can be achieved using ^{210}Pb and ^{14}C dating.

PEAT DEPOSITS

At present, the high arctic summers are believed to be too cold, brief, dry, and irregular to allow for extensive peat development (Janssens, 1990). Peat deposits did, however, accumulate in the past, when conditions were presumably warmer and/or wetter (for review see Ovendun, 1988). As noted by Brown *et al.* (1994), these fossil peat profiles may contain a suite of siliceous indicators, including diatom frustules, that can be used to interpret past environmental conditions. Because peat accumulation tends to be faster than sediment accumulation in many high arctic lakes (perhaps an order of magnitude faster in some regions; Brown *et al.*, 1994), peat deposits may potentially provide relatively high-resolution records of past environmental change. Although some work has been completed on current moss-dwelling diatoms in the Arctic (e.g., Beyens & de Bock, 1989; Hamilton *et al.*, 1994*a*; Douglas & Smol, 1995*a,b*), relatively little ecological calibration data are yet available to interpret these paleo-peat diatom assemblages, and modern analogue samples have not yet been identified.

SNOW, ICE, FROST, AND ICE CORES

The Arctic is characterized by extensive frost, ice, and snow cover. Surprisingly, even these habitats include diatoms; some possibly living as part of the snow flora (Lichti-Federovich, 1980), whilst others might be indicative of various dispersal phenomena (e.g., Lichti-Federovich, 1984, 1985, 1986; Harper, this volume). Both marine and freshwater diatom frustules are recoverable from the Greenland and arctic island ice cores (e.g., Gayley & Ram, 1984; Harwood, 1986*a,b*; Gayley *et al.*, 1989), and potentially can be used to augment environmental inferences.

MARINE/LACUSTRINE TRANSITIONS

Past patterns of glaciation and former ice sheet configurations continue to be areas of considerable controversy in high arctic regions (e.g., Blake, 1992). Paleolimnological analyses of high arctic diatom assemblages offer a powerful tool to help delineate Holocene emergence patterns from isostatic uplift (see Denys & de Wolf, this volume). For example, Hyvärinen (1969), Foged (1977), Young & King (1989), Williams (1990a), and Douglas *et al.* (1996) used diatom assemblages to document the time of lakewater freshening using sediment cores from Spitsbergen, Greenland, Devon, Baffin, and Ellesmere islands, respectively. Recently, Ng (1996) has developed a transfer function (based on diatoms present in the surface sediments of 93 Devon Island ponds) to quantitatively infer past salinity levels.

MONITORING AIRBORNE CONTAMINANTS AND LOCAL DISTURBANCES

Smol & Douglas (1996) have recently summarized some of the ways that high arctic diatoms could be used to monitor human-induced environmental changes in tundra ecosystems. These include disturbances such as airborne contaminants, such as acidic deposition (see Battarbee *et al.*, this volume), as several regions of the Arctic are susceptible to acidification. Moreover, as Inuit populations continue to increase rapidly, local disturbances (e.g., sewage inputs from communities) will continue to affect local water resources. For example, we have recently completed a study of the diatoms preserved in a sediment core from Meretta Lake, in Resolute Bay, Cornwallis Island. This lake has been receiving local sewage input since 1949. Not surprisingly, the diatom flora recorded striking successional changes at that time (Douglas & Smol, in preparation). Given the logistic problems of frequently sampling sites in arctic regions, biomonitoring tundra lakes, ponds, and rivers using diatoms would be a very cost-effective way of assessing environmental change (Smol & Douglas, 1996).

Summary

Much of the arctic landmass that is not under ice sheets is dotted with lakes and ponds. Organisms living in tundra lakes and ponds must often survive low temperatures (including total freezing of the water column in some shallow sites), extended ice and snow covers, extremes in photoperiods including long periods of continual darkness followed by 24–hour periods of daylight, and often low nutrient availability. Despite these extreme conditions, diatoms often dominate algal benthic communities in high arctic environments.

High arctic diatoms provide environmental and earth scientists with many potential applications for paleoclimatic research and for biomonitoring other types of environmental change. Some of these approaches can also be applied to Antarctic studies (see Spaulding & McKnight, this volume). For example, there is currently considerable interest in reconstructing past climatic trends in polar regions, yet many traditional proxy techniques are fraught with difficulties in these extreme environments. Because even slight climatic changes can dramatically affect the length of ice and snow cover on arctic lakes and ponds, as well as other climate-related variables (e.g., water chemistry, habitat availability), diatoms can potentially be used to track paleoclimatic histories in dated lake and pond sediment cores. Past river inflows into lakes, which are also influenced by climate, may be tracked by the deposition of lotic diatoms. Some of these analyses can be undertaken using high resolution techniques in varved sediment cores.

Other potential applications include interpreting past environmental changes from fossil peat profiles; dispersal and wind patterns from diatoms in snow, frost, and ice core samples; as well as using diatoms to reconstruct past patterns of isostatic uplift. Considerable potential exists for using diatoms to monitor anthropogenic disturbances in tundra regions, especially considering the high logistic costs of more typical monitoring programs.

Acknowledgments

Our arctic research is funded primarily by the Natural Sciences and Engineering Research Council of Canada and the Polar Continental Shelf Project. Some of this research was also funded by the Northern Studies Training Grants, and a Jennifer Robinson Memorial Scholarship, and a grant-in-aid from the Arctic Institute of North America to M. Douglas. We thank John Glew for drawing the figures. Helpful comments were provided by an anonymous reviewer and members of our labs. We also thank P. Hamilton for providing comments and some of the historical references.

References

Andrews, J. T., Evans, L. W., Williams, K. M., Briggs, W. M., Jull, A. J. T., Erlenkeuser, H., & Hardy, I. (1990). Cryosphere/ocean interactions at the margin of the Laurentide ice sheet during the Younger Dryas Chron: SE Baffin Shelf, Northwest Territories. *Paleoceanography*, **5**, 921–35.

Beyens, L., & de Bock, P. (1989). Moss dwelling diatom assemblages from Edgeøya (Svalbard). *Polar Biology*, **9**, 423–30.

Blake, W., Jr. (1989). Inferences concerning climatic change from a deeply frozen lake on Rundfjeld, Ellesmere island, Arctic Canada. *Journal of Paleolimnology*, **2**, 41–54.

(1992). Holocene emergence at Cape Herschel, east-central Ellesmere Island, Arctic Canada: Implications for ice sheet configuration. *Canadian Journal of Earth Sciences*, **29**, 1958–80.

Blake, W., Jr., M. M. Boucherle, B. Fredskild, J. A. Janssens and J. P. Smol. (1992). The geomorphological setting, glacial history and Holocene development of 'Kap Inglefield Sø', Inglefield Land, North-West Greenland. Meddelelser om Grönland, *Geosciences*, **27**, 1–42.

Bradley, R. S. (ed.), (1996). Taconite Inlet lakes project. *Journal of Paleolimnology*, **16**, 97–255.

Brown, K. M., Douglas, M. S. V., & Smol, J. P. (1994). Siliceous microfossils in a Holocene High Arctic peat deposit (Nordvestø, Northwest Greenland). *Canadian Journal of Botany*, **72**, 208–16.

Cleve, P. T. (1864). Diatomaceer från Spetsbergen. Öfversight af Kongl. *Vetenskaps-Akademiens Forhandlinger*, **24** (10), 661–9.

(1873). On diatoms from the Arctic Sea. Bihang Till Kongl. *Svenska Vetenskaps-Akademiens Handlingar*, **1**, 1–28.

(1883). Diatoms, collected during the expedition of the Vega. *Ur Vega-Expedpeditionens Vettenskapliga Iakttagelser*, **3**, 457–517.

(1896). Diatoms from Baffin Bay and Davis Strait. *Bihang Till Kongliga Svenska Vetenskaps-Akademiens Handlingar*, **22**, 1–22.

(1898). Diatoms from Franz Josef Land. *Bihang Till Kongliga Svenska Vetenskaps-Akademiens Handlingar*, **24**, 1–26.

Denys, L., & Beyens, L. (1987). Some diatom assemblages from the Angmagssalik region, south-east Greenland. *Nova Hedwigia*, **45**, 389–413.

Doubleday, N., Douglas, M. S. V., & Smol, J. P. (1995). Paleoenvironmental studies of black carbon deposition in the High Arctic: a case study from Northern Ellesmere Island. *The Science of the Total Environment*, **160/161**, 661–668.

Douglas, M. S. V. (1993). Diatom ecology and paleolimnology of high arctic ponds. PhD thesis, Queen's University, Kingston, Ontario.

Douglas, M. S. V., & Smol, J. P. (1993). Freshwater diatoms from high arctic ponds (Cape Herschel, Ellesmere Island, N.W.T.). *Nova Hedwigia*, **57**, 511–52.

(1994). Limnology of high arctic ponds (Cape Herschel, Ellesmere Island, N.W.T.). *Archiv für Hydrobiologie*, **131**, 401–34.

(1995a). Periphytic diatom assemblages from high arctic ponds. *Journal of Phycology*, **31**, 60–9.

(1995b). Paleolimnological significance of chrysophyte cysts in arctic environments. *Journal of Paleolimnology*, **13**, 79–83.

Douglas, M. S. V., Smol, J. P., & Blake, W., Jr. (1994). Marked post-18th century environmental change in high arctic ecosystems. *Science*, **266**, 416–19.

Douglas, M. S. V., Ludlam, S., & Feeney, S. (1996). Changes in diatom assemblages in Lake C2 (Ellesmere Island, Arctic Canada): Response to basin isolation from the sea and to other environmental changes, *Journal of Paleolimnology*, **16**, 217–26.

Douglas, M. S. V., Smol, J. P., & Blake, W., Jr. (1998). Paleolimnological studies of high arctic ponds: Summary of investigations at Cape Herschel, east-central Ellesmere Island. *Geological Survey of Canada*, Bulletin, (in press).

Ehrenberg, C. G. (1853). Über neue Anschauungen des kleinsten nördlichen Polarlebens. *Deutsche Akademie der Wissenschaften zu Berlin, Monatsberichte*, 1853, 522–9.

Foged, N. (1953). Diatoms from West Greenland. *Meddelelser om Grönland*, **147**, 1–86.

(1955). Diatoms from Peary Land, North Greenland. *Meddelelser om Grönland*, **194**, 1–66.

(1958). The diatoms in the basalt area and adjoining areas of Archean rock in West Greenland. *Meddelelser om Grönland*, **156**, 1–146.

(1964). Freshwater diatoms from Spitsbergen. *Tromsö Museums Skrifter*, **11**, 1–204, 22 pls.

(1972). The diatoms in four postglacial deposits in Greenland. *Meddelelser om Grönland*, **194**, 1–66.

(1973). Diatoms from Southwest Greenland. *Meddelelser om Grönland*, **194**, 1–84.

(1974). Freshwater diatoms in Iceland. *Bibliotheca Phycologia*, **15**, 1–118.

(1977). Diatoms from four postglacial deposits at Godthabsfjord, West Greenland. *Meddelelser om Grönland*, **199**, 1–64.

(1981). Diatoms in Alaska. *Bibliotheca Phycologica*, **53**: 1–31.

(1982). Diatoms in human tissues – Greenland ab. 1460 AD – Funen 1981–82 AD *Nova Hedwigia*, **36**, 345–79.

(1989). The subfossil diatom flora of four geographically widely separated cores in Greenland. *Meddelelser om Grönland (Bioscience)*, **30**, 1–75.

Gajewski, K., Garneau, M., & Bourgeois, J. C. (1995). Paleoenvironments of the Canadian High Arctic derived from pollen and plant macrofossils: Problems and potentials. *Quaternary Science Reviews*, **14**, 609–29.

Gajewski, K., Hamilton, P. B., & McNeely, R. N. (1997). A high resolution proxy-climate record from an arctic lake with laminated sediments on Devon Island, Nunavut, Canada. *Journal of Paleolimnology*, **17**, 215–25.

Gayley, R. I., & Ram, M. (1984). Observations of diatoms in Greenland ice. *Arctic*, **37**, 172–73.

Gayley, R. I., Ram, M., & Stoermer, E. F. (1989). Seasonal variations in diatom abundance and provenance in Greenland ice. *Journal of Glaciology*, **35**, 290–2.

Håkansson, H. (1988). Obituary: Niels Aage Johannes Foged 1906–1988. *Diatom Research*, **3**, 169–74.

Hamilton, P. B., & Edlund, S. A. (1994). Occurrence of *Prasiola fluviatilis* (Chlorophyta) on Ellesmere Island in the Canadian Arctic. *Journal of Phycology*, **30**, 217–21.

Hamilton, P. B., Douglas, M. S. V., Fritz, S. C., Pienitz, R., Smol, J. P., & Wolfe, A. P. (1994*b*). A compiled freshwater diatom taxa list for the Arctic and Subarctic regions of North America. In *The Proceedings of the Fourth Arctic–Antarctic Diatom Symposium (Workshop)*, ed. P. B. Hamilton, p. 85–102. Ottawa: Canadian Technical Report of Fisheries and Aquatic Sciences, 1957, 85–102.

Hamilton, P. B., Gajewski, K., & McNeely, R. N. (1998). Physical, chemical and biological characteristics of lakes from the Sidre Basin on the Fosheim Peninsula. *Geological Survey of Canada, Bulletin,* (in press).

Hamilton, P. B., Poulin, M., Prévost, C., Angell, M., & Edlund, S. A. (1994*a*). Americanarum Diatomarum Exsiccata: Fascicle II (CANA), voucher slides representing 34 lakes, ponds and streams from Ellesmere Island, Canadian High Arctic, North America. *Diatom Research*, **9**, 303–27.

Harwood, D. M. (1986*a*). Do diatoms beneath ice sheet indicate interglacials warmer than present? *Arctic*, **39**, 304–8.

(1986*b*). The search for microfossils beneath the Greenland and west Antarctic ice sheets. *Antarctic Journal of the United States*, **21**, 105–6.

Hickman, M. (1974). The epipelic diatom flora of a small lake on Baffin Island, Northwest Territories. *Arch. Protistenk.*, **116S**, 270–9.

Hobbie, J. E. (1984). Polar limnology. In *Lakes and Reservoirs. Ecosystems of the World*, ed. F. B. Taub, pp. 63–106. Amsterdam: Elsevier.

Hughen, K. A., Overpeck, J. T., Anderson, R. F., & Williams, K. M. (1996). The potential for paleoclimate records from varved arctic lake sediments: Baffin Island, Eastern Canadian Arctic. In *Paleoclimatology and Paleoceanography from Laminated Sediments*, ed. A. E. S. Kemp, vol. 116, pp. 57–71. London: Geological Society Special Publication.

Hyvärinen, H. (1969). Trullvatnet: A Flandrian stratigraphical site near Murchisinfjorden Nordaustlandet, Spitsbergen. *Geografiska Annaler*, **51A**, 42–5.

Janssens, J. A. (1990). Methods in Quaternary ecology. No. 11. Bryophytes. *Geoscience Canada*, **17**, 13–24.

Koerner, R. (1977). Devon Island ice cap: core stratigraphy and paleoclimate. *Science*, **196**, 15–18.

Lagerstedt, N. G. W. (1873). Sötvattens-Diatomaceer från Spetsergen och Beeren Eiland. *Bihang Till Kongliga Svenska Vetenskaps-Akademiens Handlingar*, **1**, 1–52.

Landers, D. H. (editor). (1995). Ecological effects of arctic airborne contaminants. *The Science of the Total Environment*, **160/161**, 1–870.

Lemmen, D. S., Gilbert, R., Smol, J. P., & Hall, R. I. (1988). Holocene sedimentation in glacial Tasikutaaq Lake, Baffin Island. *Canadian Journal of Earth Sciences*, **25**, 810–23.

Lichti-Federovich, S. (1980). Diatom flora of red snow from Isbjørneø, Carey Øer, Greenland. *Nova Hedwigia*, **33**, 395–431.

(1984). Investigations of diatoms found in surface snow from the Sydkap ice cap, Ellesmere Island, Northwest Territories. Current Research, Part A, Geological Survey of Canada, Paper 84–1A, 287–301.

(1985). Diatom dispersal phenomena: diatoms in rime frost samples from Cape Herschel, central Ellesmere Island, Northwest Territories. Current Research, Part B, Geological Survey of Canada, Paper 85–1B, 391–9.

(1986). Diatom dispersal phenomena: diatoms in precipitation samples from Cape Herschel, east-central Ellesmere Island, Northwest Territories – a quantitative assessment. Current Research, Part B, Geological Survey of Canada, Paper 86–1B, 263–9.

Lowe, C. W. (1923). Report of the Canadian Arctic Expedition 1913–18. Part A: Freshwater algae and freshwater diatoms, Southern Party 1913–1916. F. A. Acland. Printer of the King's Most Excellent Majesty: Ottawa.

Ludlam, S. D., Feeney, S., & Douglas, M. S. V. (1996). Changes in the importance of lotic and littoral diatoms in a high arctic lake over the last 191 years. *Journal of Paleolimnology*, **16**, 184–204.

Metzeltin, D., & Witkowski, A. (1996). Diatomeen der Bären-Insel. *Iconographia Diatomologica*, **4**, 1–287.

Moore, J. W. (1974*a*). Benthic algae of southern Baffin Island. I. Epipelic communities in rivers. *Journal of Phycology*, **10**, 50–7.

Moore, J. W. (1974*b*). Benthic algae of southern Baffin Island. II. The epipelic communities in temporary pools. *Journal of Ecology*, **62**, 809–19.

(1974*c*). Benthic algae of southern Baffin Island. III. Epilithic and epiphytic communities. *Journal of Phycology*, **10**, 456–62.

Nansen, F. (1897). *Farthest North*. Westminster: Archibald Constable and Company.

Ng, S. L. (1996). The paleoenvironmental record preserved in the sediments of Fish lake, Truelove Lowland, Devon Island, N.W.T., Canada. PhD thesis, The University of Western Ontario, London, Ontario.

Östrup, E. (1897*a*). Ferskvands-Diatomeer fra Öst Grönland. *Meddelelser om Grönland*, **15**, 251–90.

(1897*b*). Kyst-Diatoméer fra Grönland. *Meddelelser om Grönland*, **15**, 305–62.

(1910). Diatoms from North-East Greenland. *Meddelelser om Grönland*, **43**, 199–256.

Ovendun, L. (1988). Holocene proxy-climate data from the Canadian Arctic. Geological Survey of Canada Paper No. 88–22. Ottawa.

Petersen, J. B. (1924). Fresh water algae from the north coast of Greenland collected by the late Dr. Th. Wulff. Den II Thule Exped. til Groenlands Nordkyst (1916–18). *Meddelelser om Grönland*, **64**, 307–19.

Psenner, R., & Schmidt, R. (1992). Climate-driven pH control of remote alpine lakes and effects of acid deposition. *Nature*, **356**, 781–3.

Rigler, F. H. (1978). Limnology in the high Arctic: a case study of Char Lake. *Verhandlungen Internationale Vereingung Limnologen*, **20**, 127–40.

Roots, E. F. (1989). Climate change: high-latitude regions. *Climatic Change*, **15**, 223–53.

Ross, R. (1947). Freshwater diatomae (Bacillariophyta). In *Botany of the Canadian Eastern Arctic. Part II: Thallophyta and Bryophyta*, ed. N. Polunin, pp. 178–233. Ottawa: National Museum of Canada, Bull 97. (Biol. Ser. 26).

Rouse, W., Douglas, M., Hecky, R., Kling, G., Lesack, L., Marsh, P., McDonald, M., Nicholson, B., Roulet, N., & Smol, J. P. 1997. Effects of climate change on fresh waters of Region 2: Arctic and Sub-Arctic North America. *Hydrologic Processes*, **11**, 873–902.

Schindler, D. W. Bayley, S. E., Parker, B. R., Beaty, K. G., Cruikshank, D. R., Fee, E. J., Schindler, E. U., & Stainton, M. P. (1996). The effects of climate warming on the properties of boreal lakes and streams at the Experimental Lakes Area, northwestern Ontario. *Limnology and Oceanography*, **41**, 1004–17.

Sheath, R. G. (1986). Seasonality of phytoplankton in northern tundra ponds. *Hydrobiologia*, **138**, 75–83.

Sheath, R. G., Morgan, V., Hambrook, J. A., & Cole, K. M. (1996). Tundra stream macroalgae of North America: Composition, distribution and physiological adaptations. *Hydrobiologia*, **336**, 67–82.

Short, S., Andres, J., Williams, K., Weiner, J., & Elias, S. (1994). Late Quaternary marine and terrestrial environments, northwestern Baffin Island, Northwest Territories. *Géographie Physique et Quaternaire*, **48**, 85–95.

Skulberg, O. M. (1996). Terrestrial and limnic algae and cyanobacteria. In *A Catalogue of Svalbard Plants, Fungi, Algae and Cyanobacteria*, Part 9, ed. A. Elvebakk & P. Prestrud, pp. 383–95. Oslo: Norsk Polarinstitutt Skrifter 198.

Smol, J. P. (1983). Paleophycology of a high arctic lake near Cape Herschel, Ellesmere Island. *Canadian Journal of Botany*, **61**, 2195–204.

(1988). Paleoclimate proxy data from freshwater arctic diatoms. *Verh. Verein Internat. Limnol.*, **23**, 837–44.

Smol, J. P., & Douglas, M. S. V. (1996). Long-term environmental monitoring in arctic lakes and ponds using diatoms and other biological indicators. *Geoscience Canada*, **23**, 225–30.

Statistics Canada. (1987). *Canada Year Book 1988; A Review of Economic, Social and Political Development in Canada*. Ottawa: Statistics Canada.

Walsh, J. E. (1991). The Arctic as a bellwether. *Nature*, **352**, 19–20

Williams, K. M. (1990a). Paleolimnology of three Jackman Sound lakes, southern Baffin island, based on down-core diatom analyses. *Journal of Paleolimnology*, **4**, 203–17.

(1990b). Late Quaternary paleoceanography of the western Baffin Bay region: Evidence from fossil diatoms. *Canadian Journal of Earth Sciences*, **27**, 1487–94.

Wolfe, A. P. (1994). Late Wisconsinan and Holocene diatom stratigraphy from Amarok Lake, Baffin Island, N.W.T., Canada. *Journal of Paleolimnology*, **10**, 129–39.

(1996a). Spatial patterns of modern diatom distribution and multiple paleolimnological records from a small non-glacial Arctic lake, Baffin Island, Northwest Territories. *Canadian Journal of Botany*, **74**, 345–59

(1996b). A high resolution late-glacial and early Holocene diatom record from Baffin Island, Northwest Territories. *Canadian Journal of Earth Sciences*, **33**, 928–37.

(1998). A 6500 year diatom record from southwestern Fosheim Peninsula, Ellesmere Island, Canadian High Arctic. *Geological Survey of Canada, Bulletin,* (in press).

Wolfe, A. P., & Härtling, J. W. (1996). The late Quaternary development of three ancient tarns on southwestern Cumberland Peninsula, Baffin Island, Arctic Canada: Paleolimnological evidence from diatoms and sediment chemistry. *Journal of Paleolimnology*, **15**, 1–18.

Young R. B., & King, R. H. (1989). Sediment chemistry and diatom stratigraphy of two high arctic isolation lakes, Truelove Lowland, Devon Island, N.W.T., Canada. *Journal of Paleolimnology*, **2**, 207–25.

11 Diatoms as indicators of environmental change in antarctic freshwaters

SARAH A. SPAULDING AND DIANE M. McKNIGHT

Introduction

The Antarctic continent holds the majority of the earth's freshwater in the form of ice. However, life is dependent upon liquid water, which is scarce in Antarctica. Less than 5% of the continent is ice-free, and it is within these ice-free regions that freshwater lakes and ephemeral streams form by the melting of glacial ice. Ice-free regions are located primarily near the Antarctic coastline (Fig. 11.1). Of these regions, the 'desert oases' of East Antarctica are considered to be the coldest, driest regions on earth.

Within the limited sites where liquid freshwater is found, life is present. Cyanobacteria are the most widely distributed and abundant freshwater organisms, forming mats in lakes, ponds, and meltwater streams and on moist soils. Higher plants occur only in lower latitudes of the Antarctic Peninsula. Diatoms are found in nearly all types of Antarctic waters, and are one of few organisms that are well preserved as fossils. In this chapter, we review the investigations to date in Antarctica concerning the application of diatoms to problems of environmental change.

Antarctic diatoms

The study of freshwater algae in Antarctica began with early expeditions to the most accessible subantarctic and maritime islands of Kerguelen and South Georgia (Reinsch, 1890; Carlson, 1913). The flora of Kerguelen has been the most studied of any of the Antarctic regions (Bourrelly & Manguin, 1954; Germain & Le Cohu, 1981; Le Cohu, 1981; Le Cohu & Maillard, 1983, 1986; Riaux-Gobin, 1994). Early exploration of the continent resulted in collections of freshwater algae in general, and diatoms were often not of primary interest (West & West, 1911; Van Heurck, 1909). Although the literature is not extensive, these early studies often contain the most detailed taxonomic and ecologic work accomplished to date. Much of the work following the early expeditions consisted of species lists and reported occurrences from newly explored regions of the continent (Aleshinskaja & Bardin, 1965; Prescott, 1979; Seaburg et al., 1979). Recently, there has been renewed interested in taxonomic investigations of

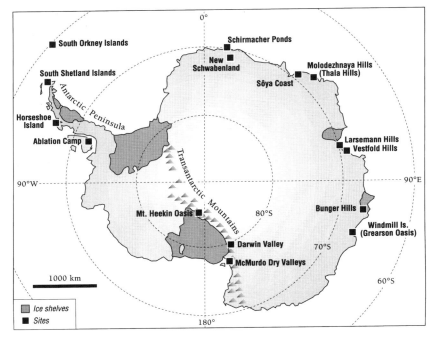

Fig. 11.1. Map of Antarctic regions containing bodies of freshwater.

Antarctic diatoms (Wasell & Håkansson, 1992; Kociolek & Jones, 1995; Spaulding & Stoermer, 1997), following the need to reconstruct historical records of environmental change in many lake regions.

Species diversity of diatoms in Antarctica is low compared to arctic regions (Douglas & Smol, this volume). Low diversity of the Antarctic flora is believed to be due, in part, to physical isolation of the continent. Antarctica is surrounded by circumpolar oceans, creating a barrier to colonization (Heywood, 1972, 1977). For example, the biota of the Antarctic Peninsula is more diverse than other regions, reflecting the proximity to the cold regions of South America. An affiliation of subantarctic and South American flora was recognized by several investigators (Frenguelli, 1924; Bourrelly & Manguin, 1954; Schmidt et al., 1990; Jones et al., 1993). For instance, many of the *Achnanthes* species found in King George Island lakes in moss habitats (Schmidt et al., 1990) were also reported in similar Chilean habitats (Krasske, 1939). Furthermore, the diatom flora of the Antarctic is quite different from that of the arctic; the polar regions have few taxa in common (Hamilton, 1994). While environmental conditions of the two poles may both be extreme, geographical isolation of Antarctica may limit the number of taxa to a greater extent than extreme environment.

Many Antarctic diatoms are aerophilic taxa, often found in aquatic habitats subject to desiccation or freezing. Other taxa are found in association with terrestrial soils and mosses (Schmidt et al., 1990). Bunt (1954) reported on diatom taxa in soils of the subantarctic Maquarie Island and characterized the types of environments where each taxon occurs. Aerophilic taxa that occur across the continent include *Navicula atomus* (Kützing) Grunow, *Diadesmis contenta* Grunow, *Luticola* spp., *Achnanthes coarctata* (Brébisson) Grunow, *Pinnularia borealis* Ehrenberg, and *Hantzschia amphioxys* (Ehrenberg) Grunow (Kawecha & Olech, 1993; Spaulding *et al.*, 1996). Other taxa have a more regional distribution, for example *Diatomella balfouriana* Greville occurs on the Antarctic peninsula (Björck *et al.*, 1991), but not East Antarctica (Spaulding *et al.*, 1996). *Muelleria peraustralis* (West & West) Spaulding & Stoermer was abundant in the McMurdo Dry Valleys and Syowa coast, but has not been found on the Antarctic Peninsula or other regions (Ko-bayashi, 1965; Spaulding & Stoermer, 1997). Diatoms that were originally described (and likely endemic taxa) from Antarctica are listed in Table 11.1.

Early studies of the Antarctic flora reported a decrease in number of algal species with increase in latitude (Hirano, 1965; Fukushima, 1970). The ratio of endemic to cosmopolitan taxa of Ross Island and several eastern Antarctic coastal regions was found to be greatest between 68 and 77 degrees south. In Antarctic Peninsula lakes, 63% of the diatoms were found to be endemic (Schmidt *et al.*, 1990). These correlations will certainly change as more taxonomic investigations are made. Further, the assumption of a direct correlation between latitude and severity of the environment may not reflect nature; there are freshwater lakes with environments similar in their extreme nature in ice-free regions of the Antarctic, regardless of latitude (Heywood, 1977). Nevertheless, areas that are conspicuously lacking in diatoms have been reported (Edward VII Peninsula (Broady, 1989) and Mt. Heekin Lakes (S. A. Spaulding, unpublished data). If will be of great interest to determine if these initial observations are supported by further investigation. Indeed, if diatoms do not occur in the highest latitudes of Antarctica may provide important insight into diatom physiology and ecology.

Diatoms as indicators of climate

Antarctica has a major influence on the climate of the Southern Hemisphere, yet the Holocene climatic record is not well known. Based on oxygen isotopes from ice cores, large-scale variations in climate correspond between Antarctica and South America (Clapperton & Sugden, 1988), but records among regions within the Antarctic are not well correlated (summarized in Zale & Karlén, 1989). In order to obtain a better record of Holocene climate, chronologies from Antarctic regions lake regions as well as from polar ice caps are valuable (Birnie, 1990; Bronge, 1992; Doran *et al.*, 1994; Appleby *et al.*, 1995). In particular, regional climatic changes are reflected in the closed-basin lakes of the McMurdo Dry Valleys, where lake-levels are controlled by the extent of glacier melt during the austral summer. In these evaporitic lakes, cool periods with low melt result in decreasing lake-levels and increased lakewater salinity, while warmer periods result in increasing lake-levels and lakewater dilution.

Table 11.1. *List of freshwater diatom taxa originally described from material of Antarctic regions (including sub-antarctic islands)*

Achnanthes abundans Bourrelly & Manguin, 1954
A. abundans var. *elliptica* Bourrelly & Manguin, 1954
A. aretasii Bourrelly & Manguin, 1954
A. bourginii Bourrelly & Manguin, 1954
A. confusa Bourrelly & Manguin, 1954
A. confusa var. *atomoides* Bourrelly & Manguin, 1954
A. delicatula var. *australis* Bourrelly & Manguin, 1954
A. germainii Bourrelly & Manguin, 1954
A. manguinii var. *elliptica* Bourrelly & Manguin, 1954
A. metakryophila Lange-Bertalot & Schmidt, 1990
A. modesta Bourrelly & Manguin, 1954
A. renei Lange-Bertalot & Schmidt, 1990
A. stauroneioides Bourrelly & Manguin, 1954

Cymbella aubertii Bourrelly & Manguin, 1954
C. kerguelenensis Germain, 1937
C. lacustris var. *australis* Bourrelly & Manguin, 1954
C. nodosa Bourrelly & Manguin, 1954
C. subantarctica Bourrelly & Manguin, 1954
Caloneis marnieri Bourrelly & Manguin, 1954

Denticula elegans var. *robusta* Bourrelly & Manguin, 1954
Diatomella hustedtii Manguin, 1954

Fragilaria vaucheriae var. *longissima* Manguin, 1954
F. vaucheriae var. *tenuis* Manguin, 1954
Frustulia pulchra Germain, 1937
F. pulchra var. *lanceolata* Manguin, 1954

Gomphonema affine var. *kerguelenense* (Manguin) Germain & Le Cohu, 1981
G. affine var. *kerguelenense* fo. *lanceolatum* Germain & Le Cohu, 1981
G. angustatum var. *aequalis* fo. *kerguelenensis* Le Cohu & Maillard, 1986
G. asymmetricum Carter, 1966 (non-Gutwinski)
G. candelariae Frenguelli, 1924
G. kerguelenensis Bourrelly & Manguin, 1954
G. kerguelenensis fo. *lanceolata* Bourrelly & Manguin, 1954
G. kerguelenensis fo. *rhomboidea* Bourrelly & Manguin, 1954
G. signyensis Kociolek & Jones, 1995
G. spatulum Carter, 1966

Luticola murrayi (West & West) Mann, 1990
L. muticopsis fo. *reducta* (West & West) Spaulding, 1996

Muelleria meridionalis Spaulding & Stoermer, 1997
M. peraustralis (West & West) Spaulding & Stoermer, 1997
  Syn: *Navicula gibbula* var. *peraustralis* (West & West) Ko-Bayashi

Navicula australomediocris Lange-Bertalot & Schmidt, 1990
N. bicephala var. *manguinii* Le Cohu & Maillard, 1986
N. bryophiloides Bourrelly & Manguin, 1954
N. bryophiloides var. *linearis* Le Cohu & Maillard, 1986
N. corrugata Bourrelly & Manguin, 1954
N. dicephala fo. *australis* Bourrelly & Manguin, 1954
N. equiornata Bourrelly & Manguin, 1954
N. glaberrima West & West, 1911
N. muticopsis fo. *evoluta* West & West (should be transferred to genus *Luticola* Mann, 1990)
N. muticopsiforme West & West, 1911 (should be transferred to genus *Luticola* Mann, 1990)
N. pseudocitrus Bourrelly & Manguin, 1954
N. shackeltoni West & West, 1911
N. spissata Bourrelly & Manguin, 1954
N. undulatistriata Bourrelly & Manguin, 1954
Neidium aubertii Bourrelly & Manguin, 1954
Nitzschia frustulum var. *kerguelenensis* Bourrelly & Manguin, 1954
N. ignorata fo. *longissima* Bourrelly & Manguin, 1954
N. westii
  Syn: *Fragilaria tenuicollis* var. *antarctica* West & West, 1911, *N. antarctica* (West & West) Fukushima

Pinnularia backebergii Bourrelly & Manguin, 1954
P. borealis var. *australis* Bourrelly & Manguin, 1954
P. borealis var. *cuneorostrata* Bourrelly & Manguin, 1954
P. circumducta Bourrelly & Manguin, 1954
P. cymatopleura West & West, 1911
P. kolbei Bourrelly & Manguin, 1954
P. microstauron var. *australis* Bourrelly & Manguin, 1954
P. microstauron var. *elongata* Bourrelly & Manguin, 1954
P. quadratarea var. *dulcicola* Bourrelly & Manguin, 1954
P. subsolaris fo. *kerguelensis* Bourrelly & Manguin, 1954

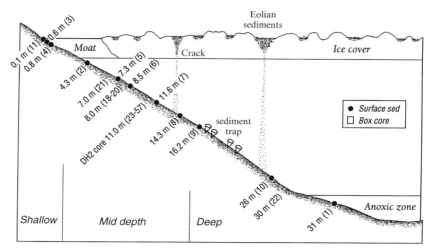

Fig. 11.2. Diagram of Lake Hoare, McMurdo Dry Valleys, East Victorialand. Horizontal axis is not to scale. Sites of surficial sediment samples, sediment cores and box cores are shown. Diagrammatic representation of perennial ice topography and eolian sediment load. 'Moat' ice may or may not melt to form open water, depending on climatic conditions.

Volume of these lakes has fluctuated greatly over the past 20 000 years (for review see Doran *et al.*, 1994). In more recent time, lake-levels have been rising since records were first kept in the early 1970s (Chinn, 1993). Increases in several lakes have been particularly dramatic, with Lake Vanda increasing in lake-level by 10 m, and Lake Wilson by 25 m (Chinn, 1993; Webster *et al.*, 1996). The record of fluctuating lake-levels has been examined primarily in terms of geochemistry of the water columns and sediments and the diatom sedimentary record in one of the McMurdo Dry Valley lakes (Lake Hoare) has been used to interpret lake history (Spaulding *et al.*, 1996).

Species composition of diatom assemblages are related to water depth in Lake Hoare, with distinct shallow water (less than 1 meter), mid-depth (4–6 m), and deep water (26–31 m) assemblages (Fig. 11.2) (Spaulding *et al.*, 1997). In the unique environment of perennial ice cover and lack of circulation of the water column, sediments record the history of a particular site rather than integration of events of the entire lake. The diatom record indicates that the current rate of lake-level increase cannot be hindcast beyond the year 1900, and that the rapid increase in lake-level is a recent event.

Nearly all Antarctic lakes are influenced by glacial ice, and many Antarctic lakes originate as proglacial lakes, on ice or at the receding edge of ice, fed by the water of glacial melt or local snowfields (Fig. 11.4) (for review see Heywood, 1977; Pickard *et al.*, 1986). As retreating ice exposes depressions, basins become filled with meltwater. Lakes range from shallow ponds a few meters wide to deep basins, covering several square kilometers. Lake regions, such as the Bunger Hills, Larsemann Hills, or Vestfold Hills contain numerous lakes of different ages and trophic status which formed following glacial retreat. The

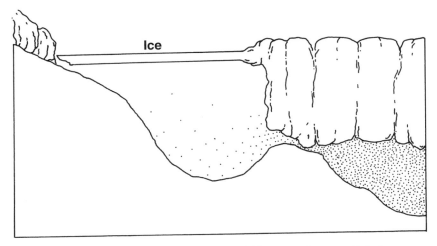

Fig. 11.3. Diagram of a glacially fed freshwater/brackish water embayment, linked to sea-water by subglacial connections. Tides result in currents within the water column, as well as mixing of sea salts.

first lakes in the Larsemann Hills were exposed nearly 9000 years BP, while those that are currently near the polar plateau have been in existence for only a few hundred years (Gillieson, 1990). Accordingly, sediments deposited early in glacial lake ontogeny are composed of sediments with little or no organic matter. As minerals are weathered within the catchment area, base cations are leached and increase in concentration in lakewater (Smol, 1988; Mäusbacher *et al.*, 1989).

Proglacial lakes are typically ultra-oligotrophic and diatoms are absent, or very few in number (Birnie, 1990; Gillieson, 1990). Differences between diatom assemblages of lakes of different ages are also evidenced in King George Island (Antarctic Peninsula) lakes (Mäusbacher *et al.*, 1989). Sediments deposited early in lake development are characterized by silty material and unstable sediments, limiting the growth of benthic mosses and diatoms. Further, early lake sediments contain higher percentages of terrestrial diatoms (e.g., *Melosira dendrophila* (Ehrenberg) Ross & Sims and *Diatomella hustedtii* Manguin) transported from the surrounding watershed. As submerged mosses became more established, the diatom assemblage in the King George Island lakes became dominated by *Achnanthes* species.

Reconstruction of lake history in relation to climate has been made in several of the Antarctic regions. Subantarctic islands are of interest because they provide a climate record in a region of open ocean, where terrestrial records are rare. Such sites are influenced by the convergence of air masses and ocean currents of the temperate and polar regions of the Southern Hemisphere (Birnie, 1990; Smith & Steenkamp, 1990). The sedimentary record on the Subantarctic island of Kerguelen shows a progression from a wet, organically enriched, acid condition to a drier, less organic, alkaline condition during the past 10000 years (Larson 1974). Schmidt *et al.* (1990) investigated diatom

S. A. SPAULDING AND D. M. MCKNIGHT

stratigraphy in two lakes on the South Shetland islands (maritime Antarctic) and found strong changes in sediment pattern and species composition correlated to the existing climatic record. It was estimated that the nearby Collins Ice Cap was retreating from 9000 to 5000 BP (Mäusbacher *et al.*, 1989), corresponding to a worldwide rising sea level and milder climate. Björk *et al.* (1991) used sediment classification, radiocarbon dating, and microfossil analyses to determine that significant changes in climate occurred in the South Shetland Islands between 3200 and 2700 years ago. Of course, the strongest support for any paleolimnological reconstruction is provided with multiple methods of analysis, in addition to the use of diatoms.

Changes in lake salinity

Because of their proximity to coastal regions, many Antarctic freshwater lakes have been influenced by the ocean. For example, several of the Bunger Hills lakes originated as marine basins or fjords in the early Holocene (Bird *et al.*, 1991; Verkulich & Melles, 1992; Melles *et al.*, 1994). Horseshoe Island lakes (Antarctic Peninsula) and Vestfold Hills lakes also have had marine origins (Pickard *et al.*, 1986; Wasell & Håkansson, 1992; Wasell, 1993). For example, the ontogeny of Ace Lake, in the Vestfold Hills, was reconstructed using the diatom record in combination with other environmental proxies (Bird *et al.*, 1991; Fulford-Smith & Sikes, 1996; Gibson & Burton, 1996). As the ice sheet retreated approximately 9000 years BP, a marine inlet became a lake as isostatic uplift raised it above sea level. The lake received freshwater as glacial meltwater flowed from the retreating ice sheet. Over the next 800 years, the lake became meromictic and supported a freshwater diatom assemblage. Then, at the time of sea level maximum approximately 6700 years BP, the diatom record indicates that seawater flooded the sill into Ace Lake and disturbed meromixis. Sediments deposited in this period were finely laminated, and contained elemental sulfur. Such features indicate that the marine input was limited in its extent and energy. The marine inflow ended 5500 years BP, and the lake again became a meromictic basin. Conditions remained stable over the next 1700 years, reflecting the conditions of the present day.

Diatoms reflect such changes from marine to brackish and freshwater conditions in both abundance and species composition. Along with increases in salinity, periodic influxes of seawater may result in an increase in nutrient concentrations and diatom abundance (Gillieson, 1990). In Ace Lake, indicator taxa have been used to interpret salinity changes within sediment core strata (Fulford-Smith & Sikes, 1996). Dramatic and clear changes in taxa provided evidence of marine incursions that had not been discovered by other paleolimnological proxies. Transfer functions for salinity have also been recently developed in lakes of Signy Island and the Vestfold Hills in order to determine past lake-level and climatic history (Roberts & McMinn, 1996*a,b*). Salinity tolerance optima for diatoms were developed from numerous Vestfold Hills lakes, with *Navicula directa* Smith, *Nitzschia cylindris* Grunow, and *Pinnularia microstauron* (Ehrenberg) Cleve representative of hypersaline, marine, and freshwater conditions, respectively. In some cases, transition from a marine environment to freshwater influences diatom species composition so heavily, that climatic influences could not be determined (Wasell & Håkansson, 1992). Such complex interactions require that sediment records be interpreted with care before drawing conclusions about climatic change.

Lacustrine environmental change

Diatom species composition also changes with lake depth (Oppenheim & Ellis-Evans, 1989; Oppenheim & Greenwood, 1990; Spaulding *et al.*, 1996). In a Signy Island lake, shallow sites are unstable environments for benthic diatoms (Oppenheim & Ellis-Evans, 1989). Substrates are disturbed year round: by wave action during the period of open water in the austral summer, and by ice scouring in winter. Deeper sites contain greater numbers of diatoms, but a greater percentage of dormant, or senescent frustules. Examination of depth profiles indicates the importance of understanding the role of light and nutrients in structuring benthic communities. These results also suggest that caution should be used in applying transfer functions in perennially ice-covered, and therefore stably-stratified lakes. Where a gradient of environmental conditions within a single lake occurs, i.e., a range of salinity in the water column of fresh to highly saline, diatom microhabitats and species composition are also variable across that gradient.

Benthic diatom ecology has been applied to interpreting the species composition of sediment cores. Several *Fragilaria* species dominated the diatom assemblages of shallow shelves in modern benthic habitats, where wave abrasion was common (Oppenheim & Ellis-Evans, 1989; Schmidt *et al.*, 1990). In sediment cores, the same *Fragilaria* species occurred in sharp zones of transition within a core (Schmidt *et al.*, 1990), and these zones were interpreted as periods of wave abrasion, caused by fluctuations in lake-level.

Use of diatoms in determining historical lake ice cover

Lakes of Antarctica range from those that are ice covered for several months of the year, to those that are perennially ice covered. Ice cover controls physical and chemical parameters (Smol, 1988; Wharton *et al.*, 1993), for example, ice cover reduces light penetration, which limits photosynthesis. Further, perennially ice-covered lakes are meromictic or amictic, and develop stable water columns (also termed perennial stratification) (Aiken *et al.*, 1991). Although ice covers impose other influences on lake chemistry, these factors are among the most important influencing diatoms.

Most notably, perennially ice-covered Antarctic lakes contain few

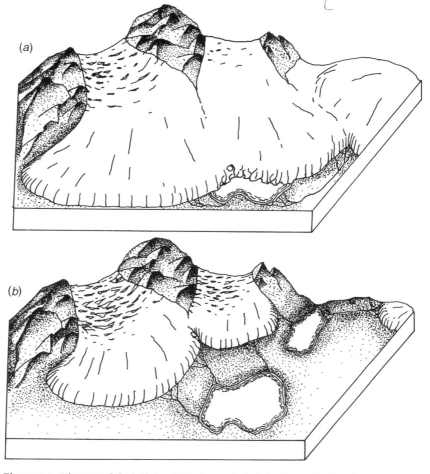

Figure 11.4. Diagram of glacial lakes. (*a*) Early proglacial lake, formed at the edge of an alpine glacier from glacial melt. (*b*) As glacier retreats, a second lake is formed. Lakes are of different ages, and potentially different trophic status following glacial retreat, despite their location in the same climatic region.

planktonic diatoms. It is believed that planktonic diatoms are infrequent because ice-covered lakes lack turbulence, which is necessary to maintain the dense silica-walled cells in suspension (Light *et al.*, 1981; Priddle *et al.*, 1986; Wharton *et al.*, 1983; Spaulding *et al.*, 1994). Diatoms collected in the plankton are often meroplanktonic, cells that are entrained in the water column having originating in benthic algal mats. Such cells have been found to be both empty frustules (dead) or cells containing chloroplasts (S. A. Spaulding, unpublished data). Meroplanktonic diatoms have been reported in other Antarctic lakes of uncertain ice cover and mixing status (Vinocur & Izaguirre, 1994). Ice-covered freshwaters linked to seawater are subject to tides, and resultant mixing of the water column (Fig. 11.3). In such lakes (e.g., Lake Miers, south of the McMurdo Dry Valleys), meroplanktonic diatoms are quite frequent (S. A. Spaulding, unpublished data). Although currents are sufficient to entrain a typically benthic flora into the water column, characteristically planktonic taxa are still absent.

Nevertheless, a limited number of planktonic taxa have been reported from Antarctic freshwaters, under ice cover (for review see Jones, 1996). Planktonic taxa include *Asterionella formosa* Hassall, *Fragilaria crotonensis* Kitton, *Cyclotella* spp., *Melosira* spp. and *Stephanodiscus minutulus* (Kützing) Cleve & Möller (Fukushima, 1961; Baker, 1967; Pankow *et al.*, 1991; Björck *et al.*, 1991). It might be expected that a planktonic flora would develop where open water is present for an extended period of time. However, even lakes that are ice free for a few months may not develop a plankton flora (Jones, 1996). This suggests that there is some other limitation to Antarctic planktonic flora; geographical isolation may play a role.

Following investigation of diatoms in arctic lakes, Smol (1983, 1988) proposed that the extent and duration of ice cover can be inferred from species composition and abundance of fossil diatoms (see Douglas & Smol, this volume). Extended periods of ice and snow cover with little development of summer 'moats' result in low production and low abundance of benthic diatoms. Taxa are characteristic of shallow waters or aerophilic taxa, and planktonic taxa are absent. During the peak of summer of slightly warmer periods, melting around the lake periphery could allow the development of a moat. Under these conditions, taxa characteristic of deeper water would become prevalent and primary production would increase. With an increase in temperature, a moat would become more extensive. Species composition would include more deep water taxa, until eventually a planktonic flora could develop. This model appears to generally be valid in Antarctic lakes. For example, diatoms in the sediments of Tiefersee and Mondsee reflect differences in the extent of moat formation during the austral summer (Schmidt *et al.*, 1990). Tiefersee, located away from a glacier develops an ice-free moat. Correspondingly, the abundance of periphytic diatoms in the sediments is greater, and also more variable than in the more persistent ice cover of Mondsee, which possesses consistently lower abundance of diatoms. Further, fluctuation in extent of moat development from year to year was interpreted based on the abundance of large pennate diatoms in sediment cores. Tiefersee showed stronger indication of an ice-free moat (more variation in periphyton abundance) than Mondsee (more constant diatom abundance). Clearly, there has been a great deal of interest in using diatoms to interpret past ice covers and there is a need for further examination of the response of diatoms to ice cover and interpreting sedimentary records.

Environmental change in surrounding watersheds

Diatoms also reflect changing environments in lake catchments as allochthonous material is deposited in lake basins. In the Antarctic, coastal lakes often experience dramatic change in nutrient concentration from the movements of marine mammals and birds. Large increases in biomass occur in lakes that have received inputs of nutrients from bird colonies (Broady, 1982, 1989; Kawecka & Olech, 1993). Twenty-one lakes on the Antarctic Peninsula, covering a gradient of extremely low nutrients to highly eutrophic systems, were examined to determine diatom community response along a nutrient gradient (Hansson & Håkansson, 1992). Most were melt-water lakes, low in productivity, but those near skua colonies received increased nutrients, while those near penguin and seal colonies received enormous influx of nutrients. Species composition was related to nutrient status, for example, *Achnanthes lanceolata* (Brébisson) Grunow and *Staurosira construens* (Ehrenberg) Williams & Round were positively correlated with phosphorus. Further efforts to quantify diatom and nutrient relations have been made using transfer functions in Signy Island lakes (Jones, 1993; Jones *et al.*, 1993; Jones & Juggins, 1995). These results were applied to sediment core assemblages, indicating that nutrient enrichment from marine bird and mammal colonies had fluctuated in Sombre Lake in recent decades, while no changes had occurred in Moss Lake. In Heywood Lake, diatom assemblages apparently did not respond to changing nutrient concentrations. The lack of response may be the result of more complex interactions that determine the diatom community; the presence of a seal colony resulted in higher turbidity and dissolved organic matter than in other lakes. Consequently, less light was available to benthic diatoms, with more of an influence than nutrient concentrations on composition and abundance (Jones & Juggins, 1995).

The species composition of diatoms may also indicate hydrologic scouring of catchment basins. During periods of increased glacial melt, meltwater carries increased suspended sediment. This suspended material appears as inorganic horizons in the downstream lake sediments. Björck *et al.* (1991) interpreted an increase in sediment accumulation rates of pollen, spores, the chlorophyte *Pediastrum*, and inorganic sediments as indicative of milder conditions (warmer summers). Concomitant with these changes, a high percentage of aerophilic taxa indicates a greater influx of allochthonous material (Schmidt *et al.*, 1990). Terrestrial diatoms associated with mosses may also reflect the development of the moss community in the surrounding watershed (Hickman & Vitt, 1973; Ohtani, 1986). For example, Schmidt *et al.* (1990) inferred that a past change in terrestrial moss habitats had occurred in the catchment based on the change in terrestrial diatom taxa in lake sediments.

Diatom species composition has been found to change in response to the input of volcanic ash in temperate regions (Birks & Lotter, 1994). Ash contains high concentrations of silica and phosphorus, nutrients that are often limiting to diatoms. In sediments of South Shetland Island lakes, peaks in the abundance of *Achnanthes subatomoides* (Hustedt) Lange-Bertalot and *Navicula seminulum* Grunow were associated with ash layers (Björk *et al.*, 1991).

Presence of freshwater taxa in marine environments

Beginning with the Dry Valleys Drilling Project in the McMurdo Valleys, freshwater diatoms have been used to interpret Antarctic paleoenvironments (Brady, 1982). Multiple glaciations of the coastal valleys of the Transantarctic Mountains, as well as fluctuations in marine ice shelves are believed to have occurred during the Pleistocene (Péwé, 1960). Marine ice contains few salts, and the surface of ice shelves contain freshwater meltpools that form from localized melt. Kellogg and Kellogg (1987) identified 53 freshwater diatom taxa in the melt pools of the Ross Ice Shelf, dominated by *Nitzschia westii* Kellogg, *Pinnularia cymatopleura* West & West, *Luticola gaussii* (Heiden) Mann, *Navicula shackletoni* West & West, *Luticola muticopsis* (Van Heurck) Mann, *Hantzschia amphioxys*, and *Cyclotella* spp. When an ice shelf melts, freshwater diatoms are deposited on marine sediments. Advances of the Ross Ice Shelf were confirmed based on the presence of freshwater diatoms (Brady, 1982). Further, more extensive freshwater deposits can be used to confirm the presence of lacustrine environments. Extensive freshwater diatom deposits were found 180 m below the present day sea level, indicating that McMurdo Sound was a freshwater oasis during the Pleistocene.

Presence of marine taxa in freshwater environments

Conversely, the marine diatoms may be deposited in freshwater sediments. The majority of Antarctic lakes are coastal, and therefore in close proximity to a source of marine diatoms. The presence of marine taxa in lacustrine sediments may indicate several types of environmental change. (i) Lakes may originate as marine basins and later become freshwater, as mentioned previously. Sediments of such lakes have a typical marine flora in the earliest strata, which may be planktonic or associated with sea-ice. (ii) Lakes may be influenced by sea spray. Presence of marine taxa in lacustrine sediments may indicate the transportation of marine diatoms, without the survival of marine taxa in freshwaters (Carlson, 1913; Negoro, 1961). Comparison of lakes that receive sea-spray with lakes that do not, showed that sea spray contains a low abundance of mostly fragmented marine diatoms (Fulford-Smith & Sikes, 1996). Alternatively, a greater influx of marine salts may change lakewater chemistry, causing an increase in growth of halophilic and alkaliphilic diatoms (Schmidt *et al.*, 1990). (iii) Lakes may be simultaneously in contact with glacial ice and marine waters, with varying marine influence. For example, primarily freshwater lakes may be tidally influenced. Although the lakewaters remain relatively fresh, tidal currents act to increase turbulence and meroplanktonic diatoms, as in Lake Miers. (iv) Freshwater lakes may be inundated with seawater following an increase in sea level, as in the Vestfold Hills (Bird *et al.*, 1991; Fulford-Smith & Sikes, 1996; Gibson & Burton, 1996). (v) Freshwater lakes may receive reworked marine sediments from the catchment (Kellogg *et al.*, 1980;

Schmidt *et al.*, 1990; Spaulding *et al.*, 1996; Fulford-Smith & Sikes, 1996). Marine deposits may be eroded by meltwater streams and deposited in lake sediments. Lacustrine strata with greater concentrations of such reworked sediments may indicate increased scouring of deposits by glacial meltwater. (vi) Freshwater lakes also contain marine diatoms transported by marine birds and from marine mammal excrement. Such animal sources of marine diatoms may be separated from other sources because animal vectors commonly result in a concomitant increase in organic carbon and nutrients (Wasell & Håkansson, 1992; Fulford-Smith & Sikes, 1996).

Diatoms in ice cores

Atmospheric transport of freshwater diatoms occurs over long distances (see Harper, this volume; Chalmers *et al.*, 1996; Gayley *et al.*, 1989; Lichti-Federovich 1984). In the Antarctic, diatoms were present in ice cores obtained from high elevations of the polar plateau, several hundred kilometers from freshwater sources (Burckle *et al.*, 1988; Kellogg & Kellogg, 1996). The diatoms were considered eolian, because no melt pools were known to occur on the surface of the ice cap. The taxa were 98% freshwater taxa; marine forms are not as likely to be transported by wind over long distances (Burckle *et al.*, 1988). Some taxa found (*Navicula muticopsis*, *N. shackletoni*, *N. mutica* var. *cohnii*, *N. deltaica* Kellogg, *Cyclotella stelligera* (Cleve) Van Heurck, *Pinnularia* sp. and *Cocconeis* sp.) are known from freshwater lakes and deposits in Antarctica. The occurrence of such freshwater taxa is intriguing because of the small source area of lacustrine deposits on the continent. Since diatoms are transported by wind and deposited on ice, it follows that they are also deposited elsewhere, including lakes (Spaulding *et al.*, 1996). The number of valves per unit mass in ice cores is very small, but wind-transported sources should be considered in interpreting lake sediment cores.

Since the work of Burckle *et al.* (1988), Kellogg and Kellogg (1996) have reported both freshwater and marine diatoms in ice cores of the Polar Plateau. These results relate to the present debate on the paleoclimate of Antarctica during the Pliocene; controversy centers on the interpretation of diatoms in the glacially deposited Sirius formation (Webb *et al.*, 1984; Denton *et al.*, 1991; Webb & Harwood, 1991; LeMasurier *et al.*, 1994). The debate centers on whether the East Antarctic Ice Sheet is stable, or warmed and melted, raising global sea level by more than 25 meters. The 'dynamic' hypothesis proposes that the Sirius deposits contain reworked marine diatoms that were deposited within basins in East Antarctica during a warm interval in the Pliocene, when the East Antarctic Ice Sheet retreated to about one-third of its present size (Webb *et al.*, 1984; Webb & Harwood, 1991). According to this hypothesis, grounded ice scoured the marine basins during a later cold period, incorporated marine sediments and diatoms, and deposited the reworked material in the Transantarctic Mountains (in the Sirius Formation). Alternatively, the 'stable' hypothesis proponents contend that the East Antarctic Ice Sheet has remained essentially unchanged for millions of years (Denton *et al.*, 1991; LeMasurier *et al.*, 1994), and that diatoms in the Sirius deposits are the result of surface contamination (Burckle & Potter, 1996). Burckle and Potter discovered Pliocene-Pleistocene diatoms within fractures of Devonian rocks, and concluded that the younger diatoms were transported by wind and deposited within the much older rocks. The presence of eolian marine and freshwater diatoms in ice cores provides evidence that diatoms are, indeed wind-transported, and may potentially be incorporated into deposits, such as the Sirius (Kellogg & Kellogg, 1996).

To resolve the debate, the transport mechanisms of diatoms into the Sirius Formation and the degree of surface 'contamination' by diatoms must be determined. If the diatoms in the Sirius Formation were incorporated into the glacial till prior to deposition, they indicate maximum ages of deposition (in the Pliocene). If diatoms in the Sirius Formation are eolian and restricted to the surface of Sirius outcrops, they indicate contamination at some time after the initial Sirius deposit. Mahood and Barron (1996) point out that controversies such as that surrounding the Sirius formation, would benefit from taxonomic scrutiny of diatom taxa that are used in biostratigraphy and environmental determination. With such detailed examination of taxa, misunderstandings of the identity, age, and source of diatoms would be far less likely.

Summary

Understanding environmental change in Antarctica is of particular importance because of the continent's major role in influencing global climate. Freshwater diatoms have been used across a range of temporal and spatial scales as indicators of environmental change in Antarctica, as well as augmenting interpretations based on other proxies. Changes in species composition and abundance of diatom assemblages are sensitive to short term environmental change such as the nutrient enrichment of inland lakes, and to longer-term change, such as the ontogeny of lakes, fjords, and uplifted marine basins. Inland lake basins have fluctuated in volume and in water chemistry, and diatoms are useful in interpreting the history of lakes and their surrounding catchments. On longer time scales, diatom records may help answer questions about the fluctuation of large ice sheets and polar ice caps.

In order to use diatoms as indicators of enviromental change, basic information on their biology is required. Such information includes adequate species descriptions, biogeographic distributions, and ecological characterizations. Unfortunately, this basic level of information is often not well known for diatoms in general (Mann & Droop, 1996), and is an even greater problem in the Antarctic (Jones, 1996). The number of Antarctic taxa is very likely to have been underestimated, because investigators have relied on the diatom literature of the Northern Hemisphere. Much of the taxonomic literature is from early expeditions, with later investigations focussing on ecology and physiology. Although these have been productive lines of research, they rely on a sound taxonomic basis (Russell & Lewis Smith, 1993). Taxonomic and ecological characterizations of diatoms is the most pressing research need, and the current limitation to their use as indicators in the Antarctic.

References

Aiken, G., McKnight, D., Wershaw, R., & Miller, L. (1991). Evidence for the diffusion of fulvic acid from the sediments of Lake Fryxell, Antarctica. In *Organic Substances and Sediments*, ed. R. Baker, pp. 75–88. Ann Arbor: Lewis.

Aleshinskaja, A. V., & Bardin, V. I. (1965). Diatomaceous flora of the Schirmacher Ponds. *Soviet Antarctic Expedition Bulletin*, **5**, 432–3.

Appleby, P. G., Jones, V. J., & Ellis-Evans, J. C. (1995). Radiometric dating of lake sediments from Signey Island (maritime Antarctic): Evidence of recent climatic change. *Journal of Paleolimnology*, **13**, 179–91.

Baker, A. (1967). Algae from Lake Miers, a solar-heated Antarctic lake. *New Zealand Journal of Botany*, **5**, 453–68.

Bird, M. I., Chivas, A. R., Radnell, C. J., & Burton, H. R. (1991). Sedimentological and stable-isotope evolution of lakes in the Vestfold Hills, Antarctica. *Paleogeography, Paleoclimatology, Paleoecology*, **84**, 109–30.

Birks, J. H. B., & Lotter, A. F. (1994). The impact of Laacher See Volcano (11 000 yr B. P.) on terrestrial vegetation and diatoms. *Journal of Paleolimnology*, **11**, 313–22.

Birnie, J. (1990). Holocene environmental change in South Georgia: Evidence from lake sediments. *Journal of Quaternary Science*, **5**, 171–87.

Björck, S., Håkansson, H., Zale, R., Karlén, W., & Jönsson, B. L. (1991). A late Holocene lake sediment sequence from Livingston Island, South Shetland Islands, with paleo-climatic implications. *Antarctic Science*, **3**, 61–72.

Bourrelly, P., & Manguin, E. (1954). Contribution à la flore algale d'eau douce des îles Kerguelen. *Mémoires de l'Institut Scientifique de Madagascar*, **5**, 7–58.

Brady, H. T. (1982). Late Cenozoic history of Taylor and Wright Valleys and McMurdo Sound inferred from diatoms in Dry Valley Drilling Project cores. In *Antarctic Geoscience* ed. C. Craddock, pp. 1123–31. Madison: University of Wisconsin Press.

Broady, P. A. (1982). Taxonomy and ecology of algae in a freshwater stream in Taylor Valley, Antarctica. *Archiv für Hydrobiologie*, **63**, 331–49.

    (1989). Survey of algae and other terrestrial biota at Edward VII Peninsula, Marie Byrd Land. *Antarctic Science*, **1**, 215–24.

Bronge, C. (1992). Holocene climatic record from lacustrine sediments in a freshwater lake in the Vestfold Hills, Antarctica. *Geografiska Annaler*, **74**, 47–58.

Bunt, J. S. (1954). A comparative account of the terrestrial diatoms of Macquarie Island. *Proceedings of the Linnean Society of New South Wales*, **79**, 34–56.

Burckle, L. H., Gayley, R. I., Ram, M., & Petit, J. R. (1988). Diatoms in Antarctic ice cores: Some implications for the glacial history of Antarctica. *Geology*, **16**, 326–9.

Burckle, L. H., & Potter, N., Jr. (1996). Pliocene–Pleistocene diatoms in Paleozoic and Mesozoic sedimentary and igneous rocks from Antarctica: A Sirius problem solved. *Geology*, **24**, 235–8.

Carlson, G. W. (1913). Süßwasseralgen aus der Antarktis, Süd-Georgien und den Falkland Inseln. *Wissenschaftliche Ergebnisse schwedische Süd-Polar-Expedition 1901–1903. Botanik*, **4**, 1–94.

Carter, J. R. (1966). Some freshwater diatoms of Tristan da Cunha and Gough Island. Report on material collected by the Royal Society expedition to Tristan da Cunha 1962 and Gough Island Scientific Survey 1956. *Nova Hedwigia*, **11**, 443–83.

Chalmers, M. O., Harper, M. A., & Marshall, W. A. (1996). *An Illustrated Catalogue of Airborne Microbiota from the Maritime Antarctic*. Cambridge: British Antarctic Survey.

Chinn, T. (1993). Physical hydrology of the dry valley lakes. In *Physical and Biogeochemical Processes in Antarctic Lakes*, Antarctic Research Series, no.59, ed. W. J. Green & E. I. Friedmann, pp. 1–51. Washington, DC: American Geophysical Union.

Clapperton, C. M., & Sugden, D. E. (1988). Holocene glacier fluctuations in South America and Antarctica. *Quaternary Science Reviews, 7*, 185–98.

Denton, G. H., Prentice, M. L., & Burckle, L. H. (1991). Cenozoic history of the Antarctic Ice Sheet. In *The Geology of Antarctica*, ed. R. J. Tingey, pp. 365–433. Oxford: Oxford University Press,

Doran, P. T., Wharton, R. A., Jr., & Lyons, W. B. (1994). Paleolimnology of the McMurdo Dry Valleys, Antarctica. *Journal of Paleolimnology, 10*, 85–114.

Frenguelli, J. (1924). Diatomeas de Tierra del Fuego. *Anales de la Sociedad Científica Argentina, 97*, 87–118, 231–66.

Fukushima, H. (1961). Diatoms from the Shin-nan Rock ice-free area, Prince Olav Coast, the Antarctic Continent. *Antarctic Record, 14*, 80–91.

(1970). Notes on the diatom flora of Antarctic inland waters. In *Antarctic Ecology*, ed. M. W. Holdgate, pp. 628–31. London: Academic Press.

Fulford-Smith, S. P., & Sikes, E. L. (1996). The evolution of Ace Lake, Antarctica, determined from sedimentary diatom assemblages. *Paleogeography, Paleoclimatology, Paleoecology, 124*, 73–86.

Gayley, R. I. Ram, M., & Stoermer, E. F. (1989). Seasonal variations in diatom abundance and provenance in Greenland ice. *Journal of Glaciology, 35*, 290–292.

Germain, H. (1937). Diatomées d'une tourbe des îles Kerguelen. *Bulletin Société Française de Microscopie, 6*, 11–17.

Germain, H., & Le Cohu, R. (1981). Variability of some features in a few species of *Gomphonema* from France and the Kerguelen Islands (South Indian Ocean). In *Proceedings of the 6th Symposium on Recent and Fossil Diatoms*. ed. R. Ross. pp. 167–77. Königstein: Koeltz.

Gibson, J. A. E., & Burton, H. R. (1996). Meromictic Antarctic lakes as recorders of climate change: the sturctures of Ace and Organic Lakes, Vestfold Hills, Antarctica. *Papers and Proceedings of the Royal Society of Tasmania, 130*, 73–8.

Gillieson, D. (1990). Diatom stratigraphy in Antarctic freshwater lakes. In *Quaternary Research in Australian Antarctica: Future Directions*. ed. D. Gillieson & S. Fitzsimmons. pp. 55–67. Canberra: University of New South Wales.

Hamilton, P. B. (ed.) (1994). *Proceedings of the Fourth Arctic–Antarctic Diatom Symposium (Workshop)*. Canadian Technical Report of Fisheries and Aquatic Sciences, No. 1957. Ottawa: Canadian Museum of Nature.

Hansson, L. A., & Håkansson, H. (1992) Diatom community response along a productivity gradient of shallow Antarctic lakes. *Polar Biology, 12*, 463–8.

Heywood, R. B. (1972). Antarctic limnology: A review. *British Antarctic Survey Bulletin 29*, 35–65.

(1977). Antarctic freshwater ecosystems: review and synthesis. In *Adaptations within Antarctic Ecosystems*, ed. G. A. Llano, pp. 801–28. Washington, DC: Smithsonian Institution.

Hickman, M., & Vitt, D. H. (1973). The aerial epiphytic diatom flora of moss species from subantarctic Campbell Island. *Nova Hedwigia, 443–58.*

Hirano, M. (1965). Freshwater algae in the Antarctic regions. In *Biogeography and Ecology in Antarctica*, ed. J. van Mieghem & P. van Oye, pp. 28–191. The Hague: Junk.

Jones, V. J. (1993). The use of diatoms in lake sediments to investigate environmental history in the Maritime Antarctic: An example from Sombre Lake, Signy Island. *Antarctic Special Topic, 91–5.*

(1996). The diversity, distribution and ecology of diatoms from Antarctic Inland Waters. *Biodiversity and Conservation, 5*, 1433–49.

Jones, V. J., & Juggins, S. (1995). The construction of a diatom-based chlorophyll a transfer function and its application at three lakes on Signy Island (maritime Antarctic) subject to differing degrees of nutrient enrichment. *Freshwater Biology, 34*, 433–45.

Jones, V. J., Juggins, S., & Ellis-Evans, J. C. (1993). The relationship between water chemistry and surface sediment diatom assemblages in maritime Antarctic lakes. *Antarctic Science,* **5,** 339–48.

Kawecka, B., & Olech, M. (1993). Diatom communities in the Vanishing and Ornithologist Creek, King George Island, South Shetlands, Antarctica. *Hydrobiologica,* **269/270,** 327–33.

Kellogg, D. E., & Kellogg, T. B. (1987). Diatoms of the McMurdo Ice Shelf, Antarctica: implications for sediment and biotic reworking. *Paleogeography, Paleoclimatology, Paleoecology,* **60,** 77–96.

(1996). Diatoms in South Pole ice: implications for eolian contamination of Sirius Group deposits. *Geology,* **24,** 115–18.

Kellogg, D. E., Stuvier, M., Kellogg, T. M., & Denton, G. H. (1980). Non-marine diatoms from late Wisconsin perched deltas in Taylor Valley, Antarctica. *Paleogeography, Paleoclimatology, Paleoecology,* **30,** 157–89.

Ko-bayashi, T. (1965). Variations of *Navicula gibbula* var. *peraustalis* (Pennate diatoms). *Antarctic Record,* **24,** 36–40.

Kociolek, J. P., & Jones, V. J. (1995). *Gomphonema signyensis* sp. nov., a freshwater diatom from maritime Antarctica. *Diatom Research,* **10,** 269–76.

Krasske, G. (1939). Zur Kieselalgenflora Südchiles. *Archiv für Hydrobiologie,* **35,** 349–468.

Larson, D. D. (1974). Paleoecological investigations of diatoms in a core from Kerguelen Islands, Southeast Indian Ocean. Institute of Polar Studies, Report no. 50, 1–61. Columbus: Ohio State University.

Le Cohu, R. (1981). Les espèces endémiques de diatomées aux îles Kerguelen. *Colloque sur les Ecosytèmes Subantarctiques,* **51,** 35–42.

Le Cohu, R., & Maillard, R. (1983). Les diatomées monoraphidées des îles Kerguelen. *Annales Limnologie,* **19,** 143–67.

(1986). Diatomées d'eau douce des îles Kerguelen (l'exclusion des Monoraphidées). *Annales Limnologie,* **22,** 99–118.

LeMasurier, W. E., Harwood, D. M., & Rex, D. C. (1994). Geology of Mount Murphy Volcanco; an 8–m.y. history of interaction between rift volcano and the West Antarctic ice sheet. *Geological Society of America Bulletin,* **106,** 265–80.

Lichti-Federovich, S. (1984). Investigation of diatoms found in surface snow from Sydkap Ice Cap, Ellesmere Island, Northwest Terriories. Current Research, Part A, Geological Survey of Canada Paper 84–1A, pp. 287–301.

Light, J. J., Ellis-Evans, J. C., & Priddle, J. (1981). Phytoplankton ecology in an Antarctic lake. *Freshwater Biology,* **11,** 11–26.

Mahood, A. D., & Barron, J. A. (1996). Comparative ultrastructure of two closely related *Thalassiosira* species: *Thalassiosira vulnifica* (Gombos) Fenner and *T. fasiculata* Harwood et Maruyama. *Diatom Research,* **11,** 283–95.

Mann, D. G., & Droop, S. J. M. (1996). Biodiversity, biogeography and conservation of diatoms. In *Biogeography of Freshwater Algae,* ed. J. Kristiansen, pp. 19–32. Dordrecht: Kluwer Academic.

Mäusbacher, R., Müller, J., & Schimidt, R. (1989). Evolution of postglacial sedimentation in Antarctic lakes (King George Island). *Zeitschrift für Geomorphologie, Neue Folge,* **33,** 219–34.

Melles, M., Verkulich, S., & Hermichen, W. D. (1994). Radiocarbon dating of lacustrine and marine sediments from the Bunger Hills, East Antarctica. *Antarctic Science,* **6,** 375–8.

Negoro, K. (1961). Diatoms from some inland waters of Antarctica (Preliminary Report). *Antarctic Record,* **11,** 872–3.

Ohtani, S. (1986). Epiphytic algae on mosses in the vicinity of Syowa Station, Antarctica. *Memoirs of the National Institute of Polar Research, Special Issue,* **44,** 209–19.

Oppenheim, D. R., & Ellis-Evans, J. C. (1989). Depth-related changes in benthic diatom assemblages of a maritime Antarctic lake. *Polar Biology,* **9,** 525–32.

Oppenheim, D. R., & Greenwood, R. (1990). Epiphytic diatoms in two freshwater maritime Antarctic lakes. *Freshwater Biology,* **24,** 303–14.

Pankow, H., Haendel, D., & Richter, W. (1991). Die Algenflora der Schirmacheroase (Ostantarktika). *Beihefte zur Nova Hedwigia, 103,* 1–197.

Péwé, T. L. (1960). Multiple glaciation in the McMurdo Sound region, Antarctica – a progress report. *Journal of Geology, 68,* 498–514.

Pickard, J., Adamson, D. A., & Heath, C. W. (1986). The evolution of Watts Lake, Vestfold Hills, East Antarctica, from Marine Inlet to Freshwater Lake. *Paleogeography, Paleoclimatology, Paleoecology, 53,* 271–88.

Priddle, J., Hawes, I., & Ellis-Evans, J. C. (1986). Antarctic aquatic ecosystems as habitats for phytoplankton. *Biological Review, 61,* 199–238.

Prescott, G. W. (1979) A contribution to a bibliography of Antarctic and Subantarctic algae. *Bibliotheca Phycologica, 45,* 1–312.

Reinsch, P. F. (1890). Die Süßwasseralgenflora von Süd-Georgien. In G. Neumeyer, *Die deutschen Expeditionen und ihre Ergebnisse, 1882–1883.* pp. 329–65.

Riaux-Gobin, C. (1994). A check-list of the *Cocconeis* species (Bacillariophyceae) in Antarctic and Subantarctic areas, with special focus on Kerguelen Islands. *Cryptogamie Algologie, 15,* 135–46.

Roberts, D., & McMinn, A. (1996a). Relationships between surface sediment diatom assemblages and water chemistry gradients in saline lakes of the Vestfold Hills, Antarctica. *Antarctic Science, 8,* 331–41.

(1996b). Paleosalinity reconstruction from saline lake diatom assemblages in the Vestfold Hills, Antarctica. *Polar Desert Ecosystem Conference Proceedings,* (in press).

Russell, S., & Lewis Smith, R. I. (1993). New significance for antarctic biological collections and taxonomic research. *Proceedings of the NIPR Symposium on Polar Biology, 6,* 152–65.

Schmidt, R., Mäusbacher, R., & Müller, J. (1990). Holocene diatom flora and stratigraphy from sediment cores of two Antarctic lakes (King George Island). *Journal of Paleolimnology, 3,* 55–74.

Seaburg, K. C., Parker, B. C., Prescott, G. W., & Whitford, L. A. (1979). The algae of southern Victorialand, Antarctica. A taxonomic and distributional study. *Bibliothecia Phycologica, 46,* Hirshberg: Cramer.

Smith, V. R., & Steenkamp, M. (1990). Climatic change and its ecological implications at a subantarctic island. *Oecologia, 85,* 14–24.

Smol, J. P. (1983). Paleophycology of a high arctic lake near Cape Herschel, Ellesmere Island. *Canadian Journal of Botany, 61,* 2195–204.

(1988). Paleoclimate proxy data from freshwater arctic diatoms. *Verh. Internat. Verein. Limnol., 23,* 837–44.

Spaulding, S. A., & Stoermer, E. F. (1997). Taxonomy and distribution of the genus *Muelleria* Frenguelli. *Diatom Research, 12,* 95–113.

Spaulding, S. A., McKnight, D. M., Smith, R. L., & Dufford, R. (1994). Phytoplankton population dynamics in perennially ice-covered Lake Fryxell, Antarctica. *Journal of Plankton Research, 16,* 527–41.

Spaulding, S. A., McKnight, D. M., Stoermer, E. F., & Doran, P. T. (1997). Diatoms in sediments of perennially ice-covered Lake Hoare, and implications for interpreting lake history in the McMurdo Dry Valleys of Antarctica. *Journal of Paleolimnology, 17,* 403–20.

Van Heurck, H. (1909). Diatomées. In *Expédition Antarctique Belge, Résultats du Voyage du S.Y. Belgica en 1897–1899. Botanique 6,* 1–126. Antwerp: Buschmann.

Verkulich, S., & Melles, M. (1992). Composition and paleoenvironmental implications of sediments in a fresh water lake in marine basins of Bunger Hills, East Antarctica. *Polarforschung, 60,* 169–80.

Vinocur, A., & Izaguirre, I. (1994). Freshwater algae (excluding Cyanophyceae) from nine lakes and pools of Hope Bay, Antarctic Peninsula. *Antarctic Science, 6,* 483–9.

Wasell, A. (1993). *Diatom stratigraphy and evidence of Holocene environmental changes in selected lake basins in the Antarctic and South Georgia.* Department of Quaternary Research Report no.23, pp. 1–15, Stockholm: Stockholm University.

Wasell, A., & Håkansson, H. (1992). Diatom stratigraphy in a lake on Horseshoe Island, Antarctica: A marine-brackish-fresh water transition with comments on the systematics and ecology of the most common diatoms. *Diatom Research, 7,* 157–94.

Webb, P. N., & Harwood, D. M. (1991). Late Cenozoic glacial history of the Ross Embayment, Antarctic. *Quaternary Science Reviews*, **10**, 215–23.

Webb, P. N., Harwood, D. M., McKelvey, B. C. Mercer, J. C., & Stott, L. D. (1984). Cenozoic marine sedimentation and ice-volume variation on the East Antarctic craton. *Geology*, **12**, 287–91.

Webster, J., Hawes, I., Downes, M., Timperley, M., & Howard-Williams, C. (1996). Evidence for regional climate change in the recent evolution of a high latitude pro-glacial lake. *Antarctic Science,* **8**, 49–59.

West, W., & West, G. S. (1911) Freshwater Algae. In *Biology*, Vol.1 *Reports on the Scientific Investigations, British Antarctic Expedition 1907–9,* ed. J. Murray. pp. 263–287. London: Heinemann.

Wharton, R. A., Jr., Parker, B. C., & Simmons, G. M., Jr. (1983). Distribution, species composition, and morphology of algal mats in Antarctic dry valley lakes. *Phycologia,* **22**, 355–65.

Wharton, R. A., Jr., McKay, C. P., Clow, G. D., & Andersen, D. T. (1993). Perennial ice covers and their influence on Antarctic lake ecosystems. In *Physical and Biogeochemical Processes in Antarctic Lakes,* Antarctic Research Series no. 59, ed. W. J. Green, & E. I. Friedmann, pp. 53–70. Washington DC: American Geophysical Union.

Zale, R., & Karlén, W. (1989). Lake sediment cores from the Antarctic Peninsula and surrounding islands. *Geografiska Annaler,* **71**, 211–20.

12 Diatoms of aerial habitats

JEFFREY R. JOHANSEN

Introduction

Although studied less than aquatic diatoms, aerial diatoms are discussed in an extensive literature. Most publications on the topic consist merely of floristic lists. Thus our understanding of aerial diatom ecology is meager. Given the brevity of the current chapter, it is not possible to list all of the pertinent literature. This paper will summarize aerial diatom studies based on floristic literature and my own work.

The most important early worker was probably Johannes Boye Petersen. Unlike many early soil phycologists, he treated diatoms with both detail and taxonomic accuracy. Petersen (1915, 1928, 1935) examined numerous aerial samples from Denmark, Iceland, and East Greenland. In all, he described 196 diatom taxa from soils, wet rocks, wet tree bark, and mosses, many of which were new to science at that time.

Other important early floristic works are those of Beger (1927, 1928), Krasske (1932, 1936, 1948), Hustedt (1942, 1949), Lund (1945), and Bock (1963). More recent studies report diatom floras associated with limestone caves, sandstone cliff faces, wet rocks, mosses, and soils. Added to these studies are numerous papers on aerial algae which discuss diatoms to some extent. Reviews on terrestrial algae have generally slighted the diatoms, although none have ignored them (Johansen, 1993; Metting, 1981; Novichkova-Ivanova, 1980; Starks *et al.*, 1981; Hoffman, 1989).

Petersen (1935) defined a number of categories for aerial algae based on their habitat type. Euaerial algae inhabit raised prominent objects that receive moisture solely from the atmosphere. Terrestrial algae are those growing on the soil. Pseudoaerial algae are those living on rocks moistened by a fairly steady source of water, such as waterfall spray, springs, or seeps. Petersen divided terrestrial algae further into three categories based on their appearance in both euaerial habitats and soils (aeroterrestrial), on soils which periodically dry out (euterrestrial), or on soils which are perpetually moist (hydroterrestrial). Finally, he differentiated surface soil algae (epiterranean) from those occurring below the surface (subterranean).

Ettl and Gärtner (1995) use aeroterrestrial more broadly than Petersen (1935), to include what he would have called aerial algae. They define two subcategories; aerophytic is used to describe any aerial habitat that is not soil, while terrestrial is used for all soils. Aerophytic is further divided by substrate type to include epiphytic (on bark, leaves, mosses), xylophytic (on wood), lithophytic (on rock), as well as lichen phycobionts. Terrestrial algae include euterrestrial (on soil), hydroterrestrial (on permanently wet soil), aeroterrestrial (on the soil surface and the transition zone to aerial habitats), and endolithic (within rock). Some confusion may arise from using aeroterrestrial for aerial and aeroterrestrial for algae in the transition zone between soil and euaerial habitats. I would recommend the retention of the term aerial algae for all non-aquatic algae, and the continued differentiation between euaerial, pseudoaerial, and terrestrial habitats. Within both euaerial and pseudoaerial classifications, the subcategories of Ettl and Gärtner (1995) for aerophytic algae are appropriate. Given that mosses often have unique diatom associations, I would recommend the use of the term bryophytic for those diatoms occurring in association with mosses, reserving epiphytic for diatoms specifically associated with vascular plants. Mosses are the confounding factor in many of these classifications, because they occur in euaerial, pseudoaerial, and terrestrial habitats. Subaerial, a term that has been used to describe soil, mosses, trees, and wet rocks (Camburn, 1982; Johansen *et al.*, 1981; Stoermer, 1962), is less descriptive and should not be used in the future without further definition.

Systematics

Numerous diatom species, varieties, and forms have been described from aerial habitats. Petersen (1915, 1928, 1935) alone described 39 new taxa. Bock (1963), Hustedt (1942, 1949), Krasske (1932, 1936, 1948), and Lund (1945) likewise published many new taxa. Recent workers have continued to find new taxa (Carter, 1971; Rushforth *et al.*, 1984; VanLandingham, 1966, 1967, and others). From literature I have examined, about 340 diatom taxa have been reported from soils, 400 diatom taxa from rock substrates, and 130 taxa from mosses. There is substantial overlap among taxa that occur in soils, on rocks, and associated with mosses, but there also appear to be taxa confined to each habitat type as well.

Aerial diatoms are systematically problematic. Many are small and difficult to study using the light microscope. A large number of naviculoid species occur in aerial habitats, but most of these are not in *Navicula sensu stricto* (Round *et al.*, 1990). Some aerial *Navicula* species have been moved to other genera, including *Luticola* (*N. mutica* Kütz. and its forms, varieties, and related species) and *Diadesmis* (*N. gallica* (Wm. Smith) Lagerst., *N. contenta* Grun., and similar forms). Other aerial groups which need systematic study are the *Caloneis bacillum* (Grun.) Cl./*C. aerophila* Bock complex and the *Cymbella microcephala* Grun./*C. falaisensis* (Grun.) Krammer et Lange-Bert. complex. *Hantzschia amphioxys* (Ehr.) Grun., a cosmopolitan taxon, demonstrates extensive variability. Some of this variability is likely genetic and the forms and varieties of this species may deserve to be reconsidered. Although many species, varieties, and forms have been described in the *Luticola* complex, there is no agreement as to what constitutes genetically based differences in forms and what proportion of morphological variability is due to environmentally induced plasticity (Bock, 1963; Lange-Bertalot & Bonik, 1978; Petersen, 1928). I have seen SEM micrographs of tropical forms of *Navicula krasskei* Hustedt and *Diadesmis contenta* (Grun.) Mann that deviate substantially from temperate forms of those taxa when observed in SEM (R. Iserentant, personal communication), even though with light microscopy they appear to fit the species well. These are only a few of the problems that exist in the systematics of aerial diatoms.

Aerial diatoms of many geographical regions have not been studied, and it is likely that new species and subspecific taxa are yet to be discovered. Areas needing attention include the tropical grassland and forest biomes, temperate grasslands, and boreal forests. Aerial diatoms of temperate forest and grassland regions of North America have been only cursorily studied (Hayek & Hulbary, 1956).

Ecology

FACTORS INFLUENCING DISTRIBUTION

Indicator status of most aerial diatom taxa is poorly known. An aerial habitat is harsh and limiting in a number of ways. This makes it difficult to determine which environmental factors are most significant in determining species distribution. Moisture availability has frequently been thought to be the most important limiting factor (Camburn, 1982), although it is probably better to consider exposure to long periods of desiccation more crucial. The question concerning moisture is whether or not the site is wet year round (truly hydroterrestrial or pseudoaerial), or wet to damp most of the year with brief periods of dryness (giving time for temporary establishment of hydrophilous species), or exposed to long periods of total dryness (hydrophilous species excluded). Several authors have categorized aerophiles according to their desiccation resistance (Ito & Horiuchi, 1991; Krasske, 1932, 1936), although none of their work was done in arid environments, where long periods of drought are even more limiting. Many of the taxa designated by these authors as xeritic do not occur in desert soils.

Associated with the effects of moisture are extremes of temperature. Aerial habitats experience considerably higher diel fluctuations of temperature than aquatic habitats, and this factor could be limiting for some species. Furthermore, soils experience greater extremes on a yearly cycle; temperatures in some desert soils exceed 50 °C in the summer and fall below freezing during winter. Elevated temperatures coincide with periods of dryness, and the severity of desiccation is even greater in climates with a hot, dry period. Despite the intuitive conjecture that temperature is an important limiting factor for diatoms, specific effects of temperature on aerial diatoms have not been tested.

It is likely that pH is nearly as critical in defining species distributions of aerial diatoms as exposure to high temperatures and desiccation. We found that pH differences of 0.5 pH units can drastically change species composition in sandstone seeps from several sites in Ohio. Limestone caves and temperate sandstone seeps have similar moisture regimes, temperature extremes, and low nutrient availability, but markedly different floras. The sandstone seeps are generally acidic (pH 3.7–6.0), while the water dripping through a limestone cave is neutral to slightly alkaline (pH 6.5–8.0). *Diadesmis gallica* Wm. Smith and *D. laevissima* (Cl.) Mann are especially confined to caves, while *D. contenta*, *Eunotia exigua* (Bréb.) Rabenh., and *N. krasskei* occur primarily on seeps.

Some taxa appear to be less sensitive to pH. *Hantzschia amphioxys* is the most conspicuous of these, with a published pH range of occurrence between 5.6 and 8.5. This, however, is more the exception than the rule. Furthermore, even taxa with wide pH ranges show distinct optima. For example, *H. amphioxys* is rare in acidic soils, but reaches high densities in neutral to slightly alkaline soils.

Substrate is also a critical limiting factor. Even given similar pH and temperature regimes, euterrestrial taxa are rarely abundant on pseudoaerial lithic substrates. For example, the cosmopolitan euterrestrial species *H. amphioxys*, *Luticola cohnii* (Hilse) Mann, *L. mutica* (Kütz.) Mann and *Pinnularia borealis* Ehr. are absent or rare in all strictly aerial and pseudoaerial lithic substrates (Camburn, 1982; Rushforth *et al.*, 1984; VanLandingham, 1964; among others). However, if mosses are present on lithic surfaces, then all of these taxa are usually present to at least some extent (Reichardt, 1985). Bryophytic diatom floras have been studied more extensively than either euterrestrial or lithophytic floras. These communities are often an agglomeration of euterrestrial and lithophytic species, along with a few predominantly bryophytic species. Algae inhabiting wooden substrates have not been studied sufficiently to determine if distinctive xylophytic associations exist (Petersen, 1928).

Most aerial diatom species are indicative of low nutrient availability. Of the 122 diatoms listed as exclusively aerial or occurring mostly in wet, moist, or temporarily dry places in Europe, Van Dam *et al.* (1994) considered only four of the taxa to be facultatively or obligately nitrogen-heterotrophic taxa, while over half of those characterized with regard to nitrogen uptake were classified as tolerating only very small concentrations of organic nitrogen. Likewise, half of the taxa scored for trophic state were oligotraphentic, with relatively few eutraphentic or hypereutraphentic species. This is not surprising, since most terrestrial, aerial, and pseudoaerial habitats have little exposure to the anthropogenic nitrogenous pollution that most surface waters in the populated world receive.

Elevated conductivity is a feature of many aerial habitats where evaporation is high, and halophilous species commonly have been collected from both soils and pseudoaerial habitats in desert environments. Indeed, *Luticola cohnii*, *L. mutica*, and *L. nivalis* (Ehr.) Mann are all species which occur at 500–1000 mg/Cl/l (Van Dam *et al.*, 1994), while *Diadesmis contenta*, *D. perpusilla* (Grun.) Mann, *Hantzschia amphioxys*, *Luticola saxophila* (Bock) Mann, *Navicula atomus* (Kütz.) Grun., *N. excelsa* Krasske, *Pinnularia borealis*, *P. obscura* Krasske and *Tryblionella debilis* Arnott, some of the most commonly occurring aerial diatoms, are all considered typical of slightly brackish waters (100–500 mg/Cl/l, Van Dam *et al.*, 1994).

Although the very term aerophilous means 'air-loving', it has not been determined whether aerial taxa indeed require high O_2 concentrations. It seems likely that such a requirement may exist for many of these species. A requirement for continuously high O_2 levels may be one characteristic that excludes some of these diatoms from aquatic environments, which may lack the consistently saturated oxygen levels that likely prevail in thin films of water in terrestrial, euaerial, and pseudoaerial habitats.

In summary, a number of environmental variables likely act in concert to restrict aerial diatom distribution. Rather than considering these species to prefer aerial habitats, it may be more accurate to consider them as most competitive under the multiple stresses that characterize these habitats, especially desiccation, temperature extremes, high conductivity, and low nutrient availability. Varying substrates and pH, while not generally considered stressors, act in conjunction with the other environmental variables to determine which species can exist in a given site.

BIOGEOGRAPHY

The importance of environmental factors in determining biogeography of aerial taxa is perhaps best illustrated by examining the diatom taxa present in habitats of varying severity. In the most extreme habitats studied (e.g., soils of the hot deserts of the American Southwest), we have found as few as two diatom taxa, *Hantzschia amphioxys* f. *capitata* O. Müller and *Luticola mutica*. In hot deserts receiving more precipitation, such as Baja California where coastal moisture is available, we find the above species together with *H. amphioxys*, *L. cohnii*, *Pinnularia borealis*, and *P. borealis* var. *rectangularis* Carlson. In higher elevation hot deserts receiving regular rain during the mild winter months we find even more species, including *Cyclotella meneghiniana* Kütz., *Cymbella minuta* Hilse, *Diatoma vulgare* Bory, *Epithemia zebra* (Ehr.) Kütz., *Fragilaria vaucheriae* (Kütz.) Peters. *Navicula exilis* Kütz. and *Navicula lanceolata* (Ag.) Kütz. (Anderson & Rushforth, 1976). Finally, in high elevation deserts, which receive both snow and winter rain, we see the addition of numerous taxa, including *Caloneis aerophila*, *Denticula elegans* f. *valida* Pedic., *Navicula contenta* f. *parallela* (Peters.) Hustedt, *Epithemia adnata* var. *minor* (Perag. et Hérib.) Patr., *L. nivalis*, and *L. paramutica* (Bock) Mann (Johansen *et al.*, 1981, 1984). When desert soils are compared to those of other climates, it becomes apparent that many species common in temperate and/or polar soils are absent or only rarely encountered, such as all

Eunotia species, most *Diadesmis* species, and most of the small naviculoid genera other than *Luticola*, such as the *Navicula atomus* and *Navicula minima* Grun. groups. Temperate soils clearly have more acidobiontic and acidophilic species, and much of the difference in floras between semi-arid shrub-steppe and temperate forest is likely due to the large variances in pH, conductivity, and nutrient availability rather than to differences in temperature extremes and moisture availability. We see large generic shifts between habitat types with only a few cosmopolitan taxa able to survive the full spectrum of environmental variability.

Interestingly, aerial diatom taxa are commonly found in the aquatic habitats in polar regions. In a study of 36 high arctic ponds on Ellesmere Island, over two thirds of the common benthic diatom taxa collected from rocks, mosses, and sediments were known aerophiles (Douglas & Smol, 1995). This suggests that many aerial diatom taxa may not owe their distribution pattern solely or even primarily to their ability to withstand desiccation. High arctic ponds are oligotrophic, unpolluted, circumneutral, cold, and highly oxygenated. The occurrence of aerial taxa in such habitats indicates that some taxa may occur in aerial habitats simply because they require highly oxygenated habitats and tolerate very low nutrient levels. Additionally, some taxa may respond to substrate type (such as moss), although this seems to be a less important predictor of occurrence. Aerial genera and species also are found in the isolated aquatic habitats of Antarctica (S. Spaulding, personal communication).

Applications

AERIAL DIATOMS AS INDICATORS

Aerial diatom species may have higher resistance to ultraviolet radiation (UVR) than aquatic species. They live in thin films of water which would be ineffective at reducing levels of UVR. Aerial algae in arid regions are exposed both to very high illuminance and UVR levels due to the dearth of shading by vascular plants. Experimental evidence indicates that stalked diatoms (*Gomphonema* and *Cymbella*) have a competitive advantage in high UVR habitats (Bothwell *et al.*, 1993). These genera are also common in the highly illuminated sandstone seep walls of the arid west (Johansen *et al.*, 1983*a,b*), even though they are quite rare on sandstone seeps in shaded temperate forests. These findings suggest a connection between UVR resistance and some aerial diatom species.

The prevalence of aerial diatoms in polar waters may be in part in response to elevated UVR resistance. If so, their relative abundance might increase with increasing levels of UVR due to destruction of ozone in the stratosphere. The apparent correlation between some aerial diatoms and elevated UVR in both desert and polar regions is intriguing. Subsequent manipulative experiments may demonstrate that aerial diatoms are useful indicators of varying levels of UVR exposure.

Aerial diatoms also may have value as clean water indicators. In a world where it becomes increasingly difficult to find unpolluted, oligotrophic lentic and lotic waters, aerial habitats may become a good source of clean water indicators. The only serious problem we have had in using aerial habitats for this purpose has been the difficulty in characterizing the levels of nutrients and pollutants in the thin film of water in these habitats. Soil chemistry can be determined, but is not equivalent to water chemistry. Lithic pseudoaerial habitats are especially problematic because often the liquid is impossible to separate from the algal film on the rock, and chemistry of the algal film or rock itself is even more misleading than soil chemistry.

AERIAL HABITATS AS STUDY SYSTEMS

Euterrestrial diatom communities, particularly in arid environments, have low species richness. They are similar over fairly extensive regions, with overlap even between floras from different ecoregions. Terrestrial habitats do not have dramatic periodic changes, such as spates or spring turnover, and so the temporal component of change is depressed. Although seasonal fluctuations in the densities of terrestrial algae are evident (Bristol-Roach, 1927; Johansen *et al.*, 1993; Petersen, 1935), the composition of the community shows little seasonal change (Lund, 1945; Johansen & Rushforth, 1985), or at least much less change than typically observed in aquatic habitats (Johansen *et al.*, 1993). Thus, in a number of ways, terrestrial diatom communities are simpler than the diatom communities of aquatic habitats.

Because of this simplicity, aerial diatom floras may serve as better experimental systems upon which to conduct manipulative experiments. Treatments applied to terrestrial habitats are in many ways easier to do, easier to confine to a given site, and less prone to natural disturbances than treatments applied to lakes and rivers. The low species diversity of aerial habitats is desirable because community changes are clearer and easier to comprehend than in aquatic habitats where stochastic factors are more prevalent. Manipulative experiments could be conducted to test the specific effects of increasing nutrients, additional moisture, UVR reduction, contamination with pesticides and herbicides, and changing pH. Such experiments might increase our ability to delimit indicators and to determine what environmental factors are most important in controlling species distribution.

Summary

The diatom community found in aerial habitats is distinctive and usually dominated by obligately aerial taxa. Exposure to desiccation, high oxygen levels, and low nutrient levels are likely factors which exclude aquatic species from aerial habitats, while periodicity of desiccation, pH, light intensity, and substrate type are likely the most important parameters in determining which aerial taxa occur at a given aerial site. Linkages between temperate aerial diatom species composition and polar diatom species distribution have been observed, and demonstrates that the environmental factors limiting known aerophilic taxa to aerial habitats may include a suite of environmental conditions apart from exposure to desiccation.

Aerial diatoms may have utility as indicators of UVR and/or low nutrient levels. Aerial diatom communities are less diverse and more temporally simple than most aquatic habitats, and thus may make good study systems for manipulative ecological research.

Given the limited extent of our understanding of aerial algae, and the potentially unique research opportunities they present, further study of euaerial, terrestrial and pseudoaerial diatoms is warranted.

Acknowledgments

The author would like to thank the many reserchers that responded to my call for references on aerial diatoms. Although the brevity of this chapter precluded the use of most of these papers, the review was better informed because of them.

References

Anderson, D. C., & Rushforth, S. R. (1976). The cryptogamic flora of desert soil crusts in southern Utah. *Nova Hedwigia*, **28**, 691–729.

Beger, H. (1927). Beiträge zur Ökologie und Soziologie der luftlebigen (atmophytischen) Kieselalgen. *Berichte der Deutschen Botanischen Gesellschaft*, **45**, 385–407.

(1928). Atmosphytische Moosdiatomeen in den Alpen. *Vierteljahrsschrift der Naturforschenden Gesellschaft in Zürich*, **73**(**Beib. 15**), 382–404.

Bock, W. (1963). Diatomeen extrem trockener Standorte. *Nova Hedwigia*, **5**, 199–254 (+3 plates).

Bothwell, M. L., Sherbot, D., Roberge, A. C., & Daley, R. J. (1993). Influence of natural ultraviolet radiation on lotic periphytic diatom community growth, biomass accrual, and species composition: Short-term versus long-term effects. *Journal of Phycology*, **29**, 24–35.

Bristol-Roach, B. M. (1927). On the algae of some normal English soils. *Journal of Agricultural Science*, **17**, 563–88.

Camburn, K. E. (1982). Subaerial diatom communities in eastern Kentucky. *Transactions of the American Microscopical Society*, **101**, 375–87.

Carter, J. (1971). Diatoms from the Devil's Hole Cave, Fife, Scotland. *Nova Hedwigia*, **21**, 657–81.

Douglas, M. S. V., & Smol, J. P. (1995). Periphytic diatom assemblages from high arctic ponds. *Journal of Phycology*, **31**, 60–9.

Ettl, H., & Gärtner, G. (1995). *Syllabus der Boden-, Luft- und Flechtenalgen*. Stuttgart: Gustav Fischer Verlag.

Hayek, J. M. W., & Hulbary, R. L. (1956). A survey of soil diatoms. *Proceedings of the Iowa Academy of Sciences*, **63**, 327–38.

Hoffman, L. (1989). Algae of terrestrial habitats. *The Botanical Review*, **55**, 77–105.

Hustedt, F. (1942). Aërofile Diatomeen in der nordwestdeutschen Flora. *Berichte der Deutschen Botanischen Gesellschaft*, **40**, 55–73.

(1949). Diatomeen von der Sinai-Halbinsel und aus dem Libanon-Gebiet. *Hydrobiolgia*, **2**, 24–55.

Ito, Y., & Horiuchi, S. (1991). Distribution of living terrestrial diatoms and its application to the paleoenvironmental analyses. *Diatom*, **6**, 23–44.

Johansen, J. R. (1993). Cryptogamic crusts of semiarid and arid lands of North America. *The Journal of Phycology*, **29**, 140–7.

Johansen, J. R., Ashley, J., & Rayburn, W. R. (1993). Effects of rangefire on soil algal crusts in semiarid shrub-steppe of the lower Columbia Basin and their subsequent recovery. *Great Basin Naturalist*, **53**, 73–88.

Johansen, J. R., & Rushforth, S. R. (1985). Cryptogamic soil crusts: Seasonal variation in algal populations in the Tintic Mountains, Juab County, Utah. *Great Basin Naturalist*, **45**, 14–21.

Johansen, J. R., Rushforth, S. R., & Brotherson, J. D. (1981). Subaerial algae of Navajo National Monument, Arizona. *Great Basin Naturalist*, **41**, 433–9.

(1983a). The algal flora of Navajo National Monument, Arizona, U.S.A. *Nova Hedwigia*, **38**, 501–53.

Johansen, J. R., Rushforth, S. R., Obendorfer, R., Fungladda, N., & Grimes, J. (1983b). The algal flora of selected wet walls in Zion National Park, Utah, USA. *Nova Hedwigia*, **38**, 765–808.

Johansen, J. R., St. Clair, L. L., Webb, B. L., & Nebeker, G. T. (1984). Recovery patterns of cryptogamic soil crusts in desert rangelands following fire disturbance. *The Bryologist*, **87**, 238–43.

Krasske, (1932). Beiträge zur kenntnis der Diatomeenflora der Alpen. *Hedwigia*, **72**, 92–134 (+ 2 plates).

(1936). Die Diatomeenflora der Moosrasen des Wilhelmshöher Parkes. *Festschrift des Vereins für Naturkunde zu Kassel zum hundertjährigen Bestehen*, 151–64 (+ 3 tables).

(1948). Diatomeen tropischer Moosrasen. *Svensk Botanisk Tidskrift*, **42**, 404–41.

Lange-Bertalot, H., & Bonik, K. (1978). Zur systematisch-taxonomischen Revision des ökologisch interessanten Formenkeises um *Navicula mutica* Kützing. *Botanica Marina*, **21**, 31–7.

Lund, J. W. G. (1945). Observations on soil algae. I. The ecology, size and taxonomy of British soil diatoms. *The New Phytologist*, **44**, 56–110.

Metting, B. (1981). The systematics and ecology of soil algae. *The Botanical Review*, **47**, 195–312.

Novichkova-Ivanova, L. N. (1980). *Soil Algae of the Sahara–Gobi Desert Region*. Leningrad: Nauka. (In Russian).

Petersen, J. B. (1915). Studier over danske aërofile Alger. *Det Kongelige Danske Videnskabernes Selskabs Skrifter, Naturvidenskabelig og Mathematisk*, **12**(7), 271–380.

Petersen, J. B. (1928). The aërial algae of Iceland. *The Botany of Iceland* **2**(8), 325–447.

(1935). Studies on the biology and taxonomy of soil algae. *Danske Botanisk Arkiv* **8**(9), 1–180.

Reichardt, E. (1985). Diatomeen an feuchten Felsen des Südlichen Frankenjuras. *Berichte Bayerische Botanische Gesellschaft*, **56**, 167–87.

Round, F. E., Crawford, R. M., & Mann, D. G. (1990). *The Diatoms*. Cambridge: Cambridge University Press.

Rushforth, S. R., Kaczmarska, I., & Johansen, J. R. (1984). The subaerial diatom flora of Thurston Lava Tube, Hawaii. *Bacillaria*, **7**, 135–57.

Starks, T. L., Shubert, L. E., & Trainor, F. R. (1981). Ecology of soil algae: a review. *Phycologia*, **20**, 65–80.

Stoermer, E. F. (1962). Notes on Iowa diatoms. II. Species distribution in a subaerial habitat. *Iowa Academy of Science, Proceedings*, **69**, 87–91.

Van Dam, H., Mertens, A., & Sinkeldam, J. (1994). A coded checklist and ecological indicator values of freshwater diatoms from the Netherlands. *Netherlands Journal of Aquatic Ecology*, **28**, 117–33.

J. R. JOHANSEN

Van Landingham, S. L. (1964). Diatoms from Mammoth Cave, Kentucky. *International Journal of Speleology*, 1, 517–39.

(1966). Three new species of *Cymbella* from Mammoth Cave, Kentucky. *International Journal of Speleology*, 2, 133–6.

(1967). A new species of *Gomphonema* (Bacillariophyta) from Mammoth Cave, Kentucky. *International Journal of Speleology*, 2, 405–6.

Part IV
Diatoms as indicators in marine and estuarine environments

13 Diatoms as indicators of coastal paleo-environments and relative sea-level change

LUC DENYS AND HEIN DE WOLF

Introduction

Since nineteenth century naturalists identified salinity as a major determinant of diatom distribution, the remains of these organisms have become popular paleoenvironmental indicators for coastal deposits. A variety of problems in coastal geology were tackled using diatom-based methods, covering fields such as stratigraphy, the study of coastal processes, paleogeography, sea-level and climate change, tectonics, natural hazard assessment and archeology. This review highlights some of the major prospects and problems of paleoenvironmental diatom research on former depositional environments and sedimentation conditions in the coastal zone, as well as its contribution to the study of relative sea-level change and other processes affecting coastal genesis. For several reasons, e.g., the early recognition of important sea-level variations and coastline changes relating to glacial/interglacial cycles, the projected impact of possible future sea-level rise on coastal lowlands, the comparability of the fossil biotic record to contemporaneous observations, and the development of high-resolution dating methods – such research has focused mainly on the Quaternary and the Holocene in particular. Relatively few studies reached further back in time (e.g., Burckle & Akiba, 1978; Harwood, 1986; Pickard *et al.*, 1986; Tawfik & Krebs, 1995). This account therefore also deals primarily with the most recent geological time window, where techniques and applications are most refined. Intended as a brief introduction only, completeness is not attempted. Some closely related topics, such as the ecology of marine–littoral diatoms, salinity calibration, estuarine settings and archeological contexts are treated more in detail elsewhere in this volume. Analytical methodology, extensively discussed by e.g., Palmer and Abbott (1986), Denys (1984), Battarbee (1986), and Gasse (1987), is not covered here, nor is the use of diatoms as biostratigraphic 'guide fossils'.

Identification and reconstruction of the paleoenvironment

Coastal geologists usually turn to diatoms to find out how particular deposits formed. Yet, the mosaic of depositional environments represented in the coastal zone and their highly dynamic character may easily turn this 'simple' question into a quest. The reasons why diatoms can nevertheless efficiently provide an answer are well-known: ubiquitous occurrence in vast numbers, high species diversity and niche specificity, substantial preservation potential of taxon-distinctive remains, and rather straightforward quantitative assessment of assemblages (Palmer & Abbot, 1986; Denys, 1984; Battarbee, 1986). The strong relationship between diatom assemblage composition and salinity along the entire gradient from fresh to hypersaline (Kolbe, 1927; Hustedt, 1957; Simonsen, 1962; Ehrlich, 1975), the substrate dependence and tidal zonation of diatom communities (e.g., Brockmann, 1950; Simonsen, 1962; Hendey, 1964), their potential to indicate sediment input from marine or fluviatile systems, and the acute sensitivity of these organisms to numerous additional environmental factors (such as hydrodynamic conditions, water and soil chemistry) further explain why diatoms are especially useful microfossils for characterizing former coastal environments. As an additional bonus, the species composition of marine–littoral diatom communities are surprisingly similar throughout the world (McIntire & Moore, 1977; Cook & Whipple, 1982). In the following sections, the three concepts used to relate fossil assemblages to specific coastal settings are outlined.

USING CONTEMPORARY ECOLOGICAL INFORMATION

Generalized autecology of species and species groups The classical method of inferring environmental conditions from fossil diatoms consists of an analysis of assemblage composition and consideration of the relevant autecological characteristics of the taxa present. Evidently, this means that taxonomic precision should be given due respect (e.g., Birks, 1994). For coastal studies in particular, this is quite demanding because taxonomy is still under development for marine–littoral diatoms, and species diversity can be extremely high. In the past, opinions differed on the degree to which abundances of individual taxa should be accounted for, and how these should be estimated (e.g., Mölder, 1943a; van der Werff, 1958; Andrews, 1972). Such decisions should be made for each investigation separately and be based on the preservation state of the assemblages as well as on the goals of the study. Qualitative (presence/absence), semi-quantitative (estimated abundance classes) and quantitative methods (determination of relative abundance or concentration by counting) all have their specific advantages and shortcomings. The shift from data evaluation by expert judgment to more objective methods has, however, directed attention to statistically reliable and reproducible approaches which are now acknowledged to be most suited for in-depth paleoenvironmental reconstructions.

Essential to the species-oriented approach is that taxa are placed in general autecological classifications, allowing the calculation of stratigraphic spectra which reflect the behavior of species groups with similar character or requirements. Implicitly, the existence of general ecological borderlines and sufficient knowledge of habitat requirements are acknowledged. To some extent, such

L. DENYS AND H. DE WOLF

borders can be drawn, but it should be noted that these remain trajectories of gradual species turnover and are of relative nature. Many taxa have a large adaptability to varying conditions, e.g., in salinity or substrate. Most commonly used in coastal studies are salinity (halobion) and life-form classifications, but relations to other habitat characteristics (such as pH, trophic conditions, occurrence along the tidal gradient or tolerance to desiccation) and information on the preservation state of the assemblage (fragmentation, corrosion) may be relevant as well. In recent years, attempts have been made to synthesize the vast amount of available data on diatom autecological characteristics into codes to facilitate the calculation of various spectra. Following the footsteps of Lowe (1974) and Beaver (1981), van Dam *et al.* (1994) summarized much of the present knowledge on 948 taxa from fresh and brackish water in this way, whereas de Wolf (1982; 698 taxa) and Denys (1991; 980 taxa) focused on taxa from the marine-littoral in particular.

In a more comprehensive exercise, Vos & de Wolf (1988*a*, 1993*a*) advocated the combination of salinity range and life form as a paleoecological key; an approach corroborated by studies on recent ecology (e.g., Simonsen, 1969; Whiting & McIntire, 1985; Kosugi, 1987). For this purpose, they consider the relative abundance of 16 ecological groups of diatoms that occur in Dutch coastal deposits. For the major salinity compartments from fresh to marine, plankton (euplankton and tychoplankton, respectively), epiphytes and benthos (epipelon, epipsammon and aerophilous species) are distinguished (Fig. 13.1). The representation of these groups then allows the identification of deposits from specific paleoenvironments. Whereas some of these are quite representative for specific conditions (e.g., brackish epiphytes for lagoons, epipsammon for more sandy subtrates, or certain bloom-producing plankton species for the subtidal nearshore zone), this is much less so for other groups. The behavior of marine tychoplankton species, such as *Cymatosira belgica* Grun. for instance, may be determined largely by sediment influx (Vos & de Wolf, 1994; Denys, 1995). If carefully used, however, identification of most paleoenvironments is rather straightforward with this method. Illustrative applications include those of Plater and Shennan (1992), Vos and de Wolf (1993*b*, 1994) and Long and Hughes (1995).

Analogue matching and inference models Pertinent paleoenvironmental conclusions as well as overall or relative estimates of paleosalinity (e.g., Lange & Wulff, 1980) or paleotidal elevation can be made by applying generalized autecological concepts to fossil diatom assemblages. There is, however, considerable potential for even more refined reconstructions on the basis of comparative field observations in the marine–littoral.

Many workers have conducted synecological research on contemporary sediment assemblages from marine-littoral (sub)environments to identify analogues for diatom taphocoenoses (e.g., Mölder, 1943*b*; Brockmann, 1950; Grohne, 1959). Studies like those of Weiss *et al.* (1978), Ireland (1987), Campeau *et al.* (1995) and Hemphill-Haley and Lewis (1995) nevertheless indicate that

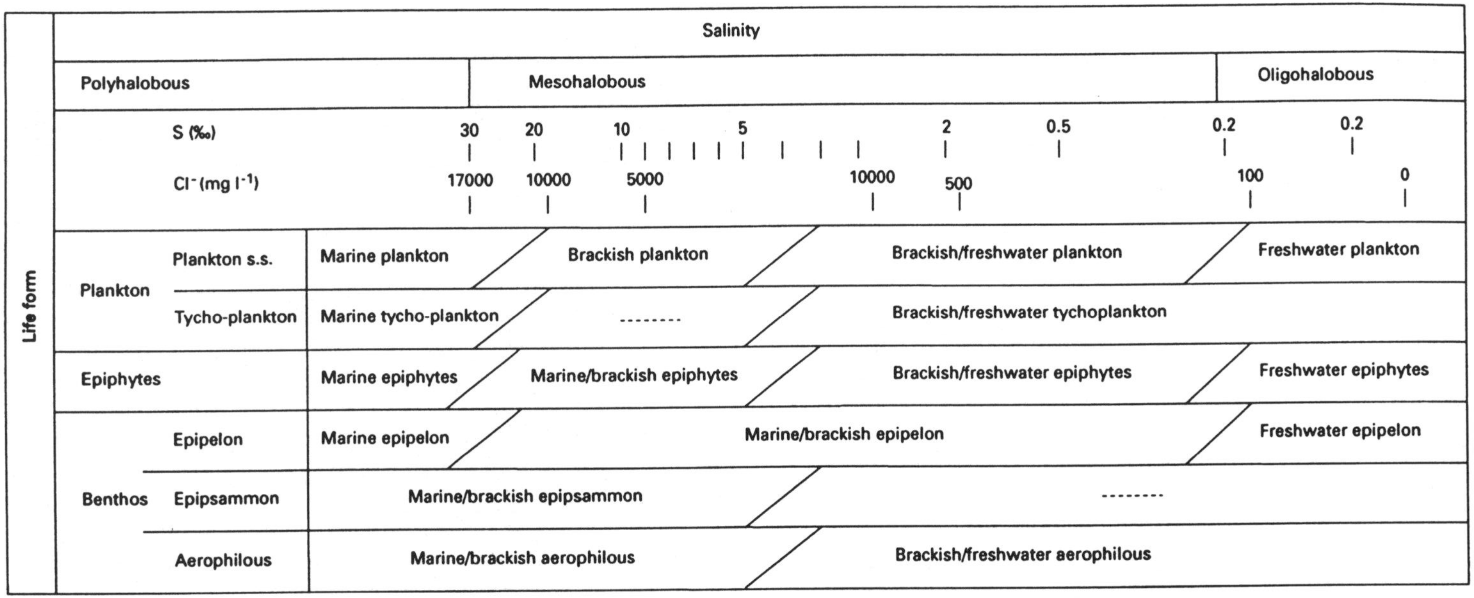

Salinity

	Polyhalobous	Mesohalobous				Oligohalobous	
S (‰)	30 20	10 5	2	0.5	0.2	0.2	0
Cl⁻ (mg l⁻¹)	17000 10000	5000	10000 500			100	

Life form

		Polyhalobous	Mesohalobous	Oligohalobous
Plankton	Plankton s.s.	Marine plankton	Brackish plankton	Brackish/freshwater plankton / Freshwater plankton
	Tycho-plankton	Marine tycho-plankton	Brackish/freshwater tychoplankton	
Epiphytes		Marine epiphytes	Marine/brackish epiphytes	Brackish/freshwater epiphytes / Freshwater epiphytes
Benthos	Epipelon	Marine epipelon	Marine/brackish epipelon	Freshwater epipelon
	Epipsammon	Marine/brackish epipsammon		
	Aerophilous	Marine/brackish aerophilous	Brackish/freshwater aerophilous	

Fig. 13.1. Ecological groups of diatoms occurring in Dutch coastal deposits (after Vos & de Wolf, 1993d).

there is still considerable need for further work here, especially in less investigated areas.

In contrast to the use of generalized autecological concepts, statistical inference models derived from actual contemporaneous species–environment relations can make optimal advantage of the ecological information that can be inferred from the assemblage, allowing quantitative inference of important parameters. A set of regional observations seems imperative in this, since hydrographic and ecological conditions differ between study areas, and community response also depends on the developmental stage or lateral extension of an ecotope (e.g., Laird & Edgar, 1992). Moreover, the paleo-setting and fossil assemblage composition must be consistent with the data on which the inference model is based. The latter requirement may often prove difficult to fulfil (cf. for instance the uncertainties on former hydrodynamic conditions and tidal amplitudes). Despite these limitations, efforts at developing this field further may be worthwhile. Juggins (1992), for instance, has shown that, using weighted average calibration, sensitive paleosalinity models can be developed for estuaries, which allow a better estimation in the brackish region than methods using halobion classifications of taxa.

Improved paleotidal reconstructions will probably be of even greater relevance to future coastal studies than more accurate paleosalinity estimates. General assemblage composition allows for the gross distinction of subtidal, lower and higher intertidal, as well as supratidal conditions in most cases. Attempts to infer paleotidal elevations more precisely on the basis of extant vertical diatom zonations have been limited though, mainly because rather few studies were directed explicitly to explore vertical distributions on soft sediments. Vertical zonation of assemblages in relation to tides results from the interaction of numerous environmental factors which are in some way connected to tidal level: light conditions, salinity, nature and stability of substrate, water movements, degree and duration of intermittent substrate desiccation, plant cover, nutrient supply, accumulation of toxic compounds, etc. (e.g., Admiraal, 1984). Spatial distribution patterns of diatom assemblages in the intertidal zone may therefore, be complex, patchy and unstable; gradations may occur, varying in vertical range according to conditions. Yet, some species and assemblages appear to be rather good indicators of specific tidal datums. Certain reference levels – mean high water spring, especially – can be recognized with reasonable precision (e.g., Haynes *et al.*, 1977; Shennan *et al.*, 1995). In a pilot study of tidal marshes in southern Oregon, Nelson and Kashima (1993) found that assemblage composition was most distinctive for the border zone between high marsh and upland, possibly allowing inference of this level with an accuracy of up to 0.2 m. Also, assemblages from mudflats, low marshes and high marshes could generally be distinguished as well. Similar observations indicating distinct marsh subzonation were reported by Hemphill-Haley (1995*a*) and Shennan *et al.* (1996). Paleotidal inference may, however, require a different approach in case species distributions show considerable overlapping or complex patterns and when transitions are more gradual (e.g.,

Oppenheim, 1988; Laird & Edgar, 1992). So far, no inference models have been developed for tidal elevation by means of weighted average calibration methods.

Taphonomic limitations As with all microfossils, there may be serious difficulties in tracing the ontogeny of fossil diatom assemblages.

Insufficient or differential preservation is common for diatom assemblages in coastal deposits. Marked differences in valve morphology, the degree of silicification, and even in the nature of the silica itself, contribute to a strongly varying resistance towards abrasion and dissolution – and hence preservation potential – for marine–littoral diatoms. The problem may be enhanced by more semi-terrestrial conditions to which high-intertidal and many supratidal facies have been exposed to. In taphocoenoses from salt marshes, reed and fen peats, often only a very small portion of the species which actually lived in these habitats remains (Brockmann, 1940; Sherrod *et al.*, 1989). Frequently, deposits from such environments become completely barren as soon as tidal influence is sufficiently reduced (Eronen *et al.*, 1987; Vos & de Wolf, 1994; Denys, 1995; Hemphill-Haley, 1995*a*). Assemblages reflecting dryer soil conditions, although quite distinctive (e.g., Sato *et al.*, 1983; Ihira *et al.*, 1985), preserve only rarely. Robust species are most likely to survive transport by tidal currents, which leads to preferential enrichment of such diatoms in high-intertidal sediments. If these species also have a morphology which enhances displacement (e.g., filamentous chains), and are most abundant in nearshore waters when extreme tides are highest and flood habitats which have low crops of autochthonous diatoms and are unfavorable for preservation, e.g., well-drained salt marshes and marginal peat areas, the apparent ecological character of the resulting assemblages will have little to do any more with the paleoenvironment. Taphocoenoses predominated by marine species as *Paralia* and *Pseudopodosira* are widespread examples of such processes (Denys, 1989, 1994; Vos & de Wolf, 1994; Hemphill- Haley, 1995*b*).

When multiple source areas contribute to a fossil assemblage, assessment of their particular roles is necessary to define the paleoenvironment at a given site by comparison to actual diatom distributions: autochthonous and allochthonous components need to be identified. The importance of recognizing transported material for tidal deposits has to be stressed. A number of guidelines have been set out for this purpose, including considerations on life form (benthic and sessile diatoms are less subjected to transport than tycho- and holoplanktonic ones), abundance and commonness of occurrence, valve preservation (corrosion, fragmentation), patterns in stratigraphic occurrence, paleogeographic setting, indicative lithological or paleontological criteria and sedimentary structures, ecological compatibility within the assemblage and consistency of paleoecological trends (Brockmann, 1940; König, 1953; Dahm, 1963; Simonsen, 1969; Mikkelsen, 1980; Beyens & Denys, 1982; Vos & de Wolf, 1988*a*, 1993*a*). Reference to present-day distributions may, of course, be equally helpful (e.g., Witkowsky, 1991; Juggins, 1992; Vos & de Wolf, 1993*b*; Hemphill-

Haley, 1995a). The nature and abundance of allochthonous material also provides additional insight into the paleogeographic setting and nature of the paleoenvironment (e.g., Brockmann, 1937; von der Brelie, 1956; Hallik, 1959; König, 1974; Barthsch-Winkler *et al.*, 1982; Vanhoorne & Denys, 1987; Vos & de Wolf, 1988b, 1994), and it may be argued that such material contributes substantially to the paleoenvironmental facies, reflecting time-averaged conditions and active transport mechanisms (Nelson & Kashima, 1993). Furthermore, small amounts of displaced marine species in non-marine deposits near the seaboard may provide elegant, and often accurately datable, evidence for periods of increased marine activity (e.g., Heyworth *et al.*, 1985; Denys, 1993). One should also bear in mind that quite different habitats may occur very close to each other in some coastal situations, giving rise to parautochthonous assemblages. Denys and Verbruggen (1989) discuss such a case where a hummocked bog was gradually replaced by mudflat.

EVIDENCE FROM THE STRATIGRAPHIC RECORD: FOSSIL DIATOM BIOFACIES

Quite often, recent analogues will not be available to guide paleoenvironmental interpretations. This may be when a coastal setting has changed considerably and the paleoecological diversity recorded by stratigraphy is no longer actually present, when species have become extinct or when taphonomic processes have made it impossible to infer the composition of the original sedimentary assemblages. Obviously, an approach using assumptions on species autecological data will also fail in cases where assemblages have been altered profoundly. In such instances, taphonomic problems may be avoided, in part, by exploring the stratigraphic sequence of the taphocoenoses and considering Walther's Law of Facies (Middleton, 1973). Facies sequence analysis, although widely used in pre-Quaternary stratigraphy and paleontology, has only rarely been applied to fossil diatom assemblages. Its use was demonstrated by Denys (1993) who characterized 11 biofacies, some with several subfacies, occurring in the Holocene deposits of the western Belgian coastal plain. Significant stratigraphical relations were identified by analyzing the frequencies of facies transitions. Together with additional information, such as data on species distributions and other environmental indications, these relations were used to attribute the various biofacies to distinct paleoenvironments. Besides highlighting some unexpected conclusions, e.g., the distinctive predominance of *Paralia* and *Pseudopodosira* in certain salt-marsh deposits, this approach resulted in a pragmatic tool for rapid paleoenvironmental assessment in the area, albeit an 'open-ended' one in need of careful handling. Also, less obvious relations were indicated between assemblage composition and the proximity of tidal channels or the degree of sediment supply. Besides the assessment of vertical facies relations, one may also consider horizontal patterns in fossil assemblage distribution from adjacent sites (Denys, 1994).

Inferring relative sea-level changes and coastal processes using diatoms

TRANSGRESSIONS, REGRESSIONS, SEA-LEVEL TENDENCIES AND SEA-LEVEL CURVES

Overall lateral shifts of the marine influence in a particular coastal area are known as transgressions and regressions. These processes can be determined from the local stratigraphies of a number of representative sites. At the site scale lateral shifts in the depositional facies and/or variations in salinity, wave and tidal energy regime, water depth or intertidal emergence, drainage conditions, and influx of marine material may all be indicative of a waning or rising marine influence. Marked directional changes of the paleoenvironmental evolution may result in stratigraphic boundary horizons: transgressive and regressive contacts. However, not all local records need to show the same picture simultaneously since many site specific factors, such as physiography, sediment supply and tidal variation, control the sense and timing of local developments. Stratigraphic evidence of such phenomena is likely to be preserved best near the inland limit of marine influence and where accretion predominates. Moreover, environmental conditions, and hence diatom assemblage composition, change most rapidly at higher tidal elevations. For diatoms, the transition from fresh to brackish water, changing substrate characteristics and transport by tidal currents appear to be most important (Whiting & McIntire, 1985; Kosugi, 1987). Lower tidal datums are much less well defined because accompanying ecological changes are less marked and sediment dynamics are higher, smearing out any potentially indicative assemblage distribution (Brockmann, 1937). Moreover, conditions may be so unfavourable for the growth of diatoms in the lower intertidal and adjacent subtidal zone that allochthonous material predominates the assemblages (Vos & de Wolf, 1988a; Hemphill-Haley, 1995a).

Studies of relative sea-level change attempt to quantify the magnitude and timing of transgressive and regressive events. Two different, hopefully complementary, routes may be followed: either inference of former sea-level stands from a variety of potential sea-level indicators – actually referring to tidal datums in most cases (van de Plassche, 1986) – resulting in shoreline displacement or time/altitude 'curves' (bands) of sea-level change, or, by qualitative analysis of 'sea-level tendencies' through time (e.g., Shennan *et al.*, 1983). The first method requires reliable sea-level index points for which the relation to a reference tide level and the age have to be obtained (Fig. 13.2). The second method avoids the demands of altitudinal precision required for constructing sea-level curves, the difficulties of relating stratigraphic phenomena to the process of relative sea-level change and the problems following from 'black-box' variables (such as changes in paleotidal range, which have been rarely quantified), by an objective study of timing and occurrence of transgressive and regressive phenomena (Fig. 13.2). Here, relative sea-level information is

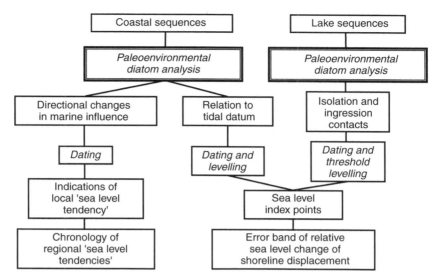

Fig. 13.2. Role of diatom research in sea level studies.

derived from the altitudinal gradients shown by indicative beds (Haggart, 1986) or the correlation of chronologies for different regions. A regional record of sea-level tendency requires numerous data-points, however. In both cases, diatoms can be used to identify the stratigraphic levels where marine influence increases or wanes and to check the value of presumed sea-level indications.

Studies of relative sea-level and coastal evolution are always heavily intertwined because sea-level change is only one of the many processes that determine the shaping of coastal areas and its effects can be isolated only with great difficulty. Obviously, the scope of paleoenvironmental diatom analysis therefore reaches beyond the topics treated below.

DIATOM CONTRIBUTIONS TO SEA-LEVEL TENDENCY ASSESSMENT

The earliest finds of deposits containing marine diatoms at inland sites (e.g., Harting, 1852; Gregory, 1855) were considered obvious evidence for changed coastline positions. In numerous further studies, changes in fossil diatom assemblages documented local or regional transgressive and regressive events associated with relative sea-level change or other coastal processes, such as barrier formation, infilling, tidal variations, etc., by illuminating the timing and nature of environmental changes and their physiographic consequences. Complex stratigraphies with 'cyclic' alternations of marine, brackish and fresh or semi-terrestrial facies have been described with much greater detail than obtainable from lithological evidence alone (e.g., Sato *et al.*, 1983; Kumano *et al.*, 1984; Huault, 1985; Haggart, 1987; Sato & Kumano, 1986; Lortie & Dionne, 1990). Shallowing of coastal waters (Lortie, 1983) and environmental changes in coastal lagoons (e.g., Robinson, 1982; Ireland, 1987; Håkansson, 1988), tidal lochs (Buzer, 1981), and estuaries (e.g., Miller, 1964; Palmer, 1978; Wendker, 1990) resulting from various coastal processes have also been documented.

Most chronologies of sea-level tendency established so far are based chiefly on the alternation of siliciclastic and more organic (peaty) beds, respectively reflecting more and less marine-influenced local conditions. The lithostratigraphic boundaries between such beds – transgressive and regressive overlaps (Tooley, 1982) – can be mapped and dated, from which regional variations in the marine influence may be deduced (as indicated by Allen, 1995, relations to sea-level variations may be very complex, however). Diatom analysis is an efficient method to establish whether the overlaps truly reflect a change in the marine (tidal) influence and whether no erosion or substantial sedimentary hiatus occurred – in which case a date from the lithological change would not represent the onset or halting of local marine activity (Haggart, 1986; Shennan, 1986; Long, 1992). The sensitivity of diatom assemblages to more subtle changes in sedimentary conditions also allows them to register less marked or long-lived local tendency changes than only those resulting in overlaps. Detailed records of the marine activity in coastal and perimarine areas can be obtained by analysis of sequences showing less lithological differentiation, making improved correlations and firmer establishment of apparent tendencies possible (Denys, 1993, 1995).

SEA-LEVEL RECONSTRUCTION

Records from open coasts In relative sea-level studies, index points must be inferred with the highest possible accuracy. Often an altitudinal error of less than a few dm is necessary. In order to define a former sedimentation level relative to a tidal datum with this precision by means of a microfossil assemblage, the latter has to refer to a very narrow range of the local tidal gradient. Estimations of water depth for sublittoral environments by means of diatoms are approximative and only of use in some cases (e.g., Lortie & Guilbault, 1984; van der Valk, 1992). Sometimes, relations to light attenuation and substrate can be applied (Whitehead & McMinn, 1997). For intertidal to supratidal situations a reasonable precision can be obtained. An overlap, such as discussed in the previous section, can provide a good sea-level index point as well. This is only so, however, if it reflects a transition from a semi-terrestrial to a high intertidal situation or vice versa, as accurate information on the position of the high water mark only results if the local water depth was negligible. If peat accumulation occurred subaquatically, e.g., in a lagoon, its relation to a tidal datum becomes too uncertain. It is relatively straightforward to check this by means of diatom analysis (Fig. 13.3). Such analyses will also define the precise level which should be dated, for the onset of a tidal regime may not concur exactly with the lithostratigraphic overlap. These features contributed substantially to the application of diatom studies in sea-level research (cf. Robinson, 1993; Zong & Tooley, 1996).

L. DENYS AND H. DE WOLF

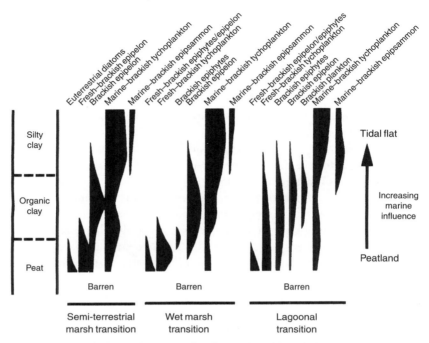

Fig. 13.3. Idealized scheme of some possible diatom assemblage shifts occurring at the transgressive overlap of a peat bed with tidal deposits under different drainage conditions. Accuracy of paleotidal inference decreases to the right. Note that the relative representation of different groups will vary according to local conditions and input of allochthonous material.

The most common process interfering with the above approaches is compaction, the amount of which is extremely difficult to estimate. Hardly ever can it be excluded, since the indicative stratum will generally be situated somewhere within a susceptible sedimentary sequence. Often organic marsh sediments or peaty horizons will be involved (also sought for as they allow precision dating by radiocarbon methods), furthering the problem. Only a few solutions are available. The occurrence of very thin basal peat strata, which developed in direct response to the sea-level rise on a consolidated substrate and were subsequently covered with intertidal sediments, probably offers the best opportunities to minimize compaction effects (van de Plassche, 1982), but again requires firm paleoenvironmental control. Denys and Baeteman (1995) used diatom evidence to select the most reliable index points from basal peats, transgressive overlaps involving transitions from semi-terrestrial to intertidal conditions (Fig. 13.3), in order to estimate the minimum height of local mean high water spring tides.

Regressive contacts and overlaps of intercalated or thicker basal peats usually yield less reliable sea-level index points because the range of possible compaction of the underlying marine and brackish sediments and/or peat will be greater. Nevertheless, within their associated error limits, they can still provide important conclusions on relative sea-level positions. Here also, diatom analysis provides an objective means of identifying the indicative horizons. For instance, Yokoyama *et al.* (1996) used the change from a marine/brackish assemblage to one with over 50% freshwater taxa in cores from the west coast of Kyushu to estimate the minimum height of the marine limit (approximating the highest position of mean high water), allowing relative mean sea level reconstructions and detection of hydro-isostatic effects.

Records from lake basins A straightforward method of reconstructing relative sea level, and hence former shoreline positions, with the help of diatoms can be applied where sedimentary basins occur in which seawater could enter only when passing over a certain threshold. In case such basins were once connected with the sea or even were part of the sea floor and have been isolated from the marine influence through crustal movements or, reversely, where ingression of freshwater lakes with seawater occurred as a result of relative sea-level rise, it is sufficient to determine the threshold elevation and the time at which isolation or ingression occurred to infer a former sea-level position, at least if local paleotidal amplitude is known (Fig. 13.2). In the diatom stratigraphy of such a basin, those levels where freshwater conditions begin or end mark the isolation or ingression event. A date from such a level yields an index point on a time/depth graph. Sea-level curves or shoreline displacement curves are obtained by combination of several such points from basins having different threshold elevations. The rationale for this type of analysis was developed early this century in Scandinavia (see Miller, 1986) in connection with the ontogeny of the Baltic and the isostatic uplift of the Scandinavian Shield since deglaciation. The method came to full exploitment with the development of radiocarbon dating (e.g., Alhonen *et al.*, 1978; Björck & Digerfeldt, 1982; Miller, 1982; Hafsten, 1983; Kjemperud, 1981*a*; Krzywinski & Stabell, 1984; Risberg *et al.*, 1996) and is now used also in other regions where Holocene isostatic rebound occurred (e.g., Williams, 1990; Pienitz *et al.*, 1991; Wasell & Håkansson, 1992; Anundsen *et al.*, 1994). Forsström *et al.* (1988) and de Wolf and Cleveringa (1995) discussed basin isolation/ingression sequences from the Eemian interglacial. Apart from the fact that the number of suitable basins providing comparable index points will be limited in areas within which differential movements remain negligible, most difficulties with this method arise from dating problems or changes in threshold elevation, e.g., by erosion. Especially when large and deep basins are involved, a certain lag between isolation or ingression and diatom response may be expected due to the euryhalinity of diatom species or physical processes such as the formation of a halocline or inflow of large amounts of freshwater (Eronen, 1974; Douglas *et al.*, 1996). In some cases, seepage or exceptional events may induce brackish conditions (e.g., Korhola, 1995). Also, it should be remembered that, in a closed basin, relative sea-level variations will be registered only if their amplitude is sufficient in relation to the altitude of the basin's threshold (de Vernal *et al.*, 1985) and if the record has not been eroded by e.g., subsequent marine incursions. Kjemperud (1981*b*) and

L. DENYS AND H. DE WOLF

Stabell (1985, 1987) discussed ecological changes and diatom response at isolation and ingression transitions in closed basins.

DIATOM STRATIGRAPHIES WITH ENVIRONMENTAL DISCORDANCIES

Not all diatom profiles from coastal deposits present gradual transitions from one paleoenvironment into another. In fact environmental discordancies frequently occur. These may be formed through natural phenomena or human interference.

Erosional unconformities, accompanied by more or less important stratigraphic hiatuses, are likely to be most common, but also most difficult to recognize. In high-energy sedimentary facies, such as of tidal channels, scouring features commonly occur. Unfortunately, diatom assemblages here tend to consist largely of transported valves, and are constantly subjected to reworking, so paleoenvironmental interpretation may not be conclusive. Characteristic sedimentary structures will be more indicative and may help to explain eventual changes in the diatom record. Where sedimentation conditions are more stable and autogenic assemblage successions occur, erosional surfaces may be identified by truncation of species curves and abrupt variations in valve concentrations, next to sedimentological characteristics. Plater and Poolton (1992) combined sediment luminescence techniques with diatom analysis to infer variations in sedimentation rates and a storm event in intertidal sediments.

Along the coast, high-energy events also bring about considerable displacement of water masses and sediment, including diatoms, from the more seaward area. In this way the contribution of epipsammic diatoms from sand flats and species living mostly on subtidal shoals (e.g., epiphytes) may increase temporarily. Often, enrichment with transport-resistent tychoplankton swept up by bottom currents is seen. As heavy storms tend to occur seasonally, deposits left by them may bear the print of the characteristic plankton assemblages of the moment as well. As such, sandy deposits occurring locally between fine-grained estuarine sediments and reed swamp peats in eastern Scotland, and identified as resulting from a major storm surge or tsunami on the North Sea, were found to contain a high frequency of *Paralia sulcata* (Ehr.) Cl. – a robust tychoplankton diatom which is abundant in the water column in the stormy season – together with sessile species from the sublittoral (Haggart, 1986, 1988). Similar observations were reported by Smith *et al.* (1992) and Hemphill-Haley (1995*b*). Hemphill-Haley (1996) discusses several case-studies where on-shore tsunami deposits were detected by means of diatom analysis. Storm activity may also be tracked from wave disturbance of offshore tidal rhytmites consisting of sand with *Delphineis minutissima* (Hust.) Simonsen and clay with *Cymatosira belgica* (de Wolf *et al.*, 1993).

Digging activities also result in discordant diatom stratigraphies. A good example is the so-called 'Spitwürfen' studied by Körber-Grohne (1967):

displaced sods of salt marsh soil. Along the North-Sea coasts, peat cutting for fuel or salt making was very widespread, resulting in extensively disturbed profiles. Diatom analysis may help in differentiating natural from such anthropogenic-related sequences (Denys & Verbruggen, 1989).

Finally, the application of diatom stratigraphies to detect abrupt tectonic events should be mentioned. Evidence for very sudden submergence of soils or emergence of intertidal deposits, as indicated by diatom records, combined with precise levelling, dating and stratigraphic control can be used to infer and map paleo-earthquakes (e.g., Bucknam *et al.*, 1992; Nelson & Kashima, 1993; Nelson *et al.*, 1996). Abrupt changes in elevation relative to tidal datums of at least several dm or transitions between assemblages from laterally non-adjacent tidal zones (e.g., from upland conditions to tidal flat without intervening salt marsh) may be interpreted similarly (Hemphill-Haley, 1995*b*; Shennan *et al.*, 1996).

Future developments

Although diatom analysis has become a standard method for interdisciplinary coastal studies with an eminent tradition in e.g., Europe, Japan and the USA, there is considerable potential for further development in large parts of the world, especially in many tropical and subtropical regions, as well as the Antarctic and Arctic.

Likewise, pre-Quaternary sea-level changes still present a major research area where diatoms may prove to be extremely useful; the more so as many of the diatoms encountered in Tertiary deposits have close extant relatives or even survived until the present. Moreover, facies analysis may be applied here to analyze paleoenvironmental conditions.

Recent datasets are being used increasingly to improve the detail and reliability of paleoenvironmental reconstructions, especially with regard to paleotidal conditions. Some other microfossils, foraminifera especially, may provide more accurate paleotidal information, but in part this may be because of differences in research emphasis. Even so, the diatom record may yield corroborative evidence or serves as a welcome subsitute. Although more precise diatom-inferred paleotidal elevation estimates are likely to contribute substantially to, for instance, paleoseismic studies, the expected gain in accuracy is largely overshadowed by additional error sources in other areas of relative sea-level research. Little is to be gained if the altitudinal errors of sea-level index points associated to sampling and, more importantly, compaction are not likewise reduced or even neglected, as is unfortunately still the case on occasion.

In most coastal areas, relative sea-level studies are plagued by considerable uncertainties on paleotidal amplitudes, which also constrain biotic inferences. Palmer and Abbott (1986) suggested a method to derive such amplitudes from diatom assemblages by comparing the elevation of coeval strata representing the upper and lower tidal limit in different cores taken along a littoral transect.

To date, this procedure has not been attempted and possible applications seem to be very limited since the method not only requires precise knowledge of compaction effects and spatial variation of paleotidal amplitude, but also hinges on the ability to pin-point the lower tidal limit. As mentioned above, there is little prospect for the latter; at this point the diatom record differs little from that of other paleoenvironmental indicators.

Finally, it should be noted that progress in coastal diatom paleoecology will greatly depend on further study of all aspects of recent diatom distributions and taxonomy, and of the sedimentation processes occurring near the seaboard. In turn, the time-perspective of the geological diatom record is likely to contribute significantly to our understanding of what is happening in some of the most dynamic and populated parts of the world.

Summary

Fossil diatom assemblages are widely used in paleoenvironmental reconstructions of coastal deposits. In this chapter, the use of autecological data, recent analogue assemblages and biofacies analysis for such purposes is briefly discussed with reference to taphonomic problems. The role of diatom analysis in relative sea-level research (the assessment of regional sea-level tendencies as well as the inference of former sea-level positions), and the study of former abrupt events (storm surges, earthquakes) and human activities in the coastal zone is reviewed. So far, applications have been directed mainly at problems of Quaternary and especially Holocene geology, where improvements in the field of paleotidal inference by means of diatoms are considered as an important research challenge for the immediate future. However, considerable potential appears to exist also for the study of pre-Quaternary sea-level changes.

References

Admiraal, W. (1984). The ecology of estuarine sediment-inhabiting diatoms. *Progress in Phycological Research*, **3**, 269–322.

Alhonen, P., Eronen, M., Nuñez, M., Salooma, R., & Uusinoka, R. (1978). A contribution to Holocene shore displacement and environmental development in Vantaa, South Finland: the stratigraphy of Lake Lammaslampi. *Bulletin of the Geological Society of Finland*, **50**, 69–79.

Allen, J. R. L. (1995). Salt-marsh growth and fluctuating sea level: implications of a simulation model for Flandrian coastal stratigraphy and peat-based sea-level curves. *Sedimentary Geology*, **100**, 21–45.

Andrews, G. W. (1972). Some fallacies on quantitative diatom paleontology. *Beihefte zu Nova Hedwigia*, **39**, 285–94.

Anundsen, K., Abella, S., Leopold, E., Stuiver, M., & Turner, S. (1994). Late-Glacial and early Holocene sea-level fluctuations in the Central Puget Lowland, Washington, inferred from lake sediments. *Quaternary Research*, **42**, 149–61.

Barthsch-Winkler, S., Ovenshine, A. T., & Kachadoorian, R. (1982). Holocene history of the estuarine area surrounding Portage, Alaska recorded in a 93 m core. *Canadian Journal of Earth Sciences*, **20**, 802–20.

Battarbee, R. W. (1986). Diatom analysis. In *Handbook of Holocene Paleoecology and Paleohydrology*, ed. B. E. Berglund, pp. 527–70. London: John Wiley.

Beaver, J. (1981). *Apparent Ecological Characteristics of some Common Freshwater Diatoms*. Don Mills: Ontario Ministry of Enviomnent.

Beyens, L., & Denys, L. (1982). Problems in diatom analysis of deposits: Allochthonous valves and fragmentation. *Geologie en Mijnbouw*, **61**, 159–62.

Birks, H. J. B. (1994). The importance of pollen and diatom taxonomic precision in quantitative paleoenvironmental reconstructions. *Review of Paleobotany and Palynology*, **83**, 107–17.

Björck, S., & Digerfeldt, G. (1982). Late Weichselian shore displacement at Hunneberg, southern Sweden, indicating complex uplift. *Geologiska Föreningens i Stockholm Förhandlingar*, **104**, 132–55.

Brockmann, C. (1937). Küstennähe und küstenferne Sedimente in der Nordsee. *Abhandlungen des Naturwissenschaftlichen Vereins zu Bremen*, **30**, 78–89.

(1940). Diatomeen als Leitfossilien in Küstenablagerungen. *Westküste*, **2**, 150–81.

(1950). Die Watt-Diatomeen des schleswig-holsteinischen Westküste. *Abhandlungen der Senckenbergischen Naturforschenden Gesellschaft*, **478**, 1–26, 6 pl.

Bucknam, R. C., Hemphill-Haley, E., & Leopold, E. B. (1992). Abrupt uplift within the past 1700 years at Southern Puget Sound, Washington. *Science*, **258**, 1611–14.

Burckle, L. H., & Akiba, F. (1978). Implications of late Neogene fresh-water sediment in the Sea of Japan. *Geology*, **6**, 123–7.

Buzer, J. S. (1981). Diatom analysis of sediments from Lough Ine, Co. Cork, Southwest Ireland. *New Phytologist*, **89**, 511–33.

Campeau, S., Héquette, A., & Pienitz, R. (1995). The distribution of modern diatom assemblages in coastal sedimentary environments of the Canadian Beaufort Sea: An accurate tool for monitoring coastal changes. *Proceedings of the 1995 Canadian Coastal Conference*, vol. 1, pp. 105–16. Dartmouth: Canadian Coastal Science and Engineering Association.

Cook, L. L., & Whipple, S. A. (1982). The distribution of edaphic diatoms along environmental gradients of a Louisiana salt marsh. *Journal of Phycology*, **18**, 64–61.

Dahm, H-D. (1963). Diagnose von Brackwassersedimenten mit Hilfe der Diatomeen. *Fortschritte in der Geologie von Rheinland und Westfalen*, **10**, 95–106.

Denys, L. (1984). Diatom analysis of coastal deposits: methodological aspects. *Bulletin van de Belgische Vereniging voor Geologie*, **93**, 291–5.

(1989). Observations on the transition from Calais deposits to surface peat in the western Belgian coastal plain – results of a paleoenvironmental diatom study. *Professional Paper Belgische Geologische Dienst*, **241**, 20–43.

(1991). A check-list of the diatoms in the Holocene deposits of the western Belgian coastal plain with a survey of their apparent ecological requirements. *Professional Paper Belgische Geologische Dienst*, **246**, 1–41.

(1993). Paleoecologisch diatomeeënonderzoek van de holocene afzettingen in de westelijke Belgische kustvlakte. Doctoral thesis. Antwerpen: Universitaire Instelling.

(1994). Diatom assemblages along a former intertidal gradient: A paleoecological study of a Subboreal clay layer (western coastal plain, Belgium). *Netherlands Journal of Aquatic Ecology*, **28**, 85–96.

(1995). The diatom record of a core from the seaward part of the coastal plain of Belgium. In *Proceedings of the Thirteenth International Diatom Symposium*, ed. M. Marino & M. Montresor, pp. 471–87. Bristol: Biopress.

Denys, L., & Baeteman, C. (1995). Holocene evolution of relative sea-level and local mean high water spring tides in Belgium – a first assessment. *Marine Geology*, **124**, 1–19.

Denys, L., & Verbruggen, C. (1989). A case of drowning – the end of Subatlantic peat growth and related paleoenvironmental changes in the lower Scheldt basin (Belgium) based on diatom and pollen analysis. *Review of Paleobotany and Palynology*, **59**, 7–36.

de Vernal, A., Lortie, G., Larouche, A., Scott, D. B., & Richard, P. J. H. (1985). Evolution d'un milieu littoral et remontée du niveau relatif de la mer à l'Holocène supérieur au nord de l'île du Cap Breton, Nouvelle-Ecosse. *Canadian Journal of Earth Sciences*, **22**, 315–23.

L. DENYS AND H. DE WOLF

de Wolf, H. (1982). Method of coding of ecological data from diatoms for computer utilization. *Mededelingen Rijks Geologische Dienst*, **36**, 95–8.

de Wolf, H., & Cleveringa, P. (1995). Eemian diatom floras in the Amsterdam glacial basin. In *Proceedings of the Thirteenth International Diatom Symposium*, ed. M. Marino & M. Montresor, pp. 489–505. Bristol: Biopress Ltd.

de Wolf, H., van der Borg, K., Cleveringa, P., de Groot, T. A. M., Meijer, T., & Westerhoff, W. E. (1993). Tidal deposits and storm depositional sequences in the western Netherlands: A paleoenvironmental reconstruction. *Terra Nova* **5** *Abstract Supplement* **1**, 616.

Douglas, M. S. V., Ludlam, S., & Freeney, S. (1996). Changes in diatom assemblages in Lake C2 (Ellesmere Island, Arctic Canada): Response to basin isolation from the sea and to other environmental changes. *Journal of Paleolimnology*, **16**, 217–26.

Ehrlich, A. (1975). The diatoms from the surface sediments of the Bardawil Lagoon (Northern Sinai) – Paleoecological significance. *Beihefte zu Nova Hedwigia*, **53**, 253–77, 3 pl.

Eronen, M. (1974). The history of the Litorina Sea and associated holocene events. *Commentationes Physico-Mathematicae*, **44**, 79–195, diagr.

Eronen, M., Kankainen, T., & Tsukada, M. (1987). Late Holocene sea-level record in a core from the Puget Lowland, Washington. *Quaternary Research*, **27**, 147–59.

Forsström, L., Aalto, M., Eronen, M., & Grönlund, T. (1988). Stratigraphic evidence for Eemian crustal movements and relative sea-level changes in eastern Fennoscandia. *Paleogeography, Paleoclimatoloy, Paleoecology*, **68**, 317–35.

Gasse, F. (1987). Les diatomées, instrument de reconstitution des paléoenvironnements. In *Géologie de la préhistoire: méthodes, techniques, applications*, réd. J. C. Miskovsky, pp. 651–67. Paris: Association pour l'Etude de l'Environnement Géologique de la Préhistoire.

Gregory, W. (1855). On a post-Tertiary lacustrine sand containing diatomaceous exuviae from Glenshira near Inverary. *Quarterly Journal of Microscopical Sciences*, **3**, 30–43.

Grohne, U. (1959). Die Bedeutung der Diatomeen zum Erkennen der subfossilen Vegetation höherer Pflanzen in Marschablagerungen. *Zeitschrift der deutschen Geologischen Gesellschaft*, **111**, 13–28.

Hafsten, U. (1983). Shore-level changes in South Norway during the last 13,000 years, traced by biostratigraphical methods and radiocarbon dating. *Norsk Geografisk Tidsskrift*, **37**, 63–79.

Haggart, B. A. (1986). Relative sea-level change in the Beauly Firth, Scotland. *Boreas*, **15**, 191–207.

(1987). Relative sea-level changes in the Moray Firth area, Scotland. In *Sea-level Changes*, ed. I. Shennan & M. J. Tooley, pp. 67–108. Oxford: Basil Blackwell.

(1988). The stratigraphy, depositional environment and dating of a possible tidal surge deposit in the Beauly Firth Area, Northeast Scotland. *Paleogeography, Paleoclimatology, Paleoecology*, **66**, 215–30.

Håkansson, H. (1988). Diatom analysis at Skateholm-Järavallen, Southern Sweden. *Acta Regiae Societatis Humaniorum Litterarum Lundensis*, **79**, 39–45.

Hallik, R (1959). Diatomeen als Anzeiger des Sedimenttransportes in der Unterelbe. *Zeitschrift der deutschen Geologischen Gesellschaft*, **111**, 29–32.

Harting, P. (1852). De bodem onder Amsterdam onderzocht en beschreven. *Verhandelingen Koninklijk-Nederlands Instituut voor Wetenschappen, Letterkunde en Schone Kunsten Klasse I*, 3e reeks, **5**, 73–232.

Harwood, D. M. (1986). Oldest record of Cainozoic glacial-marine sedimentation in Antarctica (31 Myr): results from MSSTS-1 drill-hole. *South African Journal of Science*, **82**, 516–19.

Haynes, J. R., Kiteley, R. J., Whatley, R. C., & Wilks, P. J. (1977). Microfaunas, microfloras and the environmental stratigraphy of the Late Glacial and Holocene in Cardigan Bay. *Geological Journal*, **12**, 129–58.

Hemphill-Haley, E. (1995a). Intertidal diatoms from Willapa Bay, Washington: Application to studies of small-scale sea-level changes. *Northwest Science*, **69**, 29–45.

(1995b). Diatom evidence for earthquake-induced subsidence and tsunami 300 yr ago in southern coastal Washington. *Geological Survey of America, Bulletin*, **107**, 367–78.

(1996). Diatoms as an aid in identifying late-Holocene tsunami deposits. *The Holocene*, **6**, 439–48.

Hemphill-Haley, E., & Lewis, R. C. (1995). *Distribution and Taxonomy of Diatoms (Bacillariophyta) in Surface Samples and a Two-meter Core from Winslow Marsh, Bainbridge Island, Washington*. Open-File Report 95–833. Menlo Park: US Geological Survey.

Hendey, N. I. (1964). *An Introductory Account of the Smaller Algae of British Coastal Waters*, Part V, *Bacillariophyceae (diatoms)*. London: HMSO.

Heyworth, A., Kidson, C., & Wilks, P. (1985). Late-Glacial and Holocene sediments at Clarach Bay, near Aberystwyth. *Journal of Ecology*, **73**, 459–80.

Huault, M-F. (1985). Apport des diatomées à la reconstitution des paléoenvironnements: L'exemple du Marais Vernier lors de la transgression flandrienne. *Bulletin de l'Association française pour l'Etude du Quaternaire*, **1985/4**, 209–17.

Hustedt, F. (1957). Die Diatomeenflora des Fluszsystems der Weser im Gebiet der Hansestadt Bremen. *Abhandlungen des Naturwissenschaftlichen Vereins zu Bremen*, **34**, 181–440.

Ihira, M., Maeda, Y., Matsumoto, E., & Kumano, S. (1985). Holocene sedimentary history of some coastal plains in Hokkaido, Japan. 2. Diatom assemblages of the sediments from Kushiro Moor. *Japanese Journal of Ecology*, **35**, 199–205.

Ireland, S. (1987). The Holocene sedimentary history of the coastal lagoons of Rio de Janeiro State, Brazil. In *Sea-level Changes*, ed. I. Shennan & M. J. Tooley, pp. 25–66. Oxford: Basil Blackwell.

Juggins, S. (1992). Diatoms in the Thames estuary, England: Ecology, paleoecology and salinity transfer function. *Bibliotheca Diatomologica*, **25**, 1–216.

Kjemperud, A. (1981a). A shoreline displacement investigation from Frosta in Trondsheimfjorden, Nord-Trøndelag, Norway. *Norsk Geologisk Tidsskrift*, **61**, 1–15.

(1981b). Diatom changes in sediments of basins possessing marine/lacustrine transitions in Frosta, Nord-Trøndelag, Norway. *Boreas*, **10**, 27–38.

Kolbe, R. W. (1927). Zur Ökologie, Morphologie und Systematik der Brackwasser-Diatomeen. *Pflanzenforschung*, **7**, 1–145, 3 pl.

König, D. (1953). Diatomeen aus dem Eem des Treenetales. *Schriften des Naturwissenschaftliches Vereins für Schleswig-Holstein*, **26**, 124–32.

(1974). Subfossil diatoms in a former tidal region of the Eider (Schleswig-Holstein). *Beihefte zu Nova Hedwigia*, **45**, 259–74, 4 pl.

Körber-Grohne, U. (1967). *Geobotanische Untersuchungen auf der Feddersen Wierde*. Wiesbaden: Franz Steiner Verlag.

Korhola, A. (1995). Holocene *Nitzschia scalaris* (Ehr.) W. Smith blooms in the coastal glo-lakes of Finland. In *Proceedings of the Thirteenth International Diatom Symposium*, ed. M. Marino & M. Montresor, pp. 521–9. Bristol: Biopress.

Kosugi, M. (1987). Limiting factors on the distribution of benthic diatoms in coastal regions – salinity and substratum. *Diatom*, **3**, 21–31.

Krzywinski, K., & Stabell, B. (1984). Late Weichselian sea-level changes at Sotra, Hordaland, western Norway. *Boreas*, **13**, 159–202.

Kumano, S., Sekiya, K., & Maeda, Y. (1984). Holocene sedimentary history of some coastal plains in Hokkaido, Japan. 1. Diatom assemblages of the sediments from Kutcharo Lake. *Japanese Journal of Ecology*, **34**, 389–96.

Laird, K., & Edgar, R. K. (1992). Spatial distribution of diatoms in the surficial sediments of a New England salt marsh. *Diatom Research*, **7**, 267–79.

Lange, D., & Wulff, B. (1980). Diatomeenuntersuchungen am Stechrohrkern AB 3 vom Westrand des Arkona-Beckens. *Beiträge zur Meereskunde*, **44/45**, 75–88.

Long, A. (1992). Coastal responses to changes in sea level in the East Kent Fens and southeast England, U. K. over the last 7500 years. *Proceedings of the Geologists' Association*, **103**, 187–99.

Long, A., & Hughes, P. D. M. (1995). Mid- and late-Holocene evolution of the Dungeness foreland, UK. *Marine Geology*, **124**, 253–71.

Lortie, G. (1983). Les diatomées de la mer de Goldthwait dans la région de Rivière-du-Loup, Québec. *Géographie Physique et Quaternaire*, **37**, 279–96.

Lortie, G., & Dionne, J-C. (1990). Analyse préliminaire des diatomées de la coupe de Montmagny, côte sud du Saint-Laurent, Québec. *Géographie Physique et Quaternaire*, **44**, 89–95.

Lortie, G., & Guilbault, J. P. (1984). Les diatomées et les foraminifères de sédiments marins post-glaciaires du Bas-Saint-Laurent (Québec): une analyse comparée des assemblages. *Le Naturaliste Canadien*, **111**, 297–310.

Lowe, R. L. (1974). *Environmental Requirements and Pollution Tolerance of Freshwater Diatoms*. EPA Report 670/4–74–005. Cincinnati: EPA.

McIntire, C. D., & Moore, W. W. (1977). Marine littoral diatoms: ecological considerations. In *The Biology of Diatoms*, ed. D. Werner, pp. 333–71. Berkeley: University of California Press.

Middleton, G. V. (1973). Johannes Walther's law of the correlation of facies. *Geological Society of America, Bulletin*, **84**, 979–88.

Mikkelsen, N. (1980). Experimental dissolution of Pliocene diatoms. *Nova Hedwigia*, **33**, 893–911.

Miller, U. (1964). Diatom floras in the Quaternary of the Göta River valley. *Sveriges Geologiska Undersökning Ser. Ca*, **44**, 1–69, 8 pl.

(1982). Shore displacement and coastal dwelling in the Stockholm region during the past 5000 years. *Annales Academiae Scentiarum Fennicae A. III.*, **134**, 185–211.

(1986). Ecology and paleoecology of brackish water diatoms with special reference to the Baltic Basin. In *Proceedings of the Eighth International Diatom Symposium*, ed. M. Ricard, pp. 601–611. Koenigstein: Koeltz.

Mölder, K. (1943a). Studien über die Ökologie und Geologie der Bodendiatomeen in der Pojo-Bucht. *Annales Botanici Societatis zoologicae-botanicae Fennicae Vanamo*, **18**, 1–204.

(1943b). Rezente Diatomeen in Finland als Grundlage quartärgeologischen Untersuchungen. *Geologie der Meere und Binnengewässer*, **6**, 148–240.

Nelson, A. R., & Kashima, K. (1993). Diatom zonation in southern Oregon tidal marshes relative to vascular plants, foraminifera, and sea level. *Journal of Coastal Research*, **9**, 673–97.

Nelson, A. R., Jennings, A. E., & Kashima, K. (1996). An earthquake history derived from stratigraphic and microfossil evidence of relative sea-level change at Coos Bay, southern coastal Oregon, USA. *Geological Survey of America, Bulletin*, **108**, 141–54.

Oppenheim, D. R. (1988). The distribution of epipelic diatoms along an intertidal shore in relation to principal physical gradients. *Botanica Marina*, **31**, 65–72.

Palmer, A. J. M. (1978). Diatom stratigraphy of Basin Head Harbour, Prince Edward Island. *Proceedings of the Nova Scotian Institute of Science*, **28**, 201–15.

Palmer, A. J. M., & Abbott, W. H. (1986). Diatoms as indicators of sea-level change. In *Sea-level Research, a Manual for the Collection and Evaluation of Data*, ed. O. van de Plassche, pp. 457–89. Norwich: Geo Books.

Pickard, J., Adamson, D. A., Harwood, D. M., Miller, G. H., Quilty, P. G., & Dell, R. K. (1986). Early Pliocene marine sediments in the Vestfold Hills, East Antarctica: Implications for coastline, ice sheet and climate. *South African Journal of Science*, **82**, 520–21.

Pienitz, R., Lortie, G., & Allard, M. (1991). Isolation of lacustrine basins and marine regression in the Kuujjuaq area, northern Quebec, as inferred from diatom analysis. *Géographie Physique et Quaternaire*, **45**, 155–74.

Plater, A. J., & Poolton, N. R. J. (1992). Interpretation of Holocene sea level tendency and intertidal sedimentation in the Tees estuary using sediment luminescence techniques: A viability study. *Sedimentology*, **39**, 1–15.

Plater, A. J., & Shennan, I. (1992). Evidence of Holocene sea-level change from the Northumberland coast, eastern England. *Proceedings of the Geologists' Associaton*, **103**, 201–16.

Risberg, J., Sandgren, P., & Andrén, E. (1996). Early Holocene shore displacement and evidence of irregular isostatic uplift northwest of Lake Vänern, western Sweden. *Journal of Paleolimnology*, **15**, 47–63.

Robinson, M. (1982). Diatom analysis of Early Flandrian lagoon sediments from East Lothian, Scotland. *Journal of Biogeography*, **9**, 207–21.

(1993). Microfossil analyses and radiocarbon dating of depositional sequences related to Holocene sea-level change in the Forth valley, Scotland. *Transactions of the Royal Society of Edinburgh: Earth Sciences*, **84**, 1–60.

Sato, H., & Kumano, S. (1986). The succession of diatom assemblages and Holocene sea-level changes during the last 6,000 years at Sado Island, Central Japan: the Holocene development of Lake Kamo-ko II. *Japanese Journal of Phycology*, **47**, 177–83.

Sato, H., Maeda, Y., & Kumano, S. (1983). Diatom assemblages and Holocene sea level changes at the Tamatsu site in Kobe, western Japan. *The Quaternary Research*, **22**, 77–90.

Shennan, I. (1986). Flandrian sea-level changes in the Fenland. I: The geographical setting and evidence of relative sea-level change. *Journal of Quaternary Science*, **1**, 119–54.

Shennan, I., Inness, J. B., Long, A., & Zong, Y. (1995). Holocene relative sea-level changes and coastal vegetation history at Kentra Moss, Argyll, northwest Scotland. *Marine Geology*, **124**, 43–59.

Shennan, I., Long, A. J., Rutherford, M. M., Green, F. M., Innes, J. B., Lloyd, J. M., Zong, Y., & Walker, K. J. (1996). Tidal marsh stratigraphy, sea-level change and large earthquakes, I: A 5000 year record in Washington, U. S. A. *Quaternary Science Reviews*, **15**, 1023–59.

Shennan, I., Tooley, M. J., Davis, M. J., & Haggart, B. A. (1983). Analysis and interpretation of Holocene sea-level data. *Nature*, **302**, 404–6.

Sherrod, B. L., Rollins, H. B., & Kennedy, S. K. (1989). Subrecent intertidal diatoms from St. Catherines Island, Georgia: Taphonomic complications. *Journal of Coastal Research*, **5**, 665–77.

Simonsen, R. (1962). Untersuchungen zur Systematik und Ökologie der Bodendiatomeen der westlichen Ostsee. *Internationale Revue der Gesamten Hydrobiologie, Systematische Beihefte*, **1**, 8–144, 4 pl.

Simonsen, R. (1969). Diatoms as indicators in estuarine environments. *Veröffentlichungen des Instituts für Meeresforschung Bremerhaven*, **11**, 287–91.

Smith, D. E., Firth, C. R., Turbayne, S. C., & Brooks, C. L. (1992). Holocene relative sea-level changes and shoreline displacement in the Dornoch Firth area, Scotland. *Proceedings of the Geologists' Association*, **103**, 237–57.

Stabell, B. (1985). The development and succession of taxa within the diatom genus *Fragilaria* Lyngbye as a response to basin isolation from the sea. *Boreas*, **14**, 273–86.

(1987). Changes in diatom floras in late Quaternary western and southeastern Norwegian marine and freshwater sediments: Response to basin isolation from the sea. *Nova Hedwigia*, **44**, 305–26.

Tawfik, E., & Krebs, W. N. (1995). Environment of Zeit Formation and post-Zeit section (Miocene-Pliocene) in the Gulf of Suez, Egypt. In *Proceedings of the Thirteenth International Diatom Symposium*, ed. M. Marino & M. Montresor, pp. 541–54. Bristol: Biopress.

Tooley, M. J. (1982). Sea-level changes in northern England. *Proceedings of the Geologists' Association*, **93**, 43–51.

van Dam, H., Mertens, A., & Sinkeldam, J. (1994). A coded checklist and ecological indicator values of freshwater diatoms from The Netherlands. *Netherlands Journal of Aquatic Ecology*, **28**, 117–33.

van de Plassche, O. (1982). Sea-level change and water level movements in the Netherlands during the Holocene. *Mededelingen Rijks Geologische Dienst*, **36**, 1–93.

(1986). Introduction. In *Sea-level Research, A Manual for the Collection and Evaluation of Data*, ed. O. van de Plassche, pp. 1–26. Norwich: Geo Books.

van der Valk, L. (1992). Mid- and late-Holocene coastal evolution in the beach-barrier area of the western Netherlands. Doctoral thesis. Amsterdam: Vrije Universiteit.

van der Werff, A. (1958). L'importance de la recherche sur les diatomées pour la paléobotanique. *Bulletin de la Société Botanique du Nord de la France*, **11**, 94–7.

Vanhoorne, R., & Denys, L. (1987). Further paleobotanical data on the Herzeele Formation

(Northern France). *Bulletin de l'Association française pour l'Etude du Quaternaire*, **1987**/**1**, 7–18.

von der Brelie, G. (1956). Diatomeen als Fazies-Fossilien. *Geologische Rundschau*, **45**, 84–97.

Vos P., & de Wolf, H. (1988a). Methodological aspects of paleo-ecological diatom research in coastal areas of the Netherlands. *Geologie en Mijnbouw*, **67**, 31–40.

(1988b). *Paleo-ecologisch diatomeeënonderzoek in de Noordzee en de provincie Noord-Holland in het kader van het Kustgenese Project, Taakgroep 5000*. Rapport 500 Rijks Geologische Dienst. Haarlem: Rijks Geologische Dienst.

(1993a). Diatoms as a tool for reconstructing sedimentary environments in coastal wetlands; methodological aspects. *Hydrobiologia*, **269**/**270**, 285–96.

(1993b). Reconstruction of sedimentary environments in Holocene coastal deposits of the southwest Netherlands; the Poortvliet boring, a case study of paleoenvironmental diatom research. *Hydrobiologia*, **269**/**270**, 297–306.

(1994). Paleoenvironmental research on diatoms in early and middle Holocene deposits in central North Holland (The Netherlands). *Netherlands Journal of Aquatic Ecology*, **28**, 97–115.

Wasell, A., & Håkansson, H. (1992). Diatom stratigraphy in a lake on Horseshoe Island, Antarctica: A marine-brackish-fresh water transition with comments on the systematics and ecology of the most common diatoms. *Diatom Research*, **7**, 157–194.

Weiss, D., Geitzenauer, K., & Shaw, F. C. (1978). Foraminifera, diatom and bivalve distribution in recent sediments of the Hudson estuary. *Estuarine and Coastal Marine Science*, **7**, 393–400.

Wendker, S. (1990). Untersuchungen zur subfossilen und rezenten Diatomeenflora des Schlei-Ästuars (Ostsee). *Bibliotheca Diatomologica*, **20**, 1–300, 13 pl.

Whiting, M. C., & McIntire, C. D. (1985). An investigation of distribution patterns in the diatom flora of Netarts Bay, Oregon, by correspondence analysis. *Journal of Phycology*, **21**, 655–61.

Whitehead, J. M., & McMinn, A. (1997). Paleodepth determination from Antarctic benthic diatom assemblages. *Marine Micropaleontology*, **29**, 301–18.

Williams, K. M. (1990). Paleolimnology of three Jackman Sound Lakes, southern Baffin Island, based on down-core diatom analyses. *Journal of Paleolimnology*, **4**, 203–17.

Witkowski, A. (1991). Diatoms of the Puck Bay coastal shallows (Poland, Southern Baltic). *Nordic Journal of Botany*, **11**, 689–701.

Yokoyama, Y., Nakada, M., Maeda, Y., Nagaoka, S., Okuno, J., Matsumoto, E., Sato, H., & Matsushima, Y. (1996). Holocene sea-level change and hydro-isostacy along the west coast of Kyushu, Japan. *Paleogeography, Paleoclimatology, Paleoecology*, **123**, 29–47.

Zong, Y., & Tooley, M. J. (1996). Holocene sea-level changes and crustal movements in Morecambe Bay, northwest England. *Journal of Quaternary Science*, **11**, 43–58.

14 Diatoms and environmental change in brackish waters

PAULI SNOEIJS

Introduction

Brackish waters comprise a range of exclusive habitats which can be subdivided into three major categories: transition zones between freshwater and marine habitats, transition zones between hyperhaline water and marine habitats, and inland waters (no marine water exchange, see Fritz *et al.* (this volume) with higher salinity than freshwater. The salinities of these habitats vary from relatively stable (e.g., large saline lakes) to extremely unstable in time and space (e.g., estuaries bordering tidal seas). The main salinity regulating factors are inflow of freshwater, inflow of marine water, precipitation, evaporation and ice cover. In the past many efforts have been made to classify brackish waters according to salinity and the occurrence of biological species (Segerstråle, 1959; Anonymous, 1959; den Hartog, 1964). The more detailed such classifications are, the less well they fit all brackish waters. Based on the total concentration of ionic components (‰ Salinity) generally accepted approximate limits are: limnetic (freshwater) <0.5‰, oligohaline 0.5–5‰, mesohaline 5–18‰, polyhaline 18–30‰, euhaline 30–40‰, hyperhaline >40‰ (the 'Venice System': Anonymous, 1959).

There is little universality in responses of organisms to the peculiar conditions prevailing in numerous and generally widely separated brackish water bodies. Three factors must always be taken into account when using organisms as environmental markers for salinity: (i) In environments with fluctuating salinity regimes, the species are selected more according to their ability to cope with changing salinity (euryhalinity) than to their salinity optima. (ii) In environments with stable salinity regimes, evolutionary processes have resulted in various degrees of endemism depending on the geological age and stability of the water body and the degree of isolation from other populations of the same species. (iii) Species perform differently (physiologically and ecologically) in different brackish waters because they are, besides salinity, also affected by other environmental constraints, such as alkalinity, water temperature, light regime, nutrient concentrations, degree of exposure to wave action, biotic interactions, etc. Thus, when it is observed in an estuary that a certain species has a lower salinity limit of 15‰ whereas the same species is living in the Baltic Sea down to 3‰, this may be a matter of energy allocation (through a

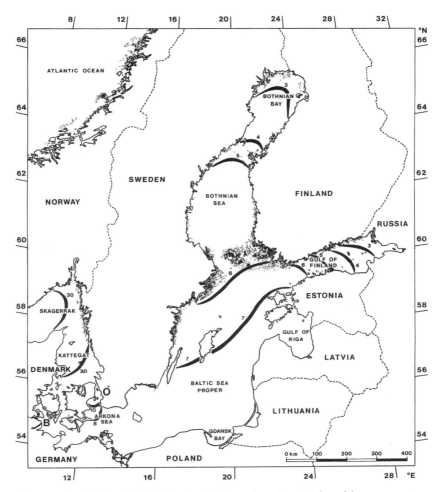

Fig. 14.1. Map showing the Baltic Sea with its five major sub-areas: the Baltic Sea proper, the Bothnian Sea, the Bothnian Bay, the Gulf of Finland and the Gulf of Riga, and adjacent waters: the Öresund (Ö), the Belt Seas (B), the Kattegat and the Skagerrak. The Skagerrak is connected to the North Sea. Lines are isohalines.

physiological adaptation), ecological preference for sites exposed to wave action (heavily exposed sites with 3‰ salinity are not found in the estuary), or an ongoing evolutionary process in the northern Baltic Sea (genetic differentiation into a Baltic ecotype). Such a multitude of possible responses based on the overall instability within and between brackish water bodies makes the use of biological species as indicators of environmental change a more complicated matter in brackish water habitats than in either marine or freshwater habitats.

This chapter presents many examples from the Baltic Sea, which is one of the world's largest brackish water bodies (surface area 377 400 km², volume 21 200 km³: Sjöberg, 1992, Fig. 14.1), and adjacent waters in the southwest

(Öresund, Belt Seas, Kattegat and Skagerrak). This area comprises an over 2000 km long transitional zone between freshwater and the North Sea (Voipio, 1981; Ketchum, 1983). The Baltic Sea receives freshwater from over 200 rivers, in total a water volume approximately equal to the net outflow from the Baltic Sea (Bergström & Carlsson, 1994). Precipitation and evaporation are more-or-less in equilibrium, and there is a limited inflow of saltier bottom water from the Kattegat over shallow and narrow thresholds. Larger inflows occur only rarely (once every 10 to 20 years), and depend on unusual air pressure and stormy weather conditions. The water turnover time of the Baltic Sea is long, ca. 35 years. The water budget keeps a stable salinity gradient intact: ca. 1–4‰ in the northern Gulf of Bothnia (= Bothnian Bay), eastern Gulf of Finland and Gulf of Riga, ca. 4–6‰ in the southern Gulf of Bothnia (= Bothnian Sea), ca. 6–8‰ in the western Gulf of Finland and the central Baltic Sea (= Baltic Sea proper), ca. 8–10‰ in the Arkona Sea, ca. 10–15‰ in the Öresund and Belt Seas, ca. 15–25‰ in the Kattegat, and ca. 25–34‰ in the Skagerrak. Inside the Baltic basin the salinity gradient is quite stable (never fluctuating by more than 1‰), but from the Öresund on northwards along the Swedish west coast, the waters become more and more estuarine (subject to salinity fluctuations), largely regulated by wind strength and direction of moving water to and from the North Sea.

The Baltic Sea is a transition area, and therefore resembles a giant estuary or rather a large threshold fjord with a series of subbasins separated from each other by sills, the largest ones being the Baltic Sea proper (including the Arkona Sea), the Bothnian Sea, the Bothnian Bay and the Gulf of Riga. The Gulf of Finland is not separated from the Baltic Sea proper by sills. There are, however, many differences: The Baltic Sea is larger and deeper than an estuary (mean ca. 55 m, maximum ca. 450 m), it has a much longer water turnover time, virtually no tidal fluctuations, and a year-round stable salinity gradient. In estuaries, salinity can vary widely on several time scales: diurnal, seasonal and irregular (Fairbridge, 1980), and the organisms living in such environments are adapted to repeated drastic salinity variations. Estuarine and salt-marsh diatoms are treated by Sullivan (this volume). This chapter will focus on diatoms that live under more stable salinity conditions.

The Baltic Sea may also be compared to other continental seas with brackish water. Other northern sea areas, such as the White Sea, Hudson Bay and the Gulf of St Lawrence, also have salinity gradients but they are affected by large tidal fluctuations and, thus, the oceanic influence is much larger, and salinity higher, than in the Baltic Sea. The intercontinental Black Sea has a narrow and shallow connection with the Mediterranean Sea, but higher salinity (ca. 16–22‰) than the Baltic Sea, although the Black Sea's salinity has been lower in the past. The Caspian Sea has no connection with the ocean (and is thus in fact a lake). It is the world's largest landlocked water body and has a salinity gradient similar to the Baltic Sea from the Volga delta out into the open waters (ca. 1–12‰); some nearly cut-off parts, e.g., the Kara-Bugaz Gulf, are hyperhaline due to evaporation. Many endemic brackish water species have evolved in the Caspian Sea, which was closed off from the ocean more than 5 million years ago (Karayeva & Makarova, 1973). This is not the case in the Baltic Sea which is very young (its present salinity conditions were established *ca.* 3000 years ago: Ignatius *et al.*, 1981), but the flora and fauna of the Baltic Sea do include many ecotypes with special adaptations to the brackish environment.

The Baltic Sea is one of the most polluted seas on earth. Since the beginning of this century it has been transformed from a clean oligotrophic to a contaminated mesotrophic water body. During this time the Baltic Sea's phosphorus load has increased about eightfold, and the nitrogen load about fourfold. This has resulted in increased pelagic primary production by an estimated 30–70% and sedimentation of organic carbon by 500–1000% (Elmgren, 1989; Jonsson & Carman, 1994). It is believed that increased production and sedimentation has caused increased oxygen consumption below the halocline, and that increased oxygen depletion in the deep waters of the Baltic Sea is a symptom of eutrophication (Nehring & Matthäus, 1991). The drainage basin is about 4.5 times as large as the sea itself and carries a human population of *ca.* 85 million in 14 countries (Sweitzer *et al.*, 1996), with industries, agriculture and forestry that discharge contaminants and nutrients to the Baltic Sea. The very limited water exchange with the ocean keeps pollutants inside the Baltic Sea for a long time. Chemical and biological degradation of pollutants are extremely slow because of low temperatures and long winters (the northern basins have a subarctic climate). Many organisms living in the Baltic Sea are salinity-stressed and need a higher rate of osmoregulation, and they may have a higher uptake of contaminants and less energy available for growth (Kautsky *et al.*, 1990; Tedengren *et al.*, 1990).

Diatoms as biological indicators in brackish water

All biological communities are dynamic. Individuals of different species are born and die, individuals of the same and of different species interact, environmental factors interact, environmental factors affect the organisms and the organisms affect the environment. All these processes go on simultaneously and continuously in different directions. I will use the term diatom community for living diatoms having interactions with each other and their biotic and abiotic environment (Round, 1981; Begon *et al.*, 1996), and the term diatom assemblage when diatoms or dead frustules are aggregated by other selection processes as well (e.g., fossil diatom assemblages, diatom assemblages on artificial substrata, etc.). When human activities change the environment, the abundances of biological species will change as a reaction to this, and community composition can be used as an indicator for the direction and rate of change caused by these activities. When using organisms as indicators, the essential aspect used is that of environmental factors affecting the survival, occurrence, abundance, growth and fecundity of organisms. The other aspects of community ecology, e.g., selective grazing, may act as drawbacks ('noise') for the indicator value, but also indirectly enforce the indicator value.

Diatom communities are sensitive indicators of environmental change. They are generally species-rich and small changes in environmental factors usually result in measurable species shifts. Compositional ecological data should, however, always be interpreted with great care and with an open mind for biotic interactions and multivariate effects. Low abundance of a particular species in a particular environment may, for example, be caused by one main environmental factor, a special combination of environmental factors, competition with other species or by being the favorite food of a predator. Conclusive ecological data on the structure and function of diatom communities should be based on the assessment of live communities only, since (allochthonous) dead frustules are widely distributed. Remnants of such communities in the fossil record of sediments (fossil diatom assemblages) are powerful tools in reconstruction of paleoenvironments (Meriläinen et al., 1983; Smol & Glew, 1992). In the same way, living diatom assemblages on artificial substrata can be used as tools for measuring environmental change without necessarily reflecting all aspects of natural communities (Snoeijs, 1991). In fossil sediments, the composition of diatom assemblages are affected by physical processes (currents, upwelling, resuspension, erosion), biological processes (bioturbation), and different degrees of dissolution of frustules of different diatom taxa. Artificial substrata are not colonized by all types of diatom life forms. Such assemblages do not reflect true species interactions, but do have indicator value for the directions and rates of change caused by changing environmental factors. The advantage of studying diatom assemblages from sediments is the large time scale that can be assessed.

Another way of using organisms for recording environmental change is to study their chemical composition in relation to the chemical composition of the environment. Analysis of one or two key taxa can give a measure of the levels and transport of pollutants in the whole ecosystem by using previously tested mathematical models. Diatoms are excellent indicators for this, given their essential role in the food web of practically all aquatic systems worldwide (Snoeijs & Notter, 1993a).

It is also expected that, when humans create new habitats by radically changing/polluting the environment, better adapted new ecotypes and species may evolve. This may happen first in small, numerous, one-celled organisms such as diatoms. In the future, when relationships between environmental factors and different clones of diatom species has been clarified, genetic markers may be used for indicating environmental change as well ('ecogenetics').

Brackish-water diatoms and salinity

HABITATS

Diatom habitats in brackish water are not different from those in other water bodies, the two main categories being pelagic (floating in the plankton or tychoplankton), and benthic (attached or associated with a substratum). Benthic diatoms can be subdivided into the sub-habitats epipelon (unattached motile forms in and on sediments), epipsammon (attached to sand grains), epiphyton (attached to plants), metaphyton (weakly motile forms living in mucilage attached to a substratum), epilithon (associated with rock surfaces), epizoon (attached to animals), ice-associated or symbiotic (Round, 1981; Round & Crawford, 1990). Each of these (sub) habitats is inhabited by a diatom flora consisting of typical taxa with typical life forms. Diatom species belonging to the same genus tend to have the same habitat and life form (Round *et al.*, 1990).

SPECIES AND POPULATIONS

Diatom taxa occurring in transitional zones between marine and freshwater habitats can be divided into two affinity groups: one containing taxa with marine affinity and one containing taxa with freshwater affinity. Round and Sims (1981) considered the distribution of diatom genera in the light of diatom evolution, and found that more than 90% of the genera are confined to either marine or freshwater habitats. Most likely this figure is actually higher, because genera previously found in both marine and freshwater habitats are currently being split, and with the new groupings more genera fit the pattern (Round *et al.*, 1990). Round & Sims (1981) considered only a few small genera to have a brackish-water distribution (*Brebissonia* Grunov, *Dimidiata* Hajos, *Scoliopleura* Grunov, *Scoliotropis* Cleve). An important aspect is how to define a 'brackish water taxon'. If this is a taxon exclusively occurring in brackish water, then there are none or extremely few in transitional zones between marine and freshwater. All diatoms living in these transitional zones should then be considered as marine or freshwater taxa with different degrees of euryhalinity (Carpelan, 1978a). Real brackish water species would then only be endemics found in land-locked brackish waters as evolutionary results of genetic bottlenecks. But if a brackish water species is a species with its abundance optimum in brackish waters, then there are many in transitional zones, all with their own typical salinity range and optimum (which however may differ in different types of brackish water habitats).

Traditionally, diatom species occurring in brackish waters have been classified into five or more groups in a so-called 'Halobion system', according to their salinity ranges. This classification system has been widely used, e.g., in the interpretation of paleoecological diatom stratigraphy where shifts in salinity are involved. The first Halobion system was established by Kolbe (1927, 1932), originally based on the diatom taxa occurring in the Sperenberger

Saltzgebiet (Germany). Later revisions were made by Hustedt (1953), Simonsen (1962), and Carpelan (1978*b*). The development of new statistical methods for analyzing complicated multivariate data (e.g., ter Braak, 1986; ter Braak & Juggins, 1993), and the intensified use of computers since the 1980s, have opened the possibility of more precise analyses and calibrations of species performances along environmental gradients. This application of diatoms is widely employed for reconstructing paleoenvironments by utilizing calibration data-sets containing individual salinity optima and ranges for each species. Such calibrations are better adapted to fit a specific geographic region or type of brackish water than the Halobion system. Salinity transfer functions and inference models have been presented by e.g., Juggins (1992) for the Thames estuary (England), and by Wilson *et al.*, 1996 for lakes in western North America, see also Cooper, Denys & de Wolf, and Fritz *et al.* in this volume. With the progress of the present intercalibration of diatom species identification in the Baltic Sea (see below), and an increasing number of ecological diatom studies carried out here, it will soon be possible to present a similar diatom calibration model for the Baltic Sea area.

In the Bothnian Sea, freshwater diatoms are the dominant epiphytes on submerged macrophytes, notably *Diatoma* spp., *Ctenophora pulchella* (Ralfs *ex* Kützing) Williams & Round, *Rhoicosphenia curvata* (Kützing) Grunov, and *Cocconeis pediculus* Ehrenberg. Snoeijs & Potapova (1998) showed that discontinuities in phenotypic characters exist between some well-known and widely distributed freshwater *Diatoma* species and Baltic populations which are generally considered conspecific. The Baltic morphotypes proved to be constant over time in the Bothnian Sea, and they were consistently different from populations in freshwater habitats. Some of the Baltic *Diatoma* forms may be defined as ecotypes (e.g., a northern Baltic form of *Diatoma moniliformis* Kützing), others may be species of their own and maybe even endemic (e.g., *Diatoma constricta* Grunov and *Diatoma bottnica* Snoeijs). This supports the hypothesis that the traditional species concept in diatoms is probably far too broad, which has been an issue for debate especially during the last decade (e.g., Ross & Mann, 1986; Mann, 1988, 1989, 1994; Round *et al.*, 1990; Mann & Kociolek, 1990; Lange-Bertalot, 1990). Biogeographical patterns that now are unclear for many diatom species, may become distinct when stable populations of ecotypes or endemic species can be recognized. A narrowing of the species concept will probably also refine methods of using (paleo-) ecological analyses of diatom communities and assemblages as markers of environmental changes; taxa with narrow ecological tolerances have higher indicator values than taxa with wide tolerances. However, defining more (and more similar) taxa will also put a still higher emphasis on the correct identification of diatom taxa.

Adaptive trade-offs seem logical in the case of the Baltic Sea. This was already recognized by Juhlin-Dannfelt (1882) for Baltic diatom taxa: 'the species living both in brackish and in fresh water do not seem to follow any rule as to their relative dimensions' (*op. cit.*, p. 13). When a freshwater species is transported in a freshwater stream to an estuary bordering a tidal marine environment, it will experience a vigorous salinity shock and die off when it meets lethal salinity. Such a freshwater species carried to the northern Baltic Sea by land-runoff meets a large habitat with extraordinary stable low salinity (hundreds of kilometers), high silicate concentrations (by the discharge of many rivers), and low biotic interactions (low diversity of macroalgae and fauna). Here some freshwater species can survive and grow, and may develop into Baltic ecotypes and endemic species genetically adapted to a certain salinity (more stenohaline). The freshwater species represented in this group are often the same species as those living in freshwaters affected by eutrophication or high conductivity, e.g., *Diatoma moniliformis*, *Cocconeis pediculus*, *Rhoicosphenia curvata* (Hofmann, 1994; Snoeijs & Potapova, 1998). Probably the Baltic diatom species that have adapted to a stable low salinity have a lower genetic variability than typical estuarine species which need to cope with a constantly changing environment. When studying genetic variation within widespread diatom species, usually unique combinations of genetic markers are found for nearly each clone, *cf.* Hoagland *et al.* (1994) who studied 147 clones of *Fragilaria capucina* Desmazières using PCR-RAPD methodology. This method may help the ecologist to group and characterize genotypes of populations, but may complicate matters for the taxonomist trying to cluster diatoms into discrete groups (species), when borders between ecotypes, varieties, subspecies, species are blurred by evolutionary activity as we notice in the northern Baltic Sea.

The floristic composition of the Baltic Sea diatoms is similar to that of estuaries elsewhere, e.g., Swedish west coast (Kuylenstierna 1989–1990), South Africa (Archibald, 1983), Western Australia (John, 1983), and to that of other inland seas, e.g., the Black Sea (Proschkina-Lavrenko, 1955). In the light of the discussion above, it is very important to illustrate diatom species used as environmental markers in different studies, preferably by both light and electron micrographs, to allow for comparisons with other areas. Intercalibration of the Baltic Sea diatoms is currently being carried out by an international working group within the framework of the 'Baltic Marine Biologists'. The results are published as illustrated guides including distributional data and taxonomic notes (Snoeijs, 1993; Snoeijs & Vilbaste, 1994; Snoeijs & Potapova, 1995; Snoeijs & Kasperoviciene, 1996; Snoeijs & Balashova, 1998). A checklist of the phytoplankton of the Baltic Sea, including the Baltic pelagic diatoms, was published by Edler *et al.* (1984).

During an average winter, the whole northern Baltic Sea is ice-covered; the mean period of ice cover for the Bothnian Bay is 120 days per year, and for the Bothnian Sea 60 days per year. Normally also the shallow coastal areas more to the south are frozen in winter. The dominant diatoms occurring in connection with this ice cover, some of them also normal constituents of the Baltic phytoplankton, belong to the Arctic flora. These diatom taxa occur in the Arctic and the Baltic Sea, and some also in the Oslofjord in southern Norway (Huttunen & Niemi, 1986; Hasle & Syvertsen, 1990; Norrman & Andersson,

1994), but they have not been reported from the Norwegian west coast. Some examples of this Arctic flora are: *Pauliella taeniata* (Grunov) Round & Basson (Syn. *Achnanthes taeniata* Grunov), *Fragilariopsis cylindrus* (Grunov) Hasle, *Melosira arctica* Dickie, *Navicula vanhoeffenii* Gran, *Navicula pelagica* Cleve, *Nitzschia frigida* Grunov, *Thalassiosira baltica* (Grunov) Ostenfeld, *Thalassiosira hyperborea* (Grunov) Hasle.

COMMUNITIES

It is the quantitative diatom community composition which is typical of brackish water with a given salinity, not the occurrence of one or two particular 'indicator' species. Each species in the community contributes information by its typical salinity range and optimum in combination with its abundance at a particular time and site. Salinity is the overruling environmental factor that determines the quantitative association of diatoms along salinity gradients, and diatom community composition can successfully reflect differences in salinity as small as 1‰ (Snoeijs, 1994a).

In the Baltic Sea area from the northern Öresund to the northern Bothnian Bay, three distributional discontinuities in the epiphytic diatom community composition are found (Snoeijs, 1992; Snoeijs, 1995). One is situated at the entrance to the Baltic Sea (at *ca.* 10‰), the second between the Baltic Sea proper and the Bothnian Sea (at *ca.* 5‰), and the third between the Bothnian Sea and the Bothnian Bay (at *ca.* 3‰). These discontinuities do not only fit with the thresholds between the different Baltic Sea subbasins, but also with subdivisions of the meso- and oligohaline zones, based on other biological observations in the Baltic: α-mesohaline (*ca.* 18–10‰), β-mesohaline (*ca.* 10–5‰), α-oligohaline (*ca.* 5–3‰), β-oligohaline (*ca.* 3–0.5‰) (Anonymous, 1959). Typical epiphytic diatoms in these different zones are: *Licmophora hyalina* (Kützing) Grunov, *Tabularia fasciculata* (C. A. Agardh) Williams & Round (whole mesohaline range), *Licmophora communis* (Heiberg) Grunov, *Licmophora oedipus* (Kützing) Grunov (α-mesohaline), *Tabularia waernii* Snoeijs (β-mesohaline), *Ctenophora pulchella* (Ralfs) Williams & Round (whole oligohaline range), *Diatoma constricta* (α-oligohaline), *Synedra ulna* (Nitzsch) Ehrenberg and *Synedra acus* Kützing (β-oligohaline). The discontinuity at 5‰ is most drastic (Fig. 14.2), and is characterized by a marked transition from an epiphytic flora dominated by diatom taxa with marine affinities (mostly little silicified species) to one dominated by taxa with freshwater affinities (mostly heavily silicified species). The transition between 5 and 7‰ represents a lower limit of salinity tolerance for many marine taxa, resulting in the absence or very low abundance of these taxa below 5‰.

Nothing general can be stated about the relationship between salinity and the primary production of diatoms as a group. This community parameter is determined more by environmental factors other than salinity, such as nutrient concentrations, temperature, light availability, exposure to water movement and substratum.

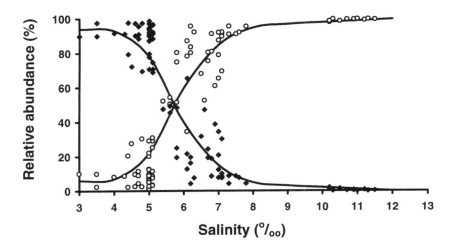

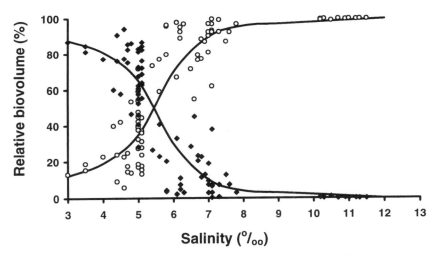

Fig. 14.2. Epiphytic diatom taxa with marine affinity (closed symbols) and freshwater affinity (open symbols) plotted against salinity (n = 88 sampling sites), based on relative abundance and relative biovolume. Adapted after Snoeijs (1995).

DIVERSITY

Macroalgal and macrofaunal species richness are extremely low in the Baltic Sea, with many marine species living at the limit of their tolerance to low salinity. No minimum for species richness is found in the Baltic Sea for epiphytic diatoms between 3 and 14‰ (Fig. 14.3(c)). Simonsen (1962) and Wendker (1990) also found that the number of species of benthic diatoms was not correlated with salinity in the Schlei Estuary (western Baltic Sea, 3–18‰). This implies that Remane's brackish-water rule based on aquatic fauna (Remane, 1940, 1958) is not valid for benthic diatoms. According to this rule, a

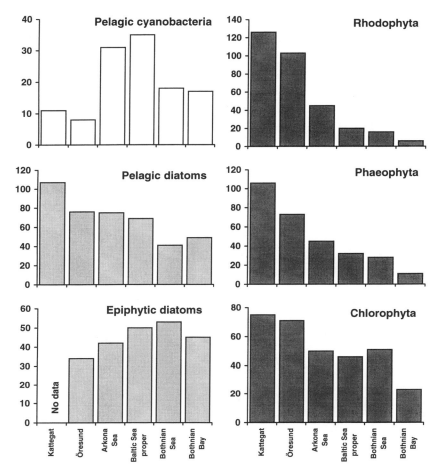

Fig. 14.3. Diversity of algae and cyanobacteria (number of species) along the Baltic Sea salinity gradient. Data for pelagic cyanobacteria and pelagic diatoms from Willén (1995: HELCOM Monitoring Programme), data for epiphytic diatoms from Snoeijs (1995), data for Rhodophyta, Phaeophyta and Chlorophyta from Nielsen *et al.* (1995: Baltic Marine Biologists, BMB).

gradual decrease occurs in the number of species as conditions become increasingly brackish with a minimum species richness at 5‰, which is the critical salinity for physiological stress in many animals. The absence of a minimum suggests that benthic diatoms as a group are not stressed by salinity in the Baltic Sea. There are few groups of organisms which, like pennate diatoms (most benthic species), to the same degree inhabit both freshwater and marine environments. These groups can thus penetrate into brackish water from both habitats, which probably explains the absence of a species minimum for epiphytic diatoms (Fig 14.3(*c*)) in the Baltic Sea. Taxa with freshwater affinities that penetrate into the Baltic Sea southward compensate for the northward loss of taxa with marine affinities. However, species richness of pelagic diatoms does show a decreasing trend with decreasing salinity

P. SNOEIJS

from the Kattegat to the northern Bothnian Bay (Fig. 14.3(*b*)); the number of species decreases successively along the salinity gradient by *ca.* 50%. The decrease in pelagic diatoms is not as drastic as that in red, brown and green macroalgae, which in the same salinity range decrease by *ca.* 95%, 90%, and 70%, respectively (Fig. 14.3(*d*)–(*f*)), or macroscopic animals (decreasing by >95%). Most probable explanations for the different diversity distributions of primary producers divisions are that red and brown algae are almost exclusively marine, and that more green macroalgal and centric diatom taxa (most of the pelagic species) occur in marine waters compared with freshwater.

PALEOECOLOGY

During deglaciation of the Weichselian ice sheet in Northwest Europe, the Baltic Sea repeatedly lost and re-established its connection with the ocean. This dynamic period lasted from *ca.* 13 000 to *ca.* 3000 years ago, and was driven by combination effects of sea level rise and crustal uplift of the Scandinavian peninsula (Ignatius *et al.*, 1981). The postglacial history of the Baltic Sea area can thus be subdivided in several typical stages: the Baltic Ice Lake (cold, fresh melting water), the *Yoldia* Sea (cold, mixture of marine water and fresh melting water), the *Ancylus* Lake (freshwater), the *Mastogloia* Sea (brackish, *ca.* 0.5–5‰), the *Litorina* Sea (brackish, *ca.* 5–25‰), the *Mya* and *Limnaea* Sea (more-or-less the present conditions, brackish). These stages can be recognized by using diatom microfossils which was already discovered more than 100 years ago (Munthe, 1892; Cleve, 1899; Alhonen, 1971; Ignatius & Tynni, 1978; Gudelis & Königsson, 1979; Miller, 1986), although the early stages are often poor in diatoms (Åker *et al.*, 1988). Of several paleosalinity indicators evaluated by Sohlenius *et al.* (1996), the diatoms were the least ambiguous in recording the Baltic Holocene history. Diatom taxa typical of different stages in the development of the Baltic Sea are given in Table 14.1. This knowledge on diatom assemblages can be applied in studies tracing past shoreline displacement in areas with substantial postglacial land uplift, e.g., Sweden and Finland (Miller, 1982; Risberg & Sandgren, 1994)

Some fossil diatom assemblage types still occur alive in the present Baltic Sea. For example, the so-called *Clypeus* flora found in sediments from the *Litorina* Sea, as well as in the Eemian Interglacial Baltic Sea of *ca.* 120 000 years ago (Grönlund 1994), is found alive in shallow sediments of the southern Bothnian Sea (5‰) with exactly the same species composition (Snoeijs, unpublished). This *Clypeus* flora consists of diatom taxa typical of brackish, shallow, lagoonal conditions, e.g., *Campylodiscus clypeus* Ehrenberg, *Anomoeoneis sphaerophora* var. *sculpta* (Ehrenberg) Krammer, *Amphora robusta* Gregory, *Tryblionella circumsuta* (Bailey) Ralfs *in* Pritchard, *Nitzschia scalaris* (Ehrenberg) W. Smith, and *Surirella striatula* Turpin. Also, the typical flora of the *Mastogloia* Sea is found epiphytic and metaphytic on macroalgae in non-exposed sites of the Bothnian Sea in summer.

Table 14.1. *Diatom taxa typical of different developmental stages of the Baltic Sea*

Approximate time period (years before present)	Developmental stage	Typical diatom species	Affinity
13000–10200	Baltic Ice Lake Cold freshwater (melting water)	*Aulacoseira islandica* ssp. *helvetica* O. Müller	F
		Aulacoseira alpigena (Grunov) Krammer	F
		Stephanodiscus rotula (Kützing) Hendey	F
		Tabellaria fenestrata (Lyngbye) Kützing	F
10200–9500	Yoldia Sea Marine water mixed with cold freshwater (melting water)	*Aulacoseira islandica* ssp. *helvetica* O. Müller	F
		Actinocyclus octonarius Ehrenberg	B
		Diploneis didyma (Ehrenberg) Ehrenberg	B
		Diploneis interrupta (Kützing) Cleve	B
		Diploneis smithii (Brébisson) Cleve	B
		Nitzschia obtusa W. Smith	B
		Thalassiosira baltica (Grunov) Ostenfeld	B
		Tryblionella navicularis (Brébisson) Ralfs	B
		Tryblionella punctata W. Smith	B
		Grammatophora oceanica Ehrenberg	BM
		Rhabdonema arcuatum (Lyngbye) Kützing	BM
		Rhopalodia musculus (Kützing) O. Müller	BM
9500–8000	Ancylus Lake Freshwater	*Aulacoseira islandica* ssp. *helvetica* O. Müller	F
		Caloneis latiuscula (Kützing) Cleve	F
		Campylodiscus noricus Ehrenberg	F
		Cocconeis disculus (Schumann) Cleve	F
		Cymatopleura elliptica (Brébisson) W. Smith	F
		Diploneis domblittensis (Grunov) Cleve	F
		Diploneis mauleri (Brun) Cleve	F
		Ellerbeckia arenaria (Ralfs) Crawford	F
		Encyonema prostratum (Berkeley) Kützing	F
		Epithemia hyndmannii W. Smith	F
		Eunotia clevei Grunov	F
		Gomphocymbella ancyli (Cleve) Hustedt	F
		Gyrosigma attenuatum (Kützing) Rabenhorst	F
		Navicula jentzschii Grunov	F

		Affinity
8000–7500	Mastogloia Sea Brackish water (0.5–5‰)	
	Stephanodiscus astraea (Ehrenberg) Grunov	F
	Stephanodiscus neoastraea Håkansson & Hickel	F
	Campylodiscus clypeus Ehrenberg	B
	Campylodiscus echeneis Ehrenberg	B
	Ctenophora pulchella (Ralfs) Williams & Round	B
	Diploneis smithii (Brébisson) Cleve	B
	Epithemia turgida (Ehrenberg) Kützing	B
	Epithemia turgida var. *westermannii* (Ehrenberg) Grunov	B
	Mastogloia braunii Grunov	B
	Mastogloia elliptica (C. A. Agardh) Cleve	B
	Mastogloia pumila (Cleve & Möller) Cleve	B
	Mastogloia smithii Thwaites	B
	Navicula peregrina (Ehrenberg) Kützing	B
	Rhoicosphenia curvata (Kützing) Grunov	B
7500–3000	Litorina Sea Brackish water (5–25‰)	
	Actinocyclus octonarius Ehrenberg	B
	Campylodiscus clypeus Ehrenberg	B
	Cocconeis scutellum Ehrenberg	B
	Coscinodiscus asteromphalus Ehrenberg	B
	Diploneis didyma (Ehrenberg) Ehrenberg	B
	Diploneis interrupta (Kützing) Cleve	B
	Chaetoceros diadema (Ehrenberg) Gran	BM
	Petroneis latissima (Gregory) A. J. Stickle & Mann	BM
	Rhabdonema arcuatum (Lyngbye) Kützing	BM
	Rhabdonema minutum Kützing	BM
	Thalassionema nitzschioides (Grunov) Grunov	BM
	Ardissonea crystallina (C. A. Agardh) Grunov	MB
	Chaetoceros mitra (J. W. Bailey) Cleve	MB
	Hyalodiscus scoticus (Kützing) Grunov	MB
	Pseudosolenia calcar-avis (Schultze) Sundström	MB

Source: From Alhonen, 1971, 1986, Ignatius & Tynni, 1978, Gudelis & Königsson, 1979, Miller & Robertsson, 1979, Robertsson, 1990, Risberg, 1991, Witkowski, 1994, Kabailiene, 1995, Hedenström, 1994. Affinities: F = freshwater taxon, B = brackish taxon, BM = brackish–marine taxon, MB = marine–brackish taxon (according to Snoeijs *et al.* 1993–1998).

Brackish water diatoms and nutrients

Changes in relative and absolute diatom abundances in pelagic and littoral systems often indicate eutrophication. This is considered mainly a combined result of shifts in the relative proportions of inorganic nutrient concentrations in the water and increased primary production. The relative proportions of nutrients change when nitrogen and/or phosphorus levels increase in land-runoff and atmospheric deposition. Silicate levels in the sea are usually little directly affected by human activity, but may be decreased by eutrophication and subsequent growth of diatoms in natural inland waters or rivers dammed for generating electrical energy. Different algal species have different nutrient requirements, but often species belonging to the same taxonomic group have similar requirements. Eutrophication effects on the composition of micro-algal communities are therefore often clearly reflected on the class or division levels. The most typical compositional change following eutrophication is usually a decrease in the relative importance of diatoms in favor of non-silicified flagellates (Smayda, 1990; Cadée & Hegeman, 1991). For example, eutrophication of the Black Sea has increased pelagic primary production many fold, and formerly dominant diatom species have been replaced by dino-flagellates, prymnesiophytes, and other non-silicious flagellates (Leppäkoski & Mihnea, 1996; Vasiliu, 1996; Humborg et al., 1997). Usually, a decrease of diatoms is regarded as negative because they have a high nutritive value for micro-, meio- and macrofauna organisms (Plante-Cuny & Plante, 1986), and very few diatom species are reported to produce toxins (Todd, 1993; Hasle et al., 1996), contrary to several groups of flagellates and cyanobacteria (Wallström, 1991; Holmqvist & Willén, 1993).

Two main processes regulate a decrease in the relative abundance of diatoms in concert with eutrophication. (i) The relative proportions of nitrogen, phosphorus and silicate in the water change, and when Si:N and/or Si:P ratios decrease, more nitrate and phosphorus remain available for the growth of non-diatom phytoplankton because silicate sets a limit to diatom growth. Diatoms have an absolute requirement for silicate, which they take up mainly as dissolved silicic acid $(Si(OH)_4)$ and incorporate as amorphous silica or poly-merized SiO_2, into a siliceous cell wall (Werner, 1977). (ii) Silicate depletion accelerates when net primary production increases because the large majority of the diatoms that initially grow with increased nitrogen and/or phosphorus sink when they die. Thus the diatoms effectively remove silicate from the photic zone, see also Schelske, this volume. Dissolution of silicate from diatom shells is slow compared with remineralization rates of organic nitrogen and phosphorus (Eppley, 1977). Silicate limitation caused by eutrophication has been reported both from marine and freshwater areas as well as from laboratory experiments (e.g., van Bennekom et al., 1975; Schelske et al., 1986; Sommer, 1994, 1996). During the last decades, numerous experiments on plankton communities under semi-natural conditions (mesocosms) with artificial eutrophication have shown that fertilization with inorganic phosphorus and/or nitrogen stimulates the dominance of flagellates over diatoms (e.g., Jacobsen *et al.*, 1995; Watanabe *et al.*, 1995; Escaravage *et al.*, 1996; Egge & Jacobsen, 1997; Suzuki *et al.*, 1997). The general cause for this dominance shift is a lower availability of silicate relative to the other nutrients with eutrophication. Diatoms can be superior competitors for nitrogen and phosphorus when silicate is available in sufficient quantities for their growth (Tilman *et al.*, 1982; Heckey & Kilham, 1988). In shallow sediments, the silicate pool of the sediment may counteract silicate limitation of epipsammon and epipelon in eutrophied areas. The relative abundance of diatoms living in and on the sandy sediments did not decrease in favor of other microalgal groups in a four-week mesocosm experiment with N and P additions (Sundbäck & Snoeijs, 1991).

In the Baltic Sea area, export of inorganic nitrogen and phosphorus is small compared to external supply or internal biological turnover (Wulff *et al.*, 1990; Sandén & Rahm, 1993). Different subbasins of the Baltic Sea each have their own typical nutrient dynamics. The Bothnian Bay is oligotrophic, and has a N:P ratio higher than the Redfield ratio, indicating phosphorus limitation (Kangas *et al.*, 1993). Phosphorus is immobilized in the sediments like in most freshwater systems (Caraco *et al.*, 1990). Silicate is never limiting because there is a continuous large riverine supply. The Bothnian Sea is a transitional area, basically oligotrophic, with a N:P ratio that about equals the Redfield ratio, in most places silicate is never limiting here either. The largest discharges of nitrogen and phosphorus take place in the southern Baltic Sea proper, the eastern Gulf of Finland and the Gulf of Riga which are meso- to eutrophic. The Baltic Sea proper has a very low N:P ratio (indicating nitrogen limitation). Here the diatom spring maximum may consume the whole winter silicate pool (Wulff *et al.*, 1990), after which diatom growth is limited. The Baltic Sea proper, the eastern Gulf of Finland and since 1992 also the Gulf of Riga may have large cyanobacteria blooms in summer (Wallström, 1991). These blooms may increase eutrophication by increased input of atmospheric N_2. In the Gulf of Riga, one of the reasons for increased cyanobacterial blooms may be that lower salinities in the deeper layers have destabilized the halocline, and consequently favored vertical transport of phosphorus to the photic zone (Kahru *et al.*, 1994).

Nutrient analyses from coastal water samples taken along the Swedish coast from the Öresund to the Bothnian Bay during the epilithic diatom spring maximum (April – May) are shown in Fig. 14.4. Silicate, nitrate + nitrite and orthophosphate concentrations show the typical north-south gradients in concert with the salinity gradient: silicate and nitrogen decrease with salinity whereas phosphorus increases, and therefore N:P and Si:P ratios decrease and the Si:N ratio does not show a trend with salinity. Wulff *et al.* (1990) and Rahm *et al.* (1996) calculated that silicate shows a decreasing trend throughout the entire Baltic Sea, and interpreted this as a further indication of increased net primary production which is attributed to eutrophication. They hypothesized that the major part of the missing amount of silicate in the trophic layer would

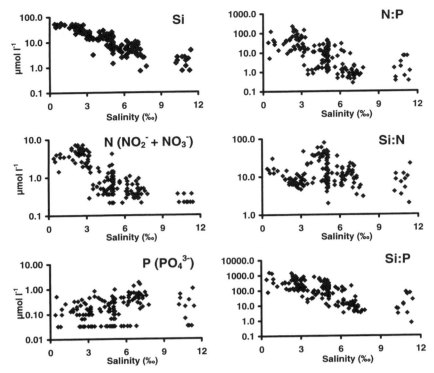

Fig. 14.4. Nutrient concentrations and nutrient quotes of coastal water samples. The samples were taken from 161 different sites along the Swedish coast from the Öresund to the Bothnian Bay during the benthic diatom spring maximum in April and May 1990 and 1991. Note that logarithmic scales are used for the nutrient quotes. Data: P. Snoeijs (previously unpublished data).

be deposited in the sediments. Diatom populations are usually mainly lost from the water column by fast sinking, while most flagellates usually disintegrate in the water column prior to deposition as slowly sinking phytodetrital material (Morris *et al.*, 1988; Passow, 1991; Heiskanen & Kononen, 1994). Increases in the present level of deposition, which are forecasted with further eutrophication in the Baltic Sea, may have a significant impact on nutrient biogeochemical cycles (Conley *et al.*, 1993, Conley & Johnstone, 1995). Wulff *et al.* (1996) assumed in a mathematical model that the process is reversible; if discharges of nitrogen and phosphorus would be lowered by 25% in the Gulf of Bothnia and by 50% in the Baltic Sea proper, silicate concentrations will increase again due to lowered primary production levels.

Rough estimations of the relative abundances of diatoms in Baltic pelagic microalgal communities for the years 1979–1988 indicate that the proportion of diatoms in the Arkona Sea was low (and that of cyanobacteria high) compared with the rest of the Baltic Sea and the Kattegat (Willén, 1995: HELCOM Monitoring Programme, open sea data). Zernova & Orlova (1996) found, in 1990, cyanobacteria-dominated phytoplankton communities as early as

February–March in the open Baltic Sea proper which they contributed to eutrophication. Unfortunately, the open sea HELCOM data are too scarce (only three to four samplings per year) for thorough interpretations of long-term trends in nutrient concentrations and phytoplankon composition (Wulff *et al.*, 1986). In some coastal areas, however, it has been noticed that diatoms have decreased in abundance and biomass. For example, long time trends in the shallow and strongly eutrophied Gulf of Riga indicate that the system might be changing from nitrogen to phosphorus, and finally, to silicate limitation with respect to diatom growth. Increases in phytoplankton primary production and diatom relative abundance since the early 1970s have in the 1990s shifted to significantly lower primary production and increased relative abundances of dinoflagellates and cyanobacteria at the expense of diatoms (Kahru *et al.*, 1994; Yurkovskis *et al.*, 1993; Balode, 1996; Yurkovskis & Kostrichkina, 1996; Poder & Jaanus, 1997).

In Baltic littoral areas there is a pronounced northwards increase in epilithic diatom biomass (towards colder climate, oligotrophic conditions and higher silicate levels). In the Bothnian Bay, cm to dm thick layers consisting of pure diatoms may cover the rocks in May (P. Snoeijs, unpublished data). In the Baltic Sea proper, epilithic filamentous green and brown algae are more successful in competition with epilithic diatoms in spring than in the Bothnian Sea and Bothnian Bay, which also may be attributed to low silicate levels and less ice scouring in the Baltic Sea proper.

The conclusion of the above paragraphs is that to be able to use living diatoms effectively as indicators for environmental change (e.g., their species composition), enough silicate must be available in the environment. Many diatom taxa known as indicators of eutrophication in freshwater occur in brackish water as well (Lange-Bertalot, 1979; Hofmann, 1994). These species have a wide conductivity tolerance, and this criterion cannot be used for indicating eutrophication in brackish water where conductivity is very high in both oligotrophic and eutrophic areas due to salt ions. Competition for nutrients between diatom species in a brackish water environment is probably mainly based on their ecological strategies such as size, shape and life form. The specific growth rate of diatoms varies with the mean cell size of the species; larger ones grow more slowly than small-celled species (Eppley, 1977). Needle-shaped diatom cells have higher surface to volume ratios than short and wide cells, and smaller cells have higher ratios than large ones (cells with a higher relative surface for uptake of nutrients have a competitive advantage). A motile life form may also be an important advantage. An experiment with sediment-living diatoms carried out in an outdoor flow-through experimental set-up on the Swedish west coast at salinity 14–24‰ showed that motile diatoms such as small *Nitzschia* species and *Cylindrotheca closterium* (Ehrenberg) Reimann & Lewin, as well as small *Amphora* species capable of rapid colonization of sand grains, were better competitors with N and P addition than other diatom taxa (Sundbäck & Snoeijs, 1991). Leskinen & Hällfors (1990) found high abundances of motile *Nitzschia* spp. between diatoms epiphytic on filamentous macroalgae, and these *Nitzschia* spp. may be considered indicators of eutrophic conditions on the southern Finnish coast (Gulf of Finland).

Deposited assemblages of dead diatom frustules in the upper sediments in the deeper water of the Baltic Sea also reflect the ongoing eutrophication process (Cato *et al.*, 1985), and recovery from eutrophication after ceased discharges in coastal areas (Korhola & Blom, 1996). Grönlund (1993) found signs of the onset of eutrophication *ca.* 200 years ago in the eastern Gotland Sea (central Baltic Sea proper), and Andrén (1994) and Witkowski (1994) *ca.* 100 years ago in the Arkona Sea and Bornholm area (southern Baltic Sea proper). Miller & Risberg (1990) and Risberg (1990) found indications for an acceleration of eutrophication which started *ca.* 20 years ago in the western Gotland Sea (north-western Baltic Sea proper). This is approximately in concert with the formation of laminated sediments in the area. The extension of bottom areas with laminated sediments in the Baltic Sea proper started to increase in the late 1960s when large inputs of organic material caused bottom oxygen deficiency and absence of bioturbation because the fauna was killed (Fonselius, 1970; Jonsson & Jonsson, 1988). The observed indications of eutrophication in the upper sediments include increased concentrations of biogenic silica, changes in diatom species composition, large increases in the abundance of *Chaetoceros* spp. resting spores, and decreases in the number of diatom taxa. The mass occurrences of the resting spores in the upper sediments can be interpreted either as a result of increased growth of pelagic *Chaetoceros* spp. (as vegetative cells) by increased access to nutrients, or as a sign of detoriated living conditions for *Chaetoceros* spp. Shifts in species composition to the advantage of autochthonous brackish or brackish–marine plankton diatoms indicate an increased growth of diatoms in the pelagic zone in the direct vicinity of the core sampling site. Diatom taxa found deposited in large amounts when the Baltic photic zone is affected by eutrophication are the centric taxa: *Actinocyclus octonarius* Ehrenberg, *Coscinodiscus asteromphalus* Ehrenberg, *Cyclotella choctawhatcheeana* Prasad, *Thalassiosira baltica*, *Thalassiosira* cf. *levanderi* van Goor, *Thalassiosira hyperborea* var. *lacunosa* (Berg) Hasle, and *Thalassiosira hyperborea* var. *pelagica* (Cleve-Euler) Hasle. These taxa are not necessarily indicators of eutrophication when they are alive. Their presence in the upper sediment layer in greater amounts than in the layers below may indicate higher primary production of pelagic diatoms in general as well as the fact that they dissolve slower than, e.g., *Skeletonema costatum* (Greville) Cleve, or vegetative *Chaetoceros* spp.

When summarizing general eutrophication trends in the Baltic Sea using diatoms as indicators, it may be concluded that effects are measurable in the southern and central part, but not in the north. The diatom records in the sediments of the southern Baltic Sea show that the eutrophication process has been going on for at least 100 years, and that it accelerated in the 1970s. In the southern Baltic Sea, eutrophication initially resulted in an increased pelagic primary production of diatoms, after which part of the silica pool has been removed from the photic zone and deposited in the sediments. Instead of nitrogen depletion, now silica depletion may occur in this area and put an end to the diatom spring bloom. It has been noted that pelagic diatoms decrease in favor of (dino)flagellates and cyanobacteria, which is expected to result in lower concentrations of biogenic silica in sediments deposited in the 1990s. In the northern Baltic Sea silicate is not a limiting nutrient, and here eutrophication probably still favors diatom growth. Local eutrophication in different sites in the Baltic Sea may result in very different diatom responses, depending on salinity, nutrient levels and quotes, but also on other environmental factors such as exposure to wave action, light availability and water temperature.

Brackish water diatoms and temperature

Water temperature is one of the principal factors regulating biological processes. Natural climatic dynamics with changing temperature regimes have always affected the biota on earth. Presently ongoing climatic change is partly related to anthropogenic activities because the burning of fossil fuel changes the carbon dioxide balance in nature (Gregory, 1988; Moore & Braswell, 1994). Increasing concentrations of carbon dioxide and other gases originating from fossil fuel burning do not hinder solar insolation, but they do trap outgoing heat radiation in the lower atmosphere which results in a temperature raise. High-latitude areas, including the Baltic Sea, are considered to be especially sensitive to the 'greenhouse effect'; here higher future temperature increases are expected to occur than in equatorial regions (Alenius, 1989; Bach, 1989). Effects of increased water temperature on living diatom communities can be studied in the discharge areas for industrial cooling water which strongly affect the local environment by thermal pollution. Deposited diatom assemblages can be used to trace past climatic changes and the results of such studies can be used for predicting expected future changes.

An important environmental factor closely related to water temperature is the occurrence of an ice cover (Leppäranta, 1989). After warm winters the phytoplankton spring bloom in the southern Baltic Sea is dominated by (dino)flagellates although some diatom species do occur, e.g., *Skeletonema costatum*, but after a cold winter the bloom is dominated by diatoms. The chain-forming diatom *Pauliella taeniata*, normally one of the phytoplankton dominants in the Baltic Sea, and dependent on ice-formation, has not occurred in the blooms in the southern Baltic Sea during the 1990s (S. Schulz, pers. comm.). *P. taeniata* is often found deposited in Baltic sediment assemblages (Hällfors & Niemi, 1975). This species indicates ice formation, and can thus be considered a marker for climatic change if ice formation on the Baltic Sea would be inhibited. In the discharge area of the Olkiluoto nuclear power plant on the Finnish west coast (Bothnian Sea), it was found that the spring maximum of *Chaetoceros wighamii* Brightwell occurred two weeks earlier than normal, and that the abundance of *Rhizosolenia minima* Levander was slightly increased (Keskitalo, 1987). However, the temperature rise did not markedly affect the species composition or total phytoplankton quantities in the discharge area. The same was found for the phytoplankton off the Forsmark nuclear power plant on the Swedish Bothnian Sea coast (Willén, 1985). On the contrary, shifts from diatoms to cyanobacteria were found in the plankton off a nuclear power plant near St. Petersburg (Russia) at the Gulf of Finland coast (Ryabova et al., 1994). This is probably explained by the nutrient status of the water: nearly oligotrophic at Olkiluoto and Forsmark and eutrophied at St. Petersburg. Large compositional changes, however, take place in the littoral algal communities in the cooling water discharge areas at Olkiluoto and Forsmark: red and brown macroalgae decrease in growth and abundance whereas green algae increase, as well as cyanobacteria in the warm season and diatoms in the cold season (Keskitalo & Heitto, 1987; Snoeijs & Prentice, 1989).

Potapova & Snoeijs (1997) showed how the life cycle in natural populations of *Diatoma moniliformis* Kützing is influenced by water temperature. This diatom is a dominant epiphyte on the green filamentous macroalga *Cladophora glomerata* (L.) Kützing in the northern Baltic Sea (salinity <6‰). Higher temperature may impose a direct selective pressure upon a population when individuals with a particular morphology are favored or disadvantaged. It may also affect a population's morphometric composition by changing the timing of the congenital life cycle. Potapova & Snoeijs (1997) followed the life cycle in two field populations of *D. moniliformis* by monthly measuring the size structure during three whole years. One of the populations was growing under natural conditions, the other one in the discharge area for the cooling water from a nuclear power plant with *ca*. 10 °C increased water temperature and increased water flow rates (increased nutrient availability). At the reference site the life cycle was well synchronized. Sexualization occurred in the cold season, after which size reduction during the vegetative life cycle could be divided into two parts. During the first part, cell volume decreased and surface to volume ratios increased. During the second part, both parameters decreased. At the reference site, the timing of the life cycle also fitted well with seasonal cycles of environmental factors. The cells of *D. moniliformis* always had highly competitive proportions, enabling fast nutrient uptake and growth (high surface to volume ratios) during the period of optimum growth in late spring (May–June). At the site affected by cooling water discharge, synchronization of the congenital life cycle was basically disrupted (Fig. 14.5). It seems probable that under natural conditions, auxosporulation is triggered by a combination of low water temperature (in the Bothnian Sea *ca*. 0–3 °C) and rapidly increasing light intensity and/or daylength in late winter–early spring. When one of these conditions is not met (at the heated site the required low temperature was absent), auxosporulation did not occur simultaneously.

The effect of anomalous water temperatures on benthic diatom community dynamics have been studied in an experimental enclosure of the southern Bothnian Sea, the Forsmark Biotest basin (Snoeijs, 1989, 1990, 1991). This research basin, on full natural scale, is the main Swedish facility for field studies on the effects of enhanced temperature on aquatic ecosystems (Snoeijs, 1994b). The basin consists of a 1 km² artificial enclosure of the Baltic Sea that

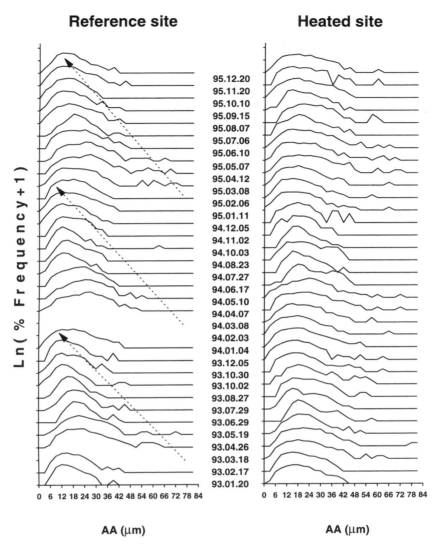

Reference site　　　　　**Heated site**

95.12.20	
95.11.20	
95.10.10	
95.09.15	
95.08.07	
95.07.06	
95.06.10	
95.05.07	
95.04.12	
95.03.08	
95.02.06	
95.01.11	
94.12.05	
94.11.02	
94.10.03	
94.08.23	
94.07.27	
94.06.17	
94.05.10	
94.04.07	
94.03.08	
94.02.03	
94.01.04	
93.12.05	
93.10.30	
93.10.02	
93.08.27	
93.07.29	
93.06.29	
93.05.19	
93.04.26	
93.03.18	
93.02.17	
93.01.20	

Ln (% Frequency + 1)

0 6 12 18 24 30 36 42 48 54 60 66 72 78 84　　　0 6 12 18 24 30 36 42 48 54 60 66 72 78 84

AA (μm)　　　　　**AA (μm)**

Fig. 14.5. Frequency distribution curves for the apical axis (AA) of the epiphyte *Diatoma moniliformis* at a reference site moderately exposed to wave action, and a site receiving cooling water discharge with a temperature anomaly of ca. 10 °C and a unidirectional flow rate of ca. 2 m/s. Monthly samples were taken during three years from January 1993 until December 1995. From Potapova & Snoeijs (1997), printed with permission from the *Journal of Phycology*.

receives brackish cooling water discharge from a nuclear power plant. Salinity is constant in the area (5‰), and the water is heated *ca.* 10 °C by the power-plant cooling water. Inside the basin, the diatoms react to higher water temperature throughout the year by an extended growing season and increased primary production. The latter effect is similar to that of eutrophication when silicate is not limiting (which is never the case in Forsmark, Snoeijs, 1989). The effects of

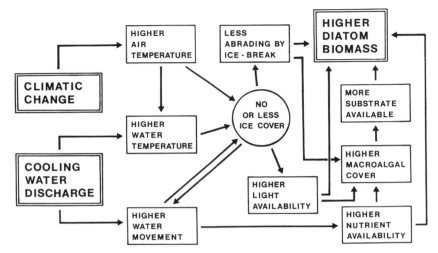

Fig. 14.6. Hypothetical synthesis of interactions between environmental factors that lead to higher diatom biomass. From Snoeijs (1990), printed with permission from the *Journal of Vegetation Science*.

anomalous temperatures on diatoms are most pronounced during the cold season. The ecological factors directly affecting the enhanced growth of the diatoms are reduced loss by reduced abrading by ice and higher production by higher availability of light, nutrients and substrate in the form of macroalgae (Fig. 14.6). Very large (dm long) epilithic diatom blooms can also be observed in spring at 0 °C, e.g., *Berkeleya rutilans* (Trentepohl) Grunov in the Bothnian Sea after a winter without ice cover (Snoeijs, 1990), or *Navicula ramosissima* (C. A. Agardh) Cleve under the ice cover in the Baltic Sea proper (Snoeijs & Kautsky, 1989). This indicates that, during the cold season, low water temperature is not a direct limiting factor for diatoms as a group. Low light intensity is only limiting around mid-winter (December–January), and inorganic nutrient concentrations are high in the same period. All these factors favor diatom growth during the cold season, in comparison with growth of other algal groups when an ice-cover is limited or absent.

The temperature anomaly inside the Forsmark Biotest basin also causes species shifts. This effect is most pronounced during the cold season, and seems to be regulated by temperature optimum for some species, but by other factors (e.g., absence of ice cover) for other species. Generally observed trends are that the diatoms enhanced by the temperature anomaly are large diatoms with colonial growth in chains (*Achnanthes brevipes* C. A. Agardh, *Melosira lineata* (Dillwyn) C. A. Agardh, *Melosira moniliformis* (O. F. Müller) C. A. Agardh), smaller diatoms with colonial growth in mucilage tubes (*Navicula perminuta* Grunov, *Nitzschia filiformis* (W. Smith) van Heurck), or large epiphytic diatoms (*Tabularia tabulata* (C. A. Agardh) Snoeijs, *Licmophora cf. gracilis* (Ehrenberg) Grunov). The species favored most are 'cold water and low light species' with growth optima around 10 °C in winter and early spring (e.g., *Melosira* spp.), and 'warm water species' with growth optima above 20 °C in late summer and autumn (e.g., *Nitzschia filiformis*). A possibly introduced species belonging to the latter category is *Pleurosira laevis* fo. *polymorpha* (Kützing) Compère. It was probably introduced during an experimental release of eels in the area; the eels were raised in aquaria in southern Europe. This large chain-forming diatom was never observed before 1989, but since 1990 it forms up to 0.5 m high colonies in the discharge area each year in September–November, especially in sites with slowly flowing water. Some diatom species do not increase or decrease greatly with temperature, but their occurrence and abundance in time changes, e.g., the epiphytic spring maximum of *Diatoma moniliformis* may occur up to two months earlier in the year (March instead of May). Few of the abundant diatom species in the area do not show any response to a raised temperature regime, one of these is the common epiphyte *Rhoicosphenia curvata*, which seems to have a very wide temperature tolerance and no clear optimum.

Brackish water diatoms and light

Reduction in the thickness of the ozone layer in the stratosphere are strongly correlated to increase in biologically damaging solar ultraviolet-B radiation (e.g., through the production of free radicals) reaching the Earth's surface (Palenik *et al.*, 1991; Madronich *et al.*, 1995). Many laboratory studies with brackish and marine microalgae have shown that ultraviolet-B radiation (UVBR, 280–320 nm) negatively affects diatom growth rates by inhibition of photosynthesis (photoinhibition), as well as inhibition of synthesis of DNA, RNA, protein and acyl lipids, dependent on the dose UVBR received by the algae (Doehler, 1989; Cullen & Lesser, 1991; Behrenfeld *et al.*, 1992; Doehler & Bierman, 1994; Lesser *et al.*, 1994). Helbling *et al.* (1992) found that ultraviolet-A and -B (UVR) radiation were more inhibitory to microplankton than to nanoplankton, and that it induced the formation of resting spores of the diatom genus *Chaetoceros*. Some species are better adapted to UVR than others, e.g., phytoplankton from tropical waters showed marked resistance to UVR as compared to Antarctic phytoplankton (Helbling *et al.*, 1992), smaller cells with greater surface area:volume ratios sustained more damage per unit DNA (Karentz *et al.*, 1991), and closely related diatom species may show marked differences in responses to UVR in the laboratory: the temperate species *Odontella sinensis* (Greville) Grunov (occurring in the Baltic Sea) is sensitive, and the Antarctic species *Odontella weissflogii* (Janisch) Grunov is insensitive (Doehler *et al.*, 1995), *Pseudo-nitzschia fraudulenta* (P. T. Cleve) Hasle and *Pseudo-nitzschia pungens* (Grunov) Hasle sensitive, and *Pseudo-nitzschia multiseries* (Hasle) Hasle insensitive (Hargraves *et al.*, 1993). Compositional changes in the diatom component of natural phytoplankton communities in the Baltic Sea attributed to UVR have not been documented, but increasing accumulation of *Chaetoceros* resting spores in the sediments as found by e.g., Andrén (1994) may be one of the effects as experimentally found by Helbling *et al.* (1992). On the

Table 14.2. *Fission products from Chernobyl measured in epilithic diatom communities in Forsmark (Sweden) one week after the Chernobyl accident (6 May 1986) in kBq/kg dry weight*

Radionuclide	Water movement			
	Quiescent	Slowly flowing	Flowing	Fast flowing
^{110m}Ag*	1	1	<1	1
^{134}Cs	14	6	3	3
^{137}Cs	25	10	6	4
^{132}Te	87	39	18	21
^{131}I	181	48	24	21
^{140}Ba	439	124	32	9
^{103}Ru	540	142	49	16
^{144}Ce	612	218	42	11
^{141}Ce	825	310	58	15
^{95}Zr	864	328	64	14
^{95}Nb	1022	339	78	18

Note:
* Discharged by both the Chernobyl accident and the Forsmark nuclear power plant.
Source: Sampling sites are arranged according to water movement conditions (from Snoeijs & Notter 1993a, Table printed with permission from the *Journal of Environmental Radioactivity*).

Swedish west coast (Skagerrak), mesocosm studies on the effects of UVBR on microbenthic sediment communities showed significantly decreased carbon fixation and lower net oxygen production (Sundbäck *et al.*, 1997). However, no changes in pigment or algal composition were found in these experiments, which may be attributed to vertical migration of epipelic diatoms as a key mechanism for avoiding damage by UVBR.

Brackish water diatoms and contaminants

Littoral diatoms were shown to be good monitoring organisms for radio-nuclides in the Forsmark area, southern Baltic Sea (Snoeijs & Notter 1993a). As monitoring organisms, they occupy a position between macroalgae and sediment. Diatoms have advantages over sediment, being part of the food chain and of responding quickly to changes in actual concentrations in the water (Table 14.2). Compared with macrophytes and animals, diatoms have higher radionuclide concentrations, less selectivity for kind of radionuclide, and the concentrations are less dependent on physiological and seasonal cycles. Radionuclide concentrations seem to be related to the algal surface area exposed to the water. Epilithic diatoms have the highest concentrations,

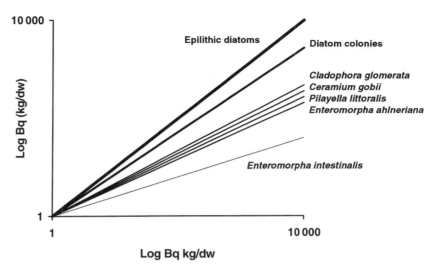

Fig. 14.7. Radionuclide concentrations in different algal samples (diatom growth forms and filamentous macroalgae) compared with concentrations in epilithic diatoms. Each pair of data consisted of one sample of epilithic diatoms and one other sample ($n = 134$ pairs). All radionuclides measured in both samples were used in the analysis. All regression lines were significant with $p < 0.001$, note that logarithmic scales are used. The diatom colonies consisted mainly of *Berkeleya rutilans* colonies in mucilage tubes. The analysis included four thin filamentous algal species: *Cladophora glomerata* (L.) Kützing, *Ceramium Gobii* Wærn, *Pilayella littoralis* (L.) Kjellman, *Enteromorpha ahlneriana* Bliding, and one filamentous algal species with wider thallus: *Enteromorpha intestinalis* (L.) Link. Adapted after Snoeijs & Notter (1993*a*).

diatoms living in mucilage tubes slightly lower, thin filamentous macroalgae still lower, and macroalgae with broader filaments the lowest (Fig. 14.7). Diatoms are at present (since 1992), used as monitoring organisms for radionuclide discharges at all four Swedish nuclear power plants. Radionuclides recycle in epilithic diatom communities to a high extent. This implies that distinct discharges of the Chernobyl-type can be traced in diatom samples for a long time. Concentrations of ^{137}Cs in epilithic diatom communities five years after the Chernobyl accident still reflected the original fallout pattern of the radioactive cloud that crossed the Swedish east coast, with mean levels of 1600 to 2100 Bq/kg dry weight in the Forsmark – Gävle region (Snoeijs & Notter, 1993*b*; Carlson & Snoeijs, 1994). New generations of diatoms still continuously take up ^{134}Cs and ^{137}Cs from the environment as these diatom communities show large yearly fluctuations in biomass, and the radiocesium isotopes only disappear from the communities according to their physical half-lives.

In the same way as for radionuclides, natural diatom communities and diatom assemblages grown on artificial substrata can be used for monitoring metals in the environment. Epilithic diatom samples taken in 1991 outside the Rönnskär industry which discharges heavy metals (Bothnian Bay, Sweden) had metal levels corresponding to the distance from the discharge point (Fig. 14.8). They seem to act as a 'sponge' attracting metal ions from the water. The

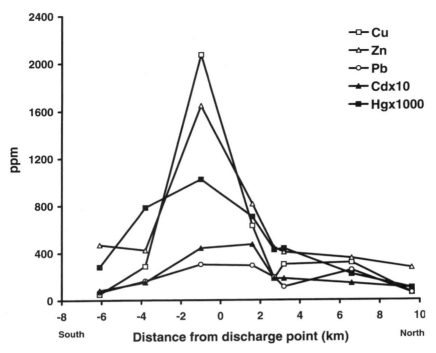

Fig. 14.8. Concentrations of copper, zinc, lead, cadmium (shown with factor ×10), and mercury (shown with factor ×1000) in epilithic diatom samples taken outside the Rönnskär industry which discharges heavy metals (Bothnian Bay, Sweden) in May 1991. Nine epilithic diatom samples taken in January 1992 at about the same distances from the discharge point outside the Forsmark nuclear power plant which discharges practically no heavy metals (Bothnian Sea, Sweden) showed the following mean values: Cu 24 ± 7 ppm, Zn 170 ± 48 ppm, Pb 31 ± 14 ppm, Cd 3 ± 1 ppm, Hg 0.15 ± 0.05 ppm. Data: P. Snoeijs (previously unpublished data).

(radioactive) metals attach mainly to the outside of the cell wall as shown by Pedersén et al. (1981) and Lindahl et al. (1983) for *Thalassiosira baltica* and *Chaetoceros wighamii* from the Baltic Sea, and only a minor portion is actually taken up into the cell. Diatoms have a thin organic casing around the silica frustule, and one of its functions may be to complex cations such as iron or aluminum for minimizing loss of silica from the frustule by dissolution (Round et al., 1990). As most radionuclides and metals are cations, it is probable that they are attached to the organic casing around the diatom frustules.

Periphyton assemblages on artificial substrata dominated by diatoms can also effectively be used as pollution indicators for toxic compounds, such as tri-n-butyl tin (TBT), in short-term toxicity tests (Molander et al., 1990; Molander & Blanck, 1992; Dahl & Blanck, 1996; Blanck & Dahl, 1996). This method, called PICT (Pollution Induced Community Tolerance), uses the selective pressure of a toxicant on sensitive species, so that the community as a whole shows an increase in tolerance for the toxicant. The increase in community tolerance is an indicator of damage to the algal community. For example, assemblages grown on artificial substrata in the field at different distances from a discharge point can, in the laboratory, be tested for PICT, chlorophyll-*a*, photosynthesis and species composition.

Summary

This chapter deals with the main environmental factors influencing diatom growth and distribution (salinity, nutrients, temperature and light) in brackish water, and how diatoms react when human society changes the conditions for life by eutrophication and climatic change. It is discussed how brackish water diatoms can, and are, being used for measuring natural dynamics and artificial changes, as well as for monitoring discharges of contaminants. Three factors must always be taken into account when using diatoms as environmental markers of salinity: (i) In environments with fluctuating salinity regimes, the species are selected more according to their ability to cope with changing salinity (euryhalinity) than to their salinity optima. (ii) In environments with stable salinity regimes, evolutionary processes have resulted in various degrees of endemism depending on the geological age and stability of the water body and the degree of isolation from other populations of the same species. (iii) Species perform differently (physiologically and ecologically) in different brackish waters because they are, besides salinity, also affected by other environmental and biological constraints. This chapter gives many examples from the Baltic Sea, which is one of the world's largest brackish water bodies.

Acknowledgments

I thank Eugene F. Stoermer, John P. Smol, Kerstin Wallström and one anonymous reviewer for valuable comments on the manuscript.

References

Åker, K., Eriksson, B., Grönlund, T., & Kankainen, T. (1988). Sediment stratigraphy in the northern Gulf of Finland. *Geological Survey of Finland, Special Paper*, **6**, 101–17.

Alenius, P. (1989). Variations of the sea temperature around the coasts of Finland. In: Conference on Climate and Water. *Publications of the Academy of Finland*, **9**, 51–62.

Alhonen, P. (1971). The stages of the Baltic Sea as indicated by the diatom stratigraphy. *Acta Botanica Fennica*, **92**, 3–17.

    (1986). Late Weichselian and Flandrian diatom stratigraphy: Methods, results and research tendencies. In *Nordic Late Quaternary Biology and Ecology*. ed. L-K. Königsson. *Striae*, **24**, 27–33.

Andrén, E. (1994). Recording environmental changes in the southern Baltic Sea – current results from a diatom study within Project ODER. In *Proceedings of the 13th International Diatom Symposium*, ed. D. Marino & M. Montresor, pp. 443–55. Bristol: Biopress Limited.

Anonymous (1959). Final Resolution of the Symposium on the Classification of Brackish Waters. *Archivio di Oceanografia e Limnologia (Supplement)*, **11**, 243–5.

Archibald, R. E. M. (1983). The diatoms of the Sundays and Great Fish rivers in the eastern Cape Province of South Africa. *Bibliotheca Diatomologica*, **1**, 1–362 + 34 pls.

Bach, W. (1989). Projected climatic changes and impacts in Europe due to increased CO_2. In Conference on Climate and Water. *Publications of the Academy of Finland*, **9**, 31–50.

Balode, M. (1996). Long-term changes of summer–autumn phytoplankton communities in the Gulf of Riga. In *Baltic Sea and Mediterranean Sea – A Comparative Ecological Approach of Coastal Environments and Paralic Ecosystems*, ed. Guelorget, O., & Lefebre, A., pp. 96–9. Montpellier: Ifremer.

Begon, M., Harper, J. L., & Townsend, C. R. (1996). *Ecology, Individuals, Populations and Communities*, 3rd edn. Oxford: Blackwell Science.

Behrenfeld, M. J., Hardy, J. T., & Lee, H. I. (1992). Chronic effects of ultraviolet-B radiation on growth and cell volume of *Phaeodactylum tricornutum* (Bacillariophyceae). *Journal of Phycology*, **28**, 757–60.

Bergström, S., & Carlsson, B. (1994). River runoff to the Baltic Sea: 1950–1990. *Ambio*, **23**, 280–7.

Blanck, H., & Dahl, B. (1996). Pollution-induced community tolerance (PICT) in mariner periphyton in a gradient of tri-n-butyltin (TBT) contamination. *Aquatic Toxicology (Amsterdam)*, **35**, 59–77.

Cadée, G. C., & Hegeman, J. (1991). Historical phytoplankton data of the Maarsdiep. *Hydrobiology Bulletin*, **24**, 111–18.

Caraco, N, Cole, J., & Likens, G. E. (1990). A comparison of phosphorus immobilization in sediments of freshwater and coastal marine systems. *Biochemistry*, **9**, 277–90.

Carlson, L., & Snoeijs, P. (1994). Radiocaesium in algae from Nordic coastal waters. In *Nordic Radioecology – The Transfer of Radionuclides through Nordic Ecosystems to Man*, ed. H. Dahlgaard, Studies in Environmental Science, vol. 62, Chapter 2.7, pp. 105–117. Amsterdam: Elsevier Science.

Carpelan, L. H. (1978a). Evolutionary euryhalinity of diatoms in changing environments. *Nova Hedwigia*, **29**, 489–526.

(1978b). Revision of Kolbe's system der Halobien based on diatoms of Californian lagoons. *Oikos*, **31**, 112–22.

Cato, I, Dale, B., & Miller, U. (1985). Mikrofossil som eutrofieringsindikator. *Report, Geological Survey of Sweden*, 16 pp. [in Swedish]

Cleve, P. T. (1899). Bidrag till Kännedom om Östersjöns och Bottniska vikens postglaciala geologi. *Sveriges Geologiska Undersökningar*, **C 180** [in Swedish]

Conley, D. J., & Johnstone, R. W. (1995). Biochemistry of N, P, and Si in Baltic Sea sediments: Response to a simulated deposition of a spring diatom bloom. *Marine Ecology Progress Series*, **122**, 265–76.

Conley, D. J., Schelske, C. L., & Stoermer, E. F. (1993). Modification of the biochemical cycle of silica with eutrophication. *Marine Ecology Progress Series*, **101**, 179–92.

Cullen, J. J., & Lesser, M. P. (1991). Inhibition of photosynthesis by ultraviolet radiation as a function of dose and dosage rate: Results for a marine diatom. *Marine Biology (Berlin)*, **111**, 183–90.

Dahl, B. & Blanck, H. (1996). Use of sand-living microalgal communities (epipsammic) in ecotoxicological testing. *Marine Ecology Progress Series*, **114**, 163–73.

den Hartog, C. (1964). Typologie des Brackwassers. *Helgoländer Wissenschaftliche Meeresuntersuchungen*, **10**, 377–90.

Doehler, G. (1989). Influence of UV-B (290–320 nm) radiation on photosynthetic carbon-14 dioxide fixation of *Thalassiosira rotula* Meunier. *Biochemie und Physiologie der Pflanzen*, **185**, 221–6.

Doehler, G., & Biermann, T. (1994). Impact of UV-B radiation on the lipid and fatty acid composition of synchronized *Ditylum brightwellii* (West) Grunow. *Zeitschrift für Naturforschung, Section C: Biosciences*, **49**, 607–14.

Doehler, G., Hoffmann, M., & Stappel, U. (1995). Patterns of proteins after heat shock and UV-B radiation of some temperate marine diatoms and the Antarctic *Odontella weissflogii*. *Botanica Acta*, **108**, 93–8.

Edler, L., Hällfors, G., & Niemi, Å. (1984). A preliminary check-list of the phytoplankton of the Baltic Sea. *Acta Botanica Fennica*, **128**, 1–26.

Egge, J. K., & Jacobsen, A. (1997). Influence of silicate on particulate carbon production in phytoplankton. *Marine Ecology Progress Series*, **147**, 219–30.

Elmgren, R. (1989). Man's impact on the ecosystem of the Baltic Sea: Energy flows today and at the turn of the century. *Ambio*, **18**, 326–32.

Eppley, R. W. (1977). The growth and culture of diatoms. In *The Biology of Diatoms*. ed. D. Werner. pp. 24–64. Botanical Monographs, Volume 13. Oxford: Blackwell Scientific Publications.

Escaravage, V., Prins, T. C., Smaal, A. C., & Peeters, J. C. H. (1996). The response of phytoplankton communities to phosphorus input reduction in mesocosm experiments. *Journal of Experimental Marine Biology and Ecology*, **198**, 55–79.

Fairbridge, R. (1980). The estuary: its definition and geodynamic cycle. In *Chemistry and Biochemistry of Estuaries*, ed. E. Olausson & I. Cato, pp. 1–35. New York: John Wiley.

Fonselius, S. (1970). Stagnant sea. *Environment*, **12**, 2–11, 40–8.

Gregory, S. (ed.) (1988). *Recent Climatic Changes*. London: Belhaven Press.

Grönlund, T. (1993). Diatoms in surface sediments of the Gotland Basin in the Baltic Sea. *Hydrobiologia*, **269/270**, 235–42.

    (1994). Lagoonal diatom flora of the Holocene Baltic *Litorina* Sea in comparison with the Eemian Baltic Sea flora. In *Proceedings of the 11th International Diatom Symposium*, ed. J. P. Kociolek, pp. 349–57. San Francisco: California Academy of Sciences.

Gudelis, V., & Königsson, L-K. (1979) (eds.). The quaternary history of the Baltic. *Acta Universitatis Upsaliensis, Symposia Universitatis Upsaliensis Annuum Quingentesimum Celebrantis*, **1**, 1–279.

Hällfors, G., & Niemi, Å. (1975). Diatoms in the surface sediment from the deep basins in the Baltic Proper and the Gulf of Finland. *Havsforskningsinstitutets Skrifter*, **240**, 71–7.

Hargraves, P. E., Zhang, J., Wang, R., & Shimizu, Y. (1993). Growth characteristics of the diatom *Pseudonitzschia pungens* and *P. fraudulenta* exposed to ultraviolet radiation. *Hydrobiologia*, **269/270**, 207–12.

Hasle, G. R., & Syvertsen, E. E. (1990). Arctic diatoms in the Oslofjord and the Baltic Sea – a bio- and paleogeographic problem? In *Proceedings of the 10th International Diatom Symposium*, ed. H. Simola, pp. 285–300. Koenigstein: Koeltz.

Hasle, G. R., Lange, C. B., & Syvertsen, E. E. (1996). A review of *Pseudo-nitzschia*, with special reference to the Skagerrak, North Atlantic, and adjacent waters. *Helgoländer Meeresuntersuchungen*, **50**, 131–75.

Heckey, R. E., & Kilham, P. (1988). Nutrient limitation of phytoplankton in freshwater and marine environments: a review of recent evidence on the effects of enrichment. *Limnology and Oceanography*, **33**, 796–822.

Hedenström, A. (1996). Preboreal shore displacement on central Södertörn peninsula, eastern middle Sweden. *Quaternaria (Stockholm), Ser. B*, **7**, 1–50.

Heiskanen, A. S., & Kononen, K. (1994). Sedimentation of vernal and late summer phytoplankton communities in the coastal Baltic Sea. *Archiv für Hydrobiologie*, **131**, 175–98.

Helbing, E. W., Villafane, V., Ferrario, M., & Holm-Hansen, O. (1992). Impact of natural ultraviolet radiation on rates of photosynthesis and on specific marine phytoplankton species. *Marine Ecology Progress Series*, **80**, 89–100.

Hoagland, K. D., Ernst, S. G., Jensen, S. I., Lewis, R. J., Miller, V. I., & Denicola, D. M. (1994). Genetic variation in *Fragilaria capucina* clones along a latitudinal gradient across North America, a baseline for detecting global climate change. In *Proceedings of the 13th International Diatom Symposium*, ed. D. Marino & M. Montresor, pp. 385–92. Bristol: Biopress Limited.

Hofmann, G. (1994). Aufwuchs-Diatomeen in Seen und ihre Eignung als Indikatoren der Trophie. *Bibliotheca Diatomologica*, **30**, 1–241.

Holmquist, E., & Willén, T. (1993). Fish mortality caused by *Prymnesium parvum*. *Vatten*, **49**, 110–5.

Humborg, C., Ittekkot, V., Cociasu, A., & von Bodungen, B. (1997). Effect of Danube River dam on Black Sea biochemistry and ecosystem structure. *Nature*, **386**, 385–8.

Hustedt, F. (1953). Die Systematik der Diatomeen in ihren Beziehungen zur Geologie und Ökologie nebst einer Revision des Halobien-Systems. *Svensk Botanisk Tidskrift*, **47**, 509–19.

Huttunen, M., & Niemi, Å. (1986). Sea-ice algae in the northern Baltic Sea. *Memoranda Societatis pro Fauna et Flora Fennica*, **62**, 58–62.

Ignatius, H., & Tynni, R. (1978). Itämeren vaiheet ja piilevätutkimus [Baltic Sea stages and diatom analysis]. *Tuurun yliopiston maaperägeologia osaston julkaisuja*, **36**, 1–26. [in Finnish, with English summary]

Ignatius, H., Axberg, S, Niemistö, L., & Winterhalter, B. (1981). Quaternary Geology of the Baltic Sea. In *The Baltic Sea*, ed. A. Voipio, pp. 54–105. Amsterdam: Elsevier Science Publishers.

Jacobsen, A., Egge, J. K., & Heimdal, B. R. (1995). Effects of increased concentration of nitrate and phosphate during a spring bloom experiment in mesocosm. *Journal of Experimental Marine Biology and Ecology*, **187**, 239–51.

John, J. (1983). The diatom flora of the Swan River estuary, Western Australia. *Bibliotheca Phycologica*, **64**, 1–359.

Jonsson, P., & Carman, R. (1994). Changes in deposition of organic matter and nutrients in the Baltic Sea during the twentieth century. *Marine Pollution Bulletin*, **28**, 417–26.

Jonsson, P., & Jonsson, B. (1988). Dramatic changes in Baltic sediments during the last three decades. *Ambio*, **17**, 158–60.

Juggins, S. (1992). Diatoms in the Thames estuary, England. Ecology, paleoecology, and salinity transfer function. *Bibliotheca Diatomologica*, **25**, 1–216.

Juhlin-Dannfelt, H. (1882). On the diatoms of the Baltic Sea. *Bihang till Kungliga Svenska Vetenskapsakademiens Handlingar*, **6**, 1–52.

Kabailiene, M. (1995). The Baltic Ice Lake and Yoldia Sea stages, based on data from diatom analysis in the central, south-eastern and eastern Baltic. *Quaternary International*, **27**, 69–72.

Kahru, M., Horstmann, U., & Rud, O. (1994). Satellite detection of increased cyanobacteria blooms in the Baltic Sea: Natural fluctuation or ecosystem change? *Ambio*, **23**, 469–72.

Kangas, P. Alasaarela, E., Lax, H. G. Jokela, S., & Storgård-Envall, C. (1993). seasonal variation of primary production and nutrient concentrations in the coastal waters of the Bothnian Bay and the Quark. *Aqua Fennica*, **23**, 165–76.

Karayeva, N. I., & Makarova, I. V. (1973). Special features and origin of the Caspian Sea diatom flora. *Marine Biology, Berlin*, **21**, 269–75.

Karentz, D., Cleaver, J. E., & Mitchell, D. L. (1991). Cell survival characteristics and molecular responses of Antarctic phytoplankton to UV radiation. *Journal of Phycology*, **27**, 326–41.

Kautsky, N., Johannesson, K., & Tedengren, M. (1990). Genotypic and phenotypic differences between Baltic and North Sea populations of *Mytilus edulis* evaluated through reciprocal transplantations. I: Growth and morphology. *Marine Ecology Progress Series*, **59**, 203–10.

Keskitalo, J. (1987). Phytoplankton in the sea area off the Olkiluoto nuclear power station, west coast of Finland. *Annales Botanici Fennici*, **24**, 281–99.

Keskitalo, J., & Heitto, L. (1987). Overwintering of benthic vegetation outside the Olkiluoto nuclear power station, west coast of Finland. *Annales Botanici Fennici*, **24**, 231–43.

Ketchum, B. H. (ed.) (1983). Estuaries and enclosed seas. *Ecosystems of the World*, **26**, 1–500.

Kolbe, R. W. (1927). Zur Ökologie, Morphologie und Systematik der Brackwasser-Diatomeen. *Pflanzenforschung*, **7**, 1–146.

(1932). Grundlinien einer allgemeinen Ökologie der Diatomeen. *Ergebnisse der Biologie*, **8**, 221–348.

Korhola, A., & Blom, T. (1996). Marked early 20th century pollution and the subsequent recovery of Toolo Bay, central Helsinki, as indicated by subfossil diatom assemblage changes. *Hydrobiologia*, **341**, 169–79.

Kuylenstierna, M. (1989–1990). *Benthic Algal Vegetation in the Nordre Älv Estuary (Swedish west Coast). Doctoral Dissertation*, Gothenburg University, Sweden. Vol. 1, Text (1990), Vol. 2, Plates (1989).

Lange-Bertalot, H. (1979). Toleranzgrenzen und Populationsdynamik benthischer Diatomeen bei unterschiedlich starker Abwasserbelastung. *Archiv für Hydrobiologie, Supplement*, **56**, 184–219.

——— (1990). Current biosystematic research on diatoms and its implications for the species concept. *Limnetica*, **6**, 12–22.

Leppäkoski, E., & Mihnea, P. E. (1996). Enclosed seas under man-induced change: a comparison between the Baltic and Black Seas. *Ambio*, **25**, 380–9.

Leppäranta, M. (1989). On climatic variations of the Baltic sea-ice conditions. In: Conference on Climate and Water. *Publications of the Academy of Finland*, **9**, 63–72.

Leskinen, E., & Hällfors, G. (1990). Community structure of epiphytic diatoms in relation to eutrophication on the Hanko peninsula, south coast of Finland. In *Proceedings of the 10th International Diatom Symposium*, ed. H. Simola, pp. 323–33. Koenigstein: Koeltz.

Lesser, M. P., Cullen, J. J., & Neale, P. J. (1994). Carbon uptake in a marine diatom during acute exposure to ultraviolet B radiation: Relative importance of damage and repair. *Journal of Phycology*, **30**, 183–92.

Lindahl, G., Wallström, K., Roomans, G. M., & Pedersén, M. (1983). X-ray microanalysis of planktic diatoms in *in situ* studies of metal pollution. *Botanica Marina*, **26**, 367–73.

Madronich, S., McKenzie, R. L., Caldwell, M. M., & Björn, L. O. (1995). Changes in ultraviolet radiation reaching the Earth's surface. *Ambio*, **24**, 143–52.

Mann, D. G. (1988). The nature of diatom species, analyses of sympatric populations. In *Proceedings of the 9th International Diatom Symposium*, ed. F. E. Round, pp. 317–27. Koenigstein: Koeltz.

——— (1989). The species concept in diatoms, evidence for morphologically distinct, sympatric gamodemes in four epipelic species. *Plant Systematics and Evolution*, **164**, 215–37.

——— (1994). The origins of shape and form in diatoms, the interplay between morphogenetic studies and systematics. In *Shape and Form in Plants and Fungi*. pp. 17–38. London: The Linnean Society.

Mann, D. G., & Kociolek, J. P. (1990). The species concept in diatoms, report on a workshop. In *Proceedings of the 10th International Diatom Symposium*, ed. H. Simola, pp. 577–83. Koenigstein: Koeltz.

Meriläinen, J., Huttunen, P., & Batterbee, R. W. (eds.) (1983). Paleolimnology. *Hydrobiologia*, **103**, 1–318.

Miller, U. (1982). Shore displacement and coastal dwelling in the Stockholm region during the past 5000 years. *Annales Academiae Scientiarum Fennicae A*, **III**, 185–211..

——— (1986). Ecology and paleoecology of brackish water diatoms with special reference to the Baltic Basin. *Proceedings of the 8th International Diatom Symposium*, ed. M. Ricard, pp. 601–11. Koenigstein: Koeltz.

Miller, U., & Risberg, J. (1990). Environmental changes, mainly eutrophication, as recorded by fossil siliceous micro-algae in two cores from the uppermost sediments of the north-western Baltic. *Beiheft zur Nova Hedwigia*, **100**, 237–53.

Miller, U., & Robertsson, A-M. (1979). Biostratigraphical investigations in the Anundsjö region, Ångermanland, northern Sweden. *Early Norrland*, **12**, 1–76.

Molander, S., & Blanck, H. (1992). Detection of pollution-induced community tolerance (PICT) in marine periphyton communities established under diuron exposure. *Aquatic Toxicology (Amsterdam)*, **22**, 129–43.

Molander, S., Blanck, H., & Söderström, M. (1990). Toxicity assessment by pollution-induced tolerance (PICT), and identification of metabolites in periphyton communities after exposure to 4,5,6–trichloroguaiacol. *Aquatic Toxicity (Amsterdam)*, **18**, 115–36.

Moore, B., & Braswell, B. H., Jr. (1994). Planetary metabolism: understanding the carbon cycle. *Ambio*, **23**, 4–12.

Morris, R. J., Niemi, Å., Niemistoe, L., & Poutanen, E. L. (1988). Sedimentary record of seasonal production and geochemical fluxes in a nearshore coastal embayment in the northern Baltic Sea. *Finnish Marine Research*, **256**, 77–94.

Munthe, H. (1892). Studier ofer Baltiska hafets qvartära historia. *Bihang till Kungliga Svenska Vetenskaps Akademins Handlingar*, **18 II**.

Nehring, D., & Matthäus, W. (1991). Current trends in hydrographic and chemical parameters and eutrophication in the Baltic Sea. *Internationale Revue der gesamten Hydrobiologie*, **76**, 297–316.

Nielsen, R., Kristiansen, A., Mathiesen, L., & Mathiesen, H. (eds.) (1995). Distributional index of the benthic macroalgae of the Baltic Sea area. *Acta Botanica Fennica*, **155**, 1–51.

Norrman, B., & Andersson, A. (1994). Development of ice biota in a temperate sea area (Gulf of Bothnia). *Polar Biology*, **14**, 531–7.

Palenik, B., Price, N. M., & Morel, F. M. M. (1991). Potential effects of UV-B on the chemical environment of marine organisms: a review. *Environmental Pollution*, **70**, 117–30.

Passow, U. (1991). Species-specific sedimentation and sinking velocities of diatoms. *Marine Biology (Berlin)*, **108**, 449–56.

Pedersén, M., Roomans, G. M., Andrén, M., Lignell, Å., Lindahl, G., Wallström, K., & Forsberg, Å. (1981). X-ray microanalysis of metals in algae – a contribution to the study of environmental pollution. *Scanning Electron Microscopy*, **1981 II**, 499–509.

Plante-Cuny, M. R., & Plante, R. (1986). Benthic marine diatoms as food for benthic marine animals. In *Proceedings of the 8th International Diatom Symposium*, ed. M. Ricard, pp. 525–37. Koenigstein: Koeltz.

Podet, T., & Jaanus, A. (1997). Nutrient concentrations and phytoplankton variables in the Gulf of Riga and Baltic proper water mixing: An attempt to test the dichotomy of limiting nutrient. *Proceedings of the 14th Baltic Marine Biologists Symposium*. Tallinn: Estonian Academy Publishers.

Potapova, M., & Snoeijs, P. (1997). The natural life cycle in wild populations of *Diatoma moniliformis* (Bacillariophyceae) and its disruption in an aberrant environment. *Journal of Phycology*, **33** (in press).

Proschkina-Lavrenko, A. I. (1955). *Diatomovye vodorosli planktona Chernogo Moria* [The plankton diatoms of the Black Sea]. Moskva: Izdatel'stvo Akademii Nauk S. S. S. R. [in Russian].

Rahm, L., Conley, D., Sandén, P., Wulff, F., & Stålnacke, P. (1996). Time series analysis of nutrient inputs to the Baltic Sea and changing DSi:DIN ratios. *Marine Ecology Progress Series*, **130**, 221–8.

Remane, A. (1940). Einführung in die zoologische Ökologie der Nord- und Ostsee. In *Die Tierwelt der Nord- und Ostsee*, Vol. 1a, ed. G. Grimpe & E. Wagler, pp. 1–238. Leipzig.

(1958). Ökologie des Brackwassers. In *Die Biologie des Brackwassers*, ed. A. Remane & C. Schlieper, pp. 1–216. Stuttgart.

Risberg, J. (1990). Siliceous microfossil stratigraphy in a superficial sediment core from the northwestern part of the Baltic proper. *Ambio*, **19**, 167–72.

(1991). Paleoenvironment and sea level changes during the early Holocene on the Södertörn peninsula, Södermanland, eastern Sweden. *Stockholm University, Department of Quaternary Research, Report*, **20**, 1–27.

Risberg, J. & Sandgren, P. (1994). Shore displacement in the Stockholm area, Sweden, during the early Holocene as recorded from diatom and magnetic analysis – a preliminary report. In *Proceedings of the 11th International Diatom Symposium*, ed. J. P. Kociolek, pp. 371–85. San Francisco: California Academy of Sciences.

Robertsson, A. M. (1990). The diatom flora of the Yoldia sediments in the Närke province, south central Sweden. *Beihefte zur Nova Hedwigia*, **100**, 255–62.

Ross, R., & Mann, D. (1986). What is a diatom species? Report on a workshop. In *Proceedings of the 8th International Diatom Symposium*, ed. M. Ricard, pp. 743–44. Koenigstein: Koeltz.

Round, F. E. (1981). *The Ecology of Algae*. Cambridge: Cambridge University Press.

Round, F. E., & Crawford, R. M. (1990). Phylum Bacillariophyta. Chapter 31 in *Handbook Of The Protoctista*, ed. L. Margulis, J. O. Corliss, M. Melkonian & D. J. Chapman, pp. 574–96. Boston: Jones & Barlett Publishers.

Round, F. E., & Sims, P. A. (1981). The distribution of diatom genera in marine and freshwater environments and some evolutionary considerations. In *Proceedings of the 6th International Diatom Symposium*, ed. R. Ross, pp. 301–20. Koenigstein: Koeltz.

Round, F. E., Crawford, R. M., & Mann, D. G. (1990). *The Diatoms – Biology and Morphology of the Genera*. Cambridge: Cambridge University Press.

Ryabova, N., Zimina, L., Zimin, V., Khajrutdinova, J., Kudryavtseva, A., & Ossipov, D. (1994). *Abstracts of the International Meeting on the Urbanization and the Protection of the Biocoenosis of the Baltic Coasts, Juodrante, Lithuania, 4–8 October 1994*. Ed. R. Volskis, pp. 12–13. UNESCO-MAB Programme.

Sandén, P., & Rahm, L. (1993). Nutrient trends in the Baltic Sea. *Environmetrics*, **4**, 75–103.

Schelske, C. L, Stoermer, E. F., Fahnenstiel, G. L., & Haibach, M. (1986). Phosphorous enrichment, silica utilization, and biochemical silica depletion in the Great Lakes. *Canadian Journal of Fisheries and Aquatic Sciences*, **43**, 407–15.

Segerstråle, S. G. (1959). Brackish water classification, a historical survey. *Archivio di Oceanografia e Limnologia (Supplement)*, **11**, 7–33.

Simonsen, R. (1962). Untersuchungen zur Systematik und Ökologie der Bodendiatomeen der westlichen Ostsee. *Internationale Revue der gesamten Hydrobiologie, Systematische Beihefte*, **1**, 1–144.

Sjöberg, B. (ed.) (1992). Sea and coast. *The National Atlas of Sweden*. Stockholm: Almqvist & Wiksell International.

Smayda, T. J. (1990). Novel and nuisance phytoplankton blooms in the sea: Evidence for a global epidemic. In *Toxic Marine Phytoplankton – Proceedings of the 4th International Conference on Toxic Marine Plankton*, ed. W. Granéli, B. Sundström, L. Edler, & D. M. Anderson, pp. 29–40.

Smol, J. P., & Glew, J. R. (1992). Paleolimnology. In *Encyclopedia of Earth System Science*, **3**, ed. W. A. Nierenberg, pp. 551–64. San Diego: Academic Press.

Snoeijs, P. (1989). Ecological effects of cooling water discharge on hydrolittoral epilithic diatom communities in the northern Baltic Sea. *Diatom Research*, **4**, 373–98.

(1990). Effects of temperature on spring bloom dynamics of epiplithic diatom communities in the Gulf of Bothnia. *Journal of Vegetation Science*, **1**, 599–608.

(1991). Monitoring pollution effects by diatom community composition: A comparison of methods. *Archiv für Hydrobiologie*, **121**, 497–510.

(1992). Studies in the *Tabularia fasciculata* complex. *Diatom Research*, **7**, 313–44.

(1993). *Intercalibration and Distribution of Diatom Species in the Baltic Sea*, vol. 1. Uppsala: Opulus Press.

(1994a). Distribution of epiphytic diatom species composition, diversity and biomass on different macroalgal hosts along seasonal and salinity gradients in the Baltic Sea. *Diatom Research*, **9**, 189–211.

(1994b). The Forsmark Biotest Basin: An Instrument for Environmental Research – Experiences of large cooling water discharges in Sweden (1969–1993) and research perspectives for the future. *Swedish Environmental Protection Agency Report*, **4374**, 1–70.

(1995). Effects of salinity on epiphytic diatom communities on *Pilayella littoralis* (Phaeophyceae) in the Baltic Sea. *Ecoscience*, **2**, 382–94.

Snoeijs, P., & Balashova, N. (1998). *Intercalibration and Distribution of Diatom Species in the Baltic Sea*. Vol. 5. Uppsala: Opulus Press.

Snoeijs, P., & Kasperoviciene, J. (1996). *Intercalibration and Distribution of Diatom Species in the Baltic Sea*. Vol. 4. Uppsala: Opulus Press.

Snoeijs, P., & Kautsky, U. (1989). Effects of ice-break on the structure and dynamics of a benthic diatom community in the northern Baltic Sea. *Botanica Marina*, **32**, 547–62.

Snoeijs, P., & Notter, M. (1993a). Benthic diatoms as monitoring organisms for radionuclides in a brackish-water coastal environment. *Journal of Environmental Radioactivity*, **18**, 23–52.

(1993b). Radiocaesium from Chernobyl in benthic algae along the Swedish Baltic Sea coast. *Swedish University of Agricultural Sciences Report SLU-REK*, **72**, 1–21.

Snoeijs, P., & Potapova, M. (1995). *Intercalibration and Distribution of Diatom Species in the Baltic Sea*. Vol. 3. Uppsala: Opulus Press.

(1998). Ecotypes or endemic species – a case study on the evolution of *Diatoma* (Bacillariophyta) taxa in the Baltic Sea. *Nova Hedwigia* (in press).

Snoeijs, P., & Prentice, I. C. (1989). Effects of cooling water discharge on the structure and dynamics of epilithic algal communities in the northern Baltic. *Hydrobiologia*, **184**, 99–123.

Snoeijs, P., & Vilbaste, S. (1994). *Intercalibration and Distribution of Diatom Species in the Baltic Sea.* Vol. 2. Uppsala: Opulus Press.

Sohlenius, G., Sternbeck, J., Andrén, E., & Westman, P. (1996). Holocene history of the Baltic Sea as recorded in a sediment core from the Gotland Deep. *Marine Geology*, **134**, 183–201.

Sommer, U. (1994). Are marine diatoms favoured by high Si:N ratios? *Marine Ecology Progress Series*, **115**, 309–15.

(1996). Nutrient competition experiments with periphyton from the Baltic Sea. *Marine Ecology Progress Series*, **140**, 161–7.

Sundbäck, K., & Snoeijs, P. (1991). Effects of nutrient enrichment on microalgal community composition in a coastal shallow-water sediment system: an experimental study. *Botanica Marina*, **34**, 341–58.

Sundbäck, K., Odmark, S., Wulff, A., Nilsson, C., & Wängberg, S-Å. (1997). Effects of enhanced UVB radiation on a marine benthic diatom mat. *Marine Biology*, **128**, 171–9.

Suzuki, K., Handa, N., Nishida, T., & Wong, C. S. (1997). Estimation of phytoplankton succession in a fertilized mesocosm during summer using high-performance liquid chromatographic analysis of pigments. *Journal of Experimental Marine Biology and Ecology*, **214**, 1–17.

Sweitzer, J., Langaas, S., & Folke, C. (1996). Land cover and population density in the Baltic Sea drainage basin: a GIS database. *Ambio*, **25**, 191–8.

Tedengren, M., André, C., Johannesson, K., & Kautsky, N. (1990). Genotypic and phenotypic differences between Baltic and North Sea populations of *Mytilus edulis* evaluated through reciprocal transplantations. III: Physiology. *Marine Ecology Progress Series*, **59**, 221–8.

ter Braak, C. J. F. (1986). Canonical correspondence analysis: A new eigenvector technique for multivariate direct gradient analysis. *Ecology*, **67**, 1167–79.

ter Braak, C. J. F., & Juggins, S. (1993). Weighted averaging partial least squares regression (WA-PLS): An improved method for reconstructing environmental variables from species assemblages. *Hydrobiologia*, **269/270**, 485–502.

Tilman, D, Kilham, S., & Kilham, P. (1982). Phytoplankton community ecology: The role of limiting nutrients. *Annual Review of Ecology and Systematics*, **13**, 349–72.

Todd, E. C. D. (1993). Domoic acid and amnesic shellfish poisoning: A review. *Journal of Food Protection*, **56**, 69–83.

van Bennekom, A. J., Gieskes, W. W. C., & Tijssen, S. B. (1975). Eutrophication of Dutch coastal waters. *Proceedings of the Royal Society of London B.*, **189**, 359–74.

Vasiliu, F. (1996). The Black Sea. In *Marine Benthic Vegetation: Recent Changes and the Effects of Eutrophication*, ed. W. Schramm & P. H. Nienhuis, pp. 435–47. Ecological Studies, Volume 123. Berlin: Springer-Verlag.

Voipio, A. (ed.) (1981). *The Baltic Sea.* Amsterdam: Elsevier Science Publishers.

Wallström, K. (1991). Ecological studies on nitrogen fixing blue-green algae and on nutrient limitation of phytoplankton in the Baltic Sea. *Acta Universitatis Upsaliensis*, **337**.

Watanabe, M., Kohata, K., Kimura, T., Takamatsu, T., Yamaguchi, S., & Ioriya, T. (1995). Generation of a *Chattonella antiqua* bloom by imposing a shallow nutricline in a mesocosm. *Limnology and Oceanography*, **40**, 1447–80.

Wendker, S. (1990). Untersuchungen zur subfossilen und rezenten Diatomeenflora des Schlei-Ästuars (Ostsee). *Bibliotheca Diatomologica*, **20**, 1–268.

Werner, D. (1977). Silicate metabolism. In *The Biology of Diatoms*, ed. D. Werner, pp. 110–49. Botanical Monographs, Volume 13. Oxford: Blackwell Scientific Publications.

Willén, T. (1985). Phytoplankton, chlorophyll a and primary production in the Biotest Basin, Forsmark, 1981–1982. *Abstracts of the 9th BMB Symposium, Turku/Åbo, Finland, 11–15 June 1985*, p. 123, Åbo Akademi.

(1995). Växtplankton i Östersjön 1979–1988, PMK Utsjöprogrammet. *Rapport Naturvårdsverket*, **4288**, Stockholm. [in Swedish]

Wilson, S. E., Cumming, B. F., & Smol, J. P. (1996). Assessing the reliability of salinity inference models from diatom assemblages: An examination of a 219–lake data set from western North America, *Canandian Journal of Fisheries and Aquatic Science*, **53**, 1580–94.

Witkowski, A. (1994). Recent and fossil diatom flora of the Gulf of Gdansk, southern Baltic Sea. *Bibliotheca Diatomologica*, **28**, 1–313.

Wulff, F., Ærtebjerg, G., Nicolaus, G., Niemi, Å, Ciszewski, P. Schulz, S., & Kaiser, W. (1986). The changing pelagic ecosystem of the Baltic Sea. *Ophelia, Supplement*, **4**, 299–319.

Wulff, F., Perttilä, M., & Rahm, L. (1996). Monitoring, mass balance calculation of nutrients and the future of the Gulf of Bothnia. *Ambio*, **6**, 28–5.

Wulff, F., Stigebrandt, A., & Rahm, L. (1990). Nutrient dynamics of the Baltic Sea. *Ambio*, **19**, 126–76.

Yurkovskis, A., & Kostrichkina, E. (1996). A long-term ecosystem response to man-made impact in the Gulf of Riga. *Abstracts of the Baltic Marine Science Conference, Rønne, Denmark, 22–26 October 1996*. p. 22.

Yurkovskis, A., Wulff, F., Rahm, L., Andrushaitis, A., & Rodriguez-Medina, M. (1993). A nutrient budget of the Gulf of Riga, Baltic Sea. *Estuarine and Coastal Shelf Science*, **37**, 113–27.

Zernova, V. V., & Orlova, Z. R. (1996). Peculiarity of phytoplankton distribution in the Baltic Sea in early spring of 1990. *Okeanologiya*, **36**, 741–6 [in Russian].

15 Applied diatom studies in estuaries and shallow coastal environments

MICHAEL J. SULLIVAN

Introduction

Diatoms are an important and often dominant component of the benthic microalgal assemblage in estuarine and shallow coastal environments. This chapter will be concerned mainly with the motile diatom assemblages of intertidal sediments in these environments and secondarily with diatom assemblages epiphytic on submerged aquatic vegetation. Admiraal (1984) provided an excellent summary of the ecology of estuarine sediment-inhabiting diatoms. A variety of topics was covered, including distribution, effects of physicochemical factors, population growth, primary production, and interactions with herbivores. The focus of the present review will be considerably narrower as only those applied studies which have utilized structural (e.g., species diversity) and/or functional (e.g., primary production rates) attributes of benthic diatom assemblages will be considered. By applied is meant studies that treat benthic diatom assemblages as tools to address concerns about larger ecosystem problems such as cultural eutrophication of estuarine and shallow coastal environments. The three diatom-related research topics that will be reviewed in this chapter include eutrophication, sediment stability, and resuspension. These topics are important because of threats posed to estuarine and shallow coastal systems by increasing nutrient levels and reduced light transmission in the water column; both may significantly impact the role of algae and other primary producers in trophic dynamics and consequently affect ecosystem health. Relevant studies conducted in the Baltic Sea will not be included as they will be part of the chapter on applied studies of diatoms in brackish waters by Snoeijs (this volume). In discussing the various studies, the names and authorities of diatom taxa will be those listed by the author(s) and no taxonomic revisions will be attempted.

For an excellent review of the physical properties of estuaries see the chapter by Cooper (this volume). Before proceeding to the actual review of specific papers that follows in this chapter, the reader may first wish to consult Round's (1979) review of estuaries from the perspective of its resident vascular plants and algae in various habitats, such as salt marshes and exposed mudflats and sandflats. An important point to be made is that the latter sedimentary environments are often described in the literature as 'unvegetated' simply because vascular plants such as *Spartina alterniflora* Loisel. or seagrasses are absent. The use of this adjective is not only misleading but also inaccurate because a diverse assemblage of cyanobacteria and eukaryotic algae thrive within such sediments. This assemblage, often termed the microphytobenthos, has recently been reviewed and referred to as a 'secret garden' by MacIntyre *et al.* (1996) and Miller *et al.* (1996). Functional properties of the benthic microalgae as an integrated whole are discussed in detail by these authors but the sedimentary diatom component is often singled out. Finally, although only intertidal and shallow subtidal habitats will be considered in this chapter, recent work by Cahoon and Cooke (1992), Cahoon and Laws (1993), and Cahoon *et al.* (1990, 1993, 1994) has shown a highly diverse and productive diatom assemblage within coastal shelf sediments. In fact, the chlorophyll (chl) *a* content of a square meter of shelf bottom may exceed that of the entire water column above, even at depths of 142 m (Cahoon *et al.*, 1990). Thus, it would appear that the contribution of benthic diatoms to coastal production and support of higher trophic levels has been greatly underestimated.

Eutrophication

Anthropogenic inputs of nutrients, mainly various forms of nitrogen and phosphorus, pose a serious threat to estuarine and shallow coastal waters (Valiela *et al.*, 1997). Although the information content of benthic diatom assemblages is high due to the large number of taxa typically present, these organisms have been little used to infer water quality or predict the effects of increased nutrient levels in these systems. Hohn (1959) was the first to employ artificial substrates in estuarine systems with the specific aim of assessing water quality. Diatom assemblages developing on glass slides in 'polluted' and 'non-polluted' areas of Galveston and Chocolate Bay, Texas exhibited very different truncated log normal curves. Species richness for diatoms, other algae, fish, protozoa, and invertebrates was drastically reduced in the 'polluted' areas of the two bays, particularly in the infamous Houston Ship Channel.

DISTRIBUTIONAL STUDIES

Admiraal (1977–1984) produced an excellent series of papers dealing with interactions between diatom populations on mudflats of the Ems–Dollard Estuary and various environmental factors to explain their spatial and temporal distribution. With regard to nutrients, Admiraal (1977*b*) showed that the K_s value for orthophosphate uptake by cultures of *Navicula arenaria* Donkin was about 0.1 μg-at P/l. This value was the same as that of several marine planktonic diatoms (see his Table 3) and indicated that phosphorus concentrations in the Ems–Dollard were not limiting to this diatom. Admiraal generalized this to the resident diatom assemblages of the intertidal flats of the Ems–Dollard;

therefore, increases in phosphorus in this body of water could not be detected using the benthic diatom assemblages.

Admiraal (1977c) then turned his attention to inorganic nitrogen. Cultures of ten diatom species isolated from mudflats of the Ems–Dollard were exposed to varying concentrations of ammonia (NH_4^+ and NH_3), nitrate, and nitrite. The majority of species grew well at high concentrations of the last two nutrients; however, ammonia concentrations higher than 0.5 μg-at N/l were strongly inhibitory to growth of the cultures, measured as chl *a*. Photosynthetic rates of cultures and field assemblages were also inhibited by this ammonia concentration and the effect was enhanced by high irradiance and high pH. Admiraal noted that tolerance of ammonia was not high amongst the species but truly sensitive species were lacking; the most tolerant diatom in the culture experiments and on the mudflats was *Navicula salinarum* Grunow. He also stated that a clear relationship between ammonia tolerance of the various species and their relative abundance on polluted and unpolluted mudflats was not obvious.

The tolerance of benthic diatom species to free sulfide (a known toxin) was also tested by Admiraal and Peletier (1979a). *Navicula salinarum*, *N. arenaria*, *Nitzschia* cf. *thermalis* (Ehrenberg) Grunow, and *Gyrosigma spencerii* (W. Smith) Cleve were relatively tolerant of free sulfide concentrations up to 6.8 mM for periods of 5 to 24 h. However, *Nitzschia closterium* (Ehrenberg) W. Smith, *N.* cf. *dissipata* (Kützing) Grunow, *N. sigma* (Kützing) W. Smith, *Surirella ovata* Kützing and *Stauroneis constricta* (W. Smith) Cleve were greatly inhibited, or killed, following 48 h of exposure to a free sulfide concentration of 0.9 mM. The high tolerance of *N. salinarum* correlated well with its dominance on polluted mudflats receiving large amounts of organic waste from potato flour and sugar mills and sewage plants bordering the Ems–Dollard Estuary. In contrast, *N. arenaria*, although relatively tolerant of free sulfide, was only dominant on well-oxygenated, relatively unpolluted sandflats. Admiraal and Peletier (1979a) concluded that the differing abilities of these taxa to tolerate free sulfide, ammonia, low salinity, anaerobic conditions, and prolonged darkness, and the capability to grow heterotrophically (see Admiraal and Peletier 1979b for uptake of organic substrates) explain their distribution in the estuary. No one factor could explain the observed distributions but a combination of them offered a logical explanation. Admiraal and Peletier (1980) came to exactly the same conclusions in studying a different mudflat in the Ems–Dollard Estuary exposed to large freshwater runoff and high concentrations of organic pollution from potato flour mills. Peletier (1996) revisited the same mudflat sampled by Admiraal & Peletier (1980). Since this time (1977–1980) the loading rates of organic waste from potato flour and cardboard have been greatly reduced. Analysis of the benthic diatom assemblages showed that the former dominants *Navicula salinarum* and *N. pygmaea* Kützing had disappeared from the mudflats as *N. flanatica* Grunow and *N. phyllepta* Kützing had become the dominant taxa. The latter two diatoms have a lower tolerance to high ammonia and free sulfide concentrations than do the former two taxa. Changes in the abundance of invertebrate grazer species were reflected in reduced chl *a* concentrations in the mudflat sediments. Hence, Admiraal's earlier work on the ecophysiology of the benthic diatom assemblages had been validated.

STUDIES UTILIZING DIVERSITY STATISTICS

Round (1981) wrote that aquatic environments can be polluted in many ways but always by agents which are not produced under natural conditions. He further stated that the best course of action is to prevent pollution wherever possible rather than attempting to determine the amount of pollution a given system can tolerate. While the prevention of pollution is a worthy goal, its attainment appears to be something that will certainly not occur in the foreseeable future. Therefore, it is critical that the 'health' of aquatic systems be constantly monitored for both their protection and our own. In recent years, studies of the diatom flora have increased in number and scope as water quality has become a subject not only of interest to aquatic biologists but also the general public. Although considerable progress has been made in utilizing diatoms to assess water quality in freshwater systems, the situation is quite the opposite in estuarine and shallow coastal environments (see Sullivan, 1986). The very few studies that have been done have differed from Admiraal's approach of considering functional responses of selected components of the diatom assemblage and relied mainly on diversity statistics. Van Dam (1982) published an informative paper on the relationship between diversity, species abundance–rank data, and water quality assessment. He pointed out the pitfalls of using diatom assemblage diversity to categorize pollution across *different* aquatic systems and stressed the importance of identifying the constituent taxa and having a knowledge of their autecology, particularly for those that were dominant in the assemblage. In a less comprehensive paper, Archibald (1972) had come to the same conclusions.

Hendey (1977) studied the diversity of diatom assemblages epiphytic on *Ceramium rubrum* Agardh along the North Coast of Cornwall, United Kingdom and its relationship to pollution by sewage. Hendey found that H' (Shannon's diversity index) (Lloyd *et al.*, 1968) was lowest for three consecutive years during summer when the summer tourist influx greatly stressed the sewage facilities; however, no control samples from nearby undeveloped areas of the coast were taken for comparative purposes. In a Massachusetts salt marsh, Van Raalte *et al.* (1976*a*) applied urea and sewage sludge containing 10% nitrogen on a long-term basis to sediments populated by the cordgrass *Spartina alterniflora*. Both organic nitrogen enrichments significantly decreased H' and S (number of species in the sample or species richness) of the edaphic (i.e., within or associated with the salt marsh sediments) diatom assemblages. The relative abundance of *Navicula salinarum* was 5–9% in control plots but increased to 20–25% in enriched plots. Sullivan (1976) studied this same microalgal assemblage beneath the same spermatophyte but in a Delaware salt marsh and added inorganic nitrogen (NH_4NO_3) and phosphorus ($CaH_4(PO_4)_2$) instead of urea or sewage sludge.

Phosphorus enrichment significantly decreased both H' and S, whereas nitrogen enrichment significantly decreased only S. Both nitrogen and phosphorus enrichment had a stimulatory effect on the relative abundance of *Navicula salinarum*; this agrees with previous work by Admiraal (1977c) and Van Raalte *et al.* (1976a) for nitrogen enrichment. Moving south to a Mississippi salt marsh, Sullivan (1981) fertilized the marsh sediments beneath the spike grass *Distichlis spicata* (Linnaeus) Greene with NH_4Cl. Responses of the edaphic diatom assemblage were quite different, in that nitrogen enrichment had virtually no effect; only the relative abundance of *Nitzschia perversa* Grunow was greatly increased by nitrogen enrichment. Comparison of the control and enriched assemblages by Stander's similarity index (SIMI) (see Sullivan, 1981) revealed that they shared between 73% and 95% of the maximum similarity possible over an annual cycle. It should be noted that in all three salt marsh studies cited above, the overstory grasses significantly increased their above-ground standing crops in response to nitrogen enrichment, and this may have been a confounding factor because of its effect on irradiance reaching the sediments.

WATER QUALITY INDICES

Although considerable progress has been made in utilizing diatoms to assess water quality in freshwater systems (see the wealth of information generated in the appropriate chapters of this book), the situation as reviewed above is quite different in estuarine and shallow coastal systems. Up until the 1970s, monitoring of water quality depended mainly on the presence of certain key species (i.e., indicator species) as indicative of polluted conditions. Round (1981, pp. 552–7) provided an excellent history of the strategy, starting with the work of Kolkwitz and Marsson (1908) at the turn of the century and the reader is referred to this for further details. As pointed out by Round and other workers, the major weakness of utilizing indicator species is that those species taken to be indicative of pollution also occur in non-polluted waters (i.e., they are not restricted to polluted waters). Therefore, the presence of such indicator species does not necessarily mean that the body of water in question is polluted, as quite the contrary may be true. Lange-Bertalot (1978, 1979) developed a more practical method to assess water quality utilizing both the specific identity and relative abundance of constituent diatom species in the assemblage for European rivers. His more important contribution, however, was to point out that the cornerstone of any index of water quality must be the *tolerance* of the individual diatom taxa to *increasing* levels of a particular pollutant. The affinity or fidelity for polluted waters of diatom taxa should not be used in such indices for reasons given above. What is essential to this approach is the species-specific limits of tolerance to *decreasing* water quality, since diatom species are usually not limited by increasing levels of water quality. Various indices based on these concepts have been proposed such as the Diatomic index (Id) of Descy (1979), the Diatom assemblage index to organic pollution (DAIpo) of Watanabe (Watanabe *et al.*, 1988), and the modified Pantle–Buck saprobic index (SI) of

Kobayasi and Mayama (1989). Recently, Prygiel et al. (1996) proposed the Practical Diatom index (IDP). Such indices have much in common. First, all are based on the relative tolerance of a diatom taxon to pollution determined empirically using ordination techniques and the subsequent assignment of each taxon to one of several pollution tolerance groups. Second, the calculation of each index yields a numerical value, which is constrained by a minimum and maximum value for the index. Therefore, this value yields information concerning the relative quality of water from which the sample was taken. Third, correlations between index values and measured parameters of water quality associated with the samples are highly significant. To date, no comparable work has been done in estuaries or shallow coastal environments. Although it should be possible to divide estuarine and marine benthic diatoms into differentiating groups of species based on their pollution tolerances, the paucity of published autecological information for these diatom taxa suggests the realization of such a system for water quality assessment in estuarine and coastal systems is a long way off. Undoubtedly, the first candidate for the most tolerant group would be *Navicula salinarum*, based on the studies of Admiraal (1977c), Van Raalte et al. (1976a), Sullivan (1976), and Peletier (1996).

FUNCTIONAL STUDIES

Other studies have utilized functional properties of the entire benthic micro-algal assemblage where diatoms were either a substantial component or completely dominant to study the effects of eutrophication. Sullivan and Daiber (1975) worked in the same spermatophyte zone as did Sullivan (1976), and found that both inorganic nitrogen and phosphorus enrichment increased the standing crop of edaphic algae as measured by chl a. Van Raalte et al. (1976b) measured primary production rates (as ^{14}C uptake) of the edaphic algae from the same plots fertilized with urea or sewage sludge described above. Primary production increased only at the highest level of nitrogen enrichment because of the shading effect of the overstory grass canopy. Light was the single most important factor limiting edaphic algal production on this Massachusetts marsh. Darley et al. (1981) enriched the sediments in both short and tall *Spartina alterniflora* zones with NH_4Cl and measured ^{14}C uptake by the edaphic algae (diatoms constituted 75–93% of total biomass) as well as their chl a levels. Both parameters increased significantly beneath the short *S. alterniflora* canopy but ammonium enrichment had little effect on the edaphic algae beneath tall *S. alterniflora*. This correlated well with the more frequent inundation of the tall form by tidal waters making nitrogen limitation here unlikely. Sundbäck and colleagues (1988, 1991a,b) have utilized benthic micro-algal assemblages from the sediments of the Skagerrak and Kattegat to investigate nutrient fluxes and succession. Nutrient enrichment of subtidal sediments (15 m) showed that the flux of inorganic nitrogen and phosphate between sediments and water column was mediated by benthic microalgae. Increasing irradiance resulted in increased photosynthetic rates and nutrient requirements for benthic microalgae preventing the release of phosphate and ammonium from the sediments. This indirect effect of benthic microalgae on nutrient fluxes was considered to be more important than the direct effect of their nutrient uptake. Furthermore, the diatom component of the benthic microalgal assemblage played a major role in controlling the flux of nutrients from the sediments to the water column. In shallow waters (0.2 m), nutrient enrichment significantly increased the standing crops of filamentous cyanobacteria, flagellates, and diatoms, but the relative dominance of these major taxonomic groups did not change. Diatoms accounted for *ca.* 50% of the benthic microalgal biomass and taxa belonging to the genera *Amphora* and *Nitzschia* within the size range 7–12 μm were most favored by the addition of inorganic nitrogen and phosphorus.

STUDIES OF SEAGRASS EPIPHYTES

Eutrophication also poses a serious threat to the world's seagrass beds (Short & Wyllie-Echeverria, 1996). Nutrient enrichment stimulates the growth of phytoplankton in the water column and microalgae epiphytic on seagrass leaves. The increased algal biomass reduces the amount of light reaching the seagrass leaves and acts as a diffusive barrier for the uptake of inorganic carbon and nutrients. Thus, light effects are really secondary effects of nutrient enrichment, the net result of which may be manifested by a decline of seagrasses through reduced growth and reproductive success. Nutrient enrichment work targeted at the epiphytic algae is on the increase, but studies providing specific information about the diatom component on seagrass leaves are few. Coleman and Burkholder (1994) enriched mesocosms with low-flow rates containing *Zostera marina* Linnaeus with nitrate. A significant increase in assemblage productivity was recorded, as measured by ^{14}C uptake. Cyanobacteria were mainly responsible for the increase 3 weeks after enrichment began and small diatoms (biovolume = 70–560 μm^3) at 6 weeks; the response of the large diatom species was variable. They concluded that nitrate enrichment has a controlling influence on species composition and dominance in epiphytic algal assemblages. Coleman and Burkholder (1995) moved their experiment to field populations of *Z. marina* in Back Sound, Beaufort, North Carolina where flow rates were much higher than in their mesocosm experiments. Nitrate enrichment did not increase total epiphytic algal production but did significantly affect some of the dominant species. Using track light–microscope–autoradiography, they showed that the species-specific production rate of *Cocconeis placentula* Ehrenberg increased, whereas that of the crustose red alga *Sahlingia subintegra* (Rosenvinge) Kornmann decreased. These data matched the significant increase and decrease in cell numbers of these two taxa, respectively. Again, they concluded that nitrate levels control species-specific productivity and community structure of the epiphytic algal assemblage of *Z. marina*. Wear *et al.* (1999) fertilized beds of the seagrasses *Halodule wrightii* Ascherson, *Syringodium filiforme* Kützing, and *Thalassia testudinum*

König in Big Lagoon, Perdido Bay, Florida with a slow-release (3–4 mo), temperature-sensitive Osmocote™ fertilizer. This fertilizer contained 19% N (as ammonium nitrate), 6% P (as anhydrous phosphoric acid), and 12% K (as anhydrous potassium hydroxide) by weight. All results for epiphytic algae were independent of the seagrass species. Enrichment caused a doubling of epiphytic algal production (^{14}C uptake) and biomass (g dry wt/cm^2) was increased to $1.4\times$ that of controls. Chl a and fucoxanthin concentrations (measured by HPLC) of the epiphytic algae tripled in response to enrichment and variation in the latter pigment explained 88% of the variation in the former pigment. Diatoms were the dominant algal group epiphytic on the three seagrasses, and accounted for virtually all of the fucoxanthin.

THERMAL POLLUTION

Before leaving the application of diatom studies to eutrophication, it is appropriate to discuss their utilization in monitoring thermal pollution here. Admiraal (1977a) demonstrated a linear relationship between the growth rate of benthic estuarine diatoms and temperature for temperatures below the optimum, and also that this response is very similar to that of planktonic diatoms. Hein and Koppen (1979) allowed diatoms to colonize styrofoam balls in the intake and discharge canals of a nuclear power plant in Forked River, New Jersey. The heated effluent caused the number of species and diversity (H') to decrease, although differences between intake and discharge benthic diatom assemblages were not statistically tested.

Sediment stability

As in terrestrial environments, the stability of sediments and their resistance to erosion (caused by tidal currents, wind-induced waves, bioturbation, and other factors) has been the subject of much research in estuaries and shallow coastal environments. Descriptive and especially experimental studies of benthic diatom assemblages have proven useful in understanding sediment dynamics in these habitats. Holland $et\,al.$ (1974) stirred unialgal cultures of six benthic diatom taxa and a mixed benthic diatom assemblage that had been allowed to develop for 12 days on different sediment types. All cultures were ineffective in stabilizing pure sand when compared to controls devoid of diatoms; however, $Navicula\ directa$ (W. Smith) Ralfs, $Hantzschia\ amphioxys$ (Ehrenberg) Grunow, and the natural benthic diatom assemblage were very effective in preventing resuspension of sediments in the 90% sand/10% silt and clay and 100% silt and clay cultures. Based on the observation that cultures producing little or no mucilage did not differ from sterile control sediments, a sediment-stabilizing mechanism based on secretion of mucilage by epipelic diatoms was proposed. The benthic diatoms were considered as effective as cyanobacterial mats for sediment stabilization.

Coles (1979) studied the relationship between intertidal sediment stability and benthic diatom populations in the Wash, a large embayment in southeast England. A year-round deposition of fine sediments (silts and clays) was found for intertidal mudflats and salt marshes but not for sandflats, where grazing was thought to be largely responsible for lack of sediment deposition. Coles found that by killing diatoms with either bleach or formalin, or by addition of the grazers *Corophium volutator* (amphipod) and *Hydrobia ulvae* (snail), the sediment surface became uncompacted followed by resuspension of the sediments by tidal scour. Conversely, an extended absence of *C. volutator* from an inner sandflat following the bleach treatments allowed the benthic diatoms to bloom, followed by the deposition of mud in the middle of the sandflat. Verification of this finding was demonstrated in the work of Daborn *et al.* (1993), who observed a 'top-down' effect of birds on sediment stability. The arrival of birds on a tidal flat in the Bay of Fundy decimated the local population of *C. volutator*; this was followed by blooms of epipelic diatoms on the flat and an increase in sediment cohesiveness. Within the salt marsh, Coles (1979) showed that, even where the spermatophyte canopy was dense, the macrophytes alone were unable to provide sufficient shelter to allow mud deposition to occur under the control of physical forces alone. The trapping and binding of fine sediment particles by copious mucilage secretion of large epipelic diatom populations permitted sediment deposition and stabilization.

Intertidal sands of Halifax Harbor, Nova Scotia exhibited heterogeneity or patchiness of diatom films caused by transposition of the bedform through the ripples (Grant *et al.*, 1986). In flume experiments, cores with diatom films lost less chl *a* than those without films. Grant *et al.* published some of the first SEMs showing that mucous strands secreted by diatoms were responsible for the binding of sand grains in the field. Bacterial cells were too small to create grain-to-grain binding. That same year, Paterson (1986) employed low temperature scanning electron microscopy (LTSEM) to produce three-dimensional images of intact benthic diatom assemblages within, and on, the surface of intertidal sediments from the Severn Estuary. Cores were subjected to an artificial tidal regime, but the normal photoperiod was maintained. Following dawn, the first cells to migrate to the surface were species of *Navicula* and *Nitzschia*. Four hours after sunrise, there was a synchronous appearance of *Scoliopleura tumida* (Brébisson ex Kützing) Rabenhorst cells at the surface. It was suggested that light acts as a migration trigger and different diatom taxa have different threshold irradiances. A massive accumulation of diatoms occurred in the top 0.2 mm of the muddy sediments; before upward and after downward migration, most of the cells were restricted to the top 1.6 mm of sediments. In a later study of these same sediments, Paterson (1989) used LTSEM in conjunction with a cohesive strength meter to demonstrate that surface stability was stronger in cores inhabited by epipelic diatoms than in cores with consolidated or unconsolidated sediments lacking diatoms. The LTSEM revealed an extensive extracellular matrix consisting mainly of mucopolysaccharides produced during diatom locomotion (i.e., migration). As in the work of Grant *et al.* (1986)

with intertidal sands, bacterial mucilage strands had no significant effect on sediment stability.

In an ingenious laboratory experiment, Paterson (1990) used glass beads 68 μm in diameter to serve as artificial sediment for natural populations of epipelic diatoms isolated from the muddy site in the Severn Estuary. Diatom biomass (as chl a) and carbohydrate were higher in non-draining (no hole in bottom of petri dish containing artificial sediments) cultures whereas sediment stability was higher in draining (hole drilled in bottom of petri dish) cultures exposed to an artificial tidal regime with 7.3 h of exposure. No chl a was detected below 2 mm, and virtually all diatoms occurred in the top 0.2 mm of 'sediment'. LTSEM showed an extensive array of cell-to-cell and cell-to-bead connections via mucilage production of the different diatom taxa. Paterson stated that excess water dissolved mucilage and as mucilage dries the binding induced by the strands becomes stronger, hence the differences observed between non-draining and draining cultures. Therefore, the water content of the sediments and hydration state of the mucilage are very important variables.

Vos *et al.* (1988) carried out an extensive set of experiments on sandy intertidal sediments of the Oosterschelde, a mesotidal estuary. The paramount importance of mucilaginous secretions of benthic diatoms for sediment stabilization that forms the central theme of work reviewed in this section was again highly evident. The authors argued that sediment stabilization consists of two factors: a cohesive and a network effect. The first effect is a result of the mucilage secretions of epipsammic and epipelic diatoms coating sediment particles improving their cohesiveness. Evidence is provided by the demonstration that grains with an organic coating and dead diatoms exhibited a 5–17% increase in the critical erosion velocity relative to grains where both the diatoms and the organic coating had been removed by H_2O_2. The network effect, the much larger of the two, is brought about when large numbers of epipelic diatoms at the sediment surface form an extensive network of mutual attachment to each other *and* the sediment particles. Network formation increased the critical erosion velocity by 25–100%. The elasticity of the mucilage (dependent on hydration state?) was considered to be important for the strength of the network and its ability to bind particles. The overall effect of 'networking' is to decrease the median grain size of the sediment as clay and silt-sized particles are bound and also a smoothing of the sediment surface as small ripples are flattened.

Underwood and Paterson (1993a,b) followed in Coles' footsteps by spraying intertidal sediments adjacent to a salt marsh in the Severn Estuary with enough formaldehyde to kill all diatoms; they then followed recolonization for 8 days. Pretreatment chl a (virtually all due to epipelic diatoms) levels returned by day 6 and final chl a concentrations were highest in sprayed plots because grazers had not yet recolonized. The mucilaginous matrix was completely lost but by day 5 the matrix was reestablished, primarily due to a large population of *Nitzschia epithemioides* Grunow at the sediment surface. Significant positive correlations were found between chl a and colloidal carbohydrates in the sediments, chl a

and critical shear strength for incipient erosion, and colloidal carbohydrates and critical shear strength. Critical shear strength was negatively correlated with water content of the sediments. The authors pointed out that the colloidal carbohydrates produced by the diatoms bind sediments and diatoms together resulting in a biofilm-enhanced sediment stability.

Yallop *et al.* (1994) studied cohesive sediments from the Severn Estuary and non-cohesive sediments from the Island of Texel, Netherlands. The cohesive sediments were characterized by thin (100 μm), unstratified biofilms totally dominated by epipelic diatoms with *Nitzschia epithemioides* accounting for 84% of the cell count. In contrast, the non-cohesive sediments were stabilized by mm-thick stratified mats where epipelic and epipsammic diatoms dominated the upper 500 μm of the mat underlain by a mass of filamentous cyanobacteria, mainly *Microcoleus chthonoplastes* Thuret. The cohesive sediments were stabilized entirely by colloidal extra polymeric substances (EPS) secreted by epipelic diatoms whereas the non-cohesive sediments were stabilized partly by an overlying layer of colloidal EPS and partly by network formation between cyanobacterial filaments.

During the present decade in the United States, efforts to restore salt marshes to their former pristine state or create new marsh by planting *Spartina alterniflora* and other appropriate species have been attempted. Once established, the obvious question is whether restored and newly-created salt marshes function in a manner similar to so-called natural marshes. To date, studies in the USA designed to answer this question have focused on the vascular plants, invertebrates, and particularly fish; the benthic microalgae have not been considered, despite their importance in the processes of sediment deposition and stabilization. Underwood (1997, 1998) followed the development of benthic microalgae over 2.5 years in a salt marsh restoration scheme on Northey Island in southeast England. The initial colonizers of newly deposited sediments following the landward displacement of a seawall bordering an eroded salt marsh were diatom biofilms. These sediment deposition events resulted in true mudflat diatom taxa being present within the newly created marsh. Only in those microalgal assemblages where the relative abundance of diatoms was >30% was there a significant linear relationship between biomass (as chl *a*) and the quantity of colloidal carbohydrate exopolymers in the surface sediments. Those managed sites where sediment compaction occurred and a cyanobacterial mat formed were characterized by a delay in the development of a typical salt marsh macrophyte community. Thus, epipelic diatom assemblages and their copious secretion of mucilaginous substances prepare the way for the eventual colonization by vascular plants and creation of new salt marsh.

Resuspension

For benthic diatom assemblages, the opposite of sediment stabilization is their resuspension into the water column along with sediment particles. In estuaries and shallow coastal waters, benthic diatoms and resuspended sediments have proven to be closely linked. Baillie and Welsh (1980) were the first to really draw attention of the scientific community to this relationship. Resuspension of mudflat sediments (33% clay, 61% silt) produced peaks of chl a during early flood and late ebb tide. Examination of diatoms in the water column revealed that 75% of frustules belonged to pennate diatoms and it was concluded that flooding tides were responsible for a net transport of epipelic diatoms from the mudflat to a salt marsh where feeding bivalves were located. Paterson (1986) has commented that the 'clouds' of diatoms observed in the water column by Baillie and Welsh probably resulted from hydration of diatom mucilages and subsequent release of these organisms by the incoming tide.

Other authors have observed large numbers of benthic diatom cells in the water column. Gallagher (1975) found that rising tidal waters entrained benthic diatoms from the surface of a Georgia salt marsh and determined that these diatoms were primarily pennate forms from the sediments. Varela and Penas (1985) suggested that resuspended diatoms were the major food source for the culture of the cockle *Cerastoderma edule* on sandflats of Ria de Arosa, northwest Spain. Virtually all microalgal cells in the water column of the shallow Peel–Harvey Estuary in southwest Australia were pennate diatoms resuspended by wind-induced currents (Lukatelich & McComb, 1986). Riaux-Gobin (1987) found the concentration of chl a could be 30 times higher at low than high tide during spring tidal cycles in North Brittany, France. Since phytoplankton were scarce at low tide and benthic diatoms abundant, she concluded that resuspension of the latter was the source of the chl a peak. Shaffer and Sullivan (1988) used K-systems analysis to show that primary production in the water column of the Barataria Estuary, Louisiana was often greatly augmented by resuspension of benthic diatom cells from the sediments. K-systems revealed that factors composed of wave height, meteorological tides, astronomical tides, and the standing crop (as chl a) and primary production of the benthic microalgae accounted for 97% of the variation in water column productivity over a 30-day period. Benthic pennate diatoms represented an average of 74% of the diatom cells in water column samples.

Delgado *et al.* (1991*a*) performed experiments directed at quantifying resuspension with sediment cores taken from the Ems estuary. As expected, the percentage of resuspended diatoms increased as the angular velocity of agitating paddles increased, up to a maximum of 45% of cells in the top 0.5 cm of sediment. However, the maximum resuspension of sediment particles was only 6%. Sediments with a high concentration of detritus and low numbers of diatoms were most vulnerable to resuspension; the highest relative resuspension of diatoms occurred when the cell density was $< 2 \times 10^5/cm^2$. Presumably, such low numbers do not allow a network of diatom mucilage and sediments to be fully developed, hence stabilization by diatom mucilage is minimal (see Vos *et al.*, 1988). Delgado *et al.* (1991*b*) isolated *Surirella ovata*, *Navicula digito-radiata* (Gregory) Ralfs, and a natural benthic diatom assemblage from the Ems Estuary and cultured them on sand. In stationary cultures there was an exponential increase in cell numbers of the unialgal cultures and mixed assemblage. In continually agitated cultures, *S. ovata* exhibited a reduction in cell number due to breakage of cells following collisions with the sand grains. *N. digito-radiata* and the natural assemblage initially lost cells, but this was followed by recovery and positive growth. The natural assemblage came to be dominated by a single, small species of *Navicula* which may have been an adaptation to avoid collisions with sand grains.

Jonge and his colleagues have written an excellent series of papers on resuspension in the Ems Estuary. Jonge (1985) defined 55 μm as the operational dividing line between mud (silt and clay) and sand particles. He found that 13% of sand grains were occupied by one or more diatom cells and that 80% of these cells were present in or on the mud coating the sand grains (20% were on bare parts of sand). Therefore, the major substratum for diatoms on tidal flats of the Ems Estuary was, in fact, mud. This led him to argue that the descriptive terms 'epipelic' and 'epipsammic' were unsuitable for classifying the major components of the estuarine diatom flora (but see Vos *et al.*, 1988). Jonge also found that the mean number of diatoms on sand grains resuspended in the water column and those resident within the tidal flats was similar (0.29 *vs.* 0.53, respectively). The identity and relative abundance of diatoms on resuspended and tidal flat sand grains was also similar, which prompted Jonge to conclude that a large exchange of sand and diatoms occurs between the tidal flats and the channels of the estuary. He further concluded that this dynamic exchange in concert with sediment sorting by currents and waves could lead to changes in species composition of tidal flat diatom assemblages.

Jonge and van den Bergs (1987) resuspended diatoms from sandy and silty sediments of the Ems Estuary with two rotating cylinders. In the sand cores, the more abundant diatoms were resuspended in two distinct groups. The second group to be resuspended was composed of *Navicula aequorea* Hustedt, *N. salinicola* Hustedt, *Opephora martyi* Héribaud, and *O. pacifica* (Grunow) Petit, diatoms which were more tightly bound to the sand grains than members of the first group (e.g., *Navicula flanatica*, *N. forcipata* Greville, *Amphora coffeaeformis* (Agardh) Kützing). Diatom species in the silty sediments were not resuspended as distinct groups. The authors concluded that resuspension and redeposition events must constitute an unending series of intensive cycles.

Jonge and van Beusekom (1992) used a large data set to calculate the contribution of benthic microalgae (virtually all pennate diatoms) to total chl *a* in the Ems Estuary. On average for the entire estuary, 11×10^3 kg chl *a* were present in the extensive tidal flats and 14×10^3 kg chl *a* in the water column. Of the latter amount, 'real' phytoplankton were responsible for 10×10^3 kg chl *a* and resuspended benthic diatoms for 4×10^3 kg chl *a*. Immediately above the tidal flats, resuspension doubled the algal biomass available to filter feeders. The implications of this for those interested in modeling trophic dynamics in estuarine systems are obvious.

Finally, Jonge and van Beusekom (1995) looked more closely at the actual resuspension event in the Ems Estuary. Mud ($<$55 μm fraction) and benthic diatoms were found to be suspended simultaneously from tidal flats. Resuspension of mud and chl a from the top 0.5 cm of sediments was a linear function of 'effective wind speed', which was the wind speed averaged over the three high water periods immediately preceding the collection of water column samples. Tidal currents appeared to have a minor effect and this factor was not considered in any of the calculations. Resuspension began in the seaward portion of the Ems Estuary at an effective wind speed of 1 m/s (2 mph) and in the innermost portion (the Dollard) at 5 m/s (11 mph). Benthic diatom networks are more developed in the latter portion. The highly significant linear regression of effective wind speed on water column chl a indicated that a wind speed of 12 m/s (27 mph) would resuspend 50% of the diatom biomass into the water column, which would eventually reach the main channels of the estuary. Again, the trophic implications of resuspended diatoms are great.

Summary

Diatom studies of an applied nature in estuaries and shallow coastal waters have been few in number, especially when compared to the situation in freshwaters and the ocean. This may be a reflection of the intermediate position such environments occupy between purely fresh and fully marine waters and their resulting highly dynamic nature. In short, the complexity of estuaries and shallow coastal waters and the continuous change in and interaction between physicochemical factors make controlled experiments difficult to carry out. Nevertheless, some excellent applied work has been done with benthic diatom assemblages, which has been arbitrarily divided into three categories within this chapter. Eutrophication is a major worldwide problem as human populations are often dense around estuaries and shallow coastal waters. The preliminary work that has been done, including what has been discussed in this chapter, has shown that diatom assemblages respond to changes in nutrient concentrations and may well emerge as a sensitive biomonitoring system of water quality in coastal areas. The paucity of autecological information on taxa from estuarine and shallow coastal systems for which no comprehensive taxonomic monograph exists makes the formulation of diatom-based water quality indices an impossibility at the present time. Such indices have proliferated in applied freshwater diatom studies. Work specifically designed to determine the tolerances of benthic marine diatoms to various forms of pollution should be given high priority.

Applied work in the area of estuarine sediment stability has convincingly shown the paramount importance of benthic diatom assemblages and their mucilaginous secretions which, at sufficient cell densities, bind the diatoms and sediment particles into a network. This network is best developed by epipelic diatoms, whose daily upward and downward migrations leave trails of sediment-binding mucilage. Two recent studies have also shown that diatom mucilages prepare the way for vascular plant colonization in restored and newly created salt marshes. Diametrically opposed to stabilization of sediments in estuaries and shallow coastal waters is the process of resuspension, whereby sediment particles with or without diatoms and unattached diatom cells enter the water column. Wind- and tidal-induced waves and currents are mainly responsible for this process. Applied work has revealed the magnitude and importance of this phenomenon. In some estuarine systems virtually all the so-called 'phytoplankton' are, in fact, resuspended benthic diatoms from the sediments. Resuspended diatoms may greatly augment the primary production of the water column and constitute an important food source for filter-feeding animals. The trophic importance of benthic diatoms, both within the sediments and resuspended in the water column, is slowly becoming a paradigm for estuarine ecologists.

Acknowledgments

This paper is a result of research sponsored in part by the National Oceanic and Atmospheric Administration, US Department of Commerce under Grant NA16RG0155, the Mississippi-Alabama Sea Grant Consortium and Mississippi State University. The US Government and the MASGC are authorized to produce and distribute reprints notwithstanding any copyright notation that may appear hereon. The views expressed herein are those of the author and do not necessarily reflect the views of NOAA or its sub-agencies.

References

Admiraal, W. (1977a). Influence of light and temperature on the growth rate of estuarine benthic diatoms in culture. *Marine Biology,* **39,** 1–9.
(1977b). Influence of various concentrations of orthophosphate on the division rate of an estuarine benthic diatom, *Navicula arenaria. Marine Biology,* **42,** 1–8.
(1977c). Tolerance of estuarine benthic diatoms to high concentrations of ammonia, nitrite ion, nitrate ion and orthophosphate. *Marine Biology,* **43,** 307–15.
(1984). The ecology of estuarine sediment-inhabiting diatoms. *Progress in Phycological Research,* **3,** 269–322.
Admiraal, W., & Peletier, H. (1979a). Sulfide tolerance of benthic diatoms in relation to their distribution in an estuary. *British Phycological Journal,* **14,** 185–96.
(1979b). Influence of organic compounds and light limitation on the growth rate of estuarine benthic diatoms. *British Phycological Journal,* **14,** 197–206.
(1980). Distribution of diatom species on an estuarine mud flat and experimental analysis of the selective effect of stress. *Journal of Experimental Marine Biology and Ecology,* **46,** 157–75.
Archibald, R. E. M. (1972). Diversity in some South African diatom associations and its relation to water quality. *Water Research,* **6,** 1229–38.
Baillie, P. W., & Welsh, B. L. (1980). The effect of tidal resuspension on the distribution of intertidal epipelic algae in an estuary. *Estuarine and Coastal Marine Science,* **10,** 165–80.
Cahoon, L. B., & Cooke, J. E. (1992). Benthic microalgal production in Onslow Bay, North Carolina, USA. *Marine Ecology Progress Series,* **84,** 185–96.

Cahoon, L. B., & Laws, R. A. (1993). Benthic diatoms from the North Carolina continental shelf: Inner and mid shelf. *Journal of Phycology, 29*, 257–63.

Cahoon, L. B., Laws, R. A., & Thomas, C. J. (1993). Viable diatoms and chlorophyll *a* in continental slope sediments off Cape Hatteras, North Carolina. *Deep-Sea Research II, 41*, 767–82.

Cahoon, L. B., Beretich, G. R., Jr., Thomas, C. J., & McDonald, A. M. (1993). Benthic microalgal production at Stellwagen Bank, Massachusetts Bay, USA. *Marine Ecology Progress Series, 102*, 179–85.

Cahoon, L. B., Redman, R. S., & Tronzo, C. R. (1990). Benthic microalgal biomass in sediments of Onslow Bay, North Carolina. *Estuarine, Coastal and Shelf Science, 31*, 805–16.

Coleman, V. L., & Burkholder, J. M. (1994). Community structure and productivity of epiphytic microalgae on eelgrass (*Zostera marina* L.) under water-column nitrate enrichment. *Journal of Experimental Marine Biology and Ecology, 179*, 29–48.

(1995). Response of microalgal epiphyte communities to nitrate enrichment in an eelgrass (*Zostera marina*) meadow. *Journal of Phycology, 31*, 36–43.

Coles, S. M. (1979). Benthic microalgal populations on intertidal sediments and their role as precursors to salt marsh development. In *Ecological Processes in Coastal Environments*, ed. R. L. Jefferies & A. J. Davy, pp. 25–42. Oxford: Blackwell.

Daborn, G. R., Amos, C. L., Brylinsky, M., Christian, H., Drapeau, G., Faas, R. W., Grant, J., Long, B., Paterson, D. M., Perillo, G. M. E., & Piccolo, M. C. (1993). An ecological cascade effect: Migratory birds affect stability on intertidal sediments. *Limnology and Oceanography, 38*, 225–31.

Darley, W. M., Montague, C. L., Plumley, F. G., Sage, W. W., & Psalidas, A. T. (1981). Factors limiting edaphic algal biomass and productivity in a Georgia salt marsh. *Journal of Phycology, 17*, 122–8.

Delgado, M., Jonge, V. N. de, & Peletier, H. (1991*a*). Experiments on resuspension of natural microphytobenthos populations. *Marine Biology, 108*, 321–8.

(1991*b*). Effect of sand movement on the growth of benthic diatoms. *Journal of Experimental Marine Biology and Ecology, 145*, 221–31.

Descy, J. P. (1979). A new approach to water quality estimation using diatoms. *Nova Hedwigia Beiheft, 64*, 305–23.

Gallagher, J. L. (1975). The significance of the surface film in salt marsh plankton metabolism. *Limnology and Oceanography, 29*, 120–3.

Grant, J., Bathmann, U. V., & Mills, E. L. (1986). The interaction between benthic diatom films and sediment transport. *Estuarine, Coastal and Shelf Science, 23*, 225–38.

Hein, M. K., & Koppen, J. D. (1979). Effects of thermally elevated discharges on the structure and composition of estuarine periphyton diatom assemblages. *Estuarine, Coastal and Shelf Science, 9*, 385–401.

Hendey, N. I. (1977). The species diversity index of some in-shore diatom communities and its use in assessing the degree of pollution insult on parts of the north coast of Cornwall. *Nova Hedwigia, Beiheft* **54**, 355–78.

Hohn, M. H. (1959). The use of diatom populations as a measure of water quality in selected areas of Galveston and Chocolate Bay, Texas. *Publications of the Institute of Marine Science, 6*, 206–12.

Holland, A. F., Zingmark, R. G., & Dean, J. M. (1974). Quantitative evidence concerning the stabilization of sediments by marine benthic diatoms. *Marine Biology, 27*, 191–6.

Jonge, V. N. de (1985). The occurrence of 'epipsammic' diatom populations: a result of interaction between physical sorting of sediment and certain properties of diatom species. *Estuarine, Coastal and Shelf Science, 21*, 607–22.

Jonge, V. N. de & van Beusekom, J. E. E. (1992). Contribution of resuspended microphytobenthos to total phytoplankton in the Ems estuary and its possible role for grazers. *Netherlands Journal of Sea Research, 30*, 91–105.

(1995). Wind- and tide-induced resuspension of sediment and microphytobenthos from tidal flats in the Ems estuary. *Limnology and Oceanography, 40*, 766–78.

Jonge, V. N. de & van den Bergs, J. (1987). Experiments on the resuspension of estuarine sediments containing benthic diatoms. *Estuarine, Coastal and Shelf Science, 24*, 725–40.

Kobayasi, H., & Mayama, S. (1989). Evaluation of river water quality by diatoms. *Korean Journal of Phycology*, **4**, 121–33.

Kolkwitz, R., & Marsson, M. (1908). Ökologie der pflanzliche Saprobien. *Berichte der Deutschen Botanischen Gesellschaft*, **26**, 505–19.

Lange-Bertalot, H. (1978). Diatomeen-Differentialarten anstelle von Leitformen: ein geeigneteres Kriterium der Gewässerbelastung. *Archiv für Hydrobiologie, Supplement 51, Algological Studies* **21**, 393–427.

(1979). Pollution tolerance of diatoms as a criterion for water quality estimation. *Nova Hedwigia, Beiheft*, **64**, 285–304.

Lloyd, M., Zar, J. H., & Karr, J. R. (1968). On the calculation of information-theoretical measures of diversity. *The American Midland Naturalist*, **79**, 257–72.

Lukatelich, R. J., & McComb, A. J. (1986). Distribution and abundance of benthic microalgae in a shallow southwestern Australian estuarine system. *Marine Ecology Progress Series*, **27**, 287–97.

MacIntyre, H. L., Geider, R. J., & Miller, D. C. (1996). Microphytobenthos: the ecological role of the 'secret garden' of unvegetated, shallow-water marine habitats. I. Distribution, abundance and primary production. *Estuaries*, **19**, 186–201.

Miller, D. C., Geider, R. J., & MacIntyre, H. L. (1996). Microphytobenthos: The ecological role of the 'secret garden' of unvegetated, shallow-water marine habitats. II. Role in sediment stability and shallow-water food webs. *Estuaries*, **19**, 202–12.

Paterson, D. M. (1986). The migratory behaviour of diatom assemblages in a laboratory tidal micro-ecosystem examined by low temperature scanning electron microscopy. *Diatom Research*, **1**, 227–39.

(1989). Short-term changes in the erodibility of intertidal cohesive sediments related to the migratory behavior of epipelic diatoms. *Limnology and Oceanography*, **34**, 223–34.

(1990). The influence of epipelic diatoms on the erodibility of an artificial sediment. In *Proceedings of the 10th International Diatom Symposium*, ed. H. Simola, pp. 345–55. Koenigstein: Koeltz Scientific Books.

Peletier, H. (1996). Long-term changes in intertidal estuarine diatom assemblages related to reduced input of organic waste. *Marine Ecology Progress Series*, **137**, 265–71.

Prygiel, J., Leveque, L., & Iserentant, R. (1996). Un nouvel indice diatomique pratique pour l'évaluation de la qualité des eaux en réseau de surveillance. *Revue des Sciences de l'Eau*, **1**, 97–113.

Riaux-Gobin, C. (1987). Phytoplancton, tripton et microphytobenthos: échanges au cours de la marée, dans un estuaire du Nord-Finistère. *Cahiers du Biologie Marine*, **28**, 159–85.

Round, F. E. (1979). Botanical aspects of estuaries. In *Tidal Power and Estuary Management*, ed. R. T. Severn, D. Dineley, & L. E. Hawker, pp. 195–213. Bristol: Scientechnica.

(1981). *The Ecology of Algae*. Cambridge: Cambridge University Press.

Shaffer, G. P., & Sullivan, M. J. (1988). Water column productivity attributable to displaced benthic diatoms in well-mixed shallow estuaries. *Journal of Phycology*, **24**, 132–40.

Short, F. T., & Wyllie-Echeverria, S. (1996). Natural and human-induced disturbance of seagrasses. *Environmental Conservation*, **23**, 17–27.

Sullivan, M. J. (1976). Long-term effects of manipulating light intensity and nutrient enrichment on the structure of a salt marsh diatom community. *Journal of Phycology*, **12**, 205–10.

(1981). Effects of canopy removal and nitrogen enrichment on a *Distichlis spicata*–edaphic diatom complex. *Estuarine, Coastal and Shelf Science*, **13**, 119–29.

(1986). Mathematical expression of diatom results: Are these 'pollution indices' valid and useful? In *Proceedings of the 8th International Diatom Symposium*, ed. M. Ricard, pp. 772–6. Koenigstein: S. Koeltz.

Sullivan, M. J., & Daiber, F. C. (1975). Light, nitrogen, and phosphorus limitation of edaphic algae in a Delaware salt marsh. *Journal of Experimental Marine Biology and Ecology*, **18**, 79–88.

Sundbäck, K., & Granéli, W. (1988). Influence of microphytobenthos on the nutrient flux between sediment and water: A laboratory study. *Marine Ecology Progress Series*, **43**, 63–9.

Sundbäck, K., & Snoeijs, P. (1991*b*). Effects of nutrient enrichment on microalgal community composition in a shallow-water sediment system: An experimental study. *Botanica Marina*, **34**, 341–58.

Sundbäck, K., Enoksson, V., Granéli, W., & Pettersson, K. (1991*a*). Influence of sublittoral microphytobenthos on the oxygen and nutrient flux between sediment and water: A laboratory continuous-flow study. *Marine Ecology Progress Series*, **74**, 263–79.

Underwood, G. J. C. (1997). Microalgal colonisation in a saltmarsh restoration scheme. *Estuarine, Coastal and Shelf Science*, **44**, 471–81.

(1998). Changes in microalgal species composition, biostabilisation potential and succession during saltmarsh restoration. *Proceedings of the Linnean Society*, in press.

Underwood, G. J. C., & Paterson, D. M. (1993*a*). Recovery of intertidal benthic diatoms after biocide treatment and associated sediment dynamics. *Journal of the Marine Biological Association of the United Kingdom*, **73**, 25–45.

(1993*b*). Seasonal changes in diatom biomass, sediment stability and biogenic stabilization in the Severn estuary. *Journal of the Marine Biological Association of the United Kingdom*, **73**, 871–87.

Valiela, I., Collins, G., Kremer, J., Lajtha, K., Geist, M., Seely, B., Brawley, J., & Sham, C. H. (1997). Nitrogen loading from coastal watersheds to receiving estuaries: new method and application. *Ecological Applications*, **7**, 358–80.

van Dam, H. (1982). On the use of measures of structure and diversity in applied diatom ecology. *Nova Hedwigia, Beiheft* **73**, 97–115.

van Raalte, C. D., Valiela, I., & Teal, J. M. (1976*a*). The effect of fertilization on the species composition of salt marsh diatoms. *Water Research*, **10**, 1–4.

(1976*b*). Production of epibenthic salt marsh algae: light and nutrient limitation. *Limnology and Oceanography*, **21**, 862–72.

Varela, M., & Penas, E. (1985). Primary production of benthic microalgae in an intertidal sand flat of the Ria de Arosa, NW Spain. *Marine Ecology Progress Series*, **25**, 111–9.

Vos, P. C., de Boer, P. L., & Misdorp, R. (1988). Sediment stabilization by benthic diatoms in intertidal sandy shoals: Qualitative and quantitative observations. In *Tide-Influenced Sedimentary Environments and Facies,* ed. P. L. de Boer, A. van Gelder & S. D. Nio, pp. 511–26, Dordrecht: D. Reidel.

Watanabe, T., Asai, K., & Houki, A. (1988). Numerical water quality monitoring of organic pollution using diatom assemblages. In *Proceedings of the 9th International Diatom Symposium,* ed. F. E. Round, pp. 123–41. Bristol: Biopress Ltd.

Wear, D. J., Sullivan, M. J., Moore, A. D., & Millie, D. F. (1999). Effects of water-column enrichment on the production dynamics of three seagrass species and their epiphytic algae. *Marine Ecology Progress Series,* in press.

Yallop, M. L., de Winder, B., Paterson, D. M., & Stal, L. J. (1994). Comparative structure, primary production and biogenic stabilization of cohesive and non-cohesive marine sediments inhabited by microphytobenthos. *Estuarine, Coastal and Shelf Science*, **39**, 565–82.

16 Estuarine paleoenvironmental reconstructions using diatoms

SHERRI R. COOPER

Introduction

Historically, there has been a lack of appreciation of the severity of human impacts on estuaries, and of how important these systems are to human society. The demand for resources and the products and residues generated as human populations grow will continue to cause cultural, economic, aesthetic and environmental problems, especially in coastal areas. Understanding the processes surrounding these problems is important for managing the continuing impacts of growing populations (National Research Council, 1993). Environmental issues relevant to estuaries include eutrophication, anoxia, harmful algal blooms, industrial pollution, loss of habitats such as wetlands and submerged aquatic vegetation, land-use effects on turbidity and sedimentation, and invasion of exotic species (National Safety Council, 1993).

Paleoecology offers powerful techniques with which to study historical changes due to human influences in depositional environments, including estuaries. A paleoecological approach makes it possible to define the naturally occurring state of an ecosystem, against which human influences can be measured (Smol, 1992). Diatoms are particularly useful not only because they are preserved in the sediment record but because they have a rapid reproductive rate and respond quickly to changes in nutrient availability and other water quality conditions. In addition, diatoms are abundant in aquatic environments, generally cosmopolitan in distribution, and have a fairly well-studied taxonomy and ecology.

Paleoecological studies in estuarine environments have lagged behind paleolimnology, in large part because of the more dynamic nature of coastal ecosystems. Estuaries are characterized by variable salinities, sediment deposition regimes, water currents, turbidity zones, and biogeochemistry of sediments. There is often mixture and transport of sediments after initial deposition, and differential silicification and preservation of diatom valves.

This chapter highlights the nature and significance of estuarine ecosystems, describes methods for using diatoms in estuarine paleoenvironmental reconstructions, and discusses issues relevant to these methods. The discussion focuses on postglacial Holocene anthropogenic influences on water quality and land use changes.

Estuaries

An estuary can be defined as a semi-enclosed coastal body of water that has a free connection with the open ocean, within which sea water is diluted with fresh water derived from land drainage (Pritchard, 1967). River mouths, coastal bays, tidal marsh systems, and sounds (behind barrier beaches) all fit this definition. Estuaries are transition zones between freshwater and marine habitats. They are most common in low-relief coastal regions including Europe and the east coast of North America. They are less extensive on uplifted coastlines such as the Pacific edge of North and South America. Although estuaries are abundant, they are geologically transient. All present-day estuaries are less than 5000 years old, the time when sea level rose to its present elevation following the last ice age (Day *et al.*, 1989).

Human populations throughout history have used estuaries as food sources, places of navigation and settlement, and repositories for waste. Although there are no specific statistics for estuaries, it is estimated that 60% of the three billion people forecasted to live in urban areas around the world by the year AD 2000 will be concentrated in settlements less than 50 miles (80.5 km) from the sea (FAO, 1993). Most of these areas include estuarine waters. Coastal urban centers are home to almost one billion people worldwide, with roughly half of the world's coasts threatened by development-related activities. Fragile coastal and estuarine ecosystems, such as wetlands, tidal flats, saltwater marshes, mangrove swamps, and the flora and fauna that depend on them, are especially endangered by urban land conversion and human uses (World Resources Institute *et al.*, 1996). In the United States, almost half of the population lives in coastal areas, a region where the majority of new development is occurring (Culliton *et al.*, 1990, 1992).

The location of estuaries depends on shoreline position, and is conditioned by geomorphology, coastal lithology, tidal range, river discharge and sediment load, climate, sea-level oscillations, tectonism, and isostasy. A coastal shore is held in delicate balance between forces that tend to move it either landward or seaward. Geomorphologic and sedimentologic changes are continuously occurring within and around estuaries. Interest in study of the geomorphology of estuaries began around the middle of the twentieth century; a comprehensive overview of geomorphology and sedimentology of estuaries is contained in Perillo (1995).

EUTROPHICATION

Eutrophication is one of the major concerns to estuarine scientists today (Neilson & Cronin, 1981; Elmgren, 1989; Jonge *et al.*, 1994; Cooper & Lipton, 1994; Hinga *et al.*, 1995; Brooks *et al.*, 1996). Worldwide, estuaries receive large amounts of nutrients from wastewaters of sewage treatment facilities and unsewered sources. In most cases, it is difficult to separate the relative impact of nutrient enrichment from point sources of sewage wastewater and industrial discharges from nonpoint sources such as agricultural and atmospheric additions and urban and rural runoff (Kennish, 1992; Puckett, 1995). Eutrophication is implicated in many coastal water quality problems, including nuisance and noxious algal blooms, hypoxic and anoxic bottom waters, fish and shellfish kills, declines in biodiversity, habitat losses and shifts in trophic dynamics. Continued population growth is likely to exacerbate eutrophication problems unless a better scientific understanding of these processes is translated into practical management strategies for the watershed and coastal waters (National Research Council, 1993).

HABITATS AND ENVIRONMENTAL HISTORY

Estuaries are biotically rich environments. The variety of habitats within an estuarine system contributes to the high productivity of these systems. The salinity gradient within an estuary is an important physical parameter that affects distributional patterns of plant and animal species. Fresh and brackish water marshes found along the shores of estuaries are some of the most productive ecosystems in the world. They are also important in terms of flood control, water quality control, and maintenance of natural chemical and biological cycles (Mitsch & Gosselink, 1986; White, 1989). In areas where the water is shallow and relatively clear, beds of submerged vascular plants grow. These habitats are home to wading birds and crustaceans and provide nursery and feeding grounds for juvenile fish. Other estuarine habitats include intertidal mud and sand flats, shallow water habitats, and deeper channels.

The sedimentary subsystem of the estuary is also very important, with its benthic organisms and unique chemistry. Within the sediments, respiration exceeds the reserves and supply of oxygen, and the sediments are typically anoxic below the water–sediment interface. Detritus from the productive zones of the estuary is decomposed, and nutrients are regenerated, recycled and stored in the sediments by the action of bacteria, fungi, protozoans, animals, and natural chemistry (Day *et al.*, 1989).

Using diatoms in estuarine paleoecology studies

The use of diatoms in estuarine paleoecology studies requires application of other paleoecological methods for collection and analysis of cores, dating of sediments, and correlation with other geochemical and biological indicators. This type of work is extremely time-intensive and requires much attention to detail.

Paleoecological indicators abound in estuarine ecosystem sediments. These indicators include pollen from terrestrial plants, pollen and seeds from wetland and submerged aquatic plants, diatoms from a variety of habitats, ostracodes, foraminifera, dinoflagellate cysts, fish otoliths and scales, charcoal, and myriad geochemical and biogeochemical parameters. Diatoms are particularly useful as indicators in estuarine systems for the same reason that they are useful in other aquatic habitats. Each species has a specific optimum and tolerances for water quality, including pH, salinity, temperature, nutrients, and light availability (turbidity). Each species identified contributes clues about the past environment, and assemblage statistics provide additional information.

DIATOMS IN ESTUARIES

There are a variety of diatom habitats within an estuary. Planktonic and tychoplanktonic species float with the currents and tides, epiphytic species grow attached to vascular plants or other algae, epipelic species migrate up and down within the upper few millimeters of muddy sediments, epilithic species are associated with rocky areas, and epipsammic species grow attached to sand grains. Intertidal marshes have their own diatom communities, inhabiting the high and low marsh (Sullivan, 1978; Hemphill-Haley, 1995a). In an estuary, freshwater diatoms may be brought in by river flow and marine species may be transported into brackish areas by tidal action, while estuarine species flourish in the productive mixing zones. Sediment samples contain all of the species that have been deposited over years of sediment accumulation, including benthic and planktonic species. Therefore, sediment samples generally give a clearer picture of diatom community structure than water samples, which may suffer from net bias and seasonality of diatom species (Hecky & Kilham, 1973). However, resuspension and transport pose problems in estuarine environments, and very delicate or lightly silicified species may be missing. Diatom communities in the coastal, estuarine and marsh environments are discussed in more detail in other chapters of this book (see Denis & de Wolf; Snoeijs & Sullivan, this volume).

SILICA CYCLING IN ESTUARIES

A certain minimum concentration of silica in solution is essential to the growth of diatoms. The few studies that include measurements of soluble silica in estuaries generally report it as a nutrient in relation to algal growth (D'Elia et al., 1983; Yamada & D'Elia, 1984; Conley & Malone, 1992). Addition of sewage effluent containing phosphorus and nitrogen contribute to eutrophication of rivers and estuaries (Ryther & Officer, 1981). Silica may become the limiting nutrient for diatom growth in eutrophic estuaries, with increasing diatom abundance and the concomitant preservation of the biogenic silica (BSi) of diatom frustules in the sediments (Conley & Malone, 1992; Dortch & Whitledge, 1992; Conley et al., 1993). Silica depletion may cause major changes in species composition and ultimately affect trophic interactions as well as other ecological and biogeochemical changes. Fraser and Wilcox (1981) and Biggs and Cronin (1981) characterize estuarine systems in general as a sink for soluble silica during low discharge and high primary productivity.

The geochemistry of silica is important to understanding preservation of diatom valves in sediments, as well as its role in the growth of diatoms. The hydration and solubility characteristics of silica are not well understood, and there is confusion in the literature concerning the silica–water interface. Most of the BSi found in the sediments of lakes and estuaries is in the form of diatom frustules, although other organisms that contribute to BSi include silicoflagellates, radiolarians and sponges. The amorphous silica of diatom frustules and other organisms appears to be much less soluble than inorganic silicas; this is true both before and after death of the organisms. There are numerous theories and explanations for this (Lewin, 1961; Siever & Scott, 1963; Iler, 1979). Chemical dissolution of diatom frustules has been found to be more prevalent in extremely alkaline or saline waters, although only minor differences were found in diatom assemblages exposed to a wide concentration of NaCl solutions in an experimental study by Barker *et al.* (1994). BSi can be a good estimate of diatom silica in sediments (Schelske *et al.*, 1986; Conley, 1988; Cooper, 1995*a*), but a linear relationship between BSi and diatom abundance is not necessarily expected because of the large size range and variable silica content of different diatom species within samples. Methods for measurement of BSi in sediments are described in Conley (1988).

VALVE PRESERVATION

Although diatoms are abundant and well preserved in estuarine sediments, they are not indestructible. Both chemical and mechanical degradation can occur under certain conditions and result in a biased preservation of species. Ecological interpretations may be hindered where finely silicified indicator species are missing due to chemical dissolution, whereas fractured or eroded valves may indicate intertidal exposure and abrasion (Hemphill-Haley, 1995*a*). On the other hand, excellent preservation of valves may indicate rapid burial and little postdepositional disturbance. Cooper (1993, 1995*a,b*) found that many lightly silicified and delicate planktonic species known to occur in Chesapeake Bay waters (such as those of *Chaetoceros, Leptocylindrus, Asterionella, Cerataulina* and *Rhizosolenia*) were not represented in sediment samples as whole valves. However, resting spores and pieces of valves representing these genera were seen. This was true for both surface sediments and deeper core sediment samples and therefore did not affect comparisons of species abundance or diversity between samples. Beasley (1987) found similar results in her study of Rehoboth Bay, Delaware. In a study of eutrophication effects on biota of the Adriatic Sea, Puškaric *et al.* (1990) also found that lightly silicified diatoms such as *Chaetoceros* and *Skeletonema* were not preserved in core samples, although they were abundant in the northern Adriatic Sea. They found that these genera were not present in surface sediments from cores, but only in sedi-

S. R. COOPER

ment trap material when incorporated in fecal pellets or other organic matrices.

Another consideration in estuarine sediment samples is the difference in silicification between freshwater and marine species. Freshwater species from fluvial inputs will generally be more heavily silicified and therefore better preserved than brackish water and marine species (Conley *et al.*, 1989). Observations on the autecology of species and the preservation of frustules will provide evidence of transport and reworking of sediments (see Chapter 13).

Methods

ESTUARINE SEDIMENTARY ENVIRONMENTS
AND CHOOSING CORING SITES

Sedimentation patterns within estuarine systems are extremely variable in both space and time (Brush, 1984, 1989; Officer *et al.*, 1984; Sanford, 1992; Dyer, 1995). Resuspension and mixing of sediments before final deposition is common in estuarine environments (Sanford, 1992; Dyer, 1995). The location of the turbidity maximum where fresh and salt waters meet may change seasonally and yearly depending on river discharge (freshwater inflow). Storms that cause high runoff from the drainage basin may flush huge amounts of sediments from rivers into estuarine areas, or scour sediments from shallow areas. Marine sediments are often transported into the mouths of estuaries by normal currents and tides (Dyer, 1995), and occasionally by storms. For these reasons, there has been doubt about the validity of using paleoecological methods within these dynamic systems.

Depositional environments in estuaries can be, and have been, identified. Many of these areas occur in deeper waters and channels (e. g., original river bed channels flooded as sea level has risen), marsh or mudflat depositional areas, and quiet lagoons (Dyer, 1995). Existing channels may be areas of good preservation and higher sedimentation rates because of the anoxic sediments found in these deeper waters that experience less bioturbation and less wind disturbance effects. Typically, these are also areas that do not require dredging, and would therefore remain primarily undisturbed. As with any sediment coring endeavor, special care needs to be taken in analyzing cores for continuous sediment accumulation through time. Prior to coring, channel incisement and morphology of estuarine bottom sediments can be investigated using seismic reflection survey equipment (Riggs *et al.*, 1992). With careful analysis, including such tools as X-raying of cores, sediment type analysis, ^{210}Pb and ^{137}Cs dating, observation of pollen and/or tephra horizons (Dugmore & Newton, 1992), and use of pollen concentrations for calculating sedimentation rates (Brush, 1989), good chronologies of sediment cores can be obtained to study the history of climate and anthropogenic influences over the past

5000 yr. Sediment cores are obtained in estuaries following the same methods as for lakes.

As an example, there are several broad sedimentation zones along a north to south transect in the Chesapeake Bay. The northern Bay sediments contain proportionately more organic carbon and less sulfur than those deposited in the middle Bay. Fluvial input is the dominant mechanism for introduction of particulate matter in this area (Helz *et al*,. 1985; Hennessee *et al*., 1986). In the mid-Bay regions, distributions of organic carbon and sulfur are more representative of marine environments, and primary production is the dominant source of organic matter (Biggs, 1970; Hennessee *et al*., 1986). The southern part of the Chesapeake Bay is characterized by sandier sediments, a greater percentage of which are of marine origin. There are also compounding factors in specific areas, such as the presence of geologically entrapped gas in sediments, that affect the bulk density and chemical properties of the sediments (Hill *et al*., 1992). This type of information is important in understanding the source and distribution of organic material found in cores, and especially when interpreting various biogeochemical parameters.

SEDIMENT DATING, SEDIMENTATION RATES, AND CORE CHRONOLOGIES

Typically, sedimentation rates within estuarine systems have increased with human land use of watershed areas. For example, the highest sedimentation rates within the mesohaline Chesapeake Bay, USA occurred during the late nineteenth century when forest clearing of the landscape for agriculture was at a historical maximum (Brush, 1984, 1986; Cooper, 1995a). For this reason, careful dating and determination of sedimentation rates is necessary to obtain a chronology for each core with which to correlate indicators such as diatoms (Table 16.1), especially when comparing anthropogenic influences with climatic and prehistorical changes. A single subsection of a core may represent as little as a seasonal signal (if sedimentation rates are very high), or as much as 100 years or more. Sedimentation rates are used to determine influx of indicators to the sediments. Pollen horizons and calculation of sedimentation rates using pollen concentrations within dated horizons of sediments have been very useful in Chesapeake Bay sediments and compare well with ^{210}Pb dating of recent sediments (Brush *et al*., 1982; Brush, 1989; Cooper & Brush, 1991). These methods are outlined in Brush *et al*. (1982).

EXTRACTION METHODS

Diatoms are plentiful in estuarine environments, and many are preserved in estuarine sediments. For example, the number of identifiable diatom valves calculated to be found in Chesapeake Bay sediments averaged about 3×10^8 valves g dry wt/sediment (Cooper, 1995a). As mentioned previously, there is some bias in preservation of whole valves related to the degree of silicification

Table 16.1. *Sedimentation rates and chronology of subsamples from core R4-50 collected in the Chesapeake Bay, USA as determined by pollen and radiocarbon methods*

Sample depth (cm)	Bulk density (g dry wt/ml)	Sedimentation rate (cm/y)	Number of years in sample	Chronology (years AD)
0–2	0.40	0.20	10	1975–1985
2–4	0.47	0.20	10	1965–1975
4–6	0.47	0.20	10	1955–1965
6–8	0.47	0.41	5	1950–1955
8–10	0.47	0.41	5	1945–1950
10–12	0.43	0.25	8	1937–1945
12–14	0.43	0.17	12	1925–1937
16–18	0.41	0.28	7	1909–1916
20–22	0.39	0.26	8	1894–1902
22–24	0.39	0.24	8	1886–1894
24–26	0.39	0.24	8	1878–1886
28–30	0.45	0.34	6	1865–1871
32–34	0.30	0.41	5	1854–1859
36–38	0.30	0.51	4	1845–1849
40–42	0.32	0.51	4	1837–1841
44–46	0.32	0.31	6	1827–1833
50–52	0.32	0.22	9	1800–1809
52–54	0.46	0.22	9	1791–1800
56–58	0.46	0.22	9	1773–1782
58–60	0.46	0.17	12	1761–1773
60–62	0.46	0.05	40	1721–1761
66–68	0.42	0.05	40	1601–1641

Source: From Cooper (1993).

of each species, and the fragility of valves. To minimize breakage during extraction procedures, it is recommended that original wet sediment be used, rather than dried or sieved material.

In order to have information on total diatoms and flux, it is important to know the volume of the wet sediment from which extractions are made, as well as the dilution of the final material, and the volume of diluted material used for each slide. Bulk density measurements are needed to calculate flux to sediment on a dry weight basis. A duplicate sample of sediment from the same depth as diatom extractions can be used to determine dry weight and bulk density. If there is not enough material for duplication, drying the sample at 60 °C is appropriate (Cooper, 1993; Parsons 1996). Dilutions to obtain slides with the appropriate density of diatoms will vary depending on site and depth within cores. In general, for estuarine sediments, an original dilution of 1/50 ml will provide at least 400 diatoms when using 0.1 ml of the diluted material per slide. Additional dilution will most likely be needed (see Cooper, 1993).

Methods for extraction of diatoms from sediments are somewhat varied, but all include steps for eliminating unwanted material from the sediment matrix. The method used in Cooper (1993), a modification of Funkhauser and Evitt (1959) produced good results. It included steps using 25% hydrogen peroxide (H_2O_2) to disperse the sample and oxidize some organics, hydrochloric acid to remove carbonates, and concentrated nitric acid with potassium dichromate to further oxidize organics. These chemical steps were followed by washing steps in distilled water, and a physical separation of larger sand particles without centrifugation. Centrifugation can also be used (Parsons, 1996) but the potential exists to damage some delicate or lightly silicified valves. Parsons (1996) experimented with several diatom extraction methods and discussed the results obtained from each. Permanent diatom slides are made using Hyrax®, Naphrax®, or other mounting media with a similar refractive index (RI ≅ 1.7).

SPECIES IDENTIFICATIONS AND REFERENCES

Good comprehensive references for identification of estuarine diatoms are few and far between. Hustedt's *Kieselalgen* (1927–1930, 1931–1959, 1961–1966) and his 1955 paper on Beaufort diatoms are very useful. Krammer and Lange-Bertalot's (1986–1991) volumes in *Süßwasserflora von Mitteleuropa* can be useful, especially in freshwater and low salinity areas, but also for some marine taxa. The recent publications of Snoeijs and Potapova (1993–1997) appears to be fairly comprehensive, and although light micrographs are not presented for all species, additional references for each species are listed. For estuarine and marine genera, Hendey (1964) and Peragallo and Peragallo (1897–1908) are helpful, among other papers and dissertations, including Cooper (1995b), Kuylenstierna (1990), Laws (1988), Bérard-Therriault et al. (1986, 1987), Cardinal et al. (1986), Poulin et al. (1984a,b, 1986), Rao & Lewin (1976), Tynni (1975–1980), Mölder & Tynni (1967–1973), Hustedt (1939, 1951, 1953), Aleem & Hustedt (1951), and Giffen (1963, 1967, 1970, 1971, 1973, 1975). For a more complete list of references on estuarine diatom taxonomy, see Cooper (1993, 1995b), and other chapters on estuarine diatoms in this book.

When conducting paleoecological studies, it is important to carefully record descriptions and take photographs of diatoms or capture digital images via video camera and image capture software. Information on size ranges of valves, striae densities, and other distinguishing features should be collected and archived, along with counts of species. References used in identification of each species should also be recorded. These notes may be critical for comparisons between studies and verification of species identifications.

S. R. COOPER

For studies of diatoms in sediment samples, sample sizes of 400–500 diatom valves have commonly been used (McIntire & Overton, 1971; Main & McIntire, 1974; Amspoker & McIntire, 1978; Sullivan, 1982; Laws, 1988). For statistical purposes, a minimum of 300 diatom valves per sample is recommended (Van Dam, 1982; Sullivan & Moncreiff, 1988; Shaffer & Sullivan, 1988). These numbers can easily be obtained from estuarine sediments, and most likely will be contained in less than 0.5 ml of wet sediment (see extraction methods).

Diatom assemblage diversity, as observed in sediment samples, may be a useful parameter, but care must be taken in determining the sedimentation rates and chronology for each core and subsample. Sediment samples representing longer time periods of deposition could be expected to exhibit higher diversity, all else being equal (Smol, 1981). On the other hand, older samples with lower sedimentation rates may show reduced diversity due to dissolution and breakage of diatom valves. Estuarine sedimentation is extremely variable. Subsamples from the same core or similar depths from two different locations may represent very different time periods, both in terms of the sedimentation rate as well as chronological age (Smol, 1981; Cooper, 1995a). This difference in sedimentation rates between samples did not appear to have a significant effect on measured diatom assemblage diversity in four cores from a transect across the mesohaline Chesapeake Bay (Cooper, 1995a).

Shannon's informational index (H′) has been shown to be an effective indicator of changes in diatom population diversity, especially within habitat comparisons as compared to between habitat comparisons, and is the most widely used diversity index in ecology (Hendey, 1977; Van Dam, 1982; Washington, 1984). Shannon's H′ is based on the total number of species as well as the evenness of species abundance. Diversity, under this definition, is an index of community structure in which rare species are structurally unimportant. Rare species may play an important role in the ecosystem, but it is a functional role rather than a significant contribution to diversity as measured by H′. Diversity is therefore a complement to information on diatom species and their autecology.

A steady decline in diversity of diatom assemblages through time, in relation to land use activity, was seen in four cores from the Chesapeake Bay (Cooper 1993, 1995a). Patten (1962), Stockner and Benson (1967), and Hendey (1977) all found that diatom community diversity (as measured by H′) decreased with eutrophication and pollution in an estuary, a lake, and the coast of Cornwall, respectively. McIntire and Overton (1971) and Wilderman (1984) also used Shannon's H′ in their studies of estuarine diatoms. Archibald (1972) reviewed various diversity indices as applied to diatom populations from rivers and cautioned against its use as the sole indicator of water quality.

Cluster analysis of diatom communities may reveal differences in distribution of species (e.g., Bruno & Lowe, 1980) or major trends in assemblage changes through time in core samples (Cooper, 1995a). Diatom assemblages preserved in the sediments of four cores from the Chesapeake Bay were remarkably more similar within time periods of similar land use, both within and between the four cores, than to assemblages within the same core dated from different land use periods (Cooper, 1993, 1995a; Cooper & Brush, 1993). This shows that the date of deposition was more important in determining diatom species presence and abundance within a sediment sample than the location within the east–west transect across the mesohaline Chesapeake Bay. Similar patterns and trends were seen in all four cores, and these were correlated to changing land use within the watershed and the concomitant changes in water quality (including eutrophication) within the Bay (Cooper, 1995a).

Multivariate statistical analyses of diatom data, along with measurements of other fossils and geochemical and physical parameters, can help elucidate what factors are most closely related to changing diatom communities. These methods are used in developing transfer functions from calibration sets (see section on calibration sets), but can also be used to compare and separate communities in relation to other parameters in sediment cores. In any paleoecological study, the more indicators and parameters that can be measured, the more information is available for corroboration of trends. Several methods of paleoecological data analysis currently in use include PCA, canonical correspondence analysis (CCA) and detrended correspondence analysis (DCA). An overview of these techniques is contained in Moser *et al.* (1996), as well as in other chapters in this book.

Weighted-averaging methodology has been used in limnology for diatom-inferred phosphorus levels (Hall & Smol, this volume) and acidification (Battarbee *et al.*, this volume). This methodology is suggested as an improved method for reconstructing environmental variables from species assemblages (ter Braak & Juggins, 1993), and has recently been used successfully in developing a salinity transfer function for the Thames Estuary (Juggins, 1992).

Studies

CALIBRATION SETS

Much progress has been made in the development of quantitative approaches to infer environmental variables from diatom assemblages (Birks, 1995). One of the most widely used tools is the surface sediment calibration set (Anderson, 1993; Charles & Smol, 1994; Moser *et al.*, 1996), which is discussed in more detail in other chapters of this book. True calibration sets have not been widely used for estuarine sediments and cores because of the within-system heterogeneity, as well as the geographical extent of these systems (as compared to lakes). In other words, estuaries do not generally occur in abundance within the same climatic and vegetational areas, and are therefore limited for use in calibration of differences between settings.

The use of a diatom calibration set for nutrients in coastal systems was explored by Anderson and Vos (1992). Much work of this kind has been done in freshwater lakes and streams (e.g., chapter by Hall & Smol, this volume; Watanabe et al., 1988; Van Dam et al., 1994; Reavie et al., 1995; Bennion et al., 1996; Reid et al., 1995). Future attempts of within-system or cross-system calibration sets for estuarine systems would certainly be worthwhile, especially in relation to land use and eutrophication, although each system will have its own unique signals due to differences in mixing, sedimentation, land use, geography, freshwater inflow, etc. Tolerance of environmental changes by different species may be useful in a cosmopolitan sense or may vary from system to system (e.g., De Sève, 1993; Cooper, 1995c). More studies on the autecology of estuarine diatoms are needed.

Surface sediment 'calibration' sets of diatom species in estuaries have identified specific physical habitats for diatoms rather than other chemical and nutrient preferences. For example, Rao and Lewin (1976) studied sediments from False Bay, Washington to identify and classify epipsammic and epipelic diatom species at different depths within sediments and distance from shore. Oppenheim (1988) characterized diatom species occurrence across physical gradients in a transect of saltmarsh, sandflat and mudflats at Berrow Flats, UK using a niche breadth statistic. Kuylenstierna (1990) studied the Nordre Älv Estuary of the Swedish west coast and identified diatoms from different habitats and salinity regimes within the estuary. Wilderman (1986) studied diatom distributions in the Severn River Estuary, USA along gradients of temperature (seasonality) and salinity.

A more quantitative approach was taken by McIntire (1973, 1978) in studies of benthic diatoms in Yaquina Estuary, Oregon. Multivariate analyses such as principal components analysis (PCA), canonical correlation analysis and cluster analysis were used to determine the relationship between the distribution of diatom species and gradients in salinity and other physical factors within the estuary. Similar studies include McIntire and Overton (1971), Main and McIntire (1974), Moore and McIntire (1977), and Whiting and McIntire (1985). Laws (1988) conducted a study of surface sediments from San Francisco Bay to characterize spatial distribution of diatom species in this estuary as related to salinity, depth, and substrate. Descriptive multivariate techniques, including Q-mode cluster analysis and PCA, were employed to analyze the data. Juggins (1992) developed a 'salinity transfer function', using weighted-averaging methodology, by analyzing surface sediments and living source communities of diatoms in the Thames Estuary. This transfer function was then used to estimate former salinity levels from subfossil diatom assemblages identified in sediment core samples with good success.

Examples of studies which have used similar techniques to analyze anthropogenic effects in estuaries include Wendker (1990), who used surface sediment calibrations of physical and chemical (including some nutrient) factors in the Schlei Estuary to compare to down-core diatom analyses. Sundbäck and Snoeijs (1991) studied the effects of nutrient enrichment on diatoms and other microalgae in a coastal shallow-water sediment system using an experimental approach. Kennett and Hargraves (1984, 1985) investigated seasonal differences in benthic diatom distributions in a Rhode Island estuary to determine differences in community composition due to stratification and chemical changes under hypoxic and anoxic water conditions.

HUMAN IMPACTS: EUTROPHICATION, LAND CLEARANCE, AND OTHER POLLUTION SOURCES

Both ecological and paleoecological studies of diatoms and other algae have been used to assess anthropogenic impacts on the environment, including eutrophication and other pollution sources (see other chapters in this volume). Specific studies of the effects of eutrophication, land use and pollution on estuarine benthic communities have been broadening, such as those of Cederwall and Elmgren (1980), Widbom and Elmgren (1988), Nilsson *et al.* (1991) and Schaffner *et al.* (1992). Studies of these same anthropogenic influences on diatom communities specifically in coastal and estuarine systems has been increasing as well (e.g., Kennett & Hargraves, 1984, 1985; Sanders & Cibik, 1985; Winkler, 1986; Marshall & Alden, 1990; Wendker, 1990; Miller & Risberg, 1990; Sundbäck & Snoeijs, 1991; Puškaric *et al.*, 1990; Cooper, 1995*a*; Rabalais *et al.*, 1996; Parsons, 1996).

Paleoecological studies of diatoms in estuaries are rare, especially when compared to other environments such as lakes and streams (Kuylenstierna, 1990). Studies that have used paleoecological techniques employing diatoms to reconstruct the effects of anthropogenic influences on estuarine systems include Brush & Davis (1984), Beasley (1987), Wendker (1990), Miller & Risberg (1990), Puškaric *et al.* (1990), Cooper (1993, 1995*a*) and Parsons (1996).

CASE STUDIES FROM MID-ATLANTIC USA

Cooper (1993, 1995*a*) has shown that diatom assemblages from subsampled sediment cores collected in the Chesapeake Bay exhibit major shifts related to land use in the watershed. These changes include a decline in diatom diversity with continued human impacts on water quality, an increase in diatom abundance with eutrophication, a shift from an assemblage with equally abundant benthic and planktonic diatoms to one dominated by planktonic diatoms, and a shift from more marine to more brackish water species. These data were clustered by Euclidean distance between diatom assemblages and correlated to major land-use changes in the watershed. The diatom data show clear trends, but are better interpreted in light of other physical parameters and indicators measured from the same sediments. These include bulk density of sediments, pollen, sedimentation rates, organic carbon, nitrogen, sulfur, biogenic silica, iron, and calculated degree of pyritization of iron (see Cooper & Brush, 1991, 1993; Cooper, 1993, 1995*a*). These indicators, and the known history of human influence on the watershed and water quality of the estuary, indicate that diatom communities have been changing in response to increased turbidity and sedimentation, increased nutrient content of the waters, increased freshwater input to the estuary, and possibly an increase in anoxic and hypoxic bottom waters. The known autecology of the diatom species that exhibit major shifts in the cores corroborate these findings, but more detailed knowledge of the ecology of different species is required to better evaluate the trophic changes within Chesapeake Bay and other estuaries.

This perspective of historical land-use effects on the Chesapeake Bay has added a new dimension to the monitoring efforts that have been in effect over the past 10 to 15 years to determine status and trends of water quality issues. During this 'short' time period of monitoring, the natural annual variability of the estuarine system has often been higher than interannual changes seen in many parameters, and trends are difficult to discern. The paleoecological approach reveals long-term trends, elucidating the effects of land clearance and increasing population on water quality and dissolved oxygen, and highlights issues of temporal and spatial scale. The data collected using paleoecological techniques have been very useful in hindcasting models of the ecosystem developed through research and monitoring, and will continue to be employed in monitoring data analyses and management decisions.

Results of previous studies of diatoms from sediment cores collected in Chesapeake Bay waters showed similar trends of increasing planktonic diatom species abundance as a result of human intervention in the watershed with a concomitant decline in turbidity-intolerant species. Diatom community diversity was found to decline with land clearance for agriculture. (Brush *et al.*, 1979; Brush & Davis, 1984).

Beasley (1987) studied the history of diatom communities in Rehoboth Bay, Delaware. She found that as nutrient input to the estuary increased as a result of land clearance for agriculture along the coast of Delaware, the diatoms showed an increase in abundance within the Bay. This was followed by a decline in diversity of the assemblages as wastewater loading increased. In the most recent sediments, she identified species characteristic of oxygen deficient waters polluted with nitrogenous compounds, indicating nitrogen pollution, and possibly declining dissolved oxygen levels in Rehoboth Bay. Beasley also determined sedimentation rates, and measured biogenic silica, organic carbon and nitrogen within sediment core samples.

For further discussion of studies of eutrophication and paleoenvironmental reconstructions using diatoms in estuaries (with special reference to the Baltic Sea), see the chapter by Snoeijs (this volume).

SEA LEVEL CHANGES/SALINITY RECONSTRUCTIONS

Diatom analysis has been widely used in paleoecological studies to reconstruct salinity changes in estuarine and coastal systems, based on a number of classification schemes (e.g., Hustedt, 1953, 1957; Simonsen, 1962). An excellent review of the history of diatom analysis to reconstruct salinity can be found in

Juggins (1992). As mentioned previously, Juggins (1992) has developed a 'transfer function' for salinity based on a calibration set from the Thames Estuary, England. Recent studies employing diatoms as indicators of salinity and sea level change include Wasell and Håkansson (1992), Nelson and Kashima (1993), Campeau *et al.* (1995), Hemphill-Haley (1995*b*) and Douglas *et al.* (1996).

Studies employing diatoms to reconstruct shoreline displacement and sea level change such as would occur with climate change, tectonics (e.g., earthquakes) or isostasy often include salinity reconstructions and habitat reconstructions. These subjects are covered in the chapter by Denys & de Wolf (this volume).

Summary

Paleoecological techniques have been underutilized in studying contemporary issues of water quality in estuarine systems around the world. These systems are the focus of concern about detrimental anthropogenic influences, and decisions are being made that will affect management options and future research. Management decisions are often based on data collected through ongoing research and monitoring programs that enable only a brief glimpse in the life of an estuary. Estuarine ecosystems are extremely variable (in both time and space), and short-term studies are not always capable of determining trends that are applicable to policy decisions. A combination of research, monitoring, and paleoecological studies can become a synergistic tool for discerning trends, causes, and consequences of watershed land-use.

In estuarine systems, diatom fossils have been used to quantify anthropogenic effects by reconstructing changes in salinity, habitat availability (e.g., benthic vs. pelagic), turbidity, sedimentation rates, nutrient availability (e.g., eutrophication), anoxia and hypoxia, and organic pollution. These changes have in turn been related to changing land-use patterns of watersheds, and can be used for hindcasting in water quality and watershed models. These data can be compared to ongoing research and monitoring results to further aid the determination of management and policy.

Paleoecological studies employing diatoms require the use of other paleoecological techniques (including site selection, core collection, dating of sediments, determination of sedimentation rates), as well as expertise in diatom taxonomy. Other indicators that can be measured in sediments will further strengthen conclusions drawn from the diatom data. Such indicators could include biogeochemical analyses of pigments, lipids, nutrients, biogenic silica, heavy metals, isotopes, and enumeration of other fossils such as dinoflagellate cysts, sponge spicules, pollen, seeds, charcoal, ostracodes and foraminifera. Multivariate statistical techniques can be used to correlate these paleoecological indicators with environmental data and this may lead to uncovering causes, and therefore consequences, of human impacts on estuaries.

References

Aleem, A. A., & Hustedt, F. (1951). Einige neue Diatomeen von der Südküste Englands. *Botaniska Notiser*, **1**, 13–20.

Amspoker, M. C., & McIntire, C. D. (1978). Distribution of intertidal diatoms associated with sediments in Yaquina Estuary, Oregon. *Journal of Phycology*, **14**, 387–95.

Anderson, N. J. (1993). Natural versus anthropogenic change in lakes: The role of the sediment record. *Trends in Ecology and Evolution*, **8**, 356–61.

Anderson, N. J., & Vos, P. (1992). Learning from the past: Diatoms as paleoecological indicators of changes in marine environments. *Netherlands Journal of Aquatic Ecology*, **26**, 19–30.

Archibald, R. E. M. (1972). Diversity in some South African diatom associations and its relation to water quality. *Water Research*, **6**, 1229–38.

Barker, P., Fontes, J-C., Gasse, F., & Druart, J-C. (1994). Experimental dissolution of diatom silica in concentrated salt solutions and implications for paleoenvironmental reconstruction. *Limnology and Oceanography*, **39**, 99–110.

Beasley, E. L. (1987). Change in the diatom assemblage of Rehoboth Bay, Delaware and the environmental implications. PhD dissertation, University of Delaware, Newark, DE.

Bennion, H., Juggins, S., & Anderson, N. J. (1996). Predicting epilimnetic phosphorus concentrations using an improved diatom-based transfer function and its application to lake eutrophication management. *Environmental Science and Technology*, **30**, 2004–7.

Bérard-Therriault, L., Cardinal, A., & Poulin, M. (1986). Les diatomées benthiques de substrats durs des eaux marines et Saumâtres du Québec 6. Naviculales: Cymbellaceae et Gomphonemaceae. *Le Naturaliste Canadien*, **113**, 405–29.

(1987). Les diatomées benthiques de substrats durs des eaux marines et Saumâtres du Québec 8. Centrales. *Le Naturaliste Canadien*, **114**, 81–103.

Biggs, R. B. (1970). Sources and distribution of suspended sediment in northern Chesapeake Bay. *Marine Geology*, **9**, 187–201.

Biggs, R. B., & Cronin, L. E. (1981). Special characteristics of estuaries. In *Estuaries and Nutrients*, ed. B. J. Neilson & L. E. Cronin, pp. 2–23. Clifton, NJ: Humana Press.

Birks, (1995). Quantitative paleoenvironmental reconstructions. In *Statistical Modelling of Quaternary Science Data: Technical Guide 5,* Quaternary Research Association, ed. D. Maddy & J. S. Brew, pp. 161–254. Cambridge.

Brooks, A., Bell, W., & Greer, J. (1996). *Our Coastal Seas, What Is Their Future?: The Environmental Management of Enclosed Coastal Seas*. Summary of an International Conference, College Park, MD: Maryland Sea Grant.

Bruno, M. G., & Lowe, R. L. (1980). Differences in the distribution of some bog diatoms: A cluster analysis. *The American Midland Naturalist* **104**, 70–9.

Brush, G. S. (1984). Patterns of recent sediment accumulation in Chesapeake Bay (Virginia-Maryland, U.S.A.) tributaries. *Chemical Geology*, **44**, 227–42.

(1986). Geology and paleoecology of Chesapeake Bay: A long-term monitoring tool for management. *Journal of the Washington Academy of Sciences*, **76**, 146–60.

(1989). Rates and patterns of estuarine sediment accumulation. *Limnology and Oceanography*, **34**, 1235–46.

Brush, G. S., & Davis, F. W. (1984). Stratigraphic evidence of human disturbance in an estuary. *Quaternary Research*, **22**, 91–108.

Brush, G. S., Davis, F. W., & Rumer, S. (1979). Biostratigraphy of Chesapeake Bay and its tributaries. *Final Report to the Environmental Protection Agency*. Philadelphia, PA: Chesapeake Bay Program.

Brush, G. S., Martin, E. A., DeFries, R. S., & Rice, C. A. (1982). Comparisons of ^{210}Pb and pollen methods for determining rates of estuarine sediment accumulation. *Quaternary Research*, **18**, 196–217.

Campeau, S., Hequette, A., & Pienitz, R. (1995). The distribution of modern diatom assemblages in coastal sedimentary environments of the Canadian Beaufort Sea: An accurate tool for monitoring coastal changes. In *Proceedings of the 1995 Canadian Coastal Conference*, **1**, 105–16. Dartmouth, Nova Scotia: The Canadian Coastal Science and Engineering Association.

Cardinal, L., Poulin, M., & Bérard-Therriault, A. (1986). Les diatomées benthiques de substrats durs des eaux marines et Saumâtres du Québec 5. Naviculales, Naviculaceae; les genres *Donkinia, Gyrosigma* et *Pleurosigma*. *Le Naturaliste Canadien*, **113**, 167–90.

Cederwall, H., & Elmgren, R. (1980). Biomass increase of benthic macrofauna demonstrates eutrophication of the Baltic Sea. *Ophelia*, Supplement 1, 287–304.

Charles, D. F., & Smol, J. P. (1994). Long-term chemical changes in lakes: Quantitative inferences from biotic remains in the sediment record. In *Environmental Chemistry of Lakes and Reservoirs. Advances in Chemistry Series* 237, ed. L. Baker, pp. 3–31. Washington DC: American Chemical Society

Conley, D. J. (1988). Biogenic silica as an estimate of siliceous microfossil abundance in Great Lake sediments. *Biogeochemistry*, **6**, 161–79.

Conley, D. J., & Malone, T. C. (1992). Annual cycle of dissolved silicate in Chesapeake Bay: Implications for the production and fate of phytoplankton biomass. *Marine Ecology Progress Series*, **81**, 121–8.

Conley, D. J., Kilham, S. S., & Theriot, E. (1989). Differences in silica content between marine and freshwater diatoms. *Limnology and Oceanography*, **34**, 205–13.

Conley, D. J., Schelske, C. L., & Stoermer, E. F. (1993). Modification of the biogeochemical cycle of silica with eutrophication. *Marine Ecology Progress Series*, **101**, 179–92.

Cooper, S. R. (1993). The history of diatom community structure, eutrophication and anoxia in the Chesapeake Bay as documented in the stratigraphic record. PhD dissertation. Department of Geography and Environmental Engineering, Johns Hopkins University, Baltimore, MD.

(1995a). Chesapeake Bay watershed historical land use: Impact on water quality and diatom communities. *Ecological Applications*, **5**, 703–23.

(1995b). Diatoms in sediment cores from the mesohaline Chesapeake Bay, U.S.A. *Diatom Research*, **10**, 39–89.

(1995c). An abundant, small brackish water *Cyclotella* species in Chesapeake Bay, U.S.A. In *A Century of Diatom Research in North America: A Tribute to the Distinguished Careers of Charles W. Reimer and Ruth Patrick*, ed. J. P. Kociolek & M. J. Sullivan, pp. 133–40. Champaign, Illinois: Koeltz Scientific Books.

Cooper, S. R., & Brush, G. S. (1991). Long-term history of Chesapeake Bay anoxia. *Science*, **254**, 992–6.

(1993). A 2,500-year history of anoxia and eutrophication in Chesapeake Bay. *Estuaries*, **16**, 617–26.

Cooper, S. R., & Lipton, D. (1994). *Mid-Atlantic Research Plan: Mid-Atlantic Regional Marine Research Program*. College Park, MD: Maryland Sea Grant.

Culliton, T. J., McDonough, J. J. III, Remer, D. G., & Lott, D. M. (1992). *Building Along America's Coast: 20 Years of Building Permits, 1970–1989*. Silver Spring, MD: National Oceanic and Atmospheric Administration.

Culliton, T. J., Warren, M. A., Goodspeed, T. R., Remer, D. G., Blackwell, C. M., & McDonough, J. J. III. (1990). *50 Years of Population Change along the Nation's Coasts 1960–2010*. Silver Spring, MD: National Oceanic and Atmospheric Administration.

Day, J. W., Hall, C. A. S., Kemp, W. M., & Yanez-Arancibia, A. (1989). *Estuarine Ecology*. New York, NY: John Wiley.

D'Elia, C. F., Nelson, D. M., & Boynton, W. R. (1983). Chesapeake Bay nutrient and plankton dynamics: III. The annual cycle of dissolved silicon. *Geochimica et Cosmochimica Acta*, **47**, 1945–55.

De Sève, M. A. (1993). Diatom bloom in the tidal freshwater zone of a turbid and shallow estuary, Rupert Bay (James Bay, Canada). *Hydrobiologia*, **269/270**, 225–33.

Dortch, Q., & Whitledge, T. E. (1992). Does nitrogen or silicon limit phytoplankton production in the Mississippi River plume and nearby regions? *Continental Shelf Research*, **12**, 1293–309.

Douglas, M. S. V., Ludlam, S., & Feeney, S. (1996). Changes in diatom assemblages in Lake C2

(Ellesmere Island, Arctic Canada): Response to basin isolation from the sea and to other environmental changes. *Journal of Paleolimnology*, **16**, 217–26.

Dugmore, A. J., & Newton, A. J. (1992). Thin tephra layers in peat revealed by X-radiography. *Journal of Archaeological Science*, **19**, 163–70.

Dyer, K. R. (1995). Sediment transport processes in estuaries. In *Geomorphology and Sedimentology of Estuaries. Developments in Sedimentology 53*, ed. G. M. E. Perillo, pp. 423–49. Amsterdam, The Netherlands: Elsevier Science Publishers BV.

Elmgren, R. (1989). The eutrophication status of the Baltic Sea: Input of nitrogen and phosphorus, their availability for plant production, and some management implications. *Baltic Sea Environment Proceedings*, **30**, 12–31.

FAO Fisheries Department. (1993). *Marine Fisheries and the Law of the Sea: A Decade of Change. FAO Fisheries Circular*, No. **853**, p. 43. Rome: Food and Agriculture Organization of the United Nations.

Fraser, T. H., & Wilcox, W. H. (1981). Enrichment of a subtropical estuary with nitrogen, phosphorus and silica. In *Estuaries and Nutrients*, ed. B. J. Neilson & L. E. Cronin, pp. 481–98. Clifton, NJ: Humana Press.

Funkhauser, J. W., & Evitt, W. R. (1959). Preparation techniques for acid insoluble micro-fossils. *Micropaleontology*, **5**, 369–75.

Giffen, M. H. (1963). Contributions to the diatom flora of South Africa. I. Diatoms of the estuaries of the Eastern Cape Province. *Hydrobiologia*, **21**, 201–65.

(1967). Contributions to the diatom flora of South Africa. III. Diatoms of the marine littoral regions at Kidd' Beach near East London, Cape Province. *Nova Hedwigia*, **13**, 245–92

(1970). New and interesting marine and littoral diatoms from Sea Point, near Cape Town, South Africa. *Botanica Marina*, **13**, 87–99.

(1971). Marine and littoral diatoms from the Gordon's Bay region of False Bay, Cape Province, South Africa. *Botanica Marina*, **14**, 1–16.

(1973). Diatoms of the marine littoral of Steensberg's Cove, St. Helena Bay, Cape Province, South Africa. *Botanica Marina*, **16**, 32–48.

(1975). An account of the littoral diatoms from Langebaan, Saldanha Bay, Cape Province, South Africa. *Botanica Marina*, **18**, 71–95.

Hecky, R. E., & Kilham, P. (1973). Diatoms in alkaline, saline lakes: ecology and geochemical implications. *Limnology and Oceanography*, **18**, 53–71.

Helz, G. R., Sinex, S. A., Ferri, K. L., & Nichols, M. (1985). Processes controlling Fe, Mn and Zn in sediments of northern Chesapeake Bay. *Estuarine, Coastal and Shelf Science*, **21**, 1–16.

Hemphill-Haley, E. (1995a). Diatom evidence for earthquake-induced subsidence and tsunami 300 yr ago in southern coastal Washington. *GSA Bulletin*, **107**, 367–78.

(1995b). Intertidal diatoms from Willapa Bay, Washington: Application to studies of small-scale sea-level changes. *Northwest Science*, **69**, 29–45.

Hendey, N. I. (1964). An introductory account of the smaller algae of British coastal waters. *Fishery Investigations Series IV. Part V. Bacillariophyceae (Diatoms)*. London, UK: Her Majesty's Stationery Office.

(1977). The species diversity index of some in-shore diatom communities and its use in assessing the degree of pollution insult on parts of the north Coast of Cornwall. *Nova Hedwigia*, Beiheft **54**, 355–78.

Hennessee, E. L., Blakeslee, P. J., & Hill, J. M. (1986). The distributions of organic carbon and sulfur in surficial sediments of the Maryland portion of Chesapeake Bay. *Journal of Sedimentary Petrology*, **56**, 674–83.

Hill, J. M., Halka, R. D., Conkwright, R. D., Koczot, K., & Park, J. (1992). Geologically con-strained shallow gas in sediments of Chesapeake Bay: Distribution and effects on bulk sediment properties. *Continental Shelf Research*, **12**, 1219–29.

Hinga, K. R., Jeon, H., & Lewis, N. (1995). Marine Eutrophication Review – Part 1: Quantifying the effects of nitrogen enrichment on phytoplankton in coastal ecosys-tems; Part 2: Bibliography with abstracts. *NOAA Coastal Ocean Program Decision Analysis Series*, No. **4**. Silver Spring, MD: NOAA Coastal Ocean Office.

Hustedt, F. (1927–1930). Die Kieselalgen Deutschlands, Österreichs und der Schweiz (three volumes). In *Dr. L. Rabenhorst's Kryptogamen-Flora von Deutschland, Österreich und der Schweiz*. Band 7, 1. Leipzig, Germany: Akademische Verlagsgesellschaft.

(1931–1959). Die Kieselalgen Deutschlands, Österreichs und der Schweiz (three volumes). In *Dr. L. Rabenhorst's Kryptogamen-Flora von Deutschland, Österreich und der Schweiz*. Band 7, 2. Leipzig, Germany: Akademische Verlagsgesellschaft.

(1939). Die Diatomeenflora des Küstengebietes der Nordsee vom Dollart bis zur Elbemündung, I. *Abhandlungen des Naturwissenschaftlichen Vereins zu Bremen*, **31**, 572–677.

(1951). Neue und wenig bekannte Diatomeen. II. *Berichte der Deutschen Botanischen Gesellschaft*, **64**, 304–14.

(1953). Die Systematik der Diatomeen in Ihren Beziehungen zur Geologie und Ökologie nebst einer Revision des Halobien-Systems. *Svensk Botanisk Tedskrift*, **47**, 509–19.

(1955). Marine littoral diatoms of Beaufort, North Carolina. *Duke University Marine Station Bulletin No.* **6**. Beaufort, NC.

(1957). Die Diatomeenflora des Fluss-Systems der Weser im Gebiet der Hansestadt Bremen. *Abhandlungen des Naturwissenschaftlichen Vereins zu Bremen*, **34**, 181–440.

(1961–1969). Die Kieselalgen Deutschlands, Österreichs und der Schweiz (three volumes). In *Dr. L. Rabenhorst's Kryptogamen-Flora von Deutschland, Österreich und der Schweiz*. Band 7, 3. Keipzig, Germany: Akademische Verlagsgesellschaft.

Iler, R. K. (1979). *The Chemistry of Silica*. New York: John Wiley.

Jonge, V. N., Boynton, W., D'Elia, C. F., Elmgren, R., & Welsh, B. L. (1994). Responses to developments in eutrophication in four different North Atlantic estuarine systems. In *Changes in Fluxes in Estuaries*, ed. K. R. Dyer & R. J. Orth, pp. 179–96. Fredensborg, Denmark: Olsen & Olsen.

Juggins, S. (1992). Diatoms in the Thames Estuary, England: Ecology, paleoecology, and salinity transfer function. In *Bibliotheca Diatomologica* 25, ed. H. Lange-Bertalot, pp. 1–216. Berlin: J. Cramer.

Kennett, D. M., & Hargraves, P. E. (1984). Subtidal benthic diatoms from a stratified estuarine basin. *Botanica Marina*, **27**, 169–83.

(1985). Benthic diatoms and sulfide fluctuations: Upper basin of Pettaquamscutt River, Rhode Island. *Estuarine, Coastal and Shelf Science*, **21**, 577–86.

Kennish, M. J. (1992). *Ecology of Estuaries: Anthropogenic Effects*. Marine Science Series. Boca Raton, FL: CRC Press, Inc.

Krammer, K., & Lange-Bertalot, H. (1986–1991). Bacillariophyceae. In *Süßwasserflora von Mitteleuropa* Band 2/1–4, ed. H. Ettl, H. Heynig & D. Mollenhauer. Stuttgart: Fischer.

Kuylenstierna, M. (1990). *Benthic algal vegetation in the Nordre Älv Estuary (Swedish west coast)* 2 volumes. PhD dissertation. University of Göteborg, Göteborg, Sweden.

Laws, R. A. (1988). Diatoms (Bacillariophyceae) from surface sediments in the San Francisco Bay estuary. *Proceedings of the California Academy of Sciences*, **45**, 133–254.

Lewin, J. C. (1961). The dissolution of silica from diatom walls. *Geochimica et Cosmochimica Acta*, **21**, 182–98.

Main, S. P., & McIntire, C. D. (1974). The distribution of epiphytic diatoms in Yaquina Estuary, Oregon (U.S.A.). *Botanica Marina*, **17**, 88–99.

Marshall, H. G., & Alden, R. W. (1990). A comparison of phytoplankton assemblages and environmental relationships in three estuarine rivers of the lower Chesapeake Bay. *Estuaries*, **13**, 287–300.

McIntire, C. D. (1973). Diatom associations in Yaquina Estuary, Oregon: A multivariate analysis. *Journal of Phycology*, **9**, 254–9.

(1978). The distribution of estuarine diatoms along environmental gradients: A canonical correlation. *Estuarine and Coastal Marine Science*, **6**, 447–57.

McIntire, C. D., & Overton, W. S. (1971). Distributional patterns in assemblages of attached diatoms from Yaquina Estuary, Oregon. *Ecology*, **52**, 758–77.

Miller, U., & Risberg, J. (1990). Environmental changes, mainly eutrophication, as recorded by fossil siliceous micro-algae in two cores from the uppermost sediments of the north-western Baltic. *Beiheft zur Nova Hedwigia*, **100**, 237–53.

Mitsch, W. J., & Gosselink, J. G. (1986). *Wetlands*. New York: Van Nostrand Reinhold.

Mölder, K., & Tynni, R. (1967–1973). Über Finnlands rezente und subfossile diatomeen I-VII. *Bulletin of the Geological Society of Finland* **39**, 199–217; **40**, 151–70; **41**, 235–51; **42**, 129–44; **44**, 141–9; **45**, 159–179.

Moore, W. W., & McIntire, C. D. (1977). Spatial and seasonal distribution of littoral diatoms in Yaquina Estuary, Oregon (U. S. A.). *Botanica Marina*, **20**, 99–109.

Moser, K. A., MacDonald, G. M., & Smol, J P. (1996). Applications of freshwater diatoms to geographical research. *Progress in Physical Geography*, **20**, 21–52.

National Research Council. (1993). *Managing Wastewater In Coastal Urban Areas*. Washington DC: National Academy Press.

National Safety Council. (1993). *Covering the Coasts: A Reporter's Guide to Coastal and Marine Resources*. Washington, DC: National Safety Council.

Neilson, B. J., & Cronin, L. E. (eds.). (1981). *Estuaries and Nutrients*. Clifton, NJ: Humana Press.

Nelson, A. R., & Kashima, K. (1993). Diatom zonation in southern Oregon tidal marshes relative to vascular plants, foraminifera, and sea level. *Journal of Coastal Research*, **9**, 673–97.

Nilsson, P., Jönsson, B., Swanberg, I. L., & Sundbäck, K. (1991). Response of a marine shallow-water sediment system to an increased load of inorganic nutrients. *Marine Ecology Progress Series*, **71**, 275–90.

Officer, C. G., Biggs, R. B., Taft, J. L., Cronin, L. E., Tyler, M. A., & Boynton, W. R. (1984). Chesapeake Bay anoxia: Origin, development, and significance. *Science*, **223**, 22–7.

Oppenheim, D. R. (1988). The distribution of epipelic diatoms along an intertidal shore in relation to principal physical gradients. *Botanica Marina*, **31**, 65–72.

Parsons, M. L. (1996). *Paleoindicators of changing water conditions in Louisiana estuaries*. PhD dissertation, Department of Oceanography and Coastal Sciences, Louisiana State University, Baton Rouge.

Patten, B. C. (1962). Species diversity in net plankton of Raritan Bay. *Journal of Marine Research*, **20**, 57–75.

Peragallo, H., & Peragallo, M. (1897–1908). *Diatomées marines de France et des districts maritimes voisins*. Grez-sur-Loing: Micrographie-Editeur.

Perillo, G. M. E. (1995). Geomorphology and sedimentology of estuaries. In *Geomorphology and Sedimentology of Estuaries. Developments in Sedimentology 53*, ed. G. M. E. Perillo, pp. 1–16. Amsterdam, The Netherlands: Elsevier Science Publishers BV.

Poulin, M., Bérard-Therriault, L., & Cardinal, A. (1984*a*). Les diatomées benthiques de substrats durs des eaux marines et Saumâtres du Québec 2. Tabellariodeae et Diatomoideae (Fragilariales, Fragilariaceae). *Le Naturaliste Canadien*, **111**, 275–95.

(1984*b*). Les diatomées benthiques de substrats durs des eaux marines et Saumâtres du Québec 3. Fragilariodeae (Fragilariales, Fragilariaceae). *Le Naturaliste Canadien*, **111**, 349–67.

(1986). Les diatomées benthiques de substrats durs des eaux marines et Saumâtres du Québec 7. Naviculales (les genres *Plagiotropis* et *Entomoneis*), Epithemiales et Surilellales. *Le Naturaliste Canadien*, **114**, 67–80.

Pritchard, D. W. (1967). Observations of circulation in coastal plain estuaries. In *Estuaries. American Association for the Advancement of Science Publication No. 83*, ed. G. H. Lauff, pp. 37–44. Washington, DC.

Puckett, L. J. (1995). Identifying the major sources of nutrient water pollution. *Environmental Science & Technology*, **29**, 408–14.

Puškaric, S., Berger, G. W., & Jorissen, F. J. (1990). Successive appearance of subfossil phytoplankton species in Holocene sediments of the Northern Adriatic and its relation to the increased eutrophication pressure. *Estuarine, Coastal and Shelf Science*, **31**, 177–87.

Rabalais, N. N., Turner, R. E., Justic, D., Dortch, Q., Wiseman, W. J. Jr., & Sen Gupta, B. K. (1996). Nutrient changes in the Mississippi River and system responses on the adjacent continental shelf. *Estuaries*, **19**, 386–407.

Rao, V. N. R., & Lewin, J. (1976). Benthic marine diatom flora of False Bay, San Juan Island, Washington. *Syesis*, **9**, 173–213.

Reavie, E. D., Hall, R. I., & Smol, J. P. (1995). An expanded weighted-averaging model for inferring past total phosphorus concentrations from diatom assemblages in eutrophic British Columbia (Canada) lakes. *Journal of Paleolimnology*, **14**, 49–67.

Reid, M. A., Tibby, J. C., Penny, D., & Gell, P. A. (1995). The use of diatoms to assess past and present water quality. *Australian Journal of Ecology*, **20**, 57–64.

Riggs, S. R., York, L. L., Wehmiller, J. F., & Snyder, S. W. (1992). Depositional patterns resulting from high-frequency Quaternary sea-level fluctuations in northeastern North Carolina. In *Quaternary Coasts of the United States: Marine and Lacustrine Systems*, eds. C. H. Fletcher & J. F. Wehmiller, pp. 141–53. SEPM (Society of Sedimentary Geology), Special Publication No. 48.

Ryther, J. H., & Officer, C. B. (1981). Impact of nutrient enrichment on water uses. In *Estuaries and Nutrients*, ed. B. J. Neilson & L. E. Cronin, pp. 247–61. Clifton, NJ: Humana Press.

Sanders, J. G., & Cibik, S. J. (1985). Reduction of growth rate and resting spore formation in a marine diatom exposed to low levels of cadmium. *Marine Environmental Research*, **16**, 165–80.

Sanford, L. P. (1992). New sedimentation, resuspension, and burial. *Limnology and Oceanography*, **37**, 1164–78.

Schaffner, L. C., Jonsson, P., Diaz, R. J., Rosenberg, R., & Gapcynski, P. (1992). Benthic communities and bioturbation history of estuarine and coastal systems: Effects of hypoxia and anoxia. *Science of the Total Environment*, Supplement, 1001–16.

Schelske, C. L, Conley, D. J., Stoermer, E. F., Newberry, T. L., & Campbell, C. D. (1986). Biogenic silica and phosphorus accumulation in sediments as indices of eutrophication in the Laurentian Great Lakes. *Hydrobiologia*, **143**, 79–86.

Shaffer, G. P., & Sullivan, M. J. (1988). Water column productivity attributable to displaced benthic diatoms in well-mixed shallow estuaries. *Journal of Phycology*, **24**, 132–40.

Siever, R., & Scott, R. A. (1963). Organic geochemistry of silica. In *Organic Geochemistry, I.*, ed. A. Breger. New York: The MacMillan Co.

Simonsen, R. (1962). Untersuchungen zur Systematik und Ökologie der Bodendiatomeen der Westlichen Ostsee. *Internationale Revue der Gesamten Hydrobiologie*, **1**, 1–144.

Smol, J. P. (1981). Problems associated with the use of 'species diversity' in paleolimnological studies. *Quaternary Research*, **15**, 209–12.

(1992). Paleolimnology: An important tool for effective ecosystem management. *Journal of Aquatic Ecosystem Health* **1**, 49–58.

Snoeijs, P., & Potapova, M. (eds.). (1993–1997). *Intercalibration and Distribution of Diatom Species in the Baltic Sea*, Volumes 1–5. Uppsala, Sweden: Opulus Press.

Stockner, J. G., & Benson, W. W. (1967). The succession of diatom assemblages in the recent sediments of Lake Washington. *Limnology and Oceanography*, **12**, 513–32.

Sullivan, M. J. (1978). Diatom community structure: Taxonomic and statistical analyses of a Mississippi salt marsh. *Journal of Phycology*, **14**, 468–75.

(1982). Distribution of edaphic diatoms in a Mississippi salt marsh: A canonical correlation analysis. *Journal of Phycology*, **18**, 130–3.

Sullivan, M. J., & Moncreiff, C. A. (1988). Primary production of edaphic algal communities in a Mississippi salt marsh. *Journal of Phycology*, **24**, 49–58.

Sundbäck, K., & Snoeijs, P. (1991). Effects of nutrient enrichment on microalgal community composition in a coastal shallow-water sediment system: An experimental study. *Botanica Marina*, **34**, 341–58.

ter Braak, C. J. F., & Juggins, S. (1993). Weighted-averaging partial least squares regression (WA-PLS): An improved method for reconstructing environmental variables from species assemblages. In *Twelfth International Diatom Symposium*, ed. H. van Dam, pp. 485–502. Dordrecht: Kluwer Academic Publishers.

Tynni, R. (1975–1980). Über Finnlands rezente und subfossile diatomeen VIII-XI. *Geological Survey of Finland Bulletin* **274, 284, 296, 312**.

Van Dam, H. (1982). On the use of measures of structure and diversity in applied diatom ecology. *Nova Hedwigia*, **73**, 97–115.

Van Dam, H., Mertens, A., & Sinkeldam, J. (1994). A coded checklist and ecological indicator values of freshwater diatoms from the Netherlands. *Netherlands Journal of Aquatic Ecology*, **28**, 117–33.

Wasell, A., & Håkansson, H. (1992). Diatom stratigraphy in a lake on Horseshoe Island, Antarctica: A marine–brackish–fresh water transition with comments on the systematics and ecology of the most common diatoms. *Diatom Research*, 7, 157–94.

Washington, H. G. (1984). Diversity, biotic, and similarity indices: A review with special relevance to aquatic ecosystems. *Water Research*, 18, 653–94.

Watanabe, T., Kazumi, A., & Akiko, H. (1988). Numerical water quality monitoring of organic pollution using diatom assemblages. In *Proceedings of the Ninth International Diatom Symposium*, ed. F. E. Round, pp. 123–41. Koenigstein: Koeltz Scientific Books.

Wendker, S. (1990). Untersuchungen zur subfossilen und rezenten Diatomeenflora des Schlei-Ästuars (Ostsee). In *Bibliotheca Diatomologica 20*, ed. H. Lange-Bertalot, pp. 1–268 with 8 plates. Berlin: J. Cramer.

White, C. P. (1989). *Chesapeake Bay: A Field Guide*. Centreville, Maryland: Tidewater Publishers.

Whiting, M. C., & McIntire, C. D. (1985). An investigation of distributional patterns in the diatom flora of Netarts Bay, Oregon, by correspondence analysis. *Journal of Phycology*, 21, 655–61.

Widbom, B., & Elmgren, R. (1988). Response of benthic meiofauna to nutrient enrichment of experimental marine ecosystems. *Marine Ecology Progress Series*, 42, 257–68.

Wilderman, C. C. (1984). The floristic composition and distribution patterns of diatom assemblages in the Severn River Estuary, Maryland. PhD dissertation, Johns Hopkins University, Baltimore, MD.

(1986). Techniques and results of an investigation into the autecology of some major species of diatoms from the Severn River Estuary, Chesapeake Bay, Maryland, U.S.A. In *Proceedings of the 8th International Diatom Symposium*, ed. M. Ricard, pp. 631–43. Koenigstein: Koeltz Scientific Books.

Winkler, M. G. (1986). Two estuaries in the Cape Cod National Seashore: Diatom documentation of environmental change. In *Fisheries and Coastal Wetlands Research*, Conference on Science in the National Parks 1986 Proceedings, vol. 6, ed. G. Larson & M. Soukup, pp. 171–84. Fort Collins, CO: Colorado State University.

World Resources Institute, United Nations Environment Programme, United Nations Development Programme & The World Bank. (1996). *World Resources: A Guide to the Global Environment 1996–97*. New York, NY: Oxford University Press.

Yamada, S. S., & D'Elia, C. F. (1984). Silicic acid regeneration from estuarine sediment cores. *Marine Ecology Progress Series*, 18, 113–18.

17 Diatoms and marine paleoceanography

CONSTANCE SANCETTA

Introduction

Diatoms have great potential for studies of marine paleoecology and paleoceanography, especially in high latitudes and coastal regions. In these settings they are diverse and abundant, usually being the dominant group in the fossil assemblage. Elsewhere, their use may be limited by the relatively poor preservation of biogenic silica. The red clays of the central oceanic gyres contain no diatoms at all; most calcareous sediments contain only fragments of a few robust forms.

The poor preservation results from two interrelated conditions. (i) Diatoms as a group are at a competitive disadvantage in conditions of low nutrient supply. In addition to the universal requirement for nitrogen and phosphorus, they will be limited by availability of silicon, and trace metals such as iron may be limiting also (Martin & Gordon, 1988; Coale *et al.*, 1996). Therefore, over large areas of the world ocean, diatoms are a minor component of the phytoplankton, and those taxa which are present are frequently very weakly-silicified. (ii) Seawater and sediment porewaters are usually undersaturated with respect to biogenic silica (Tréguer *et al.*, 1995), so that dissolution of the frustules occurs rapidly, especially when pH is relatively high, as it often is in calcareous sediments. Preservation is good only in sediments with a high component of rock particles (ice-rafted detritus, coastal sediments, volcanic ash beds) or in siliceous (diatom) oozes, which result from a combination of high rates of diatom production and exclusion of other sediment components.

Within these settings, diatoms can provide insight into a variety of environmental conditions. These are discussed individually below, with examples drawn from the literature. For each case only one or two examples are cited; given space limitations, this chapter makes no claim to be a full literature review. Useful reviews on the ecology of marine diatoms may be found in Smayda (1958) and Guillard and Kilham (1977).

Temperature

Temperature is such a basic feature of the environment that it is not surprising that attempts have been made to use the fossil diatom assemblage to estimate past surface water temperatures. Two classes of approach have been developed, both using the relative abundance (usually percent of the total assemblage) of individual taxa.

The more sophisticated approach, widely used in Recent and Pleistocene studies, involves a multivariate analysis performed on percentage composition of the entire assemblage in surface sediments (assumed to represent modern conditions) from a large number of sites and then correlates the results with modern temperature in the surface water overlying each site. The approach was pioneered by Imbrie and Kipp (1971) using factor analysis followed by regression of factor loadings against average winter (or summer) temperatures at the same sites. A variant is the use of a similarity index, in which a fossil assemblage is compared to reference (surface sediment) assemblages and the temperature for the fossil sample assumed to have been that of the modern sample it most closely resembles. These methods are appealing because they appear to provide very precise values, and several equations (e.g., winter and summer temperatures, nutrient concentration) can be derived from the same data set if one assumes that each variable being estimated is statistically independent of the others (rarely the case in reality). One may also use assemblages defined by multivariate analysis (cluster, factor, principal component, recurrent group) as general indicators of water mass conditions without specifying exact properties. For instance, an assemblage of warm-water taxa found in a boreal setting might be considered a tracer for northward extension of a western boundary current with relatively warm waters, high salinity and lower nutrient contents. A good example of this approach is the study by Koç Karpuz and Jansen (1992) using diatoms and stables isotopes of oxygen and carbon in foraminiferal calcite in cores from the Norwegian Sea. They identified a series of rapid fluctuations in sea-surface temperature (SST), implying climatic instability, between about 13 000 and 11 000 yr BP, the most recent period of deglaciation. Alternating positive and negative feedback between SST and melting of land-based ice-sheets may explain the instability of this period. Similar studies have been done in the Southern Ocean (Pichon *et al.*, 1992).

An earlier approach was developed by Kanaya and Koizumi (1966) and involves the ratio of 'warm' or 'cold' diatoms relative to either the sum of the two, or to the total assemblage (the Td index). Temperature preference is defined empirically, based on modern biogeography; cosmopolitan or very widely-ranging taxa are not included in the statistic. The ratio may be reported simply as scaled values, or the author may attempt to calibrate it by correlating the ratio values to known temperatures in some locale. This approach has the advantage of simplicity, and can be applied to assemblages containing extinct forms provided that their past biogeography is known. On the other hand, it is imprecise, and identification of taxa as 'warm' or 'cold' is subjective. For instance, the same species might be considered 'warm' in the Bering Sea but 'cold' in the temperate zone. Nevertheless, Td curves have been very useful in identifying temperature variations in the western North Pacific on a variety of time scales. Cores from the Sea of Japan show a millenial cycle of waxing and waning of the warm Tsushima Current during the Holocene (Koizumi, 1989). Td analysis has also been applied to long cores east of Japan underlying the Kuroshio Extension, and used to infer north–south shifts of the thermal front. The record clearly shows evidence for a thermal maximum during the middle Pliocene (Barron, 1992) and a series of step-like coolings thereafter (Koizumi, 1985).

   Although both approachs to temperature reconstruction are statistically valid, and have been used to produce credible results, they suffer from a few drawbacks. In the first place, it is only an assumption that a change in relative abundance of any species is related to temperature in a systematic way. The temperature tolerances of most deep-sea taxa are unknown, since they have not been cultured under controlled conditions. All taxa have some tempera-ture tolerance range, although it may be more or less broad, and experiments indicate that growth rates are highest near the maximum value of each species' range (Suzuki, 1995). Therefore, absolute abundance will be highest at some temperature and decrease to either side. However, similar experiments vary-ing other parameters (light, nutrient form and concentrations, grazing pres-sure) can also produce changes in taxon composition under constant temperature. The translation from absolute abundance of each taxon to rela-tive abundance (the only form of data available for fossils) also cannot be resolved. Suppose two species are both growing at their maximal rates; which will be *relatively* more abundant? It is more realistic to consider that each taxon is continuously solving a set of simultaneous equations in which the terms are constantly varying. Which equation is most important at any time (that is, which environmental parameter is temporarily 'controlling' the organism's success) depends upon how close the organism is to its limits in each equation. If a species has a wide range of temperature tolerance (Suzuki, 1995, for instance, found that some species had ranges up to 25 °C) and ambient temper-ature is well inside that range its abundance, relative or absolute, is probably not related to temperature but to some other parameter (e.g., perhaps the N : P ratio is dangerously low for that species but very favorable for some other species). A further complication is that in nature several environmental para-meters usually covary with temperature. Among others, temperature has a positive correlation with day length, clarity of water, stability of the water column, and salinity. Negative correlations exist with cloud cover and nutrient concentration. The result of all this ecology is that although in a broad sense it is possible to interpret a change in assemblage as having some relation to a change in temperature, the precision implied by the statistics may be mis-leading.

   A second problem with paleo-temperature estimates is the mismatch between biologic and geologic time scales. Aside from the special case of varved sediments, a sediment sample represents a mixture of many seasons and years of production. In diatomaceous deep-sea sediments, a 1 cm thick sample may include several centuries of accumulation. The analyst must assume that

C. SANCETTA

seasonal conditions of production, and their mixing ratios in the sediment, did not vary over time. To clarify this statement, consider a case in which a temperature equation has been applied to two samples and it is determined that one represents a temperature 3 °C colder than the other. What does this 'cooling' mean? Possible interpretations include: (i) year-round average temperature was colder, (ii) temperature of the productive season (spring, let us say) was colder but other seasons were not, (iii) production occurred during a different season (winter instead of spring), (iv) the cold season lasted for a longer portion of the year, (v) production occurred deeper in the water column.

All this is not to argue that temperature estimates from diatom assemblages are invalid, only to highlight the issues that must be kept in mind. Use of other fossil data may help to constrain interpretations. For instance, if a eury-thermal bloom-forming diatom co-occurs with a species of planktic foraminifer known to have a narrow temperature range, one might infer that maximum production occurred within that temperature range.

Sea-ice and the marginal ice zone

A distinctive community that includes diatoms flourishes in brine pockets at the base of sea-ice floes in both the Arctic and Antarctic (Horner & Schrader, 1982; Garrison et al., 1986). In situ studies indicate that when irradiation increases during the local spring season, there is an initial bloom within the ice, followed by a second bloom in the open water stabilized by melting of the ice (Smith & Nelson, 1986). The sea-ice algae sink out after this brief burst, and later production is dominated by different taxa (Alexander & Niebauer, 1981). There is often a transition in taxon composition with increasing distance from the ice, epontic species being most abundant close to the ice and other taxa further away (Fryxell & Kendrick, 1988; Scharek et al., 1994). The distal population probably is not derived from the ice, and represents advection from another water mass. The transition between these two types of assemblage in the sediment can serve as a tracer for the marginal ice zone (Leventer, 1992). In the Antarctic another taxon, Eucampia antarctica, is commonly found at the margin of the ice, but not in the ice itself (Fryxell & Prasad, 1990), and can serve as an alternate tracer for the ice margin. Using a ratio based on Eucampia, Kaczmarska et al. (1993) were able to document the gradual expansion of Antarctic sea-ice during the Late Pleistocene.

Conditions and rates of primary production

CHAETOCEROS SPORES AND BLOOMS

The presence of any diatoms in marine sediments indicates relatively high rates of production under conditions of moderate to high nutrient availability.

Beyond that, certain taxa are typically most abundant under particular conditions. In extreme cases, they may produce monospecific oozes, which frequently (but not always) reflect the occurrence of bloom conditions (i.e., very rapid to maximal growth rates). Chain-forming genera such as phaeoceran *Chaetoceros* and *Skeletonema* fall into this category. The latter is restricted to the coastal zone, but *Chaetoceros* spores can be common in deep sea sediments far from the continental margins. A particularly striking example occurs in the North Atlantic in a broad zone centered at 60 °N, where phaeoceran spores are common or even the dominant members of the diatom assemblage (Schrader & Koç Karpuz, 1990). The occurrence of a neritic taxon so far from land is very puzzling, as first noted by Gran (1911), who observed *Chaetoceros* blooms repeated over several years in plankton samples from the same region. He suggested that a seed population might be transported 'from the American coasts' in the Gulf Stream Extension each spring, which could then produce a bloom in the open ocean when thermal stratification is first established. Presence of the spores in the sediments is therefore an indicator both of bloom conditions and (possibly) also of strong lateral transport from neighboring shelves. In contrast, phaeocerans are essentially absent from surface waters and modern sediments of the subarctic North Pacific (Booth, 1981; Sancetta & Silvestri, 1986), a high-nutrient-low-chlorophyll (HNLC) region where rates of production and biomass are relatively low and seasonally invariant (Parsons & Lalli, 1988). The absence of *Chaetoceros* spores in the sediments is thus a reflection of the lack of bloom conditions – a rare case in which negative evidence is meaningful.

Chaetoceros spores are common in deep-sea sediments adjacent to the coastal zone of productive regions as well as in many marginal seas. In some of these systems, they have been used as indicators of seasonal upwelling, and variations in abundance during the past can be used to infer weakening of seasonal winds. For example, a 50% decrease in *Chaetoceros* flux during the last glacial interval off the coast of Oregon coincided with an equal decrease in accumulation of marine-derived organic carbon and is interpreted as reflecting a decrease of summer production and upwelling resulting from slackening of northwesterly winds (Sancetta *et al.*, 1992a). This finding agrees with predictions of an atmospheric model showing very weak or non-existent summer winds during glacial maximum, produced by the presence of the large ice sheet on North America (Kutzbach & Guetter, 1986).

An alternative approach to using individual taxa is to generate an equation to predict productivity (usually as carbon mass/area/time) using distributions of taxa in surface sediments and a regression equation, along the lines of the Imbrie–Kipp (1971) method described for temperature estimates. In this case no subjective decision is made as to the choice of taxa to be included; the equation is based on whichever statistically defined taxonomic assemblage is determined to have the highest correlation with known surface productivity. Productivity equations derived for the continental margin of Peru (Schrader, 1992) and for the eastern equatorial Pacific open ocean (Schrader *et al.*, 1993)

C. SANCETTA

indicated that primary production in this region does not show a simple linear relationship to glacial/interglacial climate cycles, a finding supported by other fossil tracers of paleoproductivity such as concentrations of biogenic silica and of organic carbon.

MONOSPECIFIC MATS AND STRATIFIED WATERS

Monospecific mats of large planktic diatoms are occasionally found in deep-sea sediments. These occur in laminated intervals with high organic content (sapropels), indicating reducing conditions. *Rhizosolenia* and *Thalassiothrix* are examples of this type. The presence of laminations indicates a consistent alternation of two conditions of production over many years, although it cannot always be determined whether the alternation is seasonal or an inter-annual cycle such as the El Niño-Southern Oscillation. The conditions under which such mats form have only been observed a few times, so that it is not certain that the same conditions always produce mats. Stable surface waters (strongly stratified, no wind mixing) appear to be necessary for the aggregates to remain at the surface. Under these conditions, rhizosolenids have been observed to exhibit positive bouyancy, rising to the surface where they form large mats or rafts up to 1 m thick (Villareal, 1988). Mild wind and wave action can concentrate them further into windrows, and the mats can remain in the surface waters for prolonged periods. Little or no grazing will occur, due to the large size of the mats. Prolonged stratification leads to nutrient exhaustion and a change to a system based on recycled nutrients, a situation in which phytoplankton such as the rhizosolenids, which harbor nitrogen-fixing endo-symbionts, can thrive (Villareal, 1992). The mats can remain in the surface waters until wind mixing or a loss of bouyancy resulting from physiologic stress causes them to sink rapidly. When they reach the sediments, they may form such a thick blanket that benthic organisms are unable to break it up (Kemp & Baldauf, 1993). The high concentration of organic matter and lack of bioturbation thus lead to reducing conditions in the surface sediments, pro-ducing sapropels.

Aggregates of rhizosolenid mats hundreds of kilometers long were seen in the eastern equatorial Pacific along a thermal front corresponding to a tropical instability wave (Yoder *et al.*, 1994). In the same region, Miocene and Pliocene sediments contain packets of laminated sapropels consisting of *Thalassiothrix* mats (Kemp & Baldauf, 1993), each packet representing hundreds of years of seasonal (or interannual) cycles. In some cases a given packet can be correlated over thousands of miles, strongly suggesting that they were formed by mecha-nisms similar to those described by Yoder *et al.* (1994). A case of laminated rhi-zosolenid mats occurs in late Pleistocene sapropels of the eastern Mediterranean (Schrader & Matherne, 1981). Here also one may infer alternat-ing cycles of stratified and well-mixed surface waters, perhaps related to density variations resulting from precipitation or Nile River influx (Sancetta, 1994).

Varved Sediments and Seasonal Variability

Varves represent a special case of laminated sediments, in which the laminae are known to be seasonal; a light–dark couplet of laminae is called a varve. The dark lamina usually represents the wet season, being dominated by silt and clay derived from runoff. The light lamina is a diatom ooze and represents the productive season. A complication occurs if lithic material can be introduced into the water column by some mechanism (e.g., wave action, a stormy period) during the productive season. Even a small amount of clay will be sufficient to darken a diatom ooze.

If the seasonal conditions in a region are known, or can be reasonably inferred, one may use the varves as tracers of seasonal and multiannual variations. In a mediterranean climate, for instance (winter wet, summer dry), the relative thickness of light and dark laminae can serve as an indicator of relative durations for the sunny and rainy seasons. Number and relative position of monospecific layers can indicate frequency of bloom events. In Saanich Inlet, British Columbia (48°30'N, 123°35'W), the varves were found to exhibit an exceptionally regular pattern consisting of a thin dark (winter) lamina, overlain by a thicker light lamina which consistently contained the following sequence: early spring bloom (*Thalassiosira* spp.), late spring bloom (*Skeletonema costata* and *Chaetoceros* vegetative cells), and late summer/fall bloom (diverse taxa, including large phaeoceran spores). The summer season was not distinguishable, consisting of the same taxa as those of late spring (Sancetta, 1989). Similar sequencing within a light lamina has been identified in sediments from the Santa Barbara Basin of southern California (Bull & Kemp, 1995) and the Guaymas Basin of the Gulf of California (Pike & Kemp, 1997). In principle, one could look for evidence of interannual or larger-scale perturbations of these trends. For instance, variations in thickness of the dark lamina might reflect higher precipitation during El Niño years; absence of one 'seasonal' sublamina could be interpreted as, e.g., a failure of upwelling during that period. Laminated sediments of the Santa Barbara and Guaymas Basins reveal distinctive changes in the relative abundance of 'warm' subtropical taxa and upwelling indicators such as *Chaetoceros* spores corresponding to the occurrence or absence of El Niño conditions (Lange *et al.*, 1990; Baumgartner *et al.*, 1985)

However, most coastal systems (where varved sediments are usually found) are more complex than Saanich Inlet. Taxon diversity is high, and for the most part, we do not know enough about the ecology of individual taxa to understand why a given species might be abundant in one year and very rare in the next. It is likely that chance plays a large role; when conditions change so as to favor production the assemblage may be simply a result of the mixture of seed populations which happened to be present at the time. In the Gulf of California, for instance, seasonal variations in circulation are fairly predictable, with strong wind-mixing and production beginning in winter and continuing into spring, and a summer season of strong thermal stratification and low, nutrient-limited production (Alvarez-Borrego & Lara-Lara, 1991). A two-year sediment-trapping program showed some repeated trends in diatom composition and flux (common *Coscinodiscus* and hyalochaetes during winter, a distinctive assemblage of small and/or weakly silicified forms during summer) but also considerable interannual variability in abundance of taxa, particularly during the spring season (Sancetta, 1995). Laminated sediments of the anoxic zone from Guaymas Basin show these trends, but there is a large degree of interannual variability: laminae of *Thalassiothrix* mats occur, some 'spring' laminae are dominated by *Chaetoceros* and others by *Skeletonema*, occasionally a normally rare taxon such as *Hemidiscus* will dominate a lamina (Sancetta, 1995, Pike & Kemp, 1997). Even were physico-chemical data available for the same time period, it is likely that no simple explanation could be found for these variations.

Lateral Transport

There are a few cases in which displaced diatoms can provide helpful paleo-environmental information. The presence of shallow-water (especially benthic) forms in deep-sea sediments indicates advection of waters from the continental shelf. The case of *Chaetoceros* spores as possible Gulf Stream tracers in the central North Atlantic was discussed above. Shelf benthic taxa are consistently more common in sediments of glacial intervals than of interglacials, from the late Pliocene off southwest Africa (Sancetta *et al.*, 1992*b*) to the late Pleistocene of the North Pacific (Sancetta *et al.*, 1992*a*), presumably a result of lower sealevel and erosion of shelf material directly into the deep sea.

Freshwater diatoms are found in marine sediments relatively near the continents. Depending on the circumstances, it may be possible to determine whether transport was by wind or water, thus inferring atmospheric circulation or fluvial input to the ocean. In the tropical Atlantic, freshwater diatoms such as *Cyclotella* spp. occur in sediments underlying the Zaire and Niger River plumes (Pokras, 1991; Van Iperen *et al.*, 1987). On the continental shelf of Portugal, the distribution of freshwater diatoms shows a close relationship to outlets of the major rivers (Abrantes, 1988). It may be possible to use freshwater diatoms to trace changes in direction of a river plume determined by shifting ocean circulation.

Freshwater diatoms also occur in sediments underlying easterly dust plumes derived from the dry lake beds of tropical Africa (Pokras, 1991). In this case the transport mode is eolian, and changes in concentration of these forms can serve as a tracer of cyclic variations in continental aridity (Pokras & Mix, 1987). Pliocene freshwater diatoms found in a late-glacial section off the west coast of North America were inferred to reflect strong easterly winds during the winter season, in agreement with predictions from a global climate model for the last glacial maximum (Sancetta *et al.*, 1992*a*).

Marine records before the Pleistocene

The examples above are almost all drawn from modern or late Pleistocene sediments, an accurate reflection of the relative concentration of effort in marine diatom paleoecology. Most of the studies on older material have been either stratigraphic or taxonomic/morphologic in nature. Typically, these studies are done on one or a few cores from a single area, examining the variations in relative abundance of indicators for 'warm' or 'cool' surface waters, presence of sea-ice, or level of productivity. Conclusions are similar to those described above for the studies of Barron (1992), Kaczmarska *et al.* (1993), and Koizumi (1985, 1989). Such studies are of value for stratigraphy and for identifying times at which distinct changes appear to have occurred. The results of such studies can be combined with those at other sites to identify regional or larger-scale trends.

The difficulty with older material lies in our progressively greater ignorance about species ecology and environmental conditions with increasing time from the present. As far back as the Pliocene, and even into the Miocene in some places, one can make inferences using either long-ranging taxa which are still extant, or assuming that an extinct form had an ecology similar to that of a morphologically similar extant form. However, as the similarity to modern assemblages decreases, it becomes progressively less valid to make such assumptions, since one is compelled to ignore large portions of the data set. If a fossil assemblage has little or no similarity to any modern one (often referred to as the 'no-analogue' problem), and the past environmental conditions at the site are otherwise unknown, there is little that can be done beyond speculation.

Marine diatoms have shown several periods of extinction and/or radiation since their appearance in the Mesozoic; Jousé (1978) defined six stages based on turnover at the genus and family level. Judging by the fossil record (notoriously incomplete), the modern pelagic diatom flora began to evolve during the Late Oligocene to Early Miocene, when several of the most diverse and abundant genera appeared or began large radiations, including *Nitzschia* and *Thalassiosira*. Paleogene and Cretaceous assemblages bear no resemblance to Neogene ones, not even at the family level. Although these ancient species undoubtedly did have individual ecologic preferences, we have no way to interpret them beyond the very general inferences that may be drawn from biogeography. Extinct taxa consistently found only in sediments known to be derived from continental margins might be considered high productivity indicators; those found only at sites formerly located in high latitudes might be 'cool' indicators, and so on.

The cause of these periods of turnover is, of course, unknown. It may well lie in biotic processes and relationships that have little to do with the physical environment. For example, relative rates of evolution of two taxa might be driven by their comparative success in competing for a critical nutrient, or in minimizing predation. A chance mutation may have had a favorable effect on physiology such as allowing more efficient uptake of a limiting nutrient (leading to higher growth rates), or more efficient use of that nutrient (lower requirement per cell, thus producing more cells). An example of the latter is the general observation that Mesozoic and Paleogene assemblages are on average more heavily silicified than modern ones, even if one allows for the effects of dissolution. Perhaps the Oligocene/Miocene turnover is related to development of a lower requirement for silica in certain families.

Summary

Fossil marine diatoms can be used as tracers of various environmental parameters. Multivariate statistics and taxon ratios are commonly used to infer sea-surface temperature from percentage data on assemblage composition. Taxa known to be closely associated with sea-ice can serve as tracers of past sea-ice cover, and shelf forms displaced into deep waters are indicators of lateral transport. Freshwater taxa provide evidence of either fluvial or eolian transport. High abundance of bloom-forming taxa reflects conditions of rapid nutrient uptake and growth rates, while aggregates of large mat-forming taxa may indicate conditions of prolonged water column stability. Laminated sediments preserve records of interannual and even seasonal variations in production. These methods have been used primarily for late Neogene sediments. Although fossil marine diatoms extend to the late Cretaceous, the ecologic requirements of ancient forms are unknown and relatively little can be achieved in the way of paleoenvironmental reconstructions.

Acknowledgments

I would like to thank the editors for giving me the opportunity to write this chapter, and John Barron, who provided helpful advice on the manuscript. The views expressed in this chapter are those of the author and do not reflect the policy of the National Science Foundation.

References

Abrantes, F. (1988). Diatom assemblages as upwelling indicators in surface sediments off Portugal. *Marine Geology*, **85**, 15–39.

Alexander, V. & Niebauer, H. J. (1981). Oceanography of the eastern Bering Sea ice-edge zone in spring. *Limnology and Oceanography*, **26**, 1111–25.

Alvarez-Borrego, S. & Lara-Lara, J. R. (1991). The physical environment and primary productivity of the Gulf of California. *American Association of Petroleum Geologists, Memoir*, **47**, 555–67.

Barron, J. A. (1992). Pliocene paleoclimatic interpretation of DSDP Site 580 (NW Pacific) using diatoms. *Marine Micropaleontology*, **20**, 23–44.

Baumgartner, T. R., Ferreira-Bartrina, V., Schrader, H., & Soutar, A. A. (1985). 20-year varve record of siliceous phytoplankton variability in the central Gulf of California. *Marine Geology*, **64**, 113–29.

Booth, B. C. (1981). Vernal phytoplankton community in the eastern subarctic Pacific: Predominant species. In *Sixth International Diatom Symposium*, ed. R. Ross, pp. 339–58. Koenigstein: J. Cramer.

Bull, D. & Kemp, A. E. S. (1995). Composition and origins of laminae in Late Quaternary and Holocene sediments from the Santa Barbara Basin. *Proceedings of the Ocean Drilling Project*, **146**, 77–87.

Coale, K., Fitzwater, S. E., Gordon, R. M., Johnson, K. S., & Barber, R. T. (1996). Control of community growth and export production by upwelled iron in the equatorial Pacific. *Nature*, **379**, 621–4.

Fryxell, G. A. & Kendrick, G. A. (1988). Austral spring microalgae across the Weddell sea-ice edge: Spatial relationships found along a northward transect during AMERIEZ 83. *Deep-Sea Research*, **35**, 1–20.

Fryxell, G. A. & Prasad, A. K. S. K. (1990). *Eucampia antarctica* var. *recta* (Mangin) stat. nov. (Biddulphiaceae, Bacillariophyceae): Life stages at the Weddell sea-ice edge. *Phycologia*, **29**, 27–38.

Garrison, D. L., Sullivan, C. W., & Akcley, S. F. (1986). Sea-ice microbial communities in Antarctica. *Bioscience*, **36**, 243–50.

Gran, H. H. (1911). Phytoplankton. In *Report of the Second Norwegian Arctic Expedition in the 'Fram' 1898–1902*, Chap. 27, 28 pp. Kristiania: A. W. Brøgger.

Guillard, R. R. L., & Kilham, P. (1977). The ecology of marine planktonic diatoms. In *The Biology of Diatoms*, ed. D. Werner, pp. 372–469. London: Blackwell.

Horner, R. & Schrader, G. C. (1982). Relative contributions of ice algae, phytoplankton, and benthic microalgae to primary production in nearshore regions of the Beaufort Sea. *Arctic*, **35**, 485–503.

Imbrie, J. & Kipp, N. G. (1971). A new micropaleontological method for quantitative paleo-climatology: Application to a Late Pleistocene Caribbean core. In *The Late Cenozoic Glacial Ages*, ed. K. K. Turekian, pp. 71–181. New Haven: Yale University Press.

Jousé, A. P. (1978). Diatom biostratigraphy on the generic level. *Micropaleontology*, **24**, 316–26.

Kaczmarska, I., Barbrick, N. E., Ehrman, J. M., & Cant, G. P. (1993). *Eucampia* index as an indi-cator of the Late Pleistocene oscillations of the winter sea-ice extent at the ODP Leg 119 Site 745B at the Kerguelen Plateau. *Hydrobiologia*, **269**, 103–12.

Kanaya, T. & Koizumi, I. (1966). Interpretation of diatom thanatocoenoses from the North Pacific applied to a study of core V20–130. *Scientific Reports of Tohoku University, Series 2*, **37**, 89–130.

Kemp, A. E. S. & Baldauf, J. G. (1993). Vast Neogene laminated diatom mat deposits from the eastern equatorial Pacific Ocean. *Nature*, **362**, 141–3.

Koç Karpuz, N. & Jansen, E. (1992). A high-resolution diatom record of the last deglaciation from the SE Norwegian Sea: Documentation of rapid climatic changes. *Paleoceanography*, **7**, 499–520.

Koizumi, I. (1985). Late Neogene paleoceanography in the western North Pacific. *Initial Reports of the Deep Sea Drilling Project*, **86**, 429–38.

  (1989). Holocene pulses of diatom growth in the warm Tsushima Current in the Japan Sea. *Diatom Research*, **4**, 55–68.

Kutzbach, J. E. & Guetter, P. J. (1986). The influence of changing orbital parameters and surface boundary conditions on climate simulations for the past 18,000 years. *Journal of Atmospheric Science*, **43**, 1726–59.

Lange, C. B., Burke, S. K. & Berger, W. H. (1990). Biological production off southern California is linked to climatic change. *Climate Change*, **16**, 319–29.

Leventer, A. (1992). Modern distribution of diatoms in sediments from the George V coast, Antarctica. *Marine Micropaleontology*, **19**, 315–32.

Martin, J. H. & Gordon, R. M. (1988). Northeast Pacific iron distributions in relation to phytoplankton productivity. *Deep-Sea Research*, **35**, 177–96.

Parsons, T. R. & Lalli, C. M. (1988). Comparative oceanic ecology of the plankton communities of the subarctic Atlantic and Pacific oceans. *Oceanography and Marine Biology Annual Reviews*, **26**, 317–59.

Pichon, J-J., Labeyrie, L. D., Barreille, G., Labracherie, M., Duprat, J., & Jouzel, J. (1992). Surface water temperature changes in the high latitudes of the southern hemisphere over the last glacial–interglacial cycle. *Paleoceanography*, **7**, 289–318.

Pike, J. & Kemp, A. E. S. (1997). Early Holocene decadal-scale ocean variability recorded in the Gulf of California laminated sediments. *Paleoceanography*, **12**, 227–38.

Pokras, E. M. (1991). Source areas and transport mechanisms for freshwater and brackish-water diatoms deposited in pelagic sediments of the equatorial Atlantic. *Quaternary Research*, **35**, 144–56.

Pokras, E. M. & Mix, A. C. (1987). Earth's precession cycle and Quaternary climatic change in tropical Africa. *Nature*, **326**, 486–7.

Sancetta, C. A. (1989). Processes controlling the accumulation of diatoms in sediments: A model derived from British Columbian fjords. *Paleoceanography*, **4**, 235–51.

    (1992). Primary production in the glacial North Atlantic and North Pacific oceans. *Nature*, **360**, 249–51.

    (1994). Mediterranean sapropels: seasonal stratification yields high production and carbon flux. *Paleoceanography*, **9**, 195–6.

    (1995). Diatoms in the Gulf of California: Seasonal flux patterns and the sediment record for the last 15,000 years. *Paleoceanography*. **10**, 67–84.

Sancetta, C. A. & Silvestri, S. (1986). Pliocene–Pleistocene evolution of the North Pacific ocean–atmosphere system, interpreted from fossil diatoms. *Paleoceanography*, **1**, 163–80.

Sancetta, C. A., Lyle, M., Heusser, L., Zahn, R., & Bradbury, J. P. (1992a). Late-glacial to Holocene changes in winds, upwelling, and seasonal production of the northern California Current system. *Quaternary Research*, **38**, 359–70.

Sancetta, C. A., Heusser, L., & Hall, M. A. (1992b). Late Pliocene climate in the southeast Atlantic: Preliminary results from a multi-disciplinary study of DSDP Site 532. *Marine Micropaleontology*, **20**, 59–75.

Scharek, R., Smetacek, V., Fahrbach, E., Gordon, L. I., Rohardt, G., & Moore, S. (1994). The transition from winter to early spring in the eastern Weddell Sea, Antarctica: Plankton biomass and composition in relation to hydrography and nutrients. *Deep-Sea Research*, **41**, 1231–50.

Schrader, H. (1992). Peruvian coastal primary paleo-productivity during the last 200,000 years. In *Upwelling Systems: Evolution Since the Early Miocene*, ed. C. P. Summerhayes, W. L. Prell, and K. C. Emeis, *Geological Society, Special Publication*, **64**, 391–409.

Schrader, H. & Koç Karpuz, N. (1990). Norwegian-Iceland Seas: Transfer functions between marine planktic diatoms and surface water temperature. In *Geological History of the Polar Oceans: Arctic Versus Antarctic*, ed. U. Bleil & J. Thiede, pp. 337–61. Amsterdam: Kluwer.

Schrader, H. J. & Matherne, A. (1981). Sapropel formation in the eastern Mediterranean Sea: Evidence from preserved opal assemblages. *Micropaleontology*, **27**, 191–203.

Schrader, H., Swanberg, N., Lycke, A. K., Paetzel, M., Schrader, T., & Schrader T. (1993). Diatom-inferred productivity changes in the eastern equatorial Pacific: The Quaternary record of ODP Leg 111, Site 677. *Hydrobiologia*, **269**, 137–51.

Smayda, T. J. (1958). Biogeographical studies of marine phytoplankton. *Oikos*, **9**, 158–91.

Smith, W. O. & Nelson, D. M. (1986). Importance of ice edge phytoplankton production in the Southern Ocean. *Bioscience*, **36**, 251–7.

Suzuki, Y. (1995). Growth responses of several diatom species isolated from various environments to temperature. *Journal of Phycology*, **31**, 880–8.

Tréguer, P., Nelson, D. M., Van Bennekom, A. J., DeMaster, D. J., Leynaert, A., & Quéguiner, B. (1995). The silica balance in the World Ocean: A reestimate. *Science*, **268**, 375–9.

van Iperen, J. M., van Weering, T. C. E., Jansen, J. H. F., & van Bennekom, A. J. (1987). Diatoms in surface sediments of the Zaire deep-sea fan (SE Atlantic Ocean) and their relation to overlying water masses. *Netherlands Journal of Sea Research*, **21**, 203–17.

Villareal, T. A. (1988). Positive bouyancy in the oceanic diatom *Rhizosolenia debyana* H. Peragallo. *Deep-Sea Research*, **35**, 1037–45.

 (1992). Marine nitrogen-fixing diatom–cyanobacteria symbioses. In *Marine Pelagic Cyanobacteria: Trichodesmium and Other Diazotrophs*, ed. E. J. Carpenter, pp. 163–75. Amsterdam: Kluwer.

Yoder, J. A., Ackleson, S. G., Barber, R. T., Flament, P. & Balch, W. M. (1994). A line in the sea. *Nature*, **371**, 689–92.

C. SANCETTA

Part V
Other applications

18 Diatoms and archeology

STEVEN JUGGINS AND NIGEL CAMERON

Introduction

One of the goals of modern archeology is to understand how past communities interacted spatially, economically, and socially with their biophysical environment (Butzer, 1982). To this end, archeologists have developed strong links with zoologists, botanists and geologists to provide information on the environment of past societies and to help understand the complex relationships between culture and environment. This chapter reviews the role of diatom analysis in such studies, and discusses how the technique can be applied at a range of spatial and temporal scales to place archeological material in its broader site, landscape and cultural context. In particular, we examine applications to the provenancing of individual archeological artefacts, the analysis of archeological sediments and processes of site formation, the reconstruction of local site environments, and the identification of regional environmental processes affecting site location and the function of site networks. We have chosen a small number of examples that best illustrate these applications; other case studies directly motivated by archeological problems may be found in recent reviews by Battarbee (1988), Mannion (1987) and Miller and Florin (1989), while diatom-based studies of past changes in sea level, climate, land-use, and water quality that are also relevant to archeological investigation are reviewed elsewhere in this volume (e.g., Bradbury; Cooper; Denys & de Wolf; Fritz *et al.*; Hall & Smol).

Analysis of archeological artefacts

The direct application of diatom analysis to archeological artefacts is best represented in the field of pottery sourcing and typology. Diatoms often survive the low temperature firing process used in the manufacture of prehistoric pottery; diatom analysis therefore offers a novel and potentially powerful method for identifying clay sources and provenancing finished pottery (Alhonen *et al.*, 1980; Jansma, 1977, 1981, 1982, 1984, 1990). Unfortunately, the method is not without problems. Diatom concentrations in pottery can be very low and valves are often poorly preserved (Håkansson & Hulthén, 1986, 1988).

In addition, diatoms may be derived from tempering material or other sources, rather than from the clay itself (Matiskainen & Alhonen, 1984; Gibson, 1986). Perhaps the main limitation is that a considerable knowledge of the distribution and diatom content of possible source clays is required to unambiguously link a pot sherd to clay deposit. However, when this information exists diatom analysis of pottery sherds can aid archeological interpretation in questions of typology, technology and transport concerning, for example, the classification of pottery types (Alhonen *et al.*, 1980), the preferential choice of pottery fabric (Alhonen & Matiskainen, 1980; Gibson, 1986; Jansma, 1981; Matiskainen & Alhonen, 1984), the movement and trade of raw materials or manufactured goods (Alhonen & Väkeväinen, 1981; Jansma, 1977, 1990), and evidence for contact between communities and their range of movement (Alhonen & Väkeväinen, 1981).

In the absence of local sources of marine clay, pottery from inland sites containing marine diatoms can provide convincing evidence for transport and trade. The converse is also true, and analysis of pottery sherds, combined with a knowledge of regional stratigraphy, can been used to demonstrate that clay or manufactured pottery was imported to island or coastal sites from the mainland or hinterland sources. For example, pottery from a number of sub-Neolithic sites on the Åland Islands in the Baltic Sea were found to contain freshwater diatoms typical of the Ancylus Lake, a freshwater lake that occupied the Baltic basin prior to the Litorina transgression (Alhonen & Väkeväinen, 1981; Matiskainen & Alhonen, 1984). Since Ancylus clays are absent from the Åland Islands, the diatom data provides clear evidence that either the finished pots or raw clays were imported from the mainland. In addition, the presence of vessels of similar style and diatom content at sites around Turku in southern Finland suggests a possible source area on the Finnish mainland.

Analysis of archeological sediments

Archeologists derive most of their basic information from excavation, and a knowledge of the origin, stratigraphy, and depositional environment of archeological sediments is fundamental to all subsequent analyses. Primary or *in situ* water-lain or terrestrial sediments generally possess a diatom flora characteristic of the conditions prevailing at the time of deposition, and should be readily distinguished from that of secondary, or reworked deposits. Diatom analysis can therefore be used to identify sediment provenance and on-site depositional environments (Miller *et al.*, 1979).

When direct deposition of marine sediments or erosion from older deposits can be discounted, the presence of marine taxa may provide supporting evidence for various economic activities, such as saltworking (Juggins, 1992a), or the collection and use of shellfish (Denys, 1992) or algal mats (Foged, 1985). At high percentages, marine and brackish diatoms provide unambiguous evidence of flood deposits on coastal sites, and fluctuations in the proportions of allochthonous marine taxa can be used to infer flooding intensity, which in turn may be related to periods of site abandonment (e.g., Groenman-van Waateringe & Jansma, 1968; Voorrips & Jansma, 1974; Jansma, 1981). One of the clearest examples of the use of diatom analysis to identify coastal inundation comes from a series of prehistoric sites along the southern Washington coast (Cole *et al.*, 1996). Occupation horizons dating from AD 1000 to AD 1700 were found preserved in a buried spruce-forest soil and overlain by a thin sand horizon and up to 1m of fine mud (Fig. 18.1). The presence of marine and brackish sandflat taxa in the sand layer was taken to indicate deposition by a tsunami (Hemphill-Haley, 1996), while diatoms in the overlying sediments suggest accumulation in intertidal mud-flat and salt marsh environments following submergence of the site (Hemphill-Haley, 1995). The evidence of rapid, earthquake-induced burial and tsunami inundation approximately 300 years ago suggests that many other sites lie buried in the present intertidal zone, and helps explain the current paucity of archeological remains in an area which supported a large native population in early historic times.

Site-based paleoenvironmental reconstructions

Knowledge of the local environment of a site is essential if its socioeconomic function is to be fully understood. Not surprisingly, diatom analysis has frequently been used to reconstruct paleoenvironments in the vicinity of coastal (e.g., Demiddele & Ervynck, 1993; Groenman-van Waateringe & Jansma, 1969; Håkansson, 1988; Jansma, 1981, 1982, 1990; König, 1983; Körber-Grohne, 1967; Miller & Robertsson, 1981; Nunez & Paabo, 1990) and inland archeological sites (Caran *et al.*, 1996; Castledine *et al.*, 1988; Jessen, 1936; Neely *et al.*, 1990; Robbins *et al.*, 1996; Straub, 1990; Wuthrich, 1971).

Estuaries in particular are often rich in archeological remains, and offer the potential for close collaboration between diatomists and archeologists. Estuarine wetlands are rich in natural resources and their navigable waters have provided routes for merchants and invaders, encouraging the development of trading and defensive settlements. Indeed, many of the world's cities have developed on estuaries because of their importance as ports. The historic and prehistoric development of these settlements is thus intimately linked to the changing estuarine environment, and diatom analysis can provide a powerful tool for studying the interface between the sea, shipping and trading routes, and the economic and military function of the shore and urban waterfronts. To this end, diatom analysis has been used to reconstruct the sedimentary environment of shipwrecks, or other isolated structures and artefacts (Cameron, 1997; Foged, 1973; Marsden *et al.*, 1989; Miller, 1995), to reconstruct local sea-level changes in areas adjacent to estuarine sites (Battarbee *et al.*, 1985; Miller & Robertsson, 1981; Wilkinson *et al.*, 1988; Wilkinson & Murphy, 1995), or to reconstruct changes in river water quality and tidal influence in urban areas

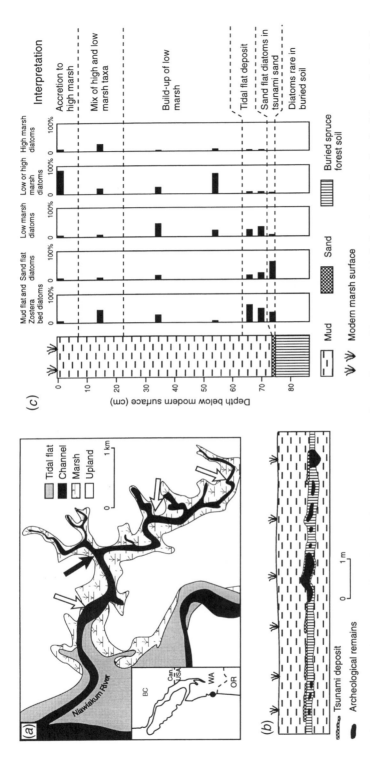

Fig. 18.1. Diagram analysis of archeological sites in the Niawiakum River Valley, southern Washington coast. (a) Site location showing the Niawiakum archeological site (solid arrow) and other sampled sedimentary sequences (hollow arrows); (b) cross-section through Niawiakum archeological site; and (c) diatom stratigraphy in vicinity of the Niawiakum archeological site. Redrawn from Cole et al. (1996) and Hemphill-Haley (1995).

(Ayers & Murphy, 1983; Boyd, 1981; Foged, 1978; Jones & Watson, 1987; Juggins, 1988, 1992b; Miller, 1982; Milne *et al.*, 1983).

One of the best examples of the collaboration between diatomists and archeologist comes from the river Thames in central London (Milne, 1985; Milne *et al.*, 1983; Juggins, 1992b). The Thames is now tidal to Teddington Weir, 30 km above the City of London, but the existence of Roman remains 4 m below the present high tide level in the outer estuary led archeologist to assume that the river was tideless in Central London during the Roman period (Willcox, 1975). Within the City, the pre-Roman foreshore now lies buried some 100 m north of the present riverbank, and a horizontal stratigraphy of wooden and stone quays and revetments spanning the last 2000 years is preserved beneath the modern waterfront (Fig. 18.2). Initial diatom analysis of foreshore sediments associated with part of the first century AD revetment at the Pudding Lane site revealed an assemblage containing *Cyclotella striata* (Kütz.) Grunow in Cleve & Grunow, a brackish water planktonic species common in the estuary today. This taxon, together with a number of marine forms, clearly demonstrates that the river was tidal in central London during the early Roman period, contrary to common belief (Milne *et al.*, 1983).

Subsequent work has examined additional foreshore samples from the first to twelfth centuries AD and refined the paleoenvironmental reconstructions using a salinity transfer function (Juggins, 1992b). Results show an initial rapid rise in mean half-tide salinity between the first and second centuries AD followed by a more gradual rise to the ninth to twelfth centuries (Fig. 18.2). Comparison of these paleosalinity estimates with evidence for changes in the tidal range suggest that the increase in salinity between the first and twelfth centuries was primarily the result of relative sea-level rise. This would have been accompanied by a substantial upstream migration of the tidal head, particularly during the early Roman period. Progressive canalization of the upper and middle estuary after the twelfth century led to a marked increase in the tidal range of the river in central London, but it appears this had little impact on the salinity regime. Any tendency for further upstream penetration of brackish water was apparently balanced by an increase in the ability of the river to impede landward flow as a result of the reduction in the tidal prism. Although more sites are needed to refine the model, these reconstructions, coupled with information on the changing river topography, provide archeologist with a preliminary framework for understanding the effects of the changing tidal regime on shipping and waterfront construction.

Regional paleoenvironmental reconstructions

In addition to reconstructing the local environment and resource space around a site, a major goal of archeology is to place sites and site networks in their landscape context, and to examine the climatic, environmental and cultural factors that have influenced regional patterns of site location and functional

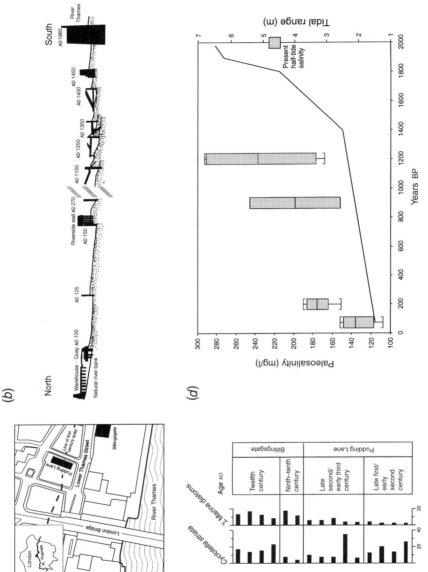

Fig. 18.2. Diatom analysis of River Thames foreshore sediments. (a) Location of Pudding Lane and Billingsgate archeological sites in central London, showing line of first century quay; (b) composite cross-section showing horizontal stratigraphy of archeological structures from pre-Roman river bank to the modern river; (c) summary diatom stratigraphy showing changes in the relative proportions of *Cyclotella striata* and total marine taxa from first to twelfth centuries; and (d) diatom-based paleosalinity estimates (boxplots) and tidal range (solid line) for the River Thames in central London for the last 2000 years. Redrawn from Milne (1985) and Juggins (1992b).

organization. Where there are appropriate sedimentary sequences in coastal or inland aquatic environments, diatom analysis offers the potential for the integration of on-site archeological evidence with off-site paleoenvironmental reconstructions. One of the clearest examples of such work comes from the Baltic coast where the interaction of isostatic land uplift and eustatic sea-level rise has given rise to a complex pattern of transgressions and regressions and an altitudinal zonation of ancient shorelines and associated archeological remains. In the Stockholm region, diatom analysis of archeological sites from Mesolithic to Medieval age reveals a complex pattern of abandonment and reoccupation according to oscillations of the Baltic strandline (Florin, 1948; Miller & Robertsson, 1981). Unfortunately, none of the archeological sites contains a continuous stratigraphic record: consequently additional diatom and ^{14}C analyses were carried out at a series of nearby lake basins with threshold altitudes that bracketed the archeological sites. Results from the lake sediments were used to identify both the timing of the final isolation of the lake from the Baltic and the maximum altitude of the various transgressive phases, allowing the archeological data to be placed in a regional framework of relative sea-level change and shore displacement (Fig. 18.3). In addition, the shoreline curve offers a tool for site prospecting by directing archeologist to the correct altitudinal contour of shorelines of different ages.

Diatom analysis from inland freshwater or saline lakes can also provide valuable information for the archeologist on regional paleohydrology and paleoclimate. However, if they are to provide more than a simple deterministic environmental 'backdrop' to archeological interpretation, such studies must consider the complex mosaic of environments within the landscape, and the possibility of asynchronous change in the quality of aquatic habitats in a regional site network. An excellent example of the integration of diatom and other paleoecological analyses into a regional archeological investigation comes from the North American Southern High Plains. The plateau of the Plains is dissected by a series of entrenched ephemeral drainage channels, or draws, that contain a rich archeological record spanning the last 12 000 years, interbedded with alluvial, lacustrine and eolian sediments (Holliday, 1985). Diatom analysis was originally used to reconstruct local depositional environments at individual sites, including the famous Clovis Paleo-Indian site on Blackwater Draw (Lohman, 1935; Patrick, 1938). More recently Winsborough (1995) has carried out diatom analysis at a network of sites as part of a wider study into regional paleoenvironments. Results show that each of the four channels examined underwent a similar evolution; from a predominantly alluvial environment during the late Pleistocene, to a complex lacustrine and marsh environment in the early Holocene, and finally to an environment dominated by eolian sedimentation in the middle Holocene. These changes were broadly synchronous between sites, indicating regional climatic control, although there were some distinct time-transgressive relationships highlighting the importance of regionally variable and/or local environmental changes (Fig. 18.4).

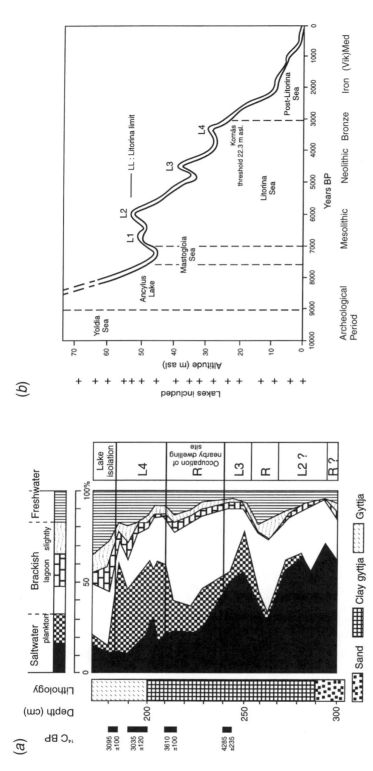

Fig. 18.3. Diatom analysis and land uplift in the Baltic. (*a*) Summary diatom stratigraphy for Körnas bog, 30 km SW of Stockholm, showing inferred Litorina transgressions and lake isolation; and (*b*) shore displacement curve for the Stockholm region derived from the Körnas bog and other diatom stratigraphies. The curve can be used to predict the altitudes of coastal dwelling sites of different ages. Redrawn from Miller & Florin (1989) and Miller & Robertsson (1981).

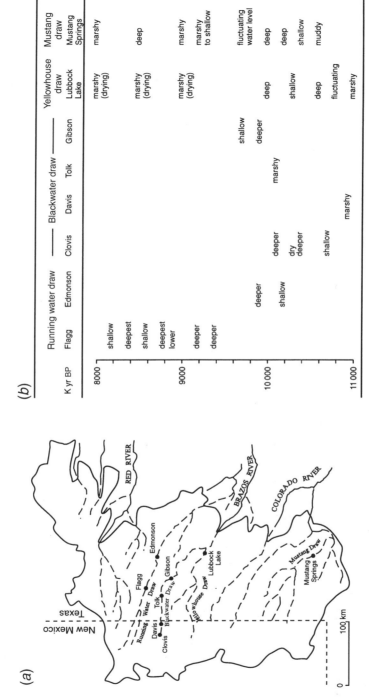

Fig. 18.4. Diatom analysis of valley fills in the Southern High Plains, North America. (*a*) Map of Southern High Plains, showing major stream channels and site locations; and (*b*) summary of late Pleistocene and early Holocene paleoenvironments inferred from diatom analysis of lake and marsh sediments. Redrawn from Winsborough (1995).

Conclusions

Despite the substantial potential of diatom analysis in archeological enquiry, it is still an underused method, and one that is poorly understood by many archeologists. As a consequence, diatom analysis has often been used *post hoc* to address poorly framed questions from inadequately sampled material, and the results relegated to specialist appendices in archeological reports. However, there are grounds for optimism: the examples above illustrate a trend in which diatomists are becoming more closely involved in archeological research and project design, leading to integrated excavation and sampling strategies that maximize the return of information. Such collaboration benefits both diatomists and archeologists, and is essential if we are to develop new methodologies that address the particular problems of preservation, taphonomy and interpretation posed by archeological material.

Summary

This chapter reviews the role of diatom analysis in archeology, and shows how the technique has been applied at a range of spatial and temporal scales to place archeological material in its broader site, landscape and cultural context. The majority of published studies have used diatom analysis to provide information on local and regional paleoenvironments, and to identify environmental factors that have influenced patterns of site location and organization. However, we also note a trend towards closer collaboration between diatomists and archeologists, leading to new applications of the technique to problems of artefact provenancing, identifying processes of site formation and abandonment, and understanding the function of archeological structures.

Acknowledgments

We are grateful to the many colleagues who sent us additional references and papers during the preparation of this chapter.

References

Alhonen, P. & Matiskainen, H. (1980). Diatom analysis from prehistoric pottery sherds – an archeological evaluation. *Proceedings of the Nordic Meeting of Diatomologists, 1980, Lammi Biological Station, Finland, May 6–7, 1980*, pp. 45–62.

Alhonen, P. & Väkeväinen, L. (1981). Diatom-analytical studies of early comb ceramic vessels in Åland. *Eripainos Suomen Museo, 1981*, 67–77.

Alhonen, P., Kokkonen, J., Matiskainen, H., & Vuorinen, A. (1980). Applications of AAS and diatom analysis and stylistic studies of Finnish Subneolithic pottery. *Bulletin of the Geological Society of Finland, 52*, 193–206.

Ayers, B. & Murphy, P. (1983). A waterfront excavation at Whitefriars Street car park, Norwich, 1979. *East Anglian Archeology Report No. 17*, 1–60.

Battarbee, R. W. (1988). The use of diatoms analysis in archeology: A review. *Journal of Archeological Science*, **15**, 621–44.

Battarbee, R. W., Scaife, R. G., & Phethean, S. J. (1985). Paleoecological evidence for sea-level change in the Bann estuary in the early mesolithic period. In *Excavations at Mount Sandel 1973–78*, ed. P. C. Woodman, pp. 111–20. Belfast: HMSO.

Boyd, P. D. A. (1981). The micropaleontology and paleoecology of medieval estuarine sediments from the Fleet and Thames in London. In *Microfossils from Recent and Fossil Shelf Seas*, ed. J. W. Neale & M. D. Brasier, pp. 274–92. Chichester: Ellis Harwood.

Butzer, K. W. (1982). *Archeology as Human Ecology*. Cambridge: Cambridge University Press.

Cameron, N. G. (1997). The diatom evidence. In *Excavations at Caldicot, Gwent: Bronze Age Paleochannels in the Lower Nedern Valley*, eds. N. Nayling & A. Caseldine, pp. 117–28. York: Council for British Archeology Research Report 108.

Caran, C., Neely, J., Winsborough, B., Sorensen, F., & Valastro, S., Jr. (1996). A late Paleo-Indian/Early-Archaic water well in Mexico. Possible oldest water-management feature in New World. *Geoarchaeology*, **11**, 1–36.

Castledine, C., Juggins, S., & Straker, V. (1988). Preliminary paleoenvironmental analysis of floodplain deposits from a section near the River Exe in Exeter, Devon. In *The Exploitation of Wetlands*, ed. P. Murphy & C. French, pp. 145–59. Oxford: BAR.

Cole, S. C., Atwater, B. F., McCutcheon, P. T., Stein, J. K., & Hemphill-Haley, E. (1996). Earthquake-induced burial of archeological sites along the southern Washington coast about AD 1700. *Geoarchaeology*, **11**, 165–77.

Denys, L. (1992). On the significance of marine diatoms in freshwater deposits at archeological sites. *Diatom Research*, **7**, 195–7.

Demiddele, H. & Ervynck, A. (1993). Diatomeeën als ecologische indicatoren in de Vlaamse archeologie: Romeins en middeleeuws Oudenburg (prov. West-Vlaanderen). *Archeologie in Vlaanderen*, **III**, 217–31.

Florin, S. (1948). Kustförskjutningen och bebyggelseutvecklingen I östra Mellansverige under senkvartär tid. II. De baltiska strandbildningarna och stenåldersboplatsen vid Dammstugan nära Katrineholm. *Geologiska Föreningens Förhandlingar*, **70**, 17–196.

Foged, N. (1973). The diatoms in a wreck from the late Middle Age, *University of Lund, Department of Quaternary Geology, Report 3*, pp. 39–45.

(1978). *The Archeology of Svendborg. No. 1. Diatom Analysis*. Odense: Odense University Press.

(1985). Diatoms in a tomb from the early bronze age. *Nova Hedwigia*, **41**, 471–82.

Gibson, A. M. (1986). Diatom analysis of clays and late Neolithic pottery from the Milfield basin, Northumberland. *Proceedings of the Prehistoric Society*, **52**, 89–103.

Groenman-van Waateringe, W. & Jansma, M. J. (1968). Diatom analysis. Appendix III to W. A. van Es, Paddepol, excavations of frustated terps, 200 BC – 250 AD. *Paleohistoria*, **14**, 286.

(1969). Diatom and pollen analysis of the Vlaardingen Creek. A revised interpretation. *Helinium*, **9**, 105–17.

Håkansson, H. (1988). Diatom analysis at Skateholm – Järavallen, southern Sweden. In *The Skateholm Project. I. Man and Environment*, ed. L. Larsson, pp. 39–45. Acta Regiae Societatis Humaniorum Litterarum Lundensis LXXIX.

Håkansson, H. & Hulthén, B. (1986). On the dissolution of pottery for diatom studies. *Norwegian Archeological Review*, **19**, 34–7.

(1988). Identification of diatoms in Neolithic pottery. *Diatom Research*, **3**, 39–45.

Hemphill-Haley, E. (1995). Diatom evidence for earthquake-induced subsidence and tsunami 300 yr ago in southern coastal Washington. *Geological Society of America Bulletin*, **107**, 367–78.

(1996). Diatoms as an aid in identifying late-Holocene tsunami deposits. *The Holocene*, **6**, 439–48.

Holliday, V. T. (1985). New data on the stratigraphy and pedology of the Clovis and Plainview sites, Southern High Plains. *Quaternary Research*, **23**, 388–402.

Jansma, M. J. (1977). Diatom analysis of pottery. In *Ex Horres: I. P.P. 1951–1976*, ed. B. L. van Beek, R. W. Brandt, & W. Groenman-van Waateringe, *Cingula*, **4**, 77–85.

(1981). Diatoms from coastal sites in the Netherlands. In *Environmental Aspects of Coasts and Islands*, ed. D. Brothwell & G. Dimbleby, pp. 145–62. Oxford: BAR.

(1982). Diatom analysis from some prehistoric sites in the coastal area of the Netherlands. *Acta Geologica Academiae Scientiarum Hungaricae*, **25**, 229–36.

(1984). Diatom analysis of prehistoric pottery. In *Proceedings of the Seventh International Diatom Symposium*, ed. D. G. Mann, pp. 529–36. Koenigstein: Koeltz Scientific Books.

(1990). Diatoms from a Neolithic excavation on the former Island of Schokland, Ijselmeerpolders, the Netherlands. *Diatom Research*, **5**, 301–9.

Jessen, K. (1936). Paleobotanical report on the Stone Age site at Newferry, County Londonderry. *Proceedings of the Royal Irish Academy*, **43C**, 31–7.

Jones, J., & Watson, N. (1987). The early medieval waterfront at Redcliffe, Bristol: A study of environment and economy. In *Studies in Paleoeconomy and Environment in South-west England*, ed. N. Balaam, B. Levitan, & V. Straker, pp. 135–62. Oxford: BAR.

Juggins, S. (1988). Diatom analysis of foreshore sediments exposed during the Queen Street Excavations, 1985. In *The Origins of the Newcastle Quayside. Excavations at Queen Street and Dog Bank*, ed. C. O'Brien, L. Bown, S. Dixon, & R. Nicholson, pp. 150–1. Monograph Series No. 3. Newcastle upon Tyne: The Society of Antiquaries of Newcastle upon Tyne.

Juggins, S. (1992*a*). Diatom analysis. In *Iron Age and Roman Salt Production and the Medieval Town of Droitwich: Excavations at the Old Bowling Green and Friar Street*, ed. S. Woodwiss, pp. microfiche 2:F4–5. Council for British Archeology Research Report No. 81.

(1992*b*). Diatoms in the Thames Estuary, England: ecology, palaeoecology, and salinity transfer function. *Bibliotheca Diatomologica*, Band 25, 216 pp.

König, D. (1983). Diatomeen des frühneolithischen Fundplatzes Siggeneben-Süd. In *Siggeneben-Süd. Ein Fundplatz der frühen Trichterbecherkultur an der holsteinischen Ostseeküste*, ed. J. Meurers-Balke, pp. 124–40. Neumünster: Karl Wachholtz.

Körber-Grohne, U. (1967). *Geobotanische Untersuchungen auf der Feddersen Wierde*. Wiesbaden: Franz Steiner Verlag.

Lohman, K. (1935). Diatoms from Quaternary lake beds near Clovis, New Mexico. *Journal of Paleontology*, **9**, 455–9.

Mannion, A. M. (1987). Fossil diatoms and their significance in archeological research. *Oxford Journal of Archeology*, **6**, 131–47.

Marsden, P., Branch, N., Evans, J., Gale, R., Goodburn, D., Juggins, S., McGail, S., Rackham, J., Tyres, I., Vaughn, D., & Whipp, D. (1989). A late Saxon logboat from Clapton, London borough of Hackney. *International Journal of Nautical Archeology and Underwater Exploration*, **18**, 89–111.

Matiskainen, H. & Alhonen, P. (1984). Diatoms as indicators of provenance in Finnish Sub-Neolithic pottery. *Journal of Archeological Science*, **11**, 147–57.

Miller, U. (1982). Shore displacement and coastal dwelling in the Stockholm region during the past 5000 years. *Annales Academiae Scientarium Fennicae*, **A. III, 134**, 185–211.

(1995). Diatoms and submarine archeology (siliceous microfossil analysis as a key to the environment of ship wrecks, harbour basins and sailing routes). In *Scientific Methods in Underwater Archeology*, ed. I. Vuorela, pp. 53–8. Rixensart, Belgium: Council for Europe.

Miller, U. & Robertsson, A. (1981). Current biostratigraphical studies connected with archeological excavations in the Stockholm Region. *Striae*, **14**, 167–73.

Miller, U. & Florin, M-B. (1989). Diatom analysis. Introduction to methods and applications. In *Geology and Paleoecology for Archaeologists*, ed. T. Hackens & U. Miller, PACT, **24**, 133–57.

Miller, U., Modig, S., & Robertsson, A. M. (1979). The Yttersel dwelling site: Method investigations. *Early Norrland*, **12**, 77–92.

Milne, G. (1985) *The Roman Port of London*. London: Batsford.

Milne, G., Battarbee, R. W., Straker, V., & Yule, B. (1983). The River Thames in London in the mid 1st century AD. *Transactions of the London and Middlesex Archeological Society*, **34**, 19–30.

Neely, J., Caran, S., & Winsborough, B. (1990). Irrigated Agriculture at Hierve el Agua, Oaxaca, Mexico. In *Debating Oaxaca Archeology*, ed. J. Marcus, pp. 115–89. Ann Arbor, Michigan: Museum of Anthropology, University of Michigan.

Nunez, M. & Paabo, K. (1990). Diatom Analysis. *Norwegian Archeological Review*, **23**, 128–30.

Patrick, R. (1938). The occurrence of flints and extinct animals in pluvial deposits near Clovis, New Mexico. Part V. Diatom evidence from the Mammoth Pit. *Proceedings of the Academy of Natural Sciences of Philadelphia*, **40**, 15–24.

Robbins, L. H., Murphy, M. L., Stevens, N. J., Brook, G. A., Ivester, A. H., Haberyan, K. A., Klein, R. G., Milo, R., Stewart, K. M., Matthiesen, D. G., & Winkler, A. J. (1996). Paleoenvironment and archeology of Drotsky's Cave: Western Kalahari Desert, Botswana. *Journal of Archeological Science*, **23**, 7–22.

Straub, F. (1990). *Hauterive-Champreveyres, 4. Diatomees et reconstitution des environments prehistorique*. Saint-Blaise: Editions du Ruau (Archeologie Neuchatoise, 10).

Voorrips, A., & Jansma, M. J. (1974). Pollen and diatom analysis of a shore section of the former Lake Wervershoof. *Geologie en Mijnbouw*, **53**, 429–35.

Willcox, G. H. (1975). Problems and posible conclusions related to the history and archeology of the Thames in the London region. *Transactions of the London and Middlesex Archeological Society*, **26**, 185–92.

Wilkinson, T. J., & Murphy, P. L. (1995). *The Archeology of the Essex Coast, Volume I: The Hullbridge Survey*. East Anglian Archeology Report No. 71. Chelmsford, UK: Essex County Council.

Wilkinson, T. J., Murphy, P., Juggins, S., & Manson, K. (1988). Wetland development and human activity in Essex estuaries during the Holocene transgression. In *The Exploitation of Wetlands*, ed. P. Murphy & C. French, pp. 213–38. Oxford: BAR.

Winsborough, B. (1995). Diatoms. In *Late Quaternary Valley Fills and Paleoenvironments of the Southern High Plains*, ed. V. Holliday, pp. 67–83. Geological Society of America, Memoir No. 186.

Wuthrich, M. (1971). Les diatomées de la station néolithique d'Auvernier (Lac de Neuchâtel). Schweizerische Zeitschrift für Hydrologie, 33, 533–52.

19 Diatoms in oil and gas exploration

WILLIAM N. KREBS

Introduction

Diatoms constitute an important rock-forming microfossil group. Unlike other microfossils, they evolved rapidly during the late Cenozoic and are found in marine, brackish water, and lacustrine sediments, and are thus useful tools for age dating and correlating sediments that accumulated in a variety of environments. Their fossil assemblages are reflective of the environments in which they lived, and their recovery in wells and outcrops can provide information on paleochemistry and paleobathymetry. They are most useful to the oil and gas industry in age dating and correlating rocks that lack or have poor recovery of useful calcareous microfossils, such as 'cold water' late Tertiary marine rocks, and sediments that accumulated in lacustrine basins and in brackish water settings.

Age dating and correlation of sedimentary rocks are important for hydrocarbon exploration in order to understand the geologic history of a basin as it relates to the formation of source rocks, reservoirs, structures, and seals. The timing of expulsion of oil and gas is a critical factor in determining the prospectivity of a basin for hydrocarbon, and necessitates an understanding of the geochronology of events. The ability to correlate subsurface rocks is a prerequisite to mapping the distribution of reservoir facies and the interpretation of sequence stratigraphy.

Environmental data pertaining to paleobathymetry and paleochemistry provide important information on the geologic history of a basin and the predictability of reservoir distribution. For example, shales barren of calcareous microfossils may have accumulated in deep water, beneath the carbonate compensation depth, or in brackish or freshwater. The incorrect interpretation will drastically affect the interpretation of a basin's evolution and the evaluation of its hydrocarbon potential. Inasmuch as fossil diatoms occur in all these settings, they can provide the critical evidence for the correct paleoenvironmental interpretation.

Applications

Cenozoic marine diatom biozonations are used within the oil and gas industry to correlate marine rocks, but at Amoco Exploration and Production, graphic

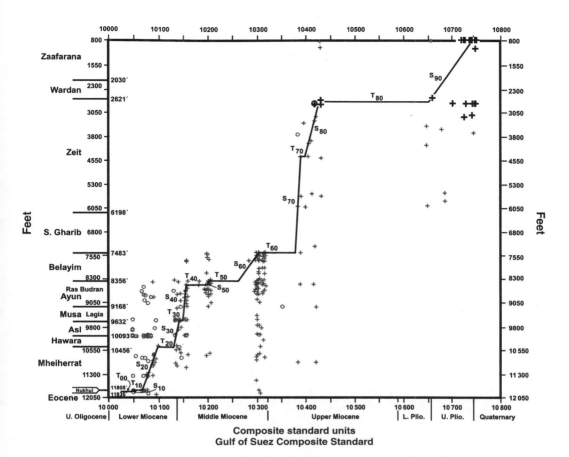

Fig. 19.1. Graphic correlation plot of the Tanka-3 well, Gulf of Suez, Egypt. The vertical axis is well depth with lithostratigraphy. Geologic time in composite standard units and epochs is represented by the horizontal axis. The symbols 'o' and '+' represent, respectively, the oldest and youngest occurrences of microfossil species in a rock section, or biostratigraphic data. The bold symbols are diatom data, and the light symbols represent calcareous nannofossil, planktonic and benthic foraminifera, pollen, spore, and dinoflagellate cyst data. The diatoms are restricted to the shallow portion of the well and provide biostratigraphic control. Note that there are nine biostratigraphic sequences (S 10 to S90) separated by nine terraces (T00 to T80). The sequences represent periods of continuous deposition, and the terraces are hiatuses in geologic time (modified from Krebs *et al.*, 1997).

correlation (Shaw, 1964; Miller, 1977; Carney & Pierce, 1995; Aurisano *et al.*, 1995) is the preferred method because it integrates all paleontological data and can resolve intrazonal hiatuses that are critical to the interpretation of sequence stratigraphy (Neal *et al.*, 1995; Krebs *et al.*, 1997). Fig. 19.1, the Tanka-3 well in the Gulf of Suez, is an example of graphic correlation. Note that diatoms are most useful in controlling the line of correlation in the upper portion of this well because they can resolve small increments of geologic time in young rocks. They are restricted to the shallowest section because they are usually destroyed by diagenesis and disappear at well depths greater than

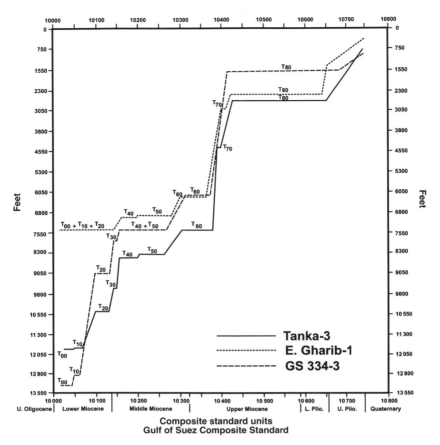

Fig. 19.2. A graphic correlation nomogram of the Tanka-3 well and two additional Gulf of Suez wells. Diatom data controlled the placement of the lines of correlation in the shallow portions of each well. Note the simplicity of age dating rock sections and correlating between wells.

about 5000 feet (1524 m). Fig. 19.2 is a nomogram of lines of correlation of three wells from the Gulf of Suez. In each case, fossil marine diatoms controlled the position of the line of correlation in the youngest section. Using the lines of correlation on the nomogram, well sections can easily be correlated to geologic time and to each other, and important stratigraphic discontinuities such as fault offsets, unconformities, and condensed sections are recognizable. Those hiatuses related to the erosion or slow deposition of rock section are essential to understanding the sequence stratigraphy of a basin.

Fossil marine diatoms are particularly useful for hydrocarbon exploration in basins characterized by thick Neogene diatomaceous rock sections, such as those along coastal California and in the Bering Sea of Alaska. In these 'cold water' Neogene sediments, calcareous microfossils are rare or have long geologic ranges, and are therefore not useful for biostratigraphy. Diatoms, however, are often abundant in these sediments, and are especially well preserved in marine basins that were 'starved' of clastic sediments. These rocks (diatomites) are common along the circum-North Pacific coast (Ingle, 1981), and one such deposit, the Monterey Formation of California, is an important oil source and reservoir. The Hondo Field, offshore of California, for example, has a fractured chert reservoir that was derived from the diagenesis of diatomaceous sediments of the Monterey Formation, and the coeval Belridge Diatomite, Kern Co., California is an oil shale (Schwartz, 1987). Diagenesis of diatomaceous sediments may have a significant impact on oil exploration by creating fractured siliceous reservoirs, and the diatomaceous sediments may themselves be important sources for petroleum (Mertz, 1984). The destruction of diatom frustules by diagenesis can be observed through the microscope (see Fig. 19.3) and noted by changes in rock properties and seismic expression (Murata & Larson, 1975; Hein *et al.*, 1978; Pisciotto, 1981; Isaacs, 1981, 1982, 1983; Grechin *et al.*, 1981; Iijima & Tada, 1981; Tada & Iijima, 1983; Compton, 1991). Both the recovery and destruction of diatoms in such basins provides useful information for the explorationist.

Although diatom frustules are usually not found beneath 1524 m in wells in their unaltered state, deeper recovery can occur when they are preserved in calcareous concretions that formed during early burial (Bramlette, 1946; Lagle, 1984; Blome & Albert, 1985). The concretions form 'closed systems' that protect the frustules from dissolution during diagenesis. In addition, pyritized diatoms can be obtained from greater well depths when recovered from well cutting residues that have been processed for foraminifera. In the North Sea, for example, beautifully preserved pyritized frustules occur at well depths as great as 3658 m and are used for biostratigraphy (M. Charnock, pers. comm. 1990). Because of their abundance in 'cold water' sediments, such recovery often occurs in well intervals that are devoid of calcareous microfossils, so the presence of pyritized diatoms can be especially important in correlating subsurface sections that are otherwise barren.

Fig. 19.4 is an example of using diatoms for paleoenvironmental interpretations in oil exploration. An outcrop section located at Wadi Abu Gaada in the Sinai, Egypt has about 170 m of massive non-calcareous mudstone that had previously been interpreted as representing a period of 'deep water' deposition beneath the carbonate compensation depth (I. Gaafar, pers. comm., 1993). Calcareous marine microfossils all but disappear at the base of this mudstone, and reappear at the top of the outcrop section. The recovery of marine and nonmarine diatoms in the subject interval, however, revealed that deposition occurred in a brackish water/lacustrine setting. This revision significantly affected the interpretation of the synrift history of the Gulf of Suez, the origin of its older evaporites, and the interpretation of its sequence stratigraphy. The evaporites, for example, clearly formed in shallow water, not in deep water brines, and the reinterpretation of this outcrop interval as a shallow water deposit had implications for reservoir sand distribution in the subsurface.

Fossil lacustrine diatoms can be recovered in wells that penetrate Neogene

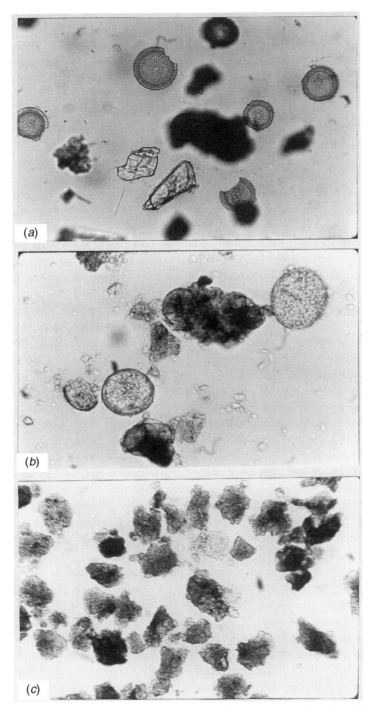

Fig. 19.3. The diagenesis and dissolution of Pliocene Glenns Ferry Formation lacustrine diatoms in the Ore-Ida -1 geothermal well, Ontario, Oregon, in the western Snake River Basin. (a) Represents a well-preserved diatom assemblage from 600 feet (183 m) in the well; (b) partial frustule dissolution at 1600 ft (488 m), and (c) frustule dissolution is complete by 1800 ft (549 m).

W. N. KREBS

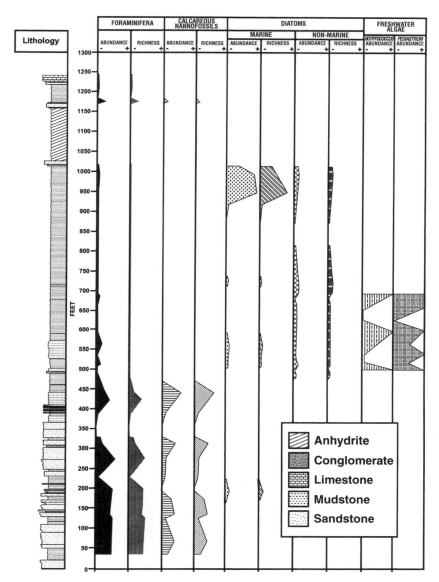

Fig. 19.4. Abundance and diversity plots (biologs) of diatoms, calcareous nannofossils, foraminifera, and freshwater algae (*Botryococcus* and *Pediastrum*) from an outcrop section at Wadi Abu Gaada, Sinai, Egypt. Note the near disappearance of marine calcareous microfossils coincides with the appearance of non-marine diatoms. The diatomaceous mudstones had previously been interpreted as deepwater marine deposits.

lake deposits and occasionally as allocthonous occurrences in Tertiary marine rocks. In the United States, lacustrine diatoms have proven useful for biostratigraphy in the Great Basin (Krebs *et al.*, 1987), particularly in the Snake River Basin of Idaho (Bradbury & Krebs, 1982) and in Carson Sink, Nevada. Both basins have thick sequences of Neogene lacustrine diatomaceous rock

Age	Formation	Characteristic diatoms
Pleistocene	Bruneau Fm.	*Stephanodiscus niagarae* *Cyclotella bodanica*
Pliocene	Glenns Ferry Fm.	*Stephanodiscus carconensis* *Cyclotella pygmaea / C. hannaites* *Cyclotella elgeri*
Late Miocene	Chalk Hills Fm.	*Mesodictyon* spp. *Cyclotella* cf. *C. elgeri*
Middle Miocene	Poison Creek Fm.	*Actinocyclus venenosus* *A. cedrus*
Middle Miocene	Sucker Creek Fm.	*Actinocyclus krasskei*

Fig. 19.5. The diatoms that characterize the lacustrine facies of five Neogene–Quaternary continental formations of the western Snake River Basin, Idaho and Oregon (Bradbury & Krebs, 1995).

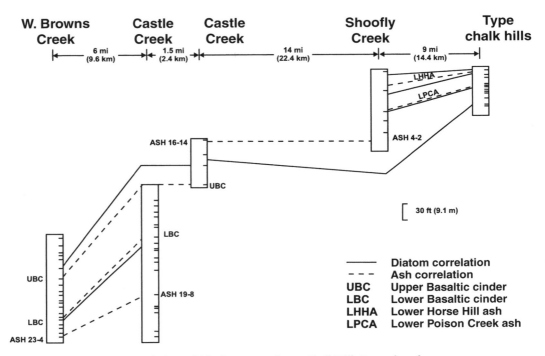

Fig. 19.6. Ash and diatom correlations within the Upper Miocene Chalk Hills Formation of the western Snake River Basin of Owyhee Co., Idaho. The dashed and solid lines represent ash and diatom correlations, respectively. The diatom correlations are based upon first and last (up-section) occurrences of key marker forms. The ash identifications and correlations were taken from Swirydczuk *et al.* (1981). The thick dashes on the right-hand margin of the columns represent stratigraphic locations of samples (Bradbury & Krebs, 1995).

W. N. KREBS

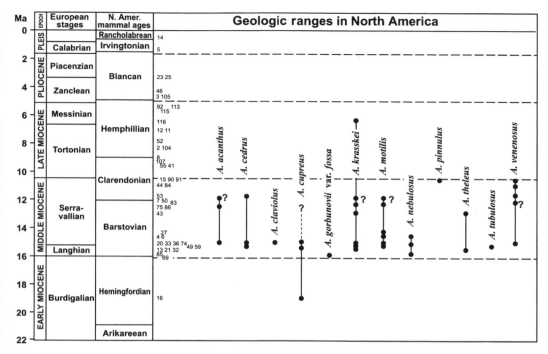

Fig. 19.7. The geologic ranges of fossil non-marine *Actinocyclus* Ehrenberg species in the western United States. The numbers refer to sites for which there are absolute ages, and the solid circles on the range line represent dated sites containing the species listed (Krebs & Bradbury, 1995).

and formations that can be characterized by their fossil diatom content. The Snake River Basin, for example, contains five formations with diatomaceous lacustrine facies that range in age from the middle Miocene to Pleistocene (Bradbury and Krebs, 1995). Each formation contains key fossil diatoms that can be used to correlate subsurface and outcrop sections within the Snake River Basin (Fig. 19.5). In addition, fossil lacustrine diatoms provide a means of making intraformational correlations. Fig. 19.6, for example, illustrates how correlations derived from the occurrences of key fossil diatoms in the lacustrine Upper Miocene Chalk Hills Formation of the western Snake River Basin are parallel to those derived from tephrochronology (Bradbury & Krebs, 1995). Diatom correlations within a single lake system, such as existed during Chalk Hills time, are possible because fluctuations in a lake's physical-chemical parameters are isochronous events that are reflected in its diatom populations.

Diatom correlations between lacustrine basins require the identification and documentation of key shared marker species that have regional time significance. For example, Fig. 19.7 represents the geologic ranges of non-marine *Actinocyclus* Ehrenberg species in the western United States (Krebs & Bradbury, 1995). Numerous occurrences of these species in Neogene lake

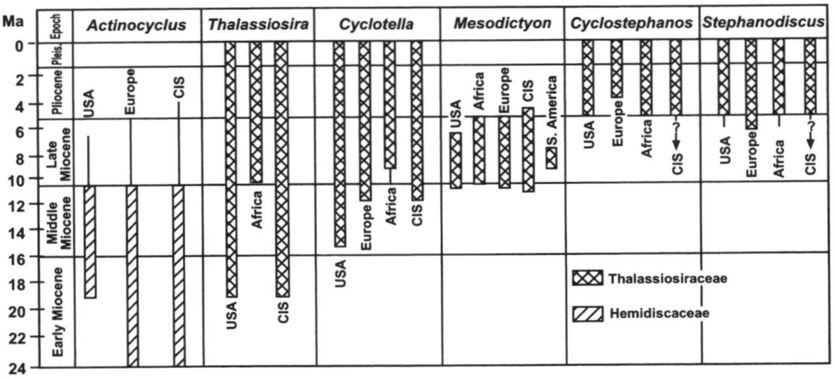

Fig. 19.8. The worldwide geologic ranges of selected non-marine diatom genera. Note the turnovers during the late middle Miocene and latest Miocene. CIS signifies the Commonwealth of Independent States of the former Soviet Union (Bradbury & Krebs, 1995).

deposits in the western United States have been calibrated to geologic time by means of associated volcanic rocks that have been radiometrically dated. In some cases, other fossil groups, such as mammals and palynomorphs, provide a means of assigning relative geologic ages to these deposits and their diatom assemblages. With sufficient data, the regional geologic ranges of key fossil lacustrine diatoms can be established and used for correlations within that area. Intercontinental correlations of Neogene lacustrine rocks based upon fossil diatoms are problematic because of the provincial nature of their geologic ranges. Nevertheless, similar global patterns of lacustrine diatom succession during the Neogene have been documented (Fig. 19.8) and can be useful for determining the relative geologic ages of lake deposits (Fourtanier, 1987; Fourtanier et al., 1993; Bradbury & Krebs, 1995).

Summary

The uses of diatoms in oil and gas exploration are varied. They are used as biostratigraphic markers for age dating and correlating marine and lacustrine rocks and as paleoenvironmental indicators. Diatoms are recovered from Tertiary marine and nonmarine sediments in cores, well cuttings, and outcrops, and are particularly useful in 'cold water' sediments that lack calcareous microfossils and in late Cenozoic brackish water and lacustrine rocks. They are rarely recovered in their unaltered state (opaline silica) at well depths greater than 1524 m because of the burial diagenesis of silica. They may occur at greater depths when preserved unaltered in calcareous concretions or when pyritized. Their destruction by burial diagensis can be observed in the microscope, and it coincides with changes in rock properties and seismic definition that have important implications for hydrocarbon exploration.

W. N. KREBS

References

Aurisano, R. W., Gamber, J. H., Lane, H. R., Loomis, E. C., & Stein, J. A. (1995). Worldwide and local composite standards: Optimizing biostratigraphic data. In *Graphic Correlation*, ed. K. O. Mann & H. R. Lane, pp. 117–30, Society of Economic Paleontologists and Mineralogists, Society for Sedimentary Geology, Special Publication no. 53.

Blome, C. D. & Albert, N. R., (1985). Carbonate concretions: An ideal sedimentary host for microfossils. *Geology*, **13**, 212–15.

Bradbury, J. P. & Krebs, W. N., (1982). Neogene and Quaternary lacustrine diatoms of the western Snake River Basin, Idaho-Oregon, U.S.A. *Acta Geologica Academiae Scientiarum Hungaricae*, **25**, 97–122.

(1995). Fossil continental diatoms: paleolimnology, evolution, and biochronology. In *Siliceous Microfossils*, ed. C. D. Blome, P. M. Whalen, & K. M. Reed (convenors). The Paleontological Society Short Courses in Paleontology, **8**, 119–38.

Bramlette, M. N. (1946). The Monterey Formation of California and the origins of its siliceous rocks. *US Geological Professional Paper* 212, 57p.

Carney, J. L. & Pierce, R. W., (1995). Graphic correlation and composite standard databases as tools for the exploration biostratigrapher. In *Graphic Correlation,* ed. K. O. Mann and H. R. Lane, pp. 23–44. Society of Economic Paleontologists and Mineralogists, Society for Sedimentary Geology, Special Publication no. 53.

Compton, J. S. (1991). Porosity reduction and burial history of siliceous rocks from the Monterey and Sisquoc Formations, Point Pedernales area, California. *Geological Society of America Bulletin*, **103**, 625–36.

Fourtanier, E. (1987). Diatomées néogènes d'Afrique; approche biostratigrahique en milieux marin (sud-ouest africain) et continental. Unpublished PhD dissertation, Ecole Normale Supérieure de Fontenay Saint-Cloud, Fontenay Saint-Cloud, France, 365 p.

Fourtanier, E., Bellier, O., Bonhomme, M. G., & Robles, I. (1993). Miocene non-marine diatoms from the western Cordilleran basins of northern Peru. *Diatom Research*, **8**, 13–30.

Grechin, V. I., Pisciotto, K. A., Mahoney, J. J., & Gordeeva, S. N. (1981). Neogene siliceous sediments and rocks off southern California, Deep Sea Drilling Project Leg 63. In *Initial Reports of the Deep Sea Drilling Project,* ed. R. J. Yeats, B. U. Haq, *et al.* vol. 63, pp. 579–93. Washington DC: US Govt. Printing Office.

Hein, J. R., Scholl, D. W., Barron, J. A., Jones, A. G., & Miller, J. (1978). Diagenesis of late Cenozoic diatomaceous deposits and formation of the bottom simulating reflector in the southern Bering Sea. *Sedimentology*, **25**, 155–81.

Iijima, A. & Tada, R. (1981). Silica diagenesis of Neogene diatomaceous and volcaniclastic sediments in northern Japan. *Sedimentology*, **28**, 185–200.

Ingle, J. C. (1981). Origin of Neogene diatomites around the North Pacific Rim. In *The Monterey Formation and Related Siliceous rocks of California,* ed. R. E. Garrison, R. G. Douglas, K. A. Pisciotto, C. M. Isaacs, & J. C. Ingle. pp. 159–79. Pacific Section, Society of Economic Paleontologists and Mineralogists, Special Publication.

Isaacs, C. M. (1981). Porosity reduction during diagenesis of the Monterey Formation, Santa Barbara coastal area, California. In *The Monterey Formation and Related Siliceous Rocks of California,* ed. R. E. Garrison, R. G. Douglas, K. A. Pisciotto, C. M. Issacs, & J. C. Ingle. pp. 257–71. Pacific Section, Society of Economic Paleontologists and Mineralogists, Special Publication.

(1982). Influence of rock compositions on kinetics of silica phase changes in the Monterey Formation, Santa Barbara area, California. *Geology*, **10**, 304–8.

(1983). Compositional variation and sequence in the Monterey Formation, Santa Barbara coastal area, California. In *Coastal Marine Sedimentation*. ed. D. K. Larue & R. J. Steel. pp. 117–32. Pacific Section, Society of Economic Paleontologists and Mineralogists, Special Publication.

Krebs, W. N., & Bradbury, J. P. (1995). Geologic ranges of lacustrine *Actinocyclus* species, western United States. *US Geological Survey Professional Paper*, **1543–B**, 53–73.

Krebs, W. N., Bradbury, J. P., & Theriot, E. (1987). Neogene and Quaternary lacustrine diatom biochronology, western USA. *Palaios*, **2**, 505–13.

Krebs, W. N., Wescott, W. A., Nummedal, D., Gaafar, I., Azazi, G., & Karamat, S. 1997. Graphic correlation and sequence stratigraphy of Neogene rocks in the Gulf of Suez. *Bulletin, Société Géologique de France,* **168**, (1), 63–71.

Lagle, C. W. (1984). Recovery of siliceous microfossils by the disaggregation of dolomite. In *Dolomites of the Monterey Formation and Other Organic-Rich Units*, R. E. Garrison, M. Kasten, & D. H. Zenger, pp. 185–94. Pacific Section, Society of Economic Paleontologists and Mineralogists, Special Publication no. 41.

Mertz, Jr., K. A. (1884). Origin of hemipelagic source rocks during the early and middle Miocene, Salinas Basin, California. *American Association of Petroleum Geologists, Bulletin,* **73** (4), 510–24.

Miller, F. X. (1977). The graphic correlation method in biostratigraphy. In *Concepts and Methods in Biostratigaphy,* ed. E. G. Kauffman & J. E. Hazel, pp. 165–86. Stroudsburg, PA: Dowden, Hutchison, and Ross.

Murata, K. J. & Larson, R. R. (1975). Diagenesis of Miocene siliceous shale, Temblor Range, California. *US Geological Survey Journal of Research,* **3**, 553–66.

Neal, J. E., Stein, J. A., & Gamber, J. H. (1995). Integration of the graphic correlation methodology in a sequence stratigraphic study: Examples from North Sea Paleogene sections. In *Graphic Correlation,* ed. K. O. Mann & H. R. Lane, pp. 95–116. Society of Economic Paleontologists and Mineralogists, Society for Sedimentary Geology, Special Publication no. 53.

Pisciotto, K. A. (1981). Diagenetic trends in the siliceous facies of the Monterey Shale in the Santa Maria region, California. *Sedimentology*, **28**, 547–71.

Schwartz, D. E. (1987). Lithology, petrophysics, and hydrocarbons in cyclic Belridge Diatomite, south Belridge oil field, Kern Co., California. In *Fourth International Congress on Pacific Neogene Stratigraphy*, ed. J. A. Barron and J. R. Blueford, abstract volume: 102. Berkeley, CA., July 29–31.

Shaw, A. B. (1964). *Time in Stratigraphy.* New York, NY: McGraw-Hill. 365 p.

Swirydczuk, K., Larson, G. P., & Smith, G. R. (1981). Volcanic ash stratigraphy of the Glenns Ferry and Chalk Hills Formations, western Snake River Plain, Idaho. *Idaho Bureau of Mines and Geology Open-File Report,* 81–1, 60 p.

Tada, R. & Iijima, A. (1983). Petrology and diagenetic changes of Neogene siliceous rocks in northern Japan. *Journal of Sedimentary Petrology,* **53**, 911–30.

20 Forensic science and diatoms

ANTHONY J. PEABODY

Introduction

The legal process has used scientific procedures for many years in its various deliberations. Some of these, for instance the DNA profiling of body fluids, are now essential and routine practice. The use of diatoms in forensic science is naturally much smaller, but in certain types of investigation, diatom taxonomy and ecology play a significant role. The diatomologist may be able to provide the investigation with evidence, which will enable the court to reach its verdict, and may be used by either the prosecution or the defence.

Drowning

The most frequent application of diatoms in forensic science is in the diagnosis of death by drowning. Drowning is a very common accidental cause of death, and thousands die each year in this fashion. The majority of these individuals die in circumstances which are not contentious, where there are witnesses, or strong indications of suicide such as a note. Where the circumstances surrounding the individual's death are less clear, then it is often important to be as certain as one can be of how death occurred.

Where the body is fresh, the pathologist may have little difficulty in reaching a verdict of drowning. However, the histopathological signs of drowning are often transient and overlaid by the grosser effects of decomposition. Additionally, in cases where an individual has been severely injured before being immersed in water, it is obviously important to determine whether death is due to these injuries or because of drowning. It is in cases such as these, where the outcome of the investigation into drowning may add years to a person's prison sentence, that the forensic scientist may be asked to assist the pathologist in reaching a decision.

The mechanisms of the drowning process have been described by Timperman (1972), and an overview of the problems associated with diatoms and drowning has been made by Peabody (1980). Some authors have referred to 'dry' drowning, which is more accurately described as death by a variety of causes whilst immersed in water, for instance, vagal inhibition, cardiac arrest or laryngeal spasm. These causes of death are not considered 'true' drowning, because death is not the result of water entering the lungs. It is in cases of drowning where water enters the lungs, that the forensic scientist can make a contribution.

When an individual drowns, water enters the lungs in increasing amounts and enters the bloodstream through ruptures in the peripheral alveoli. Any particulate material in the water is also carried in, and whilst the heart continues to pump, this material is carried through the body. There is probably some deposition of this particulate material in the capillary beds of the major organs. In most bodies of natural water, this particulate material will include diatoms, and the numbers present and their types will be a reflection of the ecology of that particular area.

Much of the pioneer work in this area was carried out in wartime Hungary, and was reported first by Incze (1942), and later by his co-workers. This early important work is fully described in Peabody (1980).

In practice, most drowning cases considered are from a freshwater location, although from time to time saltwater drownings are submitted. In most instances, the question to be answered is 'did this person drown?', rather than 'where did this person drown?'. In answering the first question, it is usually straightforward to answer the second, in the sense that any diatoms recovered from the body can be compared with diatoms from the drowning scene.

> Case. The body of an unidentified woman was found at the mouth of an
> estuary. Diatoms recovered from the body indicated that she had drowned in
> freshwater rather than in the sea or in brackish water. When her identity was
> established, her home was found to be some 60 km upstream from where she
> was found. Enquiries suggested that she had drowned near her home.

It is also common for drowning to occur in a domestic environment, particularly in the bath, or where the body of a newborn infant is found in the water of a lavatory pan following delivery. Domestic water supplies generally contain few diatoms if any (at least in the UK), and consequently the forensic scientist is unable to assist in these cases.

The essential problem in sample preparation is to recover a small number of diatoms from a large organic mass. As few as five or six diatoms might be involved, and in addition, there is the danger of contamination during the extraction procedure. To compound this, many human organs normally contain diatoms, which entered the body during life. The brain, liver and lung of 13 non-drowned subjects contained between 0–19 diatoms in 100 g samples. Most were to be found in the lung. On examination of kidney from the same subjects, much larger numbers of diatoms were recovered, up to 150 diatoms per 100 g. Common diatoms were fragments of *Coscinodiscus*, *Hantzschia amphioxys*, and *Aulacoseira granulata*. The presence of these diatoms has misled many investigators unfamiliar with diatoms, and has prompted them to suggest that this line of enquiry is unproductive. The successful investigator must be able to recognize these 'naturally occurring' diatoms for what they are, and to record those diatoms which may be present as a result of drowning.

A. J. PEABODY

Few diatomists have been concerned with the problem of diagnosing death by drowning, and some of those who have, for instance, Geissler and Gerloff (1966) and Foged (1983), are of the opinion that the diatom test is invalid. Foged found quite large numbers of diatoms in his examination of organs from non-drowned subjects. It is impossible to tell whether this might be due to contamination, but at the very least should encourage care and thought in the use of the diatom test

It is evident therefore, that this type of investigation should be made by an experienced diatomist working in conjunction with the pathologist. The diatomist assists the pathologist to make his diagnosis. The diatomist should be careful not to make a diagnosis from his or her own findings, since this will certainly impinge on the pathologist's area of expertise. It is the pathologist's concern to diagnose the cause of death.

The results of the examination are usually expressed in a form convenient to the pathologist, but bearing in mind that one's own opinions must be free standing and capable of being defended:

> ... have examined the organs from John Doe and I have found diatoms in them which could have originated from River Running at Anytown. It is my opinion therefore, that drowning should be considered as a contributory cause of death.

or:

> ... have examined the organs from John Doe and I have found no diatoms in them which could have come from River Running at Anytown. I am therefore unable to indicate whether or not John Doe has drowned.

Whilst it is possible to draw a conclusion from the presence of diatoms, the absence of diatoms cannot be taken to indicate that drowning *has not* occurred. Drowning could have occurred, but either no diatoms were infiltrated, or their numbers were so low that they were not detected.

SAMPLE PREPARATION

Many methods have been proposed for the extraction of diatoms from organs, (Peabody, 1980), and most have their drawbacks, including those used today. A method which minimizes contamination risks has been developed using 100 g samples of liver, lung and brain.

The lung is included as an indication of how many diatoms have actually reached the circulatory system. The absence of diatoms from the lung, or their presence in only low numbers will suggest that they will not be present elsewhere in the body. A large number of diatoms in the lung will indicate that water probably has reached the circulatory system, and that drowning may be confirmed by the examination of the liver and brain. An examination of lung exudate will give a preliminary indication of how many diatoms are present. The type of diatoms present is also important in that it gives an early indication as to the site of drowning.

The liver is examined because it is a well-vasculated major body organ. The brain is examined because it is remote from the lungs and preserved to some extent from damage and the exterior by the cranium.

EXTRACTION METHOD

It is essential that all laboratory ware is new, unused and cleaned before use. Likewise, all reagents must be passed through a membrane filter before use, including concentrated acid.

(i) Weigh out roughly 100 g samples of liver, lung and brain into evaporating dishes, taking account of any possible health hazards in dealing with body organs which have an unknown history. Stringent safety precautions are advised at all stages of this extraction.

(ii) Transfer the dishes to a muffle furnace and char at 250 °C for 3–5 hours.

(iii) Ash these samples overnight at 550 °C. Allow to cool. Operating below this temperature results in incomplete ashing, whilst higher temperatures result in some destruction of any diatoms present

(iv) The resultant white ash can be dissolved in conc. HCl, followed by washing in distilled water.

(v) If any organic material is evident at this point, through incomplete ashing, a standard diatom preparation using conc. H_2SO_4 and $NaNO_3$ will suffice.

(vi) Mount the washed residue in a high refractive index mountant, and examine microscopically.

Whilst not perfect, this method ensures that contamination risks are minimized, and it copes with large amounts of organic material. Parts of the process require constant attention, in particular dissolving the ash in conc. HCl. Once the samples are in the furnace, toxicological and pathological health risks are minimized, but the resultant smoke can be offensive and a possible health hazard unless ducted well away.

An enzymatic digestion method has been suggested by Ludes and Coste (1996) based on a toxicological method originally used for the extraction of drug metabolites.

Diatoms on wet clothing

When individuals run or walk through a garden pond or a stream, or indeed any body of water, particulate material in the water (including diatoms) can remain on clothing afterwards. Diatoms can be recovered from the clothing and compared with diatoms from the water under consideration. It is not unknown for other fairly resistant algal material, such as *Oedogonium* and *Phacus*, to be recovered from clothing after immersion in water containing these organisms. Diatoms however, are especially resistant and are not usually visible to the naked eye, and are therefore less likely to be removed by an alleged offender.

Case. A man gained access to a house through the garden. He stepped into a
garden pond which made his jeans wet to the knee. On arrest shortly after the
offence, he maintained that his jeans were wet because he had been in the sea.
Large numbers of freshwater diatoms of the same types as those in the garden
pond were found on his jeans. There were no marine diatoms. He received a 9
year prison sentence.

Diatoms can be removed simply by rinsing either the entire or more specifically, the affected area in water. It is useful to examine any dried material on the
garment for the presence of any other algal material. Allow the residue to settle
and examine microscopically.

Case. A man denied being the driver of a car which was found submerged in a
farm pond. He denied any connection with the pond. Another man was
found dead in the car, having failed to escape. Despite his denials, diatoms
from the farm pond were found on the clothing of the putative driver, and he
received a five year prison sentence.

The absence of diatoms from clothing, which is thought to have been through
water containing diatoms, could be significant in indicating that the wearer
has not in fact been in the water. The significance may depend on the type of
water under consideration, the numbers of diatoms in the water, and the type
of garment worn. The absence of diatoms from oilskins briefly immersed in
the sea may not be significant, whereas failure to find diatoms on trousers after
alleged immersion in a farm drainage ditch is noteworthy.

Safe ballasts

In the past, manufacturers of safes have used a variety of materials as insulation, fire proofing or protection. Some old safes have diatomaceous earth
packed between the outer casing and the inner compartment, but since this
material provides no extra security apart from being fire resistant,
manufacturers have stopped using it. However, old safes are likely to be
attacked, and should the investigators recognize the safe for what it is, valuable
evidence can be adduced (Peabody, 1971).

When such a safe is broken open, the diatomaceous earth spills out as a fine
white powder, which will settle on clothing, skin and other surfaces. It is often
invisible to the naked eye, but a small sample from a suspect surface can reveal
hundreds or thousands of diatom cells. Since these ballasts are often taken
from diatomite workings long since defunct, the chances of an individual
having this material on clothing by chance are extremely small.

Case. A safe was stolen from an office and taken to a house where it was
opened. It contained diatomaceous safe ballast, which spilled out. The safe
was then taken in a car and dumped. Three suspects had diatomaceous
material on their clothing matching the safe. The same was found in the
house of one of the suspects, and in the car of another. All three received
prison sentences.

Cases involving diatomaceous safe ballasts provide strong evidence of association relatively easily and quickly. However, diatomaceous safe ballasts are now uncommon, and are rarely submitted.

Other applications

Diatomaceous earth is used in a variety of materials which are often examined by forensic scientists, such as paint, polishes and matches. The current supplies of diatomaceous earth are restricted to a few mining sites, and consequently the same material can be used in a large number of items.

The examination of car and metal polish however, can often provide good evidence of association (Peabody, 1977).

The examination of diatoms in small flakes of white paint can sometimes provide additional evidence, but the titanium dioxide in paint must be dissolved first so that the diatoms can be seen. TiO_2 can be dissolved by boiling in conc. H_2SO_4 saturated with Na_2SO_4. Great care should be taken with this procedure.

Giving evidence

The forensic scientist must be able to communicate often complicated scientific issues to laymen who may not have any scientific background, and for whom evidence of this type is difficult to understand. This is particularly true of diatom evidence, which will almost certainly be unfamiliar to most of the individuals to whom it is presented, and who also need to understand it.

References

Foged, N. (1983). Diatoms and drowning – Once more, *Forensic Science International*, **21**, 153–9.

Geissler, U. & Gerloff, J. (1966), Das Vorkommen van Diatomeen in menschlichen Organen und in der Luft, *Nova Hedwigia*, **10**, 565–77.

Incze, G. (1942). Fremdkorper in Blutkreislauf Ertrunkener, *Zentralbl. Allg. Pathol. Palhol. Anal.*, **79**, 176.

Ludes, B. Coste, M. (1996). *Diatomées et médecine légale*. Paris: Tec & Doc Lavoisier.

Peabody, A. J. (1971), A case of safebreaking involving diatomaceous safe-ballast. *Journal of the Forensic Science Society*, **11**, 227.

(1977). Diatoms in forensic science. *Journal of the Forensic Science Society*, **17**, 81–7.

(1980). Diatoms and drowning – A review. *Medicine, Science and the Law*, **20**, 254–61.

Timperman, J. (1972). The diagnosis of drowning, – A review. *Forensic Science*, **1**, 397–409.

21 Toxic and harmful marine diatoms

GRETA A. FRYXELL AND MARIA C. VILLAC

Introduction

As White and Frady (1995) say in the Preface of their recent international directory on experts in toxic and harmful algae, 'Toxic and harmful algal blooms
present a growing global problem for fisheries, aquaculture, and public
health.' With entries from 58 countries, they list 22 countries with names and
addresses of people working with harmful diatoms and/or their toxins:
Australia (3), Canada (29), Chile (1), Croatia (1), Denmark (4), Germany (3), India
(3), Israel (1), Japan (6), Netherlands (3), New Zealand (3), Norway (4), People's
Republic of China (17), Republic of Korea (2), Romania (1), Russian Federation
(1), Spain (6), Thailand (1), Turkey (1), United Kingdom (2), United States of
America (29), and Vietnam (1), for a total of 122 workers around the world. One
such list of international specialists was compiled by Woods Hole
Oceanographic Institution Sea Grant Program in 1990 (subsequently up-
dated), so that the consequences of outbreaks of toxic and harmful algal bloom
events on fisheries and public health could be reduced. The purpose of this
summary chapter is to serve as a resource for those faced with the challenges
brought about by the changing dominant coastal diatom flora. In the study of
toxic and harmful diatoms, there are applications for fisheries, public health
institutions, mariculture, and tourism.

Harmful blooms

'A first step in applied ecology is to accurately define the present state of the
environment', according to Rowe (1996, p. 7), and progress in this endeavor has
been made. Recently, Hasle & Fryxell (1995) summarized morphological characters of toxic and harmful diatoms, and their information on the nature of
harmful events is summarized in Table 21.1, with some recent observations
added. Harmful exudates (oil, mucilage) have been known for the last half of
this century. More diatom species produce mucilage when in great abundance
than are listed in the table, and the dense ribbons of mucilage aggregates
dominated by *Pseudo-nitzschia pseudodelicatissima* (Hasle) Hasle in 1988–1989
and again in 1991 were frequently in the news from the shores of the Adriatic

Table 21.1. *Toxic and harmful diatom blooms (from Hasle & Fryxell, 1995, others below)*

Type of harm	Exudates[a]	Mechanical injury[b]	Toxin production[c]
Species	Coscinodiscus centralis C. concinnus C. wailesii Thalassiosira mala Pseudo-nitzschia delicatissima	Chaetoceros convolutus C. concavicornis Coscinodiscus wailesii Leptocylindrus spp. Tabularia affinis	Pseudo-nitzschia australis P. delicatissima P. multiseries P. pseudodelicatissima P. pungens P. seriata P. turgidula Rhizosolenia chunii
Effect	*oil:* • film on the sea surface *mucilage:* • trawls clogged/broken • clogging of gills • dense sheets	• clogging the gills • penetrating gill tissue by spines, bleeding • hypoxia • hypercapnia • shading • depletion of nutrients	*domoic acid:* neurotoxin (ASP) especially in guts, some in meat *taste:* • unmarketable mussels • mortality of mussels (long range)
Organisms affected	• water-logging of birds • mortality of birds, bivalves • anoxia • entangled on rough surfaces or organisms	• mortality of farmed finfish salmon (*Salmo salar*) lingcod (*Ophiodon elongatus*) • deterioration of *Porphyra* culture	*mortality:* • humans, birds (also rodents and monkeys) *vectors:* • mussels, scallops, razor clams, oysters, Dungeness crabs, anchovies
Locations	*oil:* North Sea *mucilage:* • Tokyo Bay, Japan • English Channel (Plymouth) • Adriatic and Tyrrhenian Seas	• northwestern coast of N. America (Washington/British Columbia) • Chilean waters • Seto Inland Sea, Japan	*ASP:* • Prince Edward Island, Canada • West Coast, USA *bitter taste:* Port Phillip Bay, Australia *target diatom spp.:* widespread

Sources: [a] Tåning (1951), Grøntved (1952), Takano (1956, 1965), Boalch & Harbour (1977), Boalch (1984), Manabe & Ishio (1991), Rinaldi et al. (1991), Nagai et al. (1994), Nagai et al. (1995), Nagai et al. (1996a).
[b] Bell (1961), Tangen (1987), Horner et al. (1990), Clement & Lembeye (1993), Rensel (1993), Taylor (1993), Nagai et al. (1996b).
[c] Parry et al. (1989), Bates (1993), Hasle (1993), Horner et al. (1997), Rhodes et al. (1996), Trainer et al. (1997), Villac et al. (1993).

and Tyrrhenian Seas (Rinaldi *et al.*, 1991). The possibility of mechanical injury to grazers or fish gills by sharply pointed spines, all oriented in one direction, has been recognized by microscopists for most of the last century and their effects on caged fish studied in the last decade (Rensel, 1993). But the knowledge that diatoms could be involved in production of a neurotoxin has only recently been proved. Although this aspect came to light during a tragic event that resulted in a crisis in a Canadian fishery, the long-term, low intoxication effects of domoic acid are also of great concern (Wood & Shapiro, 1993; Wood *et al.*, 1994). *Pseudo-nitzschia multiseries* (Hasle) Hasle was the first known to produce domoic acid (mostly during its stationary growth phase), although the larger-celled P. *australis* Frenguelli may prove to be even more dangerous (with apparent continuous domoic acid production in the field, Walz *et al.*, 1994), and three smaller-celled species also are implicated. Recently, two species have been added to the list of domoic acid producers in New Zealand, *Pseudo-nitzschia turgidula* (Hustedt) Hasle and the common P. *pungens* (Grunow) Hasle (Rhodes *et al.*, 1996; Horner *et al.*, 1997; Trainer *et al.*, 1997; Table 21.1). Microscopic identification is essential (Hasle & Fryxell, 1995), coupled with such promising recent applications as fluorescent tags (Scholin *et al.*, 1994, 1996).

Probably many high concentrations of diatoms of low specific diversity can be harmful, and the study of blooms has many applications. Even *Skeletonema costatum* (Greville) Cleve, the common inshore, temperate/boreal diatom, was associated with fish kills in fish farms during the spring bloom in north Atlantic Skagerrak/Kattegat area in 1990 (Tangen, 1991). Species in another common genus, *Leptocylindrus minimus* Gran and *L. danicus* Cleve are also believed to be harmful to sea trout (*Cynoscion regalis* (Block *et* Schneider)), Atlantic salmon (*Salmo salar* Linnaeus), and occasionally smolt of coho salmon (*Oncothynchus kisutch* (Walbaum)) in cages (Clément & Lembeye, 1993; A. Clément, personal communication). A summary listing of the diatoms reported as harmful, at least in blooms, is included in Table 21.2.

Just what is a 'bloom'? It often means high cell numbers with low species diversity in one parcel of water, but the definition of bloom used in each study must be explicitly stated if the reported observations are to provide predictive capability. Anderson (1996, p. 1) has an applied definition of harmful algal blooms as 'events where the concentration of one or several harmful algae reach levels which can cause harm to other organisms in the sea . . . or cause accumulation of algal toxins in marine organisms which eventually harm other organisms who will eat the toxic species . . .'. Qualitatively, older literature refers to a discoloration of surface water, while some recent literature has been using the term to indicate seasonal increase. Quantitatively, bloom can refer to subsurface lenses of high density of diatoms as well as to those visible at the surface. It makes a difference if the density of diatoms is measured in cell numbers, biomass calculated from chlorophyll, or cell volume, with these different measures resulting from differing time/space sampling schemes. Especially when swimming organisms, such as anchovies (*Engraulis mordax*

Table 21.2. *Diatom species considered toxic or harmful*

Cerataulina pelagica (Cleve) Hendey
Chaetoceros concavicornis Mangin
C. convolutus Castracane
C. socialis Lauder

Coscinodiscus centralis Ehrenberg
C. concinnus W. Smith
C. wailesii Gran *et* Angst

Leptocylindrus danicus Cleve
L. minimus Gran

Pseudo-nitzschia australis Frenguelli
P. delicatissima (Cleve) Heiden
P. multiseries (Hasle) Hasle
P. pseudodelicatissima (Hasle) Hasle
P. pungens (Grunow) Hasle
P. seriata (Cleve) H. Peragallo
P. turgidula (Hustedt) Hasle

Rhizosolenia chunii Karsten

Skeletonema costatum (Greville) Cleve

Tabellaria affinis (Kützing) Snoeijs

Thalassiosira mala Takano

Girard), are the filter-feeding vectors involved in concentration of harmful diatoms, it is now apparent that relative abundance of one size fraction of phytoplankton over the feeding area is a factor germane to the problem and more to the point than absolute abundance in any one parcel of water (Buck *et al.*, 1992).

In a bloom, as defined by cell numbers in one location, the biological oxygen demand is extremely high at night, as happens also after the bloom crashes and cells are deteriorating. Deaths of benthic shellfish during the bloom of *Cerataulina pelagica* (Cleve) Hendey off the northeast coast of New Zealand were attributed to anoxia induced by the bacterial decay of the diatom cells (Taylor *et al.*, 1985). The night-time catches of fish were extremely low during that extensive summer bloom, and bony fish died with gills clogged with their own mucilage and that of the diatoms, especially in the middle portions of long fishing lines that had sunk to the bottom.

Contributing factors

A second step in applied ecology is to think through hypotheses to separate correlation from contributing factors. Most of the harmful blooms discussed here have been in the spring or autumn, have no known benthic stage, and are dominated by species that are known or even relatively common in the local flora. Many bloom-forming species are found throughout the water column, although they may be concentrated near the surface. Rarely have data been available for events leading up to the bloom or those following the bloom, such as the deaths of mussels (*Mitilus edulis* Linnaeus) in Australia several months after they were noted to have a bitter taste (Parry *et al.*, 1989). Nor have data generally been obtained to ascertain long-term, low-dosage effects upon the phytoplankton community, the grazers, or the human consumers. Field sampling to obtain these data will be most effective if hypotheses are sharply focused by study of laboratory experimental results.

Many dinoflagellate and cyanobacteria blooms result from factors summarized by Paerl (1988) from freshwater and estuarine habitats: allochthonous enrichment, high irradiance, horizontal and vertical stability, and biotic factors. Diatoms may repond better to a well-mixed water column. Intrusions from airborne dust particles and some photo-induced iron forms can lead to patchy blooms in diverse habitats (Saydem, 1996). Mineral dust and sediments may serve as nutrient sources, but few of the species listed in Table 21.2 are known to produce true resting spores that settle into sediments, although at least one produces an annual sedimentary resting stage, *Coscinodiscus wailesii* Gran *et* Angst (Nagai *et al.*, 1996a). However, *C. wailesii* Gran *et* Angst came from a well-mixed water column in the spring (Boalch & Harbour, 1977). *Pseudo-nitzschia* spp. came from well-mixed water columns in the Canadian estuaries of a Prince Edward Island in the autumn (Smith *et al.*, 1990) or associated with weak upwelling events in spring, summer, and autumn, but most frequently in the autumn in Monterey Bay, California (Buck *et al.*, 1992). The Australian *Rhizosolenia chunii* Karsten bloom started when the spring stratification commenced (Parry *et al.*, 1989). Although many factors play roles in genesis of blooms, the onset of a bloom may be preceded by seeding of bloom species from lower in the water column.

Salt wedge zones of estuarine areas, nepheloid layers of continental shelves, and the nutricline in deeper water are alike in that they can serve as a residence of relict blooms and a source for seeding future phytoplankton blooms. Seeding is not limited to presence of living cells; there must be a sufficient quantity for a reliable inoculum, as the authors who have worked with maintenance of cultures can attest. Different life stages (e.g., resting spores) and slow growth phases sink to layers of different physicochemical conditions on seasonal, lunar, or even diurnal time scales, with speed of sinking sometimes increased by aggregates or floc (Kranck & Milligan, 1988; Jackson, 1990). Smetacek (1985) argued that rapid sinking of bloom diatoms out of the surface layers does not equate to their mass mortality, although some cell death certainly takes place as cells sink out of the light in the euphotic zone. Sinking is represented by Smetacek as a mechanism that can provide accumulation of an inoculum and provide a seeding strategy. The argument here is that the strategy may not be limited to seasonal time scales. Accumulation of bloom proportions can take place on shorter time scales in response to physico-chemical factors enhancing natural growth cycles, such as tidal or meteorological events.

An example of a different time/space scale takes place in small culture chambers of microtest plates (*ca.* 5 ml), where short-term cycles of a coastal diatom take place. It has been observed that planktonic *Pseudo-nitzschia* chains make long, spiraling ropes during the log-growth phase (Fryxell *et al.*, 1990; Reap & Fryxell, 1993). About 4–7 days later, when growth slows, the chains break up into singlets or doublets, which sink to the bottom of the chamber overnight. Introduction of more nutrients causes repetition of the cycle (M. E. Reap, personal communication). No resting spores have been found in *Pseudo-nitzschia*, but sequential growth phases are found at different water levels in small chambers. It may not be necessary for living cells to go into the sediments for replenishment of nutrients. Buck *et al.* (1992) found that *P. australis* was an important contributor to the downward flux of material toward the ocean floor, and Stabell and Hasle (1996) noted with special interest that the *P. pungens/P. multiseries* complex was found at one level in a core from the north-eastern Kattegat between Denmark and Sweden.

Extensive sampling across the Texas–Louisiana shelf has shown high concentration of chlorophyll *a* on the inner shelf, where high performance liquid chromatography traces showed diatoms to be the dominant class (Neuhard, 1994). On the middle shelf, near-bottom waters showed a distinct chlorophyll maximum, with fucoxanthin following the chlorophyll *a* contours to high turbidity waters within a few meters of the sediment, in what is termed the nepheloid layer. Diatoms are there, and the species are not limited to benthic species (Bontempi, 1995).

Biological paradigms

A third step in applied ecology is to provide predictive tools to provide, in this case, for warning systems of future harmful events and explain and predict toxicity in food chains. Such a warning system requires basic biological knowledge of the species and food chains involved, as well as abiotic factors. Feedback from a paradigm to the first two steps is essential.

In one paradigm, if cells sink to the base of the low-nutrient water close to the higher nutrient levels of the pycnocline, they could be maintained in large numbers. Using a transmissometer in a salinity front, Gould and Wiesenburg (1990) proved the point by finding and directly sampling a narrow, dense band of low light transmission at about 50 m depth in the Mediterranean Sea. It was formed by a lens-shaped population of gelatinous colonies of one species of *Thalassiosira* in bloom abundance in low light at the pycnocline, with a slow division rate of >0.20 doublings per day. Not only was this dense seed population concentrated and apparently growing slowly, but it had been stable long enough to have accumulated a few cells of a species of *Nitzschia* on its connecting chitin threads, in a frequently noted association. In the Pacific such a concentration, and in this case probably a flux, could explain how guts of the suspension-feeding midwater polychaete, *Poeobius meseres* Heath, could have high relative abundances of the planktonic *P. australis* seasonally (Buck *et al.*, 1992). Inshore on the US west coast, it has seemed there was a missing link in the known food chain between the toxin producers and domoic acid found in razor clams (*Siliqua patula* Dixon), since *Pseudo-nitzschia* spp. are not recognized as surf diatoms (references in Fryxell *et al.*, 1997). Surf diatoms have a sticky cell surface secretion, and flux from a coastal bloom of *Pseudo-nitzschia* sp. could stick to the surf diatoms when washed ashore and then be fed upon by razor clams (Wekell *et al.*, 1994). These examples reinforce the need to study the vertical displacement and associations in bloom formation.

Current thoughts on the onset and demise of blooms are summarized in Jumars (1993, pp. 166–171), who suggested diverse possibilities, including coagulation of aggregates and feedback mechanisms. He pointed out that diatoms with a rapid reproduction strategy (r-selected) are the bloom species. Responses are species-specific (Riebesell, 1989). Our contention is that one species can take on r-selected characters at one life stage or even one growth phase and the more complex K-selected characters at another, in a continuum similar to that found in freshwater and marine phytoplankton communities (Kilham & Kilham, 1980; Kilham & Hecky, 1988). Sinking and floating growth cycles can be found in cultures, estuaries, shelf waters, and the open ocean. It is submitted here that systematic sampling of target areas (with sampling strategy including the measurement of light transmission) of the near-bottom and nepheloid layers will reveal seed populations of dominant phytoplankton and have applications for harmful, bloom-forming species.

Summary

Many diatoms can be harmful, especially in a 'bloom', a term that must be clearly defined in each study. Toxic diatoms may have harmed some grazers, but more commonly harm has been noted to the predator that ate the grazer. Long-term injury and death have been documented in some vertebrates and invertebrates from ingestion of large numbers of some species. Mechanical injury has been caused by spines of harmful diatoms, especially to the gills of fish. Oil has accumulated from some blooms, resulting in anoxic conditions. In other cases, mucous has clogged gills. The results of each bloom have been species-specific, and identification of species remains a high priority. Bloom genesis is related to the response of the diatom (sinking, floating) to varying environmental conditions, and applied ecology can result in predictive tools.

Acknowledgments

We thank Grethe R. Hasle for many substantive contributions and Jeffrey R. Johansen and the editors of this volume for their helpful reviews. We express appreciation to Esther A. Albrecht for her translation of Tåning (1951) and to the Department of Oceanography, Texas A&M University, College Station, Texas, and also the University of Texas Marine Science Institute, Port Aransas, Texas, where some of this work was done.

References

Anderson, P. (1996). *Design and Implementation of Some Harmful Algal Monitoring Systems.* Intergovernmental Oceanographic Commission Technical Series 44, ii, 1–102, Annex VIII. UNESCO, Paris.

Bates, S. S. (1993). Domoic acid and *Pseudo-nitzschia* references. In *Domoic Acid*, Final Report of the Workshop, ed. A. M. Wood & L. M. Shapiro, pp. 13–21. ORESU-W-92–003. Oregon Sea Grant, Oregon State University.

Bell, G. R. (1961). Penetration of spines from a marine diatom into the gill tissue of lingcod (*Ophiodon elongatus*). *Nature*, **192**, 279–80.

Boalch, G. T. (1984). Algal blooms and their effects on fishing in the English Channel. *Hydrobiologia*, **116/117**, 449–52.

Boalch, G. T., & Harbour, D. S. (1977). Unusual diatom off the coast of south-west England and its effect on fishing. *Nature*, **269**, 687–8.

Bontempi, P. (1995). *Phytoplankton Distributions and Species Composition Across the Texas-Louisiana Continental Shelf During Two Flow Regimes of the Mississippi River.* MS Thesis, Texas A&M University, College Station TX. 261 pp.

Buck, K. R., Uttal-Cooke, L., Pilskaln, C. H., Roelke, D. L., Villac, M. C., Fryxell, G. A., Cifuentes, L. A., & Chavez, F. P. (1992). Autecology of the diatom *Pseudo-nitzschia australis*, a domoic acid producer, from Monterey Bay, California. *Marine Ecology Progress Series*, **84**, 293–302.

Clément, A., & Lembeye, G. (1993). Phytoplankton monitoring program in the fish farming of south Chile. In *Toxic Phytoplankton Blooms in the Sea,* ed. T. J. Smayda & Y. Shimizu, pp. 223–8. New York: Elsevier.

Fryxell, G. A., Reap, M. E., & Valencic, D. L. (1990). *Nitzschia pungens* Grunow f. *multiseries* Hasle: Observations of a known neurotoxic diatom. *Nova Hedwigia, Beih.*, **100**, 171–88.

Fryxell, G. A., Villac, M. C., & Shapiro, L. P. (1997). The occurrence of the toxic diatom genus *Pseudo-nitzschia* (Bacillariophyceae) on the West Coast of the U. S. A., 1920–1996: A review. *Phycologia*, **36**, 419–37.

Gould, R. W., Jr., & Wiesenburg, D. A. (1990). Single-species dominance in a subsurface phytoplankton concentration at a Mediterranean Sea front. *Limnology and Oceanography,* **35**, 211–20.

Grøntved, J. (1952). Investigations on the phytoplankton in the southern North Sea in May 1947. *Meddelelser fra Kommissionen for Danmarks Fiskeri- og Havundersøgelser.* Serie Plankton, **5**, 1–49.

Hasle, G. R. (1993). Nomenclatural notes on marine planktonic diatoms. The family Bacillariaceae. *Nova Hedwigia, Beih.*, **106**, 315–21.

Hasle, G. R., & Fryxell, G. A. (1995). Taxonomy of diatoms. *Manual On Harmful Marine Microalgae*, ed. G. M. Hallegraeff, D. M. Anderson, & A. D. Cembella. Intergovernmental Oceanographic Commisssion, Manuals and Guides, Chap. 17, **33**, 339–64.

Horner, R. A., Postel, J. R., & Rensel, J. E. (1990). Noxious phytoplankton blooms in western Washington waters. A review. In *Toxic Marine Phytoplankton*, ed. E. Granéli, B. Sundström, L. Edler & D. M. Anderson, pp. 171–6. New York: Elsevier.

Horner, R. A., Trainer, V. L., Postel, J. R., Ross, A. M., & Wekell, J. C. (1997). *Pseudo-nitzschia* species and domoic acid levels off the Washington coast, July 1996. Annual Meeting of the Society for Conservation Biology, Victoria, B. C. June 1997. *Abstracts*, 124.

Jackson, G. A. (1990). A model of the formation of marine algal flocs by physical coagulation processes. *Deep-Sea Research*, **37**, 1197–211.

Jumars, P. A. (1993). *Concepts in Biological Oceanography: An Interdisciplinary Primer.* New York: Oxford University Press. 348 pp.

Kilham, P., & Hecky, R. E. (1988). Comparative ecology of marine and freshwater phytoplankton. *Limnology and Oceanography*, 33(4, part 2), 776–95.

Kilham, P., & Kilham, S. S. (1980). The evolutionary ecology of phytoplankton. In *The Physiological Ecology of Phytoplankton*, ed. I. Morris, pp. 571–97. Univ. California.

Kranck, K., & Milligan, T. G. (1988). Macroflocs from diatoms: in situ photography of particles in Bedford Basin, Nova Scotia. *Marine Ecology Progress Series*, 44, 183–9.

Manabe, T., & Ishio, S. (1991). Bloom of *Coscinodiscus wailesii* and D O deficit of bloom water in Seto Inland Sea. *Marine Pollution Bulletin*, 23, 181–4.

Nagai, S., Hori, Y., Miyahara, K., Manabe, T., & Imai, I. (1994). Morphology and rejuvenation of *Coscinodiscus wailesii* Gran (Bacillariophyceae): Resting cells found in bottom sediments of Harima-Nada, Seto Inland Sea, Japan. (In Japanese, English abstract) *Nippon Suisan Gakkaishi*, 61, 179–85.

Nagai, S., Hori, Y., Miyahara, K., Manabe, T., & Imai, I. (1995). Restoration of cell size by vegetative cell enlargement in *Coscinodiscus wailesii* (Bacillariophyceae). *Phycologia* 34, 533–5.

Nagai, S., Hori, Y., Miyahara, K., Manabe, T., & Imai, I. (1996a). Population dynamics of *Coscinodiscus wailesii* Gran (Bacillariophyceae) in Harima-Nada, Seto Inland Sea, Japan. In *Harmful and Toxic Algal Blooms: Proceedings of the 7th International Conference on Toxic Phytoplankton*, ed. T. Yashumoto, Y. Oshima, & Y. Tukuyo, pp. 239–42. UNESCO, IOC, 1996.

Nagai, S., Takase, H., & Masuda, K. (1996b). The mass occurrence of the epiphytic diatom *Tabularia affinis* on Nori (*Porphyra*) in culture grounds of Hyogo Prefecture during the winter of 1995. *Hyogo Suishi Kenpo*, 33, 19–26.

Neuhard, C. A. (1994). Phytoplankton distributions across the Texas–Louisiana shelf in relation to coastal physical processes. Texas A&M University, MSc Thesis. pp. 205.

Paerl, H. W. (1988). Nuisance phytoplankton blooms in coastal, estuarine, and inland waters. *Limnology and Oceanography*, 33 (4, part 2), 823–47.

Parry, G. D., Langdon, J. S., & Huisman, J. M. (1989). Toxic effects of a bloom of the diatom *Rhizosolenia chunii* on shellfish in Port Phillip Bay, Southern Australia. *Marine Biology*, 102, 25–41.

Reap, M. E., & Fryxell, G. A. (1993). Life history and growth requirements of the domoic acid producer *Nitzschia pungens* f. *multiseries* from Galveston Bay. In *The Coastal Society Twelfth International Conference*, ed. W. M. Wise, pp. 843–51. Gloucester: The Coastal Society.

Rensel, J. E. (1993). Severe blood hypoxia of Atlantic salmon (*Salmo salar*) exposed to the marine diatom *Chaetoceros concavicornis*. In *Toxic Phytoplankton Blooms in the Sea,* ed. T. J. Smayda & Y. Shimizu, pp. 625–30. New York: Elsevier.

Rhodes, L., White, D., Syhre, M., & Atkinson, M. (1996). *Pseudo-nitzschia* species isolated from New Zealand waters: Domoic acid production *in vitro* and links with shellfish toxicity. In *Harmful and Toxic Algal Blooms*, ed. T. Yasumoto, Y. Oshima & Y. Fukuyo, pp. 155–8. Intergovernmental Oceanographic Commission of UNESCO 1996. Paris.

Riebesell, U. (1989). Comparison of sinking and sedimentation rate measurements in a diatom winter/spring bloom. *Marine Ecology Progress Series*, 54, 109–19.

Rinaldi, A., Mananari, G., Ferrari, C. R., Ghetti, A., & Vollenweider, R. A. (1991). Mucilage aggregates during 1991 in the Adriatic and Tyrrhenian Seas. *Red Tide Newsletter*, 4, 2–3.

Rowe, G. T. (1996). Azerbaijan, oil, and sustainable development on the Caspian Sea. *Quarterdeck*, Department of Oceanography, Texas A&M University, 4(3), 4–10.

Saydam, A. C. (1996). Can we predict harmful algae blooms? *Harmful Algae News*, IOC Newsletter on Toxic Algae and Algae Blooms, 15, 5–6. UNESCO.

Scholin, C. A., Buck, K. R., Britschgi, T., Cangelosi, G., & Chavez, F. P. (1996). Identification of *Pseudo-nitzschia australis* (Bacillariophyceae) using r RNA-targeted probes in whole cell and sandwich hybridization formats. *Phycologia*, 35, 190–7.

Scholin, C. A., Villac, M. C., Buck, K. R., Krupp, J. M., Powers, D. A., Fryxell, G. A., & Chavez,

Toxic and harmful marine diatoms

F. P. (1994). Ribosomal DNA sequences discriminate among toxic and non-toxic *Pseudo-nitzschia* species. *Natural Toxins*, **2**, 152–65.

Smetacek, V. S. (1985). Role of sinking in diatom life-history cycles: Ecological, evolutionary and geological significance. Marine Biology, **84**, 239–51.

Smith, J. C., Cormier, R., Worms, J., Bird, C. J., Quilliam, M. A., Pocklington, R., Angus, R., & Hanic, L. (1990). Toxic blooms of the domoic acid containing diatom *Nitzschia pungens* in the Cardigan River, Prince Edward Island, in 1988. In *Toxic Marine Phytoplankton*, ed. E. Granéli, B. Sundström, L. Edler, & D. M. Anderson, pp. 227–32. New York: Elsevier.

Stabell, B., & Hasle, G. R. (1996). Diatoms in young sediments and net samples from the Kattegat. *Pact, 50–IV*, **8**, 443–9.

Takano, H. (1956). Harmful blooming of minute cells of *Thalassiosira decipiens* in coastal water in Tokyo Bay. *Journal of the Oceanographic Society of Japan*, **12** (2), 63–67.

(1965). New and rare diatoms from Japanese marine waters-I. *Bulletin of the Tokai Reg. Fisheries Research Laboratory*, **42**, 1–10.

Tangen, K. (1987). Harmful algal blooms in Northern Europe – Their causes and effects on mariculture. *Aqua Nor*, **87**, 29–33.

(1991). A winter bloom in Skagerrak, Norway. *Red Tide Newsletter*, **4**(1), 3–4.

Taylor, F. J. R. (1993). Current problems with harmful phytoplankton blooms in British Columbia waters. In *Toxic Phytoplankton Blooms in the Sea*, ed. T. J. Smayda & Y. Shimizu, pp. 699–703. New York: Elsevier.

Taylor, F. J. R., Taylor, N. J., & Walsby, J. R. (1985). A bloom of the planktonic diatom, *Cerataulina pelagica*, off the coast of northeastern New Zealand in 1983, and its contribution to an associated mortality of fish and benthic fauna. *Internationale Revue der gesamte Hydrobiologie*, **70**, 773–95.

Trainer, V. L., Wekell, J. C., Horner, R. A., Hatfield, C. L., & Stein, J. E. (1997). Domoic acid production by *Pseudo-nitzschia pungens*. *Proceedings of the Eighth International Conference on Toxic Phytoplankton*. In press.

Tåning, A. V. (1951). Olieforurening af havet og massedød af fugle. *Naturens Verden*, **35**, 34–43.

Villac, M. C., Roelke, D. L., Villareal, T. A., & Fryxell, G. A. (1993). Comparison of two domoic acid-producing diatoms: A review. *Hydrobiologia*, **269/270**, 213–24.

Walz, P. M., Garrison D. L., Graham W. M., Cattey M. A., Tjeerdema R. S., & Silver M. W. (1994). Domoic acid-producing diatom blooms in Monterey Bay, California: 1991–1993. *Natural Toxins* **2**, 271–9.

Wekell, J. C., Gauglitz, E. J. Barnett, H. J., Hatfield, C. L., Simons, D., & Auyes, D. (1994). Occurrence of domoic acid in Washington State razor clams (*Siliqua patula*) during 1991–1993. *Natural Toxins*, **2**, 197–205.

White, A. W. & Frady, T. L. (ed.) (1995). *Harmful Algae: An International Directory of Experts in Toxic and Harmful Algae and Their Effects on Fisheries and Public Health*. IOC Document 1015, UNESCO 1995. (English only) Paris, France. pp. 127.

Wood, A. M. & Shapiro, L. M. (ed.) (1993). *Domoic Acid*. Final Report of the Workshop, ORESU-W-92–003. Oregon Sea Grant, Oregon State University. pp. 21.

Wood, A. M., Shapiro, L. M., & Bates, S. S. (ed.) (1994). *Domoic Acid*. Final Report of the Workshop, ORESU-W-94–001. Oregon Sea Grant, Oregon State University. pp. 22.

22 Diatoms as markers of atmospheric transport

MARGARET A. HARPER

Introduction

In 1846 Charles Darwin wrote:

> 'On the 16th of January (1833), when the Beagle was ten miles off the N. W. end of St. Jago (Cape Verde Islands), some very fine dust was found adhering to the under side of the horizontal wind-vane at the mast-head; it appeared to have been filtered by the gauze from the air as the ship lay inclined to the wind. The wind had been for twenty-four hours previously E. N. E., and hence, from the position of the ship, the dust probably came from the coast of Africa'.

Darwin sent some of this dust to Ehrenberg (1844) who found diatoms. Wind-blown material is widespread (Pye, 1987) and the study of aeolian diatoms can indicate past changes in atmospheric circulation. The most abundant aeolian diatoms are freshwater ones blown from North Africa into the inter-tropical Atlantic, where they often outnumber the marine ones (Pokras & Mix, 1985).

Most diatoms grow in water bodies where the surface tension film protects them from light winds. Stronger winds pick up water and diatoms which can act as nuclei for raindrops and so often drop out. Schlichting (1964) trapped more soil algae from updrafts on cloudy days; but he caught few live diatoms. Marshall and Chalmers (1997) only trapped broken cells of *Pinnularia borealis* Ehrenberg; when alive these motile diatoms avoid dry surfaces. Diatom remains from dry fossil deposits and algal flakes from dried-up water bodies are more readily picked up by the wind.

Funnels, filters and sticky surfaces are used to trap wind-blown material. Their effectiveness depends on their aerodynamics and exposure. Diatoms are rare in most traps, and, when present, are not always recognized. Biologists have studied the dispersal of some freshwater algae by culturing wind and animal transported material (Kristiansen, 1996). Marine diatoms occasionally can be wind dispersed; as *Chaetoceros*, and three benthic genera grew in cultures exposed 35 meters above the intertidal zone (Lee & Eggleston, 1989). Atmospheric scientists are interested in particles and aerosols, palynologists in disease spores and pollen floras, and forensic workers in airborne contaminants (diatoms; Geissler & Gerloff, 1966).

Long distance transport of African diatoms

North African winds pick up about 500 million tonnes of dust each year (D'Almeida, 1986). Satellite images show that most of this dust is carried west by trade winds into the Atlantic Ocean (Moulin *et al.*, 1997). The latitude of the main dust plume changes, from around 5° N in winter to 20° N in summer. The summer plume comes from the Sahara desert. The winter plume comes from wetter areas in the upper Niger and Lake Chad and so contains more diatom remains (Pokras & Mix, 1985). Several workers (see Gasse *et al.*, 1989) have traced this dust, using *Aulacoseira* (ex *Melosira*) abundances in ocean sediments. However, *Aulacoseira* remains can come from major rivers, for instance, the Zaire (Congo) River carries them 1000 km into the Gulf of Guinea (Gasse *et al.*, 1989). Pokras (1991) compared the distribution of taxa and he found *Stephanodiscus* fragments like those of *Aulacoseira* occured under the winter dust plume. However, *Cyclotella striata* (Kützing) Grunow had an entirely fluvial distribution in the Zaire fan (Pokras, 1991). Gasse *et al.* (1989) comment on the lack of diversity in freshwater floras from dust and ocean sediments, compared with those from African rivers and diatom deposits on land. Diatoms are often fragmented in dust storms and partly dissolved while settling through the ocean. Thick-walled *Aulacoseira* survive this process more often than *Stephanodiscus*. Both are more common in Plio-Pleistocene than younger deposits of Lake Chad (Gasse *et al.*, 1989).

Moulin *et al.* (1997) analyzed 11 years of satellite data, and found that dustier years occur when there is a deep atmospheric pressure low off Iceland. Others have related increases in terrigenous material in cores to stronger winds, northeast trade winds (Pye, 1987) or southerly trade winds (Stabell, 1986). Evidence on land differs, Rognon and Coudé-Gaussen (1996) found westerly surface winds indicated by paleodunes in Northwest Africa. In winter, surface northeasterly winds extend down to 5° N where they meet the wetter southwesterly winds. Dust is also carried west above the turbulent convergence zone by the mid-level (height 1–4 km) African Easterly Jet (DeMenocal *et al.*, 1993). Pokras and Mix (1985) found peaks of diatom abundance in cores from 3° N to 1° S and related these to summer insolation at 30° N. Variations in summer insolation are controlled by 19 000–23 000 year precessional changes in the Earth's orbit. DeMenocal *et al.* (1993) confirmed this periodicity over the last million years for a core at 1° S. Other dust particles showed 100 000 and 41 000 year glacial periodicities, which change the prevalence of dry winds in North Africa (DeMenocal *et al.*, 1993). Deflation removes diatom sources which are restored by humid periods (Pokras & Mix, 1985). Less is known about a southern plume from the Namib desert, which deposits *Aulacoseira* off Angola (Pokras & Mix, 1985).

Red rain, with dust from North Africa, is common in the Mediterranean and has been known to occur in England (Pye, 1987). Seyve and Fourtanier (1985) collected Saharan dust in southern France; it included diatoms coated with red clay, but not all wind-blown material is clay coated (Game, 1964). They found mainly aerophilic diatom species (*Hantzschia amphioxys* (Ehrenberg) Grunow, *Navicula* aff. *mutica* Kützing), with *Aulacoseira granulata* (Ehrenberg) Simonsen forming only 11% of the population. Håkansson and Nihlén (1990) found *A. granulata* was more abundant in dust from northeastern Africa sampled in Crete, Italy and Greece. They said *Cyclostephanos* cf. *damasii* (Hustedt) Stoermer & Håkansson in the dust flora indicated it was African. Sahel diatoms are even blown to America, for instance *A. granulata* was trapped on a Barbados sea cliff in Saharan dust (Delany *et al.*, 1967). The same species was caught at 3000 m above Florida, in air from the Atlantic (Maynard, 1968).

Polar transport of diatoms – Arctic

Gayley *et al.* (1989) studied core samples from the Greenland ice-cap and identified *A. granulata* and *Stephanodiscus niagarae* Ehrenberg. Most of these diatoms were from layers deposited in spring time. In spring, anticyclones block westerly winds, directing them north to Greenland. These winds originate from seasonally dry areas of North America and carry *Ambrosia* pollen to Spitsbergen (Bourgeois *et al.*, 1985). Changes in diatom concentrations downcore in ice parallel those of 0.38–1.31 μm microparticles (Gayley *et al.*, 1989); diatom valves are larger, but flake-like and perforated, so they are more readily airborne. Fewer valves reach snow 500 km North of Alaska (Darby *et al.*, 1974).

Polar transport of diatoms – Antarctic

Most diatoms in continental Antarctica grow in ephemeral moist patches, meltwater streams, or meltwater moats (margins of ice-covered lakes). Planktonic diatoms sink under ice and therefore are very rare in freshwater. No live diatoms were found in cyanobacterial mats south of Minna Bluff (78°35'; C. Howard-Williams, NIWA pers. comm. 1994). Remains from further south are usually air transported, or fossil. Kellogg and Kellogg (1996) found both marine and freshwater taxa of aeolian origin in ice cores from the South Pole. Planktonic *Cyclotella stelligera* (Cleve) Van Heurck was the most common species, with 184 per litre in one sample. Each concentration peak could come from a single source area and be transported by the same wind event, but, snow can blow around for years before compacting. Maximum concentrations in their South Pole ice cores were lower than Gayley *et al.* (1989) found in Greenland. This agrees with Thompson and Moseley-Thompson's (1981) microparticle data, where they find glacial and postglacial concentrations are greater for Greenland than Antarctica.

Marine diatoms wind-transported to the South Pole are evidence that sparse marine diatoms found in the Sirius tillites on the Trans-Antarctic Mountains could be wind-blown. The original hypothesis was that they were carried there from inland seas by ice sheet growth, which would mean East

Antarctica was deglaciated in the Pliocene (Webb *et al.*, 1984). Other evidence the diatoms are aeolian is they mainly occur in the surface tillite (Stroeven *et al.*, 1996). The few found further inside the tillite could have been reworked down micro-cracks (Barrett *et al.*, 1997). Plio-Pleistocene species also occur in cracks in ancient rocks (Burckle & Potter, 1996) and in non-Sirius surface material (Barrett *et al.*, 1997). Wind-blown diatoms, including fossils, could have reached the tillites any time since their Miocene deposition. Most dust reached central Antarctica during glacial periods, from Patagonia (Grousset *et al.*, 1992) where *Aulacoseira* is abundant (Izaguirre, 1990). Fragments of *Aulacoseira* are relatively common (all diatoms are exceedingly sparse less than 10 per gram) in tillite at Mt. Feather (Barrett *et al.*, 1997). Scherer (1991) found diatoms including Pleistocene marine species under a glacial icestream and concluded the West Antarctic ice sheet was smaller at times. Burckle *et al.* (1997) argue their diatoms are wind-blown and concentrated under the ice by basal melting. Scherer's (1991) microclasts of *Mesodictyon* (a Miocene freshwater genus) need not be wind-blown as the West Antarctic Ice Sheet was not fully formed in the Miocene (Barrett, 1996).

Subpolar wind-blown diatoms

Marine diatoms are usually rare in air except near beaches (Geissler & Gerloff, 1966; Delany *et al.*, 1967; Lee & Eggleston, 1989) or on stormy seas (Pokras & Mix, 1985). Maritime polar regions differ, *Fragilariopsis cylindrus* (Grunow) Kreiger (syn. *Nitzschia cylindrus* (Grunow) Hasle) was often dominant in traps on Signy Island (sub-Antarctic) and was abundant in rime frost on Ellesmere Island (sub-Arctic, Lichti-Federovich, 1985). These diatoms came from dense growths on sea ice. Sub-Arctic snow 50 km inland contained 17 marine taxa (total 110 taxa, Lichti-Federovich, 1984). Traps on Signy Island, 1 km inland, collected 36 marine taxa (total 43 taxa, Chalmers *et al.*, 1996). Nearly all diatom remains were small (area less than $10 \times 30 \, \mu$m) with the few larger ones being elongated or flimsy; no *Aulacoseira* was present. This flora is of local origin, unlike wind-blown floras in central Antarctica.

Other instances of long distance transport

Exposed diatomites are readily wind-blown. *Cyclotella elgeri* Hustedt and *Stephanodiscus carconensis* (Eulenstein) Grunow from Cascade diatomites were blown into the Gulf of California (Sancetta *et al.*, 1992). *Melosira sulcata* var. *siberica* Grunow (syn. *Paralia siberica* (A. Schmidt) Crawford & Sims) and *Stephanopyxis* from Tethys diatomite (north and east of the Black Sea) reached Finland (Tynni, 1970). Southeasterly winds blow dust from the Black Sea area to Scandinavia (Pye, 1987). There are unrecognized instances of diatoms being blown a long way. Dust is blown out of other continents besides Africa (Pye,

1987). Exotic taxa are often ignored, but with checks for contamination, diatomists could recognize more instances of aerial transport.

Summary

Darwin first recognized that freshwater diatoms were blown from Africa into the Atlantic. Changes in abundance of aeolian *Aulacoseira* in cores from off Northwest Africa show African monsoon cycles follow the Earth's 19 000 to 23 000 year orbital periodicity. Freshwater diatoms blown from North America are found in Greenland ice. Marine and freshwater diatoms are blown to the South Pole, and floras in ice cores indicate their deposition by particular sets of weather events. Subpolar airborne material is unusual as marine diatoms are often more abundant than terrestrial, most being small diatoms from sea ice. Many wind-blown diatoms come from diatomites. Diatom remains behave like smaller dust particles ($<1.4\ \mu$m diameter), as most are porous flakes.

Acknowledgments

Thanks to David Walton, William Marshall and Matthew Chalmers for the opportunity to analyse British Antarctic Survey trap data; and to Clement Prevost of the Terrain Sciences Division, Natural Resources of Canada, John Carter and others in the School of Earth Sciences, Victoria University of Wellington for help with this chapter.

References

Barrett, P. J. (1996) Antarctic paleoenvironment through Cenozoic times – A review. *Terra Antarctica*, 3, 103–19.

Barrett, P. J., Bleakley, N. L., Dickinson, W. W., Hannah, M. J.. & Harper, M. A. (1997). Distrbution of siliceous microfossils on Mount Feather, Antarctica, and the age of the Sirius group. In *The Antarctic Region, Geological Evolution and Processes*, ed. C. A. Ricci, pp. 763–70. Siena: Terra Antarctica Publication.

Bourgeois, J. C., Koerner, R. M., & Alt, B. T. (1985). Airborne pollen: A unique air mass tracer, its influx into the Canadian high Arctic. *Annals of Glaciology*, 7, 109–16.

Burckle, L. H. & Potter, J. R. (1996). Pliocene–Pleistocene diatoms in Paleozoic and Mesozoic sedimentary and igneous rocks from Antarctica: A Sirius problem solved. *Geology*, 24, 235–8.

Burckle, L. H., Kellogg, D. E., Kellogg, T. B., & Fastook, J. L. (1997). A mechanism for the emplacement and concentration of diatoms in glaciogenic deposits. *Boreas*, 26, 55–60.

Chalmers, M. O., Harper, M. A., & Marshall, W. A. (1996). *An Illustrated Catalogue of Airborne Microbiota from the Maritime Antarctic*. Cambridge, UK: British Antarctic Survey. 175 pp.

D'Almeida, G. A. (1986). A model for Saharan dust transport. *Journal of Climate and Applied Meteorology*, 25, 903–16.

Darby, D. A., Burckle, L. H., & Clark, D. L. (1974). Airborne dust on Arctic pack ice, its composition and fallout rate. *Earth and Planetary Science Letters*, 24, 166–72.

Darwin, C. E. (1846). An account of the fine dust which often falls on vessels in the Atlantic Ocean. *Quarterly Journal of the Geological Society* (London), **2**, 26–30.

Delany, A. C., Delany, A. C., Parkin, D. W., Griffin, J. J., Goldberg, E. D., & Reimann, B. E. F. (1967). Airborne dust collected at Barbados. *Geochemica et Cosmochimica Acta*, **31**, 885–909.

DeMenocal, P. B., Ruddiman, W. F., & Pokras, E. M. (1993). Influences of high- and low-latitude processes on African terrestrial climate: Pleistocene eolian records from equatorial Atlantic, Ocean Drilling Program Site 633. *Paleoceanography*, **8**, 209–42.

Ehrenberg, C. G. (1844). Einige vorläufige Resultate der Untersuchunden der von der Südpolreise des Captain Ross. *Bericht über die zur Bekanntmachungen geeigneter Verhandlungen der königlich Preussischen Akademie der Wissenschaften zu Berlin*, **1844**, 182–207.

Game, P. M. (1964). Observations on a dustfall in the Eastern Atlantic, February, 1962. *Journal of Sedimentary Petrology*, **34**, 355–9.

Gasse, F., Stabell, B., Fourtanier, E., & van Iperen, Y. (1989). Freshwater diatom influx in intertropical Atlantic: Relationships with continental records from Africa. *Quaternary Research*, **32**, 229–43.

Gayley, R. I., Ram, M., & Stoermer E. F. (1989). Seasonal variations in diatom abundance and provenance in Greenland ice. *Journal of Glaciology*, **35**, 290–2.

Geissler, U. & Gerloff, J. (1966). Das Vorkommen von Diatomeen in menschlichen Organen und in der Luft. *Nova Hedwigia*, **10**, 565–77.

Grousset, F. E., Biscaye, P. E., Revel, M., Petit, J-R., Pye, K., Joussaume, S., & Jouzel, J. (1992). Antarctic (Dome C) ice-core dust at 18 k.y. B. P.: Isotopic constraints on origins. *Earth and Planetary Science Letters*, **111**, 175–82.

Håkansson, H. & Nihlén, T. (1990). Diatoms of eolian deposits in the Mediterranean. *Archiv für Protistenkunde*, **138**, 313–22.

Izaguirre, I. (1990). Comparative analysis of the phytoplankton of six lentic environments from the province of Chubut (Argentina). *Physis (Buenos Aires)*, B, **48**, 7–23.

Kellogg, D. E., & Kellogg, T. B. (1996). Diatoms in South Pole ice: Implications for eolian contamination of Sirius Group deposits. *Geology*, **24**, 115–18.

Kristiansen, J. (1996). Dispersal of freshwater algae – A review. *Hydrobiologia*, **336**, 151–7.

Lee, T. F. & Eggleston, P. M. (1989). Airborne algae and cyanobacteria. *Grana*, **28**, 63–6.

Lichti-Federovich, S. (1984). Investigation of diatoms found in surface snow from the Sydkap ice cap, Ellesmere Island, Northwest Territories. *Current Research, Geological Survey of Canada*, **84-1A**, 287–301.

—— (1985). Diatom dispersal phenomena: diatoms in rime frost samples from Cape Herschel, central Ellesmere Island, Northwest Territories. *Current Research, Geological Survey of Canada*, **85-1B**, 391–9.

Marshall, W. A. & Chalmers, M. O. (1997). Airborne dispersal of Antarctic terrestrial algae and cyanobacteria. *Ecography*, **20**, 585–94.

Maynard, N. G. (1968). Significance of airborne algae. *Zeitschrift für Allgemeine Mikrobiologie*, **8**, 225–6.

Moulin, C., Lambert, C. E., Dulac, F., & Dayan, U. (1997). Control of atmospheric export of dust from North Africa by the North Atlantic Oscillation. *Nature*, **387**, 691–4.

Pokras, E. M. (1991). Source areas and transport mechanisms for freshwater and brackish-water diatoms deposited in pelagic sediments of the equatorial Atlantic. *Quaternary Research*, **35**, 144–56.

Pokras, E. M. & Mix, A. C. (1985). Eolian evidence of spatial variability of late Quaternary climates in Tropical Africa. *Quaternary Research*, **24**, 137–49.

Pye, K. (1987). *Aeolian Dust and Dust Deposits*. London: Academic Press, 334 pp.

Rognon, P., & Coudé-Gaussen, G. (1996). Paleoclimates of Northwest Africa (28°–35°N) about 18,000 yr. B. P. based on continental eolian deposits. *Quaternary Research*, **46**, 118–26.

Sancetta, C., Lyle, M., Heusser, L., Zahn, R., & Bradbury, J. P. (1992). Late-Glacial to Holocene changes in winds, upwelling and seasonal production of the northern Californian current system. *Quaternary Research*, **38**, 359–70.

Scherer, R. P. (1991). Quaternary and Tertiary microfossils from beneath the Ice Stream B:

Evidence for a dynamic West Antarctic ice sheet history. *Paleogeography, Paleoclimatology, Paleoecology*, **90**, 395–412.

Schlichting, H. E., Jr. (1964). Meteorological conditions affecting the dispersal of airborne algae and protozoa. *Lloydia*, **27**, 64–78.

Seyve, C. & Fourtanier, E. (1985). Contenu microfloristique d'un sédiment éolian actuel. *Bulletin – Centre pour Recherche et Exploration-Production Elf-Aquitaine*, **9**, 137–54.

Stabell, B. (1986). Variations of diatom flux in the eastern equatorial Atlantic during the last 400,000 years. *Marine Geology*, **72**, 305–23.

Stroeven, A. P., Prentice, M. L., & Klemen J. (1996). On marine microfossil transport and pathways in Antarctica during the late Neogene: Evidence from the Sirius Group at Mount Fleming. *Geology*, **24**, 727–30.

Thompson, L. G. & Moseley-Thompson, E. (1981). Microparticle concentration variations linked with climatic change: Evidence from polar ice cores. *Science*, **212**, 812–14.

Tynni, R. (1970). Diatoms from a dust stained snowfall in 1969. *Geology*, **21**, 79–81.

Webb, P. N., Harwood, D. M., McKelvey, B. C., Mercer, J. H., & Stott, L. D. (1984). Cenozoic marine sedimentation and ice-volume variation on the East Antarctic craton. *Geology*, **12**, 287–91.

23 Diatomite

DAVID M. HARWOOD

Introduction

Diatomite is a porous, lightweight sedimentary rock resulting from accumulation and compaction of diatom remains (class Bacillariophyceae). The delicate shell or frustule of diatoms, which gives diatomite many of its useful properties, is composed of amorphous opaline silica ($SiO_2 \cdot nH_2O$). Most diatoms fall within the 10 μm to 100 μm size range, although some are as large as 1 mm (Tappan, 1980). It is estimated that 1 cubic inch of diatomite may contain 40 to 70 million diatoms (Crespin, 1946). While the specific gravity (density) of diatom frustules is nearly twice that of water, the perforations and open structure of the frustule renders diatomite a considerably lower effective density (between 0.12 g/cm³ and 0.25 g/cm³) and high porosity (from 75 to 85%), able to absorb and hold up to 3.5 times its own weight in liquid (Cleveland, 1966).

Diatomite of varying quality has been deposited in freshwater environments since at least the Eocene ($\sim$50 million years ago) and in marine environments since the Late Cretaceous ($\sim$80 million years ago). The purity of diatomites depends on the presence and amount of both clastic particles (silt and clay) and of organic materials, which limit the utility of diatomite in industrial applications. Some high commercial-grade diatomite contains up to 90% SiO_2, with minor occurrence of calcium carbonate, volcanic glass, and terrigenous particles (Cressman, 1962; Cummins, 1960). Diatom-bearing rocks with a higher terrigenous component (e.g., diatomaceous shale or siliceous shale) are commonly interbedded with diatomite. The association of many oil fields with diatomite or diatom-bearing shale indicates that the high lipid-oil content in the diatoms is a likely source for petroleum. Early names for diatomite include tripoli, kieselghur, and infusorial earth.

Origin of diatomite

Diatomite accumulates in areas where the rate of deposition of diatom frustules greatly exceeds that of other sediment components (Berger, 1970; Heath, 1974; Barron, 1987). In the oceans, this requirement is met in areas where cold, deep, nutrient-rich water is brought to the surface by wind or currents and

Fig. 23.1. Outcrop of a freshwater diatomite deposit from the Nightingale Quarry. This deposit contains an abundance of the genus *Aulacoseira*. (Photo by E. F. Stoermer.)

diatom production and sedimentation of diatom frustules is high. Areas of biogenic silica accumulation occur today in the equatorial Pacific Ocean, the circum-Antarctic, the North Pacific, and in areas where surface currents draw water away from the margin of continents (western Africa, Peru, etc.). Analysis of accumulation rates of diatomaceous sediment since the Cretaceous indicates considerable variation in the timing and distribution of siliceous deposits in the World's oceans (Leinen, 1979; Baldauf & Barron, 1990; Barron, 1987), with periods of maximum opal accumulation during the Late Cretaceous, middle Eocene and late Miocene. Heath (1974) estimated that up to 90% of the biogenic silica deposited in the ocean may be laid down in estuaries and nearshore basins, but diatomite does not accumulate here, as the diatoms are diluted by terrigenous sediment particles. Similarly, nonmarine diatomite is commonly only associated with lakes where clastic sediment input from the surrounding land is low. The survival of non-marine diatomite requires protection from erosion and diagenesis resulting from contact with corrosive alkaline or silica-deficient pore waters (Conger, 1942; Lohman, 1960).

Diatomite beds may be laminated or massive, and occur in tabular or lenticular beds that range from several centimeters to tens of meters in thickness (Fig. 23.1). Total thickness of an interbedded diatomite-bearing sequence may be in excess of 300 meters (Taylor, 1981), although this sequence may include chert and clay-rich intervals of little industrial value. Laminated diatomites may reflect a general absence of bioturbation in anoxic basins, seasonal variation in sediment input and biogenic deposition, and/or rapid deposition of diatom mats (Kemp & Baldauf, 1993). Due to dissolution in the water column, it is estimated that only 1% to 10% of the diatom frustules produced at the surface waters reach the bottom of the ocean (Lisitzin, 1972; Calvert, 1974).

Preservation of the opaline silica in diatom frustules depends on: (i) local conditions of alkalinity (pH<9 is favorable); (ii) the presence of dissolved silica in pore water; (iii) an association with volcanic ash (Taliaferro, 1933); and (iv) interstratification with lithologies that may limit permeability of corrosive pore waters. With progressive time, increasing heat (above ~50 °C) and depth of burial (depending on thermal gradient), the amorphous, opaline silica ($SiO_2 \cdot nH_2O$, commonly called opal-A) of diatom frustules is progressively transformed to anhydrous silica or porcellanite (SiO_2 – a disordered form of crystobalite and tridymite, referred to as opal C-T), and finally to quartz (SiO_2) as chert. This transformation results in a significant decrease in porosity and increase in density and hardness (Bramlette, 1946; Issacs, 1981).

History and use of diatomite

The unique properties and siliceous composition of diatomite were recognized long before diatoms were identified as the source of these rocks. The 'floating bricks' of Antiquity and vases made of diatomite are known from the time of the Greeks (Ehrenberg, 1842). Reconstruction of the dome of Hagia Sofia Church in Constantinople used diatomite as a building material as early as AD 432. In 1836, C. Fischer reported his microscopic observations of diatoms in kieselghur from a peat deposit to Ehrenberg, who was impressed with this discovery and spread the news throughout Europe (Ehrenberg, 1836). This initiated much activity by Ehrenberg and others into the study of, and search for, siliceous sediments around the world. This led to discovery, in 1837, of the important freshwater diatomite deposit of Quaternary age at Luneburger Heide district near Hannover, Germany. This was the leading deposit in the world, until full-scale operations of the marine deposit at Lompoc, California began after world war one. Soon after the reports by Ehrenberg, Bailey (1839) reported the first deposit of diatomite in the US and predicted the occurrence of other diatomites in many environments. Many new deposits were discovered in the middle to late nineteenth century, including that at Lompoc, California in 1888. The interest in diatomite and the search for additional applications led to production of numerous descriptive studies of diatomite, diatom deposits and industrial uses (Wahl, 1876; Card & Dun, 1897; Moss, 1898; Bigot, 1920; Eardley-Wilmont, 1928; Calvert, 1929; Hendey, 1930; Krczil, 1936; Kawashimma & Shiraki, 1941–1946; Conger, 1942).

The characteristics that make diatomite most suitable for industrial uses are low density, high porosity, low thermal conductivity, high melting point (1400 °C to 1750 °C depending on impurities), solubility only in strong alkaline solutions and hydrofluoric acid, and being chemically inert (Durham, 1973). These properties have been applied in 300 to 500 commercial applications

(Schroeder, 1970; Durham, 1973). Principal uses include filtration, insulation, fine abrasion, absorption, building materials, mineral fillers, as a pesticide, catalysts, carriers, coatings, food additive, and anti-caking agent. Diverse industries make use of diatomite – food, beverage, pharmaceutical, chemical, agricultural, paint, plastics, paper, construction, dry cleaning, recreation, sewage treatment, among many others (Cummins, 1975).

The first commercial application, and perhaps the most significant use of diatomite was in the production of dynamite, as discovered in 1867 by Alfred Nobel. Nitroglycerin was previously discovered by A. Sobrero in 1847, but it was a dangerous material that could not be controlled. Absorption of nitroglycerin within the porous diatomite, or kieselguhr, from the Luneburger Heide deposit in Germany, provided a means for safe transport. Dynamite made from diatomite is no longer used, except in special applications that require a softer 'bang'. It is clear that, without this discovery of dynamite by Nobel, the advancement of railroads across the US and other countries, the construction of canals for shipping, tunnels, dams, highways, and the extraction of the coal and raw materials that fueled the industrial revolution would not have occurred at the same pace. The progress during this time period impacts us yet today, and enables the honoring of the human spirit and discovery in arts, science, literature and peace with the Nobel Prizes.

Today, filtration is the main application for diatomite (73% is used for this purpose) followed by use as a filler (14%). The principal markets for filtration are in alcoholic beverages (beer and wine) and in filtering sugars, oils and water treatment. The use of diatomite in filtration will likely be reduced by the increasing use of membrane filters, as the disposal of diatomite-filtrate waste in landfill sites will become more expensive and may possibly be prohibited in the future. Although alternate materials may be substituted for diatomite in the filtration process (expanded perlite, asbestos, and silica sand), diatomite is in most cases a superior material. There may be new applications of diatomite as solid filters in biotechnology, with potential for other new applications. Concern for crystalline silica as a health hazard may limit the utilization of diatomite in paint in the future.

The stratigraphic record of diatoms in the geological record provides an extremely useful tool for the determination of age and environmental conditions of the past. High rates of diatom evolution and extinction during the Paleogene and Cretaceous afford a high-resolution time scale to date geologic events. Strelnikova (1990) estimated that 1.63 new genera were evolving every million years, and 1.29 genera were going extinct during these periods. Higher rates probably occurred associated with and following the explosion of the pennate diatoms in the Neogene (Tappan, 1980). Reviews of the biostratigraphic utility of diatoms can be found in Jouse (1978), Strelnikova (1992), Barron and Baldauf (1995), and Harwood and Nikolaev (1995). Although much research is still needed to document the stratigraphic record of the diatoms in the older records, the last 30 million years of the marine diatom record is sufficiently well known to warrant close study of specific taxa and begin to

Fig. 23.2. Mining operations of the Lompoc Diatomite, an upper Miocene marine diatomite. Excavation and transport vehicles are more than 6 meters high. (Photo by E. F. Stoermer.)

address the details of diatom evolution, paleobiogeography, and paleoceanography (see papers by Yanagisawa & Akiba, 1990, Yanagisawa, 1996). Diatoms will continue to play a major role in the interpretation of paleoenvironmental change due to the ecological limits and restricted geographic distribution of many diatom taxa (Gersonde, 1990).

Production of diatomite

World production of diatomite was estimated at 1.5 million tonnes in 1995, with the United States as the largest producer of diatomite (670 tonnes), followed by France (250 tonnes), the former Soviet Union (120 tonnes), Mexico (90 tonnes) and Korea (70 tonnes) (USGS Mineral Commodity Summaries, 1996). The upper Miocene diatomite deposit at Lompoc, Santa Barbara County, California is the largest producing deposit in the world, representing more than 60% of the US total production (Fig. 23.2). World resources of crude diatomite provide an adequate supply for the foreseeable future. The US Geological Survey Minerals Yearbook (available on the worldwide web) indicates that diatomite is a small player in the overall economics of minerals with US production at a value of $175 million, compared to $30 billion for gold, copper and aluminium and approximately $60 billion for cement.

Most diatomite mining operations are surface excavations from open pits, with selection of individual strata of different grades for specific treatment and application. The raw materials are crushed, sorted and dried, and often

D. M. HARWOOD

calcined in kilns to limit the insoluble impurities and remove organic material. Control of dust in the mining process is adequate due to the high moisture content of the raw materials. Silica dust production during processing is controlled by enclosing these areas.

Summary

Diatomite has a long history of use by humans, which spans more than two centuries. The ability of this porous, yet relatively strong material, to safely contain and protect nitroglycerin impacted the rate of advancement of our civilization through the use of dynamite in the construction of our transportation systems and mining. The distinctive properties of high absorption, low density and high porosity have led to the use of diatomite in diverse applications, with the greatest application today in filtration processes and as a lightweight filler. One benefit of diatom-bearing deposits that is often overlooked is their potential as source rocks for the generation of petroleum.

Diatomites that have the greatest utility in industrial applications result from: (i) an environmental condition that was conducive to the growth of diatoms; (ii) a depositional setting that reduced or prevented the input of volcanic, terrigenous, and biogenic carbonate; and (iii) a subsequent history of shallow burial and limited diagenesis or erosion. These conditions are known from marine and nonmarine settings around the globe, with most economic deposits coming from the Miocene to Quaternary.

Acknowledgments

Photographs were generously provided by Gene Stoermer. The paper benefited from the wealth of information collected in the collection of essays, notes, bibliography and history of many aspects of diatomite and diatoms available in Terra Diatomacea from Johns-Manville (Cummins, 1975).

References

Bailey, J. W. (1839). On fossil infusoria discovered in peat-earth at West Point, N.Y. *American Journal of Science*, Ser. I, 35, 118–24.

Baldauf, J. G. & Barron, J. A. (1990). Evolution of biosiliceous sedimentation patterns – Eocene through Quaternary: Paleoceanographic response to polar cooling. In *Geological History of the Polar Oceans: Arctic Versus Antarctic*, ed. U. Bleil & J. Thiede, pp. 575–607, Dordrecht: Kluwer Academic Publishers.

Barron, J. A. (1987). Diatomite: Environmental and geological factors affecting its distribution. In *Siliceous Sedimentary Rock-Hosted Ores and Petroleum*, ed. J. R. Hein, pp. 164–78, New York: Van Nostrand Reinhold Co.

Barron, J. A. & Baldauf, J. G. (1995). Cenozoic marine biostratigraphy and applications to paleoclimatology and paleoceanography. In *Siliceous Microfossils*, ed. C. D. Blome *et al.* (convenors), Paleontological Society Short Courses in Paleontology, 8, 107–18.

Berger, W. H. (1970). Biogenous deep-sea sediments – Fractionation by deep-sea circulation. *Geological Society of America Bulletin*, **81**, 1385–401.

Bigot, A. (1920). *Industrie des silices d'infusoires et de diatomées*. Rev. Ingénieur et Index Technique. **26**, 302–25.

Bramlette, M. N. (1946). *The Monterey Formation of California and the Origin of its Siliceous Rocks*, US Geological Survey Professional Paper 212, 57pp.

Calvert, R. (1929). *Diatomaceous Earth*. American Chemical Society Monograph 52, Chemical Catalogue Co., New York, 250pp.

Calvert, S. E. (1974). *Deposition and Diagenesis of Silica in Marine Sediments*, Special Publications of the International Association of Sedimentology, No. 1, pp. 273–99.

Card, G. W. & Dun, W. S. (1897). The diatomaceous earth deposits of New South Wales. *Geological Records of NSW*, **5**, 128–48.

Cleveland, G. B. (1966). Diatomite. In *Mineral and Water Resources of California. California Division of Mines and Geology Bulletin*, **191**, 151–58.

Conger, P. S. (1942). Accumulation of diatomaceous deposits. *Journal of Sedimentary Petrology*, **12**, 55–66.

Crespin, I. (1946). Diatomite. *Mineral Resources of Australia, Summary Report* No. 12, pp. 1–31.

Cressman, E. R. (1962). Non-detrital siliceous sediments. In *Data of Geochemistry*, 6th edn. US Geological Survey Professional Paper 440–T, pp. T1–T23.

Cummins, A. B. (1960). Diatomite. In *Industrial Minerals and Rocks*, 3rd edn. New York: American Institute of Mining, Metallurgical, and Petroleum Engineers, pp. 303–19.

Cummins, A. B. (1975). *Terra Diatomacea*, Denver, Colorado: Johns-Manville, 246 pp.

Durham, D. L. (1973). *Diatomite*, US Geological Survey Professional Paper 820, pp. 191–5.

Eardley-Wilmont, W. L. (1928). *Diatomite: Its Occurrence, Preparation and Uses*. Publication 691, Canada: Department of Mines, 182pp.

Ehrenberg, C. G. (1836). Über Fossile Infusions-thiere. *Bericht Berlin Akademie*, **1836**, 50–4.
(1842). Über die wie Kork auf Wasser schwimmenden Mauersteine der Alten Griechen und Roemer. *Bericht Berlin Akademie*, 1842, Bd. 8, pp.132–7.

Gersonde, R. (1990). The paleontological significance of fossil diatoms from the high-latitude oceans. In *Polar Marine Diatoms*, ed. L. K. Medlin & Priddle, Cambridge: British Antarctic Survey, pp. 57–63.

Harwood, D. M. & Nikolaev, N. A. (1995). Cretaceous diatoms: Morphology, taxonomy, biostratigraphy. In *Siliceous Microfossils*, C. D. Blome *et al.* (convenors), Paleontological Society Short Courses in Paleontology, **8**, 81–106.

Heath, G. R. (1974). Dissolved silica and deep-sea sediments. In *Studies in Paleo-Oceanography*, ed. W. W. Hay, Society of Economic Paleontologists and Mineralogists Special Publication, 20, pp. 77–93.

Hendey, N. I. (1930). Diatomite: its analysis and use in pharmacy. *Quarterly Journal of Pharmacy*, **3**, 390–407.

Issacs, C. M. (1981). Porosity reduction during diagenesis of the Monterey Formation, Santa Barbara coastal area, California. In *The Monterey Formation and Related Siliceous Rocks of California*, ed. R. E. Garrison & R. G. Douglas, Los Angeles: Pacific Section, Society of Economic Paleontologists and Mineralogists, pp.257–72.

Jouse, A. P. (1978). Diatom biostratigraphy on the generic level. *Micropaleontology*, **24**, 316–26.

Kawashimma, C & Shiraki, Y. (1941 to 1946). Fundamental studies of Japanese diatomaceous earths and their industrial applications. *Journal of the Japanese Ceramic Association*. Series of 18 parts from Part I (1941) in vol. 49, pp. 14–25 to Part XVIII (1946) in vol. 54, pp. 52–56.

Kemp, A. E. S. & Baldauf, J. G. (1993). Vast Neogene laminated mat deposits from the eastern equatorial Pacific Ocean. *Nature*, **362**, 141–3.

Krczil, F. (1936). Kieselguhr. Inre Gerwinnung, Veredlung and Anwendung. Sammlung *Chem. u Chem. Tech. Vortrage Heft* 32, Stuttgart, 197pp.

Leinen, M. A. (1979). Biogenic silica accumulation in the central equatorial Pacific and its implications for Cenozoic paleoceanography. *Geological Society of America Bulletin*, 90, 1310–76.

Lisitzin, A. (1972). Sedimentation in the World Ocean. *Society of Economic Paleontologists and Mineralogists, Special Publication*, **17**, 218 pp.

Lohman, K. E. (1960). The ubiquitous diatom – A brief survey of the present state of knowledge. *American Journal of Science*, **258–A**, 180–91.

Moss, J. (1898). Kieselghur and other infusorial earths. *Proceedings of the British Pharmaceutical Conference for 1898*, 337–45.

Schroeder, H. J. (1970). Diatomite in mineral facts and figures. *US Bureau of Mines Bulletin*, pp. 967–75.

Strelnikova, N. I. (1990). Evolution of diatoms during the Cretaceous and Paleogene periods. In *Proceedings of the 10th International Diatom Symposium, Joensuu, Finland 1988*, ed. H. Simola, pp. 195–204, Koenigstein, Germany, O. Koeltz Scientific Books.

(1992). *Paleogene Diatom Algae*. St. Petersburg: St. Petersburg University Press. 313 pp.

Taliaferro, N. L. (1933). The relation of volcanism to diatomaceous and associated siliceous sediments. *California University Publications, Geological Sciences*, **23**, 1–56.

Tappan, H. (1980). *The Paleobiology of Plant Protists*. San Francisco: W. H. Freeman, 1028pp.

Taylor, G. C. (1981). California's diatomite industry. *California Geology*, **34(9)**, 183–192.

US Geological Survey (1996). *Diatomite, Mineral Commodity Summaries 1996*. Washington, DC: US Government Printing Office. 2pp.

Wahl, W. H. (1876). Infusorial Earth and its uses. *Quarterly Journal Science, London,* New Series, **6**, 336–51.

Yanagisawa, Y. (1996). Taxonomy of the genera *Rossiella*, *Bogorovia* and *Koizumia* (Cymatosiraceae, Bacillariophyceae). *Nova Hedwigia, Beiheft*, **12**, 273–81.

Yanagisawa, Y. & Akiba, F. (1990). Taxonomy and phylogeny of the three marine diatom genera, *Crucidenticula*, *Denticulopsis* and *Neodenticula*. *Bulletin of the Geological Survey of Japan*, **41**, 197–301.

Part VI
Conclusions

24 Epilogue: a view to the future

EUGENE F. STOERMER AND JOHN P. SMOL

At this point it is appropriate, if not necessarily wise, to attempt a brief view into the future. By its very nature, scientific research is not usually kind to those who would attempt to plan it, or even forecast its future direction.

Be this as it may, it seems to us that some of the immediate future directions in applied diatom studies seem almost foreordained. It is very clear that a good deal of effort needs to be devoted to the formalities of taxonomy and nomenclature, which have been sadly neglected for the past century. Great strides have been made very recently in the alpha level taxonomy of diatoms (e.g., Lange-Bertalot & Metzelin, 1996). It has also become much more common for diatomists to document their work in published iconographs (e.g., Douglas & Smol, 1993; Cumming *et al.*, 1995) followed by deposition of properly vouchered material from major studies. Even more promising, the application of modern systematic techniques to diatoms (e.g., Kociolek & Stoermer, 1989; Kociolek *et al.*, 1989; Theriot & Stoermer, 1984; Williams, 1985) is becoming more and more established. However, it is also true that relatively few diatomists are formally trained systematists. The very increase in interest in diatom taxonomy, particularly that part fueled by practical applications, has left behind it a virtual morass of nomenclatorial problems. Unfortunately, many well-intentioned attempts to alleviate the situation have resulted in inappropriate synonymies, conservations, and circumscriptions, which only serve to further complicate it

It is unfortunately true that the recent past has not been kind to institutions, such as the Academy of Natural Sciences, Philadelphia, and the Natural History Museum, London, which have been the traditional guardians of taxonomic propriety and the training ground of diatom taxonomists. It seems incongruous that the capacity of these institutions has been diminished precisely at the time they are most needed. The number of university laboratories training diatom taxonomists has certainly increased, but few of these have the history, and attendant specimen and literature collections, held by the few major museums that have active diatom collections. Indeed the majority of university laboratories specializing in diatom studies are not deeply embedded in the structure of their parent institutions, and tend to be short lived. It is rare that such programs, despite their transient fame or notoriety, outlive their founders, particularly in North America. Thus, it appears to us that the greatest and growing limitation on application of diatom studies to real world problems lies not in the field itself, but in academic infrastructure which should be present to support it.

Infrastructure problems extend beyond the confines of academia. Governmental policies of the recent past in most countries of the world have attempted economies through the constraints of 'directed research' and 'client oriented' programs. While such approaches may have affected economies in fields where applications are well based in fundamental science, they have proven devastating to areas of research where discovery and synthesis are far from complete. Fortunately, at least in the United States, the necessity to support fundamental research in systematics is being increasingly recognized through programs such as the US National Science Foundation's Partnerships for Enhancing Expertise in Taxonomy (PEET), and other programs aimed at understanding biodiversity.

Adopting the hopeful outlook that fundamental infrastructure problems can be resolved or conquered, where may the greatest advances in applied diatom studies occur?

From our viewpoint it seems quite clear that the development and sophistication of the tools of statistical analysis currently overreach our ability to produce the basic data necessary for their full exercise. It is rare that an applied study is based on enumerations that fully expose the population structure of the community or assemblage under investigation (Patrick *et al.*, 1954) and even more rare that enumerations are sufficiently detailed to yield statistically reliable abundance estimates for more that a few members of the assemblage (Pappas & Stoermer, 1996). Given the apparent great power of recent work, even given these limitations, how much more could be revealed through simply extending population analyses? Testing this proposition demands more and better trained analysts, and the resources necessary to support them.

Not all these resources are necessarily more of the same. There is certainly reason to hope that the same technology which has allowed great gains in data analysis may also eventually provide additional power and certainty in producing diatom population data. Relatively simple statistical techniques are available to test hypotheses concerning species relationships (Julius *et al.*, 1997). Image analysis techniques have been successfully applied to diatoms (e.g., Stoermer & Ladewski, 1982). While early dreams of fully automated systems (Cairns *et al.*, 1977) have so far foundered on the grim realities of producing a machine system sufficiently sophisticated to solve the problems of specimen inconsistencies (broken, overlapping, abnormal, etc.), which seem so trivial to the human brain, but demand literally reams of computer-readable code to circumvent, the fundamental principles are sound, and development will yield to application of resources and effort. At the very minimum, image analysis techniques now available can provide aid in resolving population differences and help with some tasks critical to correct identifications (Stoermer, 1996).

E. F. STOERMER AND J. P. SMOL

As we pointed out in the prologue to this book, there remains a great need for experimental evidence on diatom physiology. Cellular level capabilities and processes are known for relatively few species, and virtually none of the known species occur in oligotrophic environments. Most of our current physiological assumptions in applied studies are derived from community level experiments, with their attendant difficulties in certain interpretation. While this type of study will undoubtedly continue to furnish valuable clues, the most pressing current need is for carefully controlled studies of single populations.

There is also a remaining need for better knowledge of life cycles and life history processes. The success of specific diatom populations may depend crucially upon their ability to move, grow stalks, orient to light, etc. It is still uncommon to see these factors, even if recognized, explicitly supported by observations and measurements in many interpretative studies. Life history strategies may also be critical to interpreting diatom population changes in some systems (e.g., Stoermer, 1993). Edlund and Stoermer (1997) estimate that there is published information concerning any aspect of their life for only about 200 diatom taxa. Unfortunately, this very minor fraction of the total number of species does not include many of the species dominant in our most widespread and important aquatic habitats.

It is also our sense of the situation that applied studies employing modern diatom communities have not yet utilized the rich information potentially available from work in paleontology and phytogeography. These subsets of the grand systematic endeavor are assumed in work on higher organisms, but it is not at all clear that most diatomists have yet broken free of the predominately Eurocentric view of diatom distribution and evolution brought about by the history of the field. Rich opportunities are being realized by those who have ventured out of the conventional mode of thought, and even greater opportunities await.

Although we have tended to emphasize challenges remaining in the field, in the hope this will challenge further research, we should probably end by saying that we know of few fields of research with greater opportunity, intellectual challenge, and social relevance. The authors in this volume have certainly laid out research agendas which can be broadened and refined. It should be equally clear that many great works remain to be done.

References

Cairns, J. Jr., Dickson, K. L., & Slocomb, J. (1977). The ABC's of diatom identification using laser holography. *Hydrobiologia*, **54**, 7–16.

Cumming, B. F., Wilson, S. E., Hall, R. I., & Smol, J. P. (1995). Diatoms from Lakes in British Columbia (Canada) and their relationship to lakewater salinity nutrients and other limnological variables. *Bibliotheca Diatomologica*, **31**. E. Stuttgart: Schweizerbart'sche Verlagsbuchhandlung, 207 pp.

Douglas, M. S. V., & Smol, J. P. (1993). Freshwater diatoms from high Arctic ponds (Cape Herschel, Ellesmere Island, N.W.T.). *Nova Hedwigia*, **57**, 511–52.

Edlund, M. B., & Stoermer, E. F. (1997). Ecological, evolutionary, and systematic significance of diatom life histories. *Journal of Phycology*, **33**, 897–918.

Julius, M. L., Estabrook, G. F., Edlund, M. B., & Stoermer, E. F. (1997). Recognition of taxonomically significant clusters near the species level, using computationally intense methods, with examples from the *Stephanodiscus niagarae complex*. *Journal of Phycology*, **33**, 1049–54.

Kociolek, J. P., & Stoermer, E. F. (1989). Phylogenetic relationships and evolutionary history of the diatom genus Gomphoneis. *Phycologia*, **28**, 438–54.

Kociolek, J. P., Theriot, E. C., & Williams, D. M. (1989). Inferring diatom phylogeny: A cladistic perspective. *Diatom Research*, **4**, 289–300.

Lange-Bertalot, H., & Metzeltin, D. (1996). Indicators of Oligotrophy, 800 taxa representative of three ecologically distinct lake types, carbonate buffered – oligodystrophic – weakly buffered soft water. *Iconographia Diatomologica – Annotated Diatom Micrographs, **Volume 2**, Ecology–Diversity–Taxonomy* ed. H. Lange-Bertalot, Königstein, Koeltz Scientific Books, 390 pp.

Pappas, J. L., & Stoermer, E. F. (1996). Formulation of a method to count number of individuals representative of number of species in algal communities. *Journal of Phycology*, **32**, 693–6.

Patrick, R., Hohn, M. H., & Wallace, J. H. (1954). A new method for determining the pattern of the diatom flora. *Notulae Naturae, Academy of Natural Sciences of Philadelphia*, **259**, 1–12.

Stoermer, E. F. (1993). Evaluating diatom succession: Some peculiarities of the Great Lakes case. Journal of Paleolimnology, **8**, 71–83.

(1996). A simple, but useful, application of image analysis. *Journal of Paleolimnology*, **15**, 111–13.

Stoermer, E. F., & Ladewski, T. B. (1982). Quantitative analysis of shape variation in type and modern populations of *Gomphoneis herculeana*. *Nova Hedwigia, Beiheft*, **73**, 347–86.

Theriot, E. C., & Stoermer, E. F. (1984). Principal component analysis of Stephanodiscus: Observations on two new species from the *Stephanodiscus niagarae* complex. *Bacillaria*, **7**, 37–58.

Williams, D. M. (1985). Morphology, taxonomy and inter-relationships of the ribbed araphid diatoms from the genera *Diatoma* and *Meridion* (Diatomaceae: Bacillariophyta). *Bibliotheca Diatomologica*, **8**, E. Stuttgart: Schweizerbart'sche Verlagsbuchhandlung, 238 pp.

Glossary and acronyms

A : C ratio: Index of lake trophic status based on the number of frustules of planktonic araphid to centric diatoms. No longer widely used.

accretion: A gradual increase brought about by natural forces over a period of time; the upward growth of the sedimentary column due to accumulation of matter (= vertical accretion).

acid waters: Streams and lakes with a pH value less than 7.

acidified waters: Streams and lakes that have become increasingly acidic through time.

acidobiontic: Diatoms that have their greatest abundance at pH values less than 5.5.

acidophilous: Diatoms that are most abundant at pH values below pH 7.

adnate diatoms: Diatoms that grow flat, tightly attached to substrates.

aerophilic: 'Air loving'; term used for diatoms occurring commonly in subaerial environments, often living exposed to air, and not totally submerged under water (= aerophilous).

aerophilous: 'Air loving'; term used for diatoms occurring commonly in subaerial environments, often living exposed to air, and not totally submerged under water (= aerophilic).

afforestation: The planting or natural regrowth of trees, often of conifer forests on land previously covered by heathland and moorland.

alkalibiontic: Diatoms that are most abundant at pH values above pH 7.0, and assumed to require this condition for growth.

alkaliphilic: Diatoms that prefer habitats in which the pH is greater than 7.0 (= alkaliphilous).

alkaliphilous: Diatoms that prefer habitats in which the pH is greater than 7.0 (= alkaliphilic).

allochthonous: Material introduced into aquatic environments from external (usually terrestrial) sources. The term may be used in contrast to autochthonous, which refers to material that originates within the aquatic environment.

alpha-mesohaline: Salinity between 10 and 18 ppt.

alpha-oligohaline: Salinity between 3 and 5 ppt.

alluvial: Relating to the products of sedimentation by rivers or estuaries.

amictic: Lakes that do not undergo vertical mixing. Includes lakes that possess a perennial ice cover that prevents wind-induced mixing. Such lakes are limited primarily to Antarctica, but are also rarely found in the High Arctic and at high elevations.

amorphous opaline silica: A non-crystalline isotropic mineral ($SiO_2 \cdot nH_2O$). It is often found in the siliceous skeletons of various aquatic organisms, including diatoms.

analogue matching: A technique in which a similarity or dissimilarity index is used to assess if 'fossil' assemblages are represented in a modern 'training set' (i.e., are the fossil assemblages represented by modern assemblages?). The analogue-matching technique can also be used to estimate limnological conditions for 'fossil' samples by assuming that the 'best' matches, or the average of a specified number of 'best' matches (or weighted average, where the percent similarity/dissimilarity is used as the weight), of 'fossil' samples with 'modern' samples provides an adequate estimate of the environmental conditions of the 'fossil' sample.

ANC: Acid Neutralizing Capacity; the equivalent capacity of a solution to neutralize strong acids.

anoxia: The condition of zero oxygen. Anoxic waters contain no dissolved oxygen.

apparent root mean squared error: An estimate of the predictive ability of a model based on samples that were used to derive that model, in contrast to error derived from an independent dataset. This error is calculated from the sum of the observed minus inferred values, divided by the total number of samples.

aquifer: A body of rock that contains sufficient saturated permeable material to conduct groundwater.

attributes: Characteristics of assemblages that are measured to assess biotic integrity and the environmental stressors affecting biotic integrity (see metrics).

autecological: Adjective referring to the ecological conditions in which a specific species (taxon) occurs.

autecological indices: A quantitative inference about environmental conditions in a habitat based on the species composition of organisms in the habitat and the ecological condition in which those species are usually found.

autecology: The study of the relationships between individual organisms or taxa and their environment.

autochthonous: Formed or produced in the place where it is found (as opposed to allochthonous).

auxosporulation: The formation of an auxospore, which is a special cell, usually a zygote, produced by diatoms which expands to the near the maximum size of a given species, thus compensating for the cell size diminution that occurs during vegetative cell division.

baymouth bar: A sandbar formed across the opening of a coastal bay, caused by current transport of sand along the coast. This process eventually results in formation of a lake as the bay becomes isolated from the larger body of water adjacent to it.

benthic: Organisms attached to the substrata (e.g., rocks, sand, mud, logs, and plants) or otherwise living on the bottoms of aquatic ecosystems (see periphyton).

benthos: Organisms living on the bottom of aquatic systems, either attached to substrates or moving along the bottom.

beta-mesohaline: Salinity between 5 and 10 ppt.

beta-oligohaline: Salinity between 0.5 and 3 ppt.

Bhalme and Mooley Drought Index: An index of drought intensity calculated from monthly precipitation measurements from individual meteorological stations.

biofilm: A concentration of motile diatom cells in the upper 2 mm of sediments connected to one another by mucilaginous secretions.

biogenic: Formed by living organisms.

biogenic silica: Also called 'biogenic opal'. $SiO_2 \cdot nH_2O$, the mineral of which diatom frustules are composed, as well as the siliceous components of radiolarians, chrysophyte algae, sponge spicules, etc.

biogeochemistry: The science that deals with the relation of earth chemicals to plant and animal life.

biomass assay: One of many measurements of biomass. Chlorophyll *a*, ash-free dry mass, cell numbers, and biovolume are some of the assays used to estimate algal biomass.

biomass: Mass of biological components of a habitat or ecosystem.

biotic integrity: The similarity of community attributes (structural and functional characteristics) at sites being assessed compared to attributes of communities which were historically or currently considered natural for the region. Sometimes defined as 'the capability of supporting a balanced, integrated, adaptive community of organisms having a species composition, diversity, and functional organization comparable to that of natural habitats of the region'.

bioturbation: The mixing of sediment by organisms.

black ice: Transparent ice, which effectively transmits light (as opposed to white ice).

blooms: Large increases in numbers of planktonic organisms, sometimes associated with eutrophication and seasonal temperature changes.

bootstrapping: A computer intensive statistical resampling procedure that randomly generates 'new' datasets (e.g., 1000), with replacement, that are the same size as the original dataset. The predictive ability of a model is based on estimates derived on samples when they do not form part of the randomly generated dataset.

brackish: Saline water with a concentration between freshwater and sea water.

brine composition: The major ions (cations and anions) that characterize water; usually used in reference to saline water.

Bronze Age: A historical period of human activity that occurred *ca.* 1500 to 500 BC.

BSi: Biogenic silica

C : P ratio: Index of lake trophic status based on the numbers of planktonic centric diatom species versus pennate species.

calcined: When a substance has been heated to the point at which it loses its water component.

CA: Correspondence Analysis.

Canonical Correspondence Analysis: A constrained ordination technique, that uses a weighted-average algorithm, in which ordination axes are constrained to be linear combinations of the supplied environmental variables.

carbonaceous particles: Spheroidal fly-ash particles generated by the high temperature combustion of coal and oil in power stations; can be identified in soils and in lake sediments.

CCA: Canonical Correspondence Analysis

chert: A hard, compact, dense, sedimentary rock that is composed of microcrystalline to cryptocrystalline quartz grains or amorphous silica (opal). Usually exhibits conchoidal fracture.

circumneutral: Diatoms that have their greatest abundance at pH values around pH 7.

CLAG: Critical Loads Advisory Group.

clastic: Sedimentary material derived from the erosion of pre-existing rocks.

closed-basin lake: Lake with no outlet. Water may flow into the lake by a variety of sources (streams, direct precipitation on lake surface, ground water inflows), but in closed-basin lakes the only way that water can leave is by evaporative processes. Therefore, closed-basin lakes tend to contain higher concentrations of solutes. (See evaporative lake.)

cluster analysis: Statistical techniques that group assemblages of organisms based on characteristics of those assemblages, usually species composition, and one of many mathematical means of comparing assemblages.

coeval: Occurring at the same time.

composite diversity: A characterization of species composition that includes both richness and evenness of the species in the habitat or sample.

conspecific: Belonging to the same species.

Correspondence Analysis: An ordination technique that uses a weighted average algorithm to maximize the dispersion of species or sites in low dimensional space.

corrie lakes: Lakes formed in small, amphitheater-shaped, glacially eroded basins in mountain regions.

Cretaceous: The period in Earth history spanning the interval from approximately 140 to 65 million years ago, popularly known as the age of the dinosaurs. Mass extinctions mark the end of the Cretaceous.

critical loads: The values ascribed to the levels of acid deposition (or other pollutants) that soils and surface waters can tolerate before detrimental effects occurs.

cryophilic: Literally cold loving. Applied to organisms which thrive in cold environments.

cryoturbation: The mixing of sediments as a result of freezing and thawing.

cyanobacteria: Unicellular, colonial, and filamentous bacteria containing chlorophyll *a* and evolving oxygen during photosynthesis; also called blue-green algae.

DAIpo: Diatom Assemblage Index to Organic Pollution.

DAR: Diatom Accumulation Rate.

DCA: Detrended Correspondence Analysis.

deflate: An erosion process that removes unconsolidated sediment by wind.

deflation: The sorting, lifting, and removal of loose, dry, fine- grained particles by wind.

dendroecology: The science that uses annual tree-rings to assess past growth rates, from which past climatic and other environmental variables can be inferred.

desert oasis: As used here, a term applied to the McMurdo Dry Valleys, Antarctica, where precipitation is less than 10 cm per year. In these regions, lakes form from glacial meltwaters. Such lakes may be the only liquid water in otherwise dry (or ice-covered) terrain.

Detrended Correspondence Analysis: The detrended form of Correspondence Analysis. Detrending is a mathematical technique used to remove the 'arch' or 'horseshoe effect' on the second axis, which is a mathematical artifact.

DI: Diatom-inferred (as in DI-TP = diatom-inferred total phosphorus concentration)

diagenesis: Postdepositional physical, biological, and chemical alteration of sediments.

diastrophism: The process or processes by which the crust of the Earth is deformed, producing continents, ocean basins, plateaus and mountains, flexures and folds of strata, and faults.

Diatom Assemblage Index to Organic Pollution (DAIpo): Mathematical index of water quality based on the relative abundances of diatom taxa in a sample.

diatomaceous earth: See diatomite.

Diatom Inferred Trophic Index: Developed by Agbeti and Dickman (1989) and used to estimate changes in lake trophic status based on diatom assemblages. The index is based on multiple regression analyses of log-transformed percent abundances of diatom species that were divided into six indicator categories: (i) oligotrophic; (ii) oligomesotrophic; (iii) mesotrophic; (iv) mesoeutrophic; (v) eutrophic; and, (vi) eurytopic.

Diatomic Index (Id): A mathematical index of water quality based on the relative abundances of diatom taxa in a sample.

diatomite: A porous, lightweight sedimentary rock resulting from accumulation and compaction of diatom remains.

DIC: Dissolved Inorganic Carbon.

dimictic: Two seasonal periods of circulation occurring in a lake. Typical in temperate lakes.

Discriminant Analysis: A form of multivariate analysis used to determine significant groups.

dissimilarity index: A metric used to assess how dissimilar two or more sites are; common metrics include Chi-square, chord, Euclidean distances, etc.

DITI: Diatom Inferred Trophic Index.

diversity: A characterization of species composition of a habitat or sample that may include richness (number of species) and/or evenness (their relative abundance) of species.

DOC: Dissolved Organic Carbon.

ecological integrity: The similarity of ecosystem attributes (structural and functional characteristics) at sites being assessed, compared to attributes of communities which were historically or currently considered natural for the region. Ecological integrity is distinguished from biotic integrity because the former includes physical habitat integrity and water quality characteristics, as well as biotic integrity of assemblages (see biotic integrity).

ecogenetics: A combination of the disciplines ecology and genetics.

ecotone: A narrow transition zone or region that separates two or more regional bio-zones.

ecotope: A geographical unit of ecological homogeneity characterized by a specific combination of abiotic conditions.

ecotype: A subset of individuals within a species with a characteristic ecology.

edaphic: In phycological terms, adjective describing algae within or associated with marine intertidal sediments. Also refers to soil conditions.

edge-effect: Refers to the bias introduced into estimates of species optima when lake surveys do not sample the entire range of environmental conditions that species inhabits. Edge-effects refer to truncated species responses at the ends of environmental gradients (e.g., at very low or high values).

Eemian: The most recent interglaciation; named for marine deposits along a small stream in the eastern Netherlands. The term is often used as equivalent to marine oxygen isotope stage 5e, about 125 100 years ago.

El Niño: Literally the child – refers to the characteristic timing, near Christmas, of the manifestations of ENSO events on the South American coast.

EMAP: Environmental Monitoring and Assessment Program (US EPA).

embayment: A bay.

endemic: Organisms that are only found within a certain (often limited) region.

endosymbiont organism: Living within the cells of another organism, often in a mutually beneficial arrangement (for example, algae in coral polyps).

ENSO (El Niño Southern – Ocean Oscillation): A climate feature of the Pacific Ocean characterized by east-west shifts of the tropical pressure gradient, occurring every 4–7 years. During ENSO events, waters of the eastern Pacific are abnormally warm.

Eocene: The next to the earliest epoch of the Tertiary; comprising the time between about 54 and 38 million years ago.

eolian: Any material or organism that may be carried by the wind. Also spelled aeolian.

EPA: Environmental Protection Agency.

epilithic: Adjective describing algae attached to rocks, cement, glass slides, or similar hard surfaces.

epilithon: Algae growing attached to rocks and similar hard surfaces.

epipelic: Adjective describing algae that live on, or in, mud.

epipelon: The highly motile community living on soft substrates, especially mud.

epiphytic: Adjective describing algae attached to other algae or aquatic plants.

epiphyton: Algae growing attached to other algae or aquatic plants.

epipsammic: Adjective describing algae attached to sand grains.

epipsammon: Algae growing attached to sand grains.

epizoon: Algae living attached to animals.

epontic: Occurring in or on the basal layer of sea-ice.

estuary: A semi-enclosed coastal body of water which has a free connection with the open sea and within which sea water is measurably diluted with fresh water derived from land drainage.

euplankton: The 'true' plankton; not actively swimming organisms; normally occurring in the water column; opposed to tychoplankton, which occurs intermittently in the water column after being swept up from the bottom.

euplanktonic: Truly planktonic organisms that spend their entire life floating in the open water.

euhaline: Salinity of 30–40 ppt.

euryhaline: Referring to an organism with a relatively broad tolerance range for salinity.

eurytopic: Organisms that live within a broad range of environmental parameters, whether it is temperature, light, nutrient regime, etc.

eustatic: A large-scale change in sea level; not a relative change resulting from local coastal subsidence or elevation.

eustasy: Large-scale changes in sea level due to rise and fall of the ocean, not to that of the land.

eutrophication: The process of becoming more eutrophic. Eutrophic waters are rich in dissolved nutrients, often from anthropogenic fertilization (for example, sewage outlows or agricultural runoff). Eutrophic waters may experience seasonal oxygen deficiency.

evaporitic lake: See closed-basin lake.

evenness: A measure of the similarity in numbers of organisms of each species in a habitat or sample.

exopolymers: Substances (typically carbohydrates) secreted into the water column by algae.

exotic species: Foreign species that are introduced, and are not native.

extant: Organisms that are still living (i.e., not extinct).

extinct: The disappearance of species or larger biological units.

facies sequence analysis: Analysis of the occurrence of facies changes in order to improve understanding of the relations between different facies and the conditions leading to their formation.

Factor Analysis: A form of multivariate analysis in which operations are performed on a correlation or covariance matrix. Eigenvectors are extracted so as to explain as much of the original variance as possible, and each eigenvector is expressed as a factor which can be thought of as representing a theoretical end-member sample; the original samples can be expressed quantitatively as a mixture of factors. Synonymous with Principal Components Analysis.

fluvial: Produced by, coming from, or related to a river.

Foraminifera: Member of the class Sarcodina; unicellular organisms that secrete an external test of calcium carbonate or of cemented particles, usually consisting of a series of chambers.

foreshore: The part of a beach that becomes covered and uncovered by water during the process of tidal rise and fall.

frustule: The amorphous silicon cell wall of a diatom. It is composed of two halves, called valves, that are connected together by a girdle band or cingulum. The distinctive markings and shapes of frustules are used to identify the genera and species of diatoms.

Gaussian Logit Regression (GLR): Mathematical technique that attempts to fit either a Gaussian (bell shaped) regression curve or an increasing or decreasing monotonic curve to species abundance data generated from a training set.

Gaussian response: A normal distribution model.

generic level indices: Indices that use results of identifying genera and usually counting abundances of genera in samples.

geomorphology: The study of the classification, description, nature, origin, and development of landforms and their relationships to underlying structures, and the history of geologic changes as recorded by these surface features.

GLR: Gaussian Logit Regression.

glycerol: An alcohol composed of a backbone of three carbon atoms, each carrying a hydroxyl group; used in the synthesis of lipids.

graben: A linear block of the Earth's crust that has been dropped down along faults relative to rocks on either side. A depression produced by subsidence between two normal faults.

greenhouse gases: Atmospheric gases which contribute to the so-called Greenhouse Effect. These gases are transparent to incoming solar radiation but opaque to outgoing infrared radiation. Example gases include carbon dioxide, water vapor, methane, and chloro-fluorocarbons (CFCs).

gyre: Any roughly circular or elliptical region of water with a relatively stagnant stratified central region, surrounded by clockwise or counter-clockwise currents; usually refers to the large subtropical central areas of the oceans.

H′: Shannon – Weiner Diversity Index.

halobion: Referring to the biota living in salt water; a system for classifying organisms according to their salinity tolerances.

halocline: Depth zone within which salinity changes maximally.

Holocene: The most recent epoch of the Quaternary. Approximately the last 10 000 years of Earth history.

halophilic: Organisms that prefer, or are most abundant in, habitats that contain a relatively high concentration of salts.

holoplankton: Not actively swimming organisms, completing their entire life cycle in the water column.

humic waters: Streams and lakes with high concentrations of dissolved organic carbon (DOC).

humification: Formation of humus, the organic brown or black portion of soils resulting from decomposition of plant and animal material.

Hyalochaete: One of two subgenera of the genus *Chaetoceros*, characterized by robust spine-like extensions of the frustule (setae) and an absence of resting spores; most species are deep-water forms (pelagic).

hydro-isostasy: Crustal adjustment to loading and unloading attributed to water.

hypereutrophic: Highly nutrient enriched waters.

hypersaline: Water with an ionic concentration > 50 g/l.

hyposaline: Water with an ionic concentration between 3 and 20 g/l.

hypoxic: Waters deficient in dissolved oxygen (< 2 mg O_2/l).

IBI: Index of Biotic Integrity.

ice core: A core of ice taken from the surface of an ice cap or glacier, generally to document past global climatic or environmental change.

ice-wedge polygons: Polygon-shaped depressions delimited by ice-wedges which form in permafrost. They often fill up with water to form ponds, which result from the expansion of the freeze–thaw cycle. Common in tundra regions of the Arctic.

IDP: Practical Diatom Index.

Index alpha: An index proposed by Nygaard to infer lakewater pH from diatom species composition.

Index of Biotic Integrity: Quantitative scale for relating effects of human activity to responses of organisms in ecosystems (see biotic integrity).

Index of Disturbance: A measure of the extent to which lakes have been disturbed by human activities, based on diatoms in sediment cores. The Index of Disturbance is calculated using species richness, diversity, detrended correspondence analysis scores, and inferred [TP], [Cl], and transparency values.

indifferent: Organisms that tolerate a wide range of environmental values, such as pH.

inference model: A mathematical model, or transfer function, that estimates the value of an environmental variable (e.g., lake-water phosphorus concentration) as a function of biological data (e.g., diatom assemblages).

infusorial earth: Obsolete synonym of diatomite.

ingression: The entering of the sea at a given place.

Inter-Tropical Convergence Zone: The boundary zone where the northeast trade winds of the Northern Hemisphere meet the southeast trade winds of the Southern Hemisphere.

interglacial: The time and climatic conditions that separates two glacial or ice age periods.

intertidal: Between low and high tide.

intertidal zone: Region of coastal zone between limits of low and high tide.

isostasy: The condition of equilibrium, comparable to floating, of the units of the Earth's lithosphere above the asthenosphere. Isostatic compensation and correction occur to maintain this equilibrium, causing relative elevation changes at the Earth's surface.

isostatic rebound: Rising of the Earth's crust in response to loss of the weight of glacial ice sheet, much like an inflated ball returns to its original shape once pressure is taken off of it.

isostatic uplift: Apparent sea level change resulting after deglaciation, when land masses gradually rebounded after being released from the pressure of the huge ice sheets.

ITCZ: Inter-Tropical Convergence Zone.

jack-knifing: The simplest form of cross validation (also known as 'leave-one-out' validation) for estimating the root mean squared error of prediction, where the reconstruction procedure is applied n times using a training set of size (n-1). In each of the n predictions, one sample is left out, and it is from these samples that the predictive ability of the calibration is evaluated.

Kieselghur: A German word meaning diatomite.

lake trophic status: Refers to the nutrient and productivity status of a lake; for example lakes with high nutrient content and high productivity are termed eutrophic, lakes of low nutrients and productivity are termed oligotrophic.

Last Glacial Maximum: The time of the maximum advance of the most recent Pleistocene glaciers, commonly dated at about 18 000 years before present.

lead-210 (^{210}Pb) dating: Geochronological tool for dating recent (e.g., last 150 years) sediments. ^{210}Pb, with a half-life of approximately 22 years, is part of the radium decay series.

lentic taxa: Taxa living in standing water, such as lakes and ponds.

LGM: Last Glacial Maximum.

LIA: Little Ice Age.

liming: The addition of limestone, usually in finely powdered form, to the catchments or surfaces of lakes and streams to mitigate the effects of acidification.

linear regression: A statistical technique that shows the numerical relationship between a response (or dependent) variable and an explanatory (or independent) variable for variables that are linearly related.

Little Ice Age: An interval of more frequent cold conditions dated from approximately AD 1300–1850, originally described from western Europe and characterized by the advance of mountain glaciers.

littoral: The nearshore area of a lake, typically defined as the area where rooted aquatic macrophytes can grow. Often delimited by the depth of the photic zone.

Lotic Index: An index based on the ratio of lotic diatoms to the total number of benthic diatoms in a given habitat.

lotic taxa: Taxa living in running waters, such as streams and rivers.

LTMP: Long-Term Monitoring Program.

Maar: Monogenetic volcanic craters, generally less than 2 km in diameter, formed by phreatomagmatic explosions and subsequent collapse.

macrophyte: Macroscopic forms of aquatic vegetation, commonly referred to as aquatic weeds.

MAGIC: Model of Acidification of Groundwater in Catchments; a dynamic computer model used to estimate acidification.

mangrove swamps: Brackish-water coastal wetlands of tropical and subtropical areas that are usually dominated by shrubby halophytes such as mangroves, and are partly inundated by tidal flow.

mean bias: The mean difference between inferred and measured values within the set of training lakes; used to estimate the tendency of inference models to over- or underestimate measured values.

Medieval Warm Period: An interval originally described in Europe of more frequent warm conditions, dated from approximately AD 900 to 1300.

meromictic: Lakes in which mixing, or circulation, of the entire water body is incomplete on an annual basis. Such lakes possess an upper water mass that circulates, separated from a bottom, non-circulating portion by a strong concentration gradient.

meroplanktonic: Organisms that spend only a part of their life cycle in the plankton. Diatoms that grow primarily in the benthos, but are passively entrained in the water column are termed meroplanktonic.

mesic: Moist.

mesohaline: Salinity of 5–18 ppt.

mesosaline: Waters with an ionic concentration between 20 and 50 g/l.

Mesozoic: The era of Earth history spanning the interval from approximately 250 to 65 million years ago, encompassing the Triassic, Jurassic and Cretaceous periods, during which time the land fauna was dominated by reptiles.

metalimnion: A region of rapid temperature decline with depth in a thermally stratified lake. Separates the upper waters (epilimnion) from the lower waters (hypolimnion) and commonly acts as a barrier for nutrients released in the bottom of the lake.

metaphyton: Algae found in the littoral (shallow water) region of a lake that are neither attached nor truly planktonic.

metrics: Literally a measure. As applied here, attributes of assemblages that change in response to human alterations of watersheds.

microphytobenthos: The assemblage of microalgae (mostly diatoms and cyanobacteria) living in the benthos.

Miocene: The first epoch of the Neogene period, and fourth epoch of the Triassic, spanning the interval from 22 to 5 million years ago. During the Miocene, glaciation developed on Antarctica and continental climates became drier, with large areas converted from forests to grasslands.

moat: The portion of open water between a floating ice pan in the middle of a lake and the shore. Common in high arctic lakes.

modern analogue: A surface sediment species assemblage that has similar taxonomic composition to a fossil sample.

Modified Pantle Buck Saprobic Index (SI): A mathematical index of water quality based on the relative abundances of diatom taxa in a sample.

monomictic: One period of seasonal circulation occurring in a lake.

monsoon: A wind system that changes direction seasonally, blowing from one direction in summer and the opposite direction in winter; associated with major seasonal shifts in precipitation, bringing heavy rains when blowing onshore.

moss epiphyte: Periphytic organism living on moss.

mucilage: Polysaccharides secreted by diatoms and other algae to facilitate movement in and/or for attachment to sediments; also called mucopolysaccharides.

multimetric indices: Indices that are composed of two or more metrics.

multiple linear regression: Statistical technique that is used to show the relationship between a response or dependent variable and a number of explanatory or independent variables.

multivariate analysis: Simultaneous analysis of data with high dimensionality.

Neogene: The period of Earth history spanning the interval from 22 million years ago to the present, encompassing the Miocene, Pliocene, and Pleistocene epochs.

neogenesis: The formation of new minerals, as by diagenesis or metamorphism.

Neolithic Age: The new stone age; historical period of human activity that occurred *ca*. 5000–3000 BC.

neritic: Occurring in shallow marine waters; usually refers to waters overlying the continental shelves.

Oligocene: The last epoch of the Paleogene period, spanning the interval from approximately 38 to 22 million years ago.

oligohaline: Salinity of 0.5–5 ppt.

Ontario Trophic Status Model: An empirical mass-balance model for estimating past, present, and future lake water phosphorus concentration, presented by Hutchinson *et al*. (1991).

ooze: Fine-grained sediment dominated by a single type of microfossil (for example, a diatom ooze).

optimum: A measure of the 'preferred' environmental conditions of a taxon, usually with respect to a particular environmental variable; the value of that environmental variable at which the taxon is most abundant.

ordination: A collective term for statistical techniques that attempt to arrange sites in low-dimensional space based on their species composition (e.g., a dataset consisting of many sites and species can effectively be summarized by one or more ordination axes).

osmoregulation: The processes that enable a cell or an organism to maintain the osmotic concentration of internal fluids within some narrow range, despite fluctuations in the external medium.

oxygen isotope stratigraphy: A stratigraphy based on the ratio of stable isotopes of oxygen (^{16}O and ^{18}O) preserved in the carbonate fraction of, typically, invertebrates. The isotopic ratios are used, for example, to infer the amount of ice on land (as glaciers) and so documents the waxing and waning of glacial periods.

paleoenvironmental proxy: Indicator that can be used to infer past environmental conditions. These may include microfossils, such as diatoms, as well as indicators such as pollen, sediment grain size, etc.

Paleogene: The period of Earth history spanning the interval from approximately 65 to 22 million years ago. During this time mammal faunas underwent rapid radiation and dominated the land areas.

paleolimnological: Pertaining to the history of a lake; examination of the proxy indicators within sediments in order to infer past environments.

paleomagnetic chronology: A stratigraphy or chronology based on the ages of the reversals of the Earth's magnetic field.

paleotidal: Refers to former tidal conditions.

Palmer Cell: A microscope slide adapted to hold 0.1 ml of water in a shallow well under a coverglass. The depth of the well is shallow enough that the short working distance of a 40 × objective will reach the bottom of the well where algae settle.

paludification: The waterlogging and subsequent development of peats on mineral soils.

palynology: The study of pollen and spores. A micropaleontotogical method based on the identification and counting of fossil pollen and spore types that have been preserved in lake and mire deposits. It allows the reconstruction of past vegetation.

parautochthonous: Intermediate between autochthonous and allochthonous; said of a mixed assemblage of which the autochthonous and allochthonous parts cannot be discerned.

passive samples: Samples included in ordinations that have no influence on the position of other samples (i.e., active samples) in the analysis. For example, fossil samples with no corresponding environmental data.

PCA: Principal Components Analysis

pelagic: Occurring in the deep ocean beyond the continental shelves, or in the deep parts of a lake.

periglacial lake: A lake originating on, or in immediate contact with, a glacier.

periphyton: Microorganisms attached to substrata (e.g., rocks, sand, mud, logs, and plants).

permafrost: Permanently frozen ground.

Phaeoceran: One of two subgenera of the genus *Chaetoceros*, characterized by thin spine-like extensions of the frustule (setae). Many species form spores and are coastal forms (neritic).

phytoplankton: Photosynthetic organisms suspended in the water and not capable of determining their own position in the water column, particularly their horizontal position. Some of these organisms can migrate vertically in the water column, but their position is largely determined by currents.

PIRLA: Paleoecological Investigation of Recent Lake Acidification.

plankton: Bacteria, algae, protozoa, fungi, and small animals suspended in the water and not capable of determining their own position in the water column, particularly their horizontal position. Some of these organisms can migrate vertically in the water column, but their position in largely determined by currents.

playa lake: A shallow intermittent lake in an arid or semi-arid region, which is filled only during wet periods and then subsequently dries via evaporation, usually leaving deposits of soluble salts.

Pleistocene: The first of two geological epochs within the Quaternary period, for the time from approximately 1.6 million years ago to 10 000 years ago, or the Holocene.

pleniglacial: Full glacial; maximum of a glaciation.

Pliocene: The fifth and last epoch of the Tertiary. Generally considered to comprise the time between approximately 5 and 2 million years ago.

pluvial lake: A lake formed in a period of exceptionally heavy rainfall; in the

Pleistocene epoch the term refers to a lake formed during a moist interval that is now either extinct or exists as a remnant.

POC: Particulate Organic Carbon

polyhaline: Salinity of 18–30 ppt.

practical diatom index (IDP): A mathematical index of water quality based on the relative abundances of diatom taxa in a sample.

Praetiglian: A European time-stratigraphic term that refers to cold climate environments and sediments deposited during the late Pliocene (*ca.* 2.4 million years ago) and preceding the Tiglian warm conditions.

Principal Components Analysis: An ordination technique which uses a weighted sum algorithm to maximize the dispersion of species or sites in low dimensional space. Synonymous with Factor Analysis.

proline: A non-polar amino acid.

pyrite: An iron sulfide compound that is found in soils and sediments, and is formed in the absence of oxygen.

quasibiennial oscillation: A self-generating, dynamic relation in the atmosphere that appears as an oscillation in temperature, pressure, and other climatic and oceanographic parameters occurring roughly every 2 years.

r^2: Coefficient of determination.

radiocarbon (^{14}C) dating: A dating technique that utilizes the ratio between the unstable (^{14}C) and stable (^{12}C) isotopes of carbon preserved in once-living organic materials. The technique can produce reliable dates back to about 30 000 years ago.

RDA: Redundancy Analysis

Redfield Ratio: The approximate Molar ratios of carbon, nitrogen, and phosphorus found in living phytoplankton (C:N:P of 106:16:1).

Redundancy Analysis: A constrained ordination technique, that uses a weighted-sum algorithm, in which ordination axes are constrained to be linear combinations of supplied environmental variables (see Jongman *et al.,* 1995).

regression: (i) A statistical technique that describes the dependence of one variable on another (cf. correlation, which assesses the relationship between two variables); (ii) When used in the context of training sets, the 'regression' step refers to the estimation of species parameters (e.g., optima, tolerance) from the species abundances in the training set; (iii) Retreat of the sea from a land area.

Remane's brackish-water rule: A model of the generalized penetration of marine, freshwater, and brackish-water animals into an estuary in relation to salinity. Diversity of marine and freshwater animals are shown as a percentage of species diversity in each source habitat. Diversity of brackish-water animals is shown as a subdivision of marine animals. At high salinity there is a pronounced minimum in the total animal diversity, with *ca.* 20% of the species number of freshwater habitats. Named after A. Remane.

rhizosolenid: Any one of several planktonic genera characterized by large, elongate, pencil-shaped cells.

richness: The number of species (or higher specified taxonomic level, e.g., number of genera) in a habitat or sample.

RMSE: Root Mean Squared Error.

RMSEP: Root Mean Squared Error of Prediction.

S: The number of algal taxa (species and their varieties) in a specified count.

salinity: The total quantity of dissolved salts, usually expressed as the sum of the ionic concentration of the four major cations (sodium, magnesium, calcium, and potassium) and four major anions (carbonate, bicarbonate, sulfate, and chloride) in mass or milliequivalents per liter.

Sangamon: Stratigraphic name referring to the last interglacial stage in North America. The term is a name for a paleosol (buried soil) widespread in the central United States. The term generally refers to the period between 125 000 and 70 000 years ago that encompasses both warm and cool periods.

sapropel: A fine-grained aquatic sediment rich in organic material, often of a greenish or black color.

sea-ice algae: Algae living in brine pockets within sea-ice, frozen in when the ice forms in fall, and released to produce a large pulse of primary production when ice melts the following spring.

seagrass: Flowering plant rooted in marine sediments belonging to several families, but not members of the grass family.

semi-terrestrial: Transitional between terrestrial and aquatic; only seasonally/periodically submersed.

senescent: Aging or damaged cells that have a decreased rates of physiological processes, such as photosynthesis and respiration, compared to cells of actively growing populations.

Shannon Diversity Index (H'): Informational index used to measure diversity of algal assemblages; expressed as 'bits' ($\log_2$) or 'nits' ($\log_e$) per individual.

SI: Modified Pantle Buck Saprobic Index.

SIMI: Stander's Similarity Index.

similarity index: A coefficient expressing the similarity of two multivariate samples, usually a ratio based on cumulative differences calculated between abundance of each variable in the two samples.

smectites: Dioctahedral (montmorillonite) and trioctahedral (saponite) clay minerals that possess high cation-exchange capacities.

sound: A long and/or broad inlet of the ocean, generally with its larger part extending roughly parallel to the shore.

species level indices: Indices that are based on identifying taxa to the species level and usually counting abundances of species in samples.

Spermatophyta: Vascular plant that reproduces by seeds; most commonly used to refer to flowering plants or angiosperms.

stalked diatoms: Diatoms that are attached to substrates by mucilaginous stalks.

Stander's Similarity Index: Index used to compare the statistical similarity of two samples based on the species present and their relative abundances.

stenohaline: Referring to an organism with a relatively narrow tolerance range for salinity.

subaquatic: Situated under water.

sublittoral: In marine systems, the shore area between low tide and a depth of *ca.* 100 m.

subsaline: Water with an ionic concentration between 0.5 to 3 g/l.

subtidal: Below the low tide level.

sun-spot cycles: An 11.1-year cycle in the number of spots (faculae) on the surface of the sun.

supratidal: The shore area just above high tide level.

surface-sediment calibration: A technique whereby contemporary limnological characteristics of a suite of lakes are related to the species composition of surface sediments (upper *ca.* 0.5–3 cm) in order to estimate the range and optima of taxa relative to some environmental gradient(s). Also called 'training sets'.

SWAP: Surface Water Acidification Program.

synecology: The study of the relationships between biotic communities and their environment.

taphocoenosis: The fossil assemblage as it results from all taphonomic processes (death, burial, diagenesis, discovery); the stage following the thanatocoenosis (death assemblage).

taphonomy: The branch of paleoecology concerned with the processes that affect material before its permanent burial; in the case of fossil organisms, the processes that alter the composition of the death assemblage relative to the living assemblage.

TD Index: A ratio based on the percentage of 'warm' diatom specimens relative to the total percentage of 'warm' and 'cold' specimens in a sample; used as a semi-quantitative indicator of relative temperature differences between samples.

tectonism: Synonym of diastrophism, which is defined as a general term for all movement of the Earth's crust produced by tectonic processes, including the formation of ocean basins, continents, plateaus, and mountain ranges.

tephra: A volcanic ash layer.

tephrochronology: A chronological technique that utilizes the presence of volcanic ashes of known age in deposits to ascribe a date to the deposit.

terrestrial: Land-based.

terrigenous: Derived from the land.

Tertiary: The earlier (between about 60 and 2 million years ago) of the two geologic periods of the Cenozoic era.

thermokarst lakes: Periglacial features formed in areas of permafrost, where ground ice has thawed and filled with water. These depressions can result from climate change, fire, vegetation change, etc.

tidal datum: Chart datum based on a phase of the tide.

tidal flats: Tracts of wet, low-lying, level land that are inundated regularly by ocean tides.

TN: Total Nitrogen.

tolerance: The measure of the breadth of a species response curve, or its ecological amplitude, with respect to an environmental variable.

TP: Total Phosphorus.

training set: A survey of biotic assemblages preserved in the surface sediments of lakes with an associated set of contemporaneous environmental measurements. Used to generate transfer functions or inference models. Synonymous with 'surface sediment calibration sets'.

transfer function: A mathematical function that describes the relationships between biological species and environmental variables that allow the past values of an environmental variable (e.g., pH, salinity) to be inferred from the composition of a fossil assemblage.

transgression: Flooding of land area by a rise in relative sea level resulting in an onshore migration of the high-water mark.

tripoli: A light-colored powdery sedimentary rock that forms as a result of weathering of chert. This term was originally applied to diatomite deposits, however this is now considered to be an incorrect usage.

trona: A white sodium-based mineral, which occurs in columnar layers or masses in saline deposits: $Na_2(CO_3)$, $Na(HCO_3) \cdot 2H_2O$.

TROPH 1: A diatom-based trophic index developed by Whitmore to estimate changes in lake trophic status. The index is based on the ratio of diatoms indicating high trophic status to those indicating low trophic conditions.

TSM: Ontario Ministry of Environment & Energy's Trophic Status Model.

tychoplankton: Organisms occurring intermittently in the water column as well as at times living in benthic communities.

USEPA: United States Environmental Protection Agency.

unimodal response: The expected non-linear response of a biological species to an environmental variable along an environmental gradient. The abundance of a species is expected to be at its maximum at the center of its range.

UV: Ultraviolet radiation.

UV-B: Ultraviolet B radiation. The B component of the spectral band has a wavelength of 280 to 320 nm. This wavelength is the most damaging to biological forms.

valves: The glass cell wall of a diatom (frustule) is composed of two halves, which are called valves. These valves often separate when the protoplasm is oxidized out of cells.

varve: A pair of contrasting sediment laminae representing accumulation during two seasons of a single year (for example, a light summer layer and a dark winter one).

Venice System: A final resolution from a Symposium held in Venice, Italy, 1958, on the Classification of Brackish Waters, and published in 1959.

WA: Weighted Average.

WA-PLS: Weighted Averaging Partial Least Squares Regression and Calibration.

WACALIB: A computer program developed by Line *et al.* (1994) that is used to develop transfer functions (e.g., Weighted Averaging Regression and Calibration, and other techniques).

Walther's Law of Facies: Only those facies and facies areas can be superimposed primarily which can be observed beside each other at the present time; consequently, vertical facies successions reflect horizontal spatial relations.

weighted averaging: A technique used to estimate either: (i) the optimum of a taxon (weighted-averaging regression) based on measured values of environmental variables from the lakes in a training set, where the weight is proportional to the species abundance; or (ii) an environmental variable from the species composition of a sample, based on estimates of species parameters from a training set, where species are weighted relative to their abundance (weighted-averaging calibration).

wet mounts: Entrapment of a suspension of microorganisms under a coverglass on a microscope slide. One method for wet mounts enables quantitative assessments of algae with 1000× oil immersion lens by placing a small volume of concentrated algal suspension (20 ml) evenly under a coverglass, and sealing the suspension under the coverglass with a coating of fingernail polish or resin around the edge of the coverglass and onto the microscope slide.

white ice: Ice that appears white in color due to entrapment of air or other materials during melting and refreezing. This type of ice is far less transparent to light than black ice.

Würm: The last glacial stage of the Pleistocene in the Alps, equivalent in age to the Weichselian in northern Europe, the Devensian in Britain, and the Wisconsinian of North America.

Younger Dryas: A cold interval at the end of the most recent Pleistocene deglaciation, dated from 12 600 to 11 450 calendar years before present.

zeolite: A generic term for a large group of hydrous aluminosilicates that have sodium, calcium, and potassium as their chief metals.

Index

A:C ratio, 136
abberant diatoms, 24–5
acidification, 85–117
acidobiontic, 86
acidophilous, 86
adnate diatoms, 233–4
AFDM, *see* ash-free dry mass
aerial habitats, 264–71
aerophilic, 246–7, 264–71
afforestation, 105
alkalibiontic, 86
alkaliphilic, 86
alkaliphilous, 86
aluminum, 97–9
alpha-mesohaline, 306
alpha-oligohaline, 306
alpine, 205–20
analogue matching, 279–82
anoxia, 157
Antarctic, 245–59, 269, 377, 431–2
archeology, 104, 389–98
Arctic, 205–21, 227–40, 254, 269, 377, 431–2
Arctic Front, 207
artificial substrates, 15–16, 323–5, 341
ash-free dry mass, 16–18
atmospheric transport, 429–33
autecological indices, 11, 21–4, 278–9

bays, 195
baymouth bar, 195
beta-mesohaline, 306
beta-oligohaline, 306
Bhalme and Mooley Drought Index, 66–7
bioassay, 18–19, 131
biofilm, 5
biogenic silica, 73–83, 157, 177, 356, 379
biogeography, 175, 268–9, 375
biomass, 5, 16
biomass assay, 18–19
biotic integrity, 16, 26–30
birds, 255

bootstrapping, 46–7, 94–5, 141
boreal, 205
brackish, 298–325
brine composition, 43, 50
Bronze Age, 105, 147
BSi, *see* biogenic silica
building materials, 438

C:P ratio, *see* Centrales:Pennales ratio
CA, *see* Correspondence Analysis
CALIBRATE, 23, 139
canals, 148, 184
Canonical Correspondence Analysis, 89–91,
    138, 210–11, 362
CCA, *see* Canonical Correspondence
    Analysis
cell size, 219, 320
Centrales:Pennales ratio, 136
Cenozoic, 402, 410
Chaoborus, 98, 100
Chernobyl, 322–5
chironomids, 157
climate change, 41–72, 205–21, 227–40
closed-basin lakes, 41–72, 247
clothing, 416–17
cluster analysis, 21, 361–2
coastal environments, 48, 277–91, 334–48,
    374
contaminants, 239, 322–5
correlation, 402–10, 439
Correspondence Analysis, 20
counting protocols, 23–4
craters, 176
Cretaceous, 382, 383, 436, 437, 439
critical loads, 110–11
cryoturbation, 235
cyanobacteria, 73, 128, 129, 135, 157, 245, 317,
    335, 340

DAIpo, *see* Diatom Assemblage Index to
    Organic Pollution

466